AF606546

VOLUME FIVE HUNDRED AND FORTY

METHODS IN ENZYMOLOGY

Reconstituting the Cytoskeleton

METHODS IN ENZYMOLOGY

VOLUME FIVE HUNDRED AND FORTY

METHODS IN ENZYMOLOGY

Reconstituting the Cytoskeleton

Edited by

RONALD D. VALE

Department of Cellular and Molecular Pharmacology
and the Howard Hughes Medical Institute
University of California, San Francisco
California, USA

ELSEVIER

AMSTERDAM • BOSTON • HEIDELBERG • LONDON
NEW YORK • OXFORD • PARIS • SAN DIEGO
SAN FRANCISCO • SINGAPORE • SYDNEY • TOKYO
Academic Press is an imprint of Elsevier

AP

Academic Press is an imprint of Elsevier
525 B Street, Suite 1800, San Diego, CA 92101-4495, USA
225 Wyman Street, Waltham, MA 02451, USA
Radarweg 29, PO Box 211, 1000 AE Amsterdam, The Netherlands
The Boulevard, Langford Lane, Kidlington, Oxford, OX5 1GB, UK
32 Jamestown Road, London NW1 7BY, UK

First edition 2014

ISBN: 978-0-12-397924-7
ISSN: 0076-6879

Printed and bound in United States of America
14 15 16 17 11 10 9 8 7 6 5 4 3 2 1

CONTENTS

Section II
Polymer Nucleation and Regulation

Section III
Molecular Motor Ensembles on Natural and Engineered Cargoes

Section V
Cell Extract Systems

CONTRIBUTORS

Anna Akhmanova
Division of Cell Biology, Faculty of Science, Utrecht University, Padualaan, Utrecht, The Netherlands

Jawdat Al-Bassam
Molecular Cellular Biology, University of California Davis, California, USA

Charles L. Asbury
Department of Physiology and Biophysics, University of Washington School of Medicine, Seattle, Washington, USA

Pradeep Barak
Department of Biological Sciences, Tata Institute of Fundamental Research, Colaba, Mumbai, Maharashtra, India

Peter Bieling
Department of Bioengineering and Biophysics Group, University of California-Berkeley, Berkeley, and Department of Cellular and Molecular Pharmacology, University of California-San Francisco, San Francisco, California, USA

Sue Biggins
Division of Basic Sciences, Fred Hutchinson Cancer Research Center, Seattle, Washington, USA

Laurent Blanchoin
Institut de Recherches en Technologies et Sciences pour le Vivant, iRTSV, Laboratoire de Physiologie Cellulaire et Végétale, CNRS/CEA/INRA/UJF, Grenoble, France

Rajaa Boujemaa-Paterski
Institut de Recherches en Technologies et Sciences pour le Vivant, iRTSV, Laboratoire de Physiologie Cellulaire et Végétale, CNRS/CEA/INRA/UJF, Grenoble, France

Marie-France Carlier
Laboratoire d'Enzymologie et Biochimie Structurales, CNRS, Gif-sur-Yvette, France

Brian T. Castle
Department of Biomedical Engineering, University of Minnesota, Minneapolis, Minnesota, USA

Baoyu Chen
Department of Biophysics and Howard Hughes Medical Institute, University of Texas Southwestern Medical Center at Dallas, Dallas, Texas, USA

Yuk-Kwan Choi
Division of Life Science, The Hong Kong University of Science and Technology, Hong Kong, China

Pamela E. Constantinou
Departments of Chemistry and Bioengineering, Rice University, Houston, Texas, USA

Michael R. Diehl
Departments of Chemistry and Bioengineering, Rice University, Houston, Texas, USA

Serge Dmitrieff
Cell Biology and Biophysics Unit, European Molecular Biology Laboratory, Heidelberg, Germany

Marileen Dogterom
FOM Institute AMOLF, Science Park, Amsterdam, The Netherlands

Jonathan W. Driver
Department of Physiology and Biophysics, University of Washington School of Medicine, Seattle, Washington, and Departments of Chemistry and Bioengineering, Rice University, Houston, Texas, USA

David G. Drubin
Department of Molecular and Cell Biology, University of California, Berkeley, California, USA

Alok Kumar Dubey
Department of Biological Sciences, Tata Institute of Fundamental Research, Colaba, Mumbai, Maharashtra, India

Christine M. Field
Department of Systems Biology, Harvard Medical School, Boston, and The Marine Biological Laboratory, Woods Hole, Massachusetts, USA

Daniel A. Fletcher
Department of Bioengineering and Biophysics Group, University of California-Berkeley, Berkeley, California, USA

Franck J. Fourniol
London Research Institute, Cancer Research UK, London, United Kingdom

Rémi Galland
Institut de Recherches en Technologies et Sciences pour le Vivant, iRTSV, Laboratoire de Physiologie Cellulaire et Végétale, CNRS/CEA/INRA/UJF, Grenoble, France

Margaret Gardel
Department of Physics, Institute for Biophysical Dynamics, and James Franck Institute, University of Chicago, Chicago, Illinois, USA

Melissa K. Gardner
Department of Genetics, Cell Biology & Development, University of Minnesota, Minneapolis, Minnesota, USA

Christopher P. Garnham
Cell Biology and Biophysics Unit, National Institute of Neurological Disorders and Stroke, Bethesda, Maryland, USA

Jeff Gelles
Department of Biochemistry, Brandeis University, Waltham, Massachusetts, USA

Yale E. Goldman
Pennsylvania Muscle Institute, and Department of Physiology, Perelman School of Medicine, University of Pennsylvania, Philadelphia, Pennsylvania, USA

Bruce L. Goode
Department of Biology and Rosenstiel Basic Medical Sciences Research Center, Brandeis University, Waltham, Massachusetts, USA

Brian S. Goodman
Department of Cell Biology, Harvard Medical School, Boston, Massachusetts, USA

Ilya Grigoriev
Division of Cell Biology, Faculty of Science, Utrecht University, Padualaan, Utrecht, The Netherlands

Aaron C. Groen
Department of Systems Biology, Harvard Medical School, Boston, and The Marine Biological Laboratory, Woods Hole, Massachusetts, USA

Christophe Guérin
Institut de Recherches en Technologies et Sciences pour le Vivant, iRTSV, Laboratoire de Physiologie Cellulaire et Végétale, CNRS/CEA/INRA/UJF, Grenoble, France

Adam G. Hendricks
Pennsylvania Muscle Institute, and Department of Physiology, Perelman School of Medicine, University of Pennsylvania, Philadelphia, Pennsylvania, USA

Lisa Henry
Department of Biophysics and Howard Hughes Medical Institute, University of Texas Southwestern Medical Center at Dallas, Dallas, Texas, USA

Erika L.F. Holzbaur
Pennsylvania Muscle Institute, and Department of Physiology, Perelman School of Medicine, University of Pennsylvania, Philadelphia, Pennsylvania, USA

Florian Huber
FOM Institute AMOLF, Science Park, Amsterdam, The Netherlands

Peter J. Hume
Department of Pathology, University of Cambridge, Cambridge, United Kingdom

Daniel Humphreys
Department of Pathology, University of Cambridge, Cambridge, United Kingdom

Keisuke Ishihara
Department of Systems Biology, Harvard Medical School, Boston, and The Marine Biological Laboratory, Woods Hole, Massachusetts, USA

D. Kenneth Jamison
Departments of Chemistry and Bioengineering, Rice University, Houston, Texas, USA

Antoine Jégou
Laboratoire d'Enzymologie et Biochimie Structurales, CNRS, Gif-sur-Yvette, France

Gijsje H. Koenderink
FOM Institute AMOLF, Science Park, Amsterdam, The Netherlands

Vassilis Koronakis
Department of Pathology, University of Cambridge, Cambridge, United Kingdom

Liedewij Laan
Fas Center for Systems Biology, Harvard University, Cambridge, Massachusetts, USA

Duck-Yeon Lee
Biochemistry Core, National Heart, Lung and Blood Institute, Bethesda, Maryland, USA

Tai-De Li
Department of Bioengineering and Biophysics Group, University of California-Berkeley, Berkeley, California, USA

Roop Mallik
Department of Biological Sciences, Tata Institute of Fundamental Research, Colaba, Mumbai, Maharashtra, India

Alphée Michelot
Physics of the Cytoskeleton and Morphogenesis Group, institut de Recherches en Technologies et Sciences pour le Vivant, iRTSV, LPCV/CNRS/CEA/INRA/UJF, Grenoble, France

Timothy J. Mitchison
Department of Systems Biology, Harvard Medical School, Boston, and The Marine Biological Laboratory, Woods Hole, Massachusetts, USA

Hiroaki Mizuno
Laboratory of Single-Molecule Cell Biology, Tohoku University Graduate School of Life Sciences, Sendai, Miyagi, Japan

R. Dyche Mullins
Department of Cellular and Molecular Pharmacology, University of California-San Francisco, San Francisco, California, and Howard Hughes Medical Institute, Chevy Chase, Maryland, USA

Michael Murrell
Departments of Biomedical Engineering and Materials Science and Engineering, University of Wisconsin, Madison, Wisconsin, USA

François Nédélec
Cell Biology and Biophysics Unit, European Molecular Biology Laboratory, Heidelberg, Germany

Phuong A. Nguyen
Department of Systems Biology, Harvard Medical School, Boston, and The Marine Biological Laboratory, Woods Hole, Massachusetts, USA

David J. Odde
Department of Biomedical Engineering, University of Minnesota, Minneapolis, Minnesota, USA

Shae B. Padrick
Department of Biophysics and Howard Hughes Medical Institute, University of Texas Southwestern Medical Center at Dallas, Dallas, Texas, USA

Natalie A. Petek
Department of Cellular and Molecular Pharmacology, University of California, San Francisco, California, USA

Andrew F. Powers
Department of Physiology and Biophysics, University of Washington School of Medicine, Seattle, Washington, USA

Louis S. Prahl
Department of Biomedical Engineering, University of Minnesota, Minneapolis, Minnesota, USA

Magdalena Preciado López
FOM Institute AMOLF, Science Park, Amsterdam, The Netherlands

Céline Pugieux
Cell Biology and Biophysics Unit, European Molecular Biology Laboratory, Heidelberg, Germany

Robert Z. Qi
Division of Life Science, The Hong Kong University of Science and Technology, Hong Kong, China

Ashim Rai
Department of Biological Sciences, Tata Institute of Fundamental Research, Colaba, Mumbai, Maharashtra, India

Priyanka Rai
Department of Biological Sciences, Tata Institute of Fundamental Research, Colaba, Mumbai, Maharashtra, India

Samara L. Reck-Peterson
Department of Cell Biology, Harvard Medical School, Boston, Massachusetts, USA

Arthur Rogers
Departments of Chemistry and Bioengineering, Rice University, Houston, Texas, USA

Antonina Roll-Mecak
Cell Biology and Biophysics Unit, National Institute of Neurological Disorders and Stroke, and Center for Biophysics, National Heart, Lung and Blood Institute, Bethesda, Maryland, USA

Guillaume Romet-Lemonne
Laboratoire d'Enzymologie et Biochimie Structurales, CNRS, Gif-sur-Yvette, France

Michael K. Rosen
Department of Biophysics and Howard Hughes Medical Institute, University of Texas Southwestern Medical Center at Dallas, Dallas, Texas, USA

Sophie Roth
FOM Institute AMOLF, Science Park, Amsterdam, The Netherlands

Krishna K. Sarangapani
Department of Physiology and Biophysics, University of Washington School of Medicine, Seattle, Washington, USA

Benjamin A. Smith
Department of Biochemistry, Brandeis University, Waltham, Massachusetts, USA

Michel O. Steinmetz
Laboratory of Biomolecular Research, Paul Scherrer Institut, Villigen, Switzerland

Cristian Suarez
Institut de Recherches en Technologies et Sciences pour le Vivant, iRTSV, Laboratoire de Physiologie Cellulaire et Végétale, CNRS/CEA/INRA/UJF, Grenoble, France

Thomas Surrey
London Research Institute, Cancer Research UK, London, United Kingdom

Katarzyna Tarnawska
Cell Biology and Biophysics Unit, European Molecular Biology Laboratory, Heidelberg, Germany

Todd Thoresen
Department of Physics, Institute for Biophysical Dynamics, and James Franck Institute, University of Chicago, Chicago, Illinois, USA

Manuel Théry
Institut de Recherches en Technologies et Sciences pour le Vivant, iRTSV, Laboratoire de Physiologie Cellulaire et Végétale, CNRS/CEA/INRA/UJF, Grenoble, France

Annapurna Vemu
Cell Biology and Biophysics Unit, National Institute of Neurological Disorders and Stroke, Bethesda, Maryland, USA

Naoki Watanabe
Laboratory of Single-Molecule Cell Biology, Tohoku University Graduate School of Life Sciences, Sendai, Miyagi, Japan

PREFACE: THE ROLE OF RECONSTITUTION IN CYTOSKELETON RESEARCH

Reconstitution is an experimental strategy that aims to recapitulate a cellular activity outside of a cell so that it can be studied in a "test tube"-like environment. Being able to manipulate molecules *in vitro* provides an unparalleled opportunity to dissect mechanism in ways that cannot be easily achieved in a living cell. Many of the most important discoveries in cell and molecular biology in the twentieth century utilized reconstitution; examples include the classic mechanistic studies of DNA replication, ubiquitin-mediated protein degradation, protein insertion into the endoplasmic reticulum, and vesicular trafficking to the cell surface.

Reconstitution also has had a rich history in the cytoskeleton field. A notable early success was Albert Szent-Gyorgyi's *in vitro* reconstitution of the contraction of actomyosin threads in the early 1940s (Szent-Györgyi, 1942), which allowed him to dissect the molecules involved in muscle contraction. For cilia and flagella motility, Summers and Gibbons (1971) demonstrated microtubule sliding in permeabilized axonemes, which helped to understand how dynein motors drive axonemal beating. In the 1980s, Sheetz and Spudich (1983) directly observed the motility of myosin-coated beads along actin cables from dissected *Nitella* cells; shortly thereafter, the Spudich lab demonstrated myosin motility on purified actin filaments. At a similar time, several *in vitro* motility assays were developed for studying the molecular motors that drive the transport of membrane organelles in nerve cells (Vale, Schnapp, Reese, & Sheetz, 1985). Collectively, these *in vitro* assays were essential stepping stones for the single-molecule assays that soon followed. *In vitro* motility assays, single-molecule measurement techniques, and protein expression in bacteria or insect cells opened the doors to the study of many molecular motors that could not be easily studied in living cells. Thus, reconstitution was essential for broadening the investigation of cellular motility beyond the classical systems of muscle and cilia.

Understanding the dynamic properties of cytoskeletal polymers also was aided by reconstitution strategies. Shinya Inoue first observed the dynamic nature of the cytoskeletal fibers that comprise the mitotic spindle (Inoue, 1953). More than two decades later, the polymerization and depolymerization of purified tubulin were demonstrated *in vitro* when the correct buffer conditions were identified (Weisenberg, 1972). Later, Mitchison and

Kirschner (1984) demonstrated that microtubules grow slowly followed by catastrophic shrinkage. This unusual behavior (termed "dynamic instability") was first observed with purified tubulin and then later confirmed to occur in living cells. For the actin cytoskeleton, the first principles of polymerization of purified actin were derived by Oosawa and colleagues in the early 1960s (Kasai, Asakura, & Oosawa, 1960), and later Pollard and coworkers (Woodrum, Rich, & Pollard, 1975) developed a more detailed understanding of subunit addition to the two ends of an actin filament.

In addition to work on purified proteins, reconstitution strategies have been developed for complex systems of cytoskeletal filaments. For example, using frog egg extracts, Murray and colleagues (Shamu & Murray, 1992) observed mitotic spindle formation and anaphase movement of chromosomes, a complex system involving the orchestrated actions of hundreds of proteins. Heald et al. (1996) followed by demonstrating that DNA-coated beads can replace natural chromosomes in inducing the formation of a mitotic spindle. Success in reconstituting actin-based cell motility arguably has been even more dramatic. Actin-based polymerization at the leading edge is believed to drive cell migration. *Listeria*, a pathogenic bacteria, hijacks this actin polymerization system to propel itself through cytoplasm, and this process was reconstituted in Xenopus egg extracts (Theriot, Rosenblatt, Portnoy, Goldschmidt-Clermont, & Mitchison, 1994). Several years later in 1999, Carlier and colleagues (Loisel, Boujemaa, Pantaloni, & Carlier, 1999) demonstrated that a defined cocktail of actin, Arp2/3, and additional actin regulatory proteins can propel *Listeria in vitro*, and 2 years later, they replaced the *Listeria* itself with artificial beads.

Even in our modern era of "omics"-driven research, reconstitution is alive and well, and indeed flourishing as never before in cytoskeletal research. Building upon the earlier work described above, investigators are now pushing to reconstitute more complex systems and develop more sophisticated assays. Combining *in vitro* reconstitution with *in vivo* studies of living cells/organisms is proving to be a powerful overall approach for elucidating the roles and mechanisms of cytoskeletal proteins.

This book captures the excitement and progress on reconstituting cytoskeletal processes. It begins with new methods of assaying polymer dynamics, including powerful computational methods (Prahl et al.) and ways of modulating experimental conditions using microfluidics (Carlier et al.). In addition to the classic systems of actin and microtubules, the *in vitro* polymerization of actin-like proteins from bacteria can now be studied as well (Petek and Mullins). The nucleation and dynamics of cytoskeletal filaments are controlled by numerous regulatory proteins that interact with actin and microtubules, and these mechanisms can be dissected by *in vitro* experimentation. The nucleation step is controlled by large, multi-subunit protein complexes. In this volume, Chen et al. describe how to prepare the WAVE complex (a regulator of Arp2/3 and actin nucleation) using a very clever protein expression method that may prove generally useful for other multi-subunit protein machines as well. Our understanding of tubulin nucleation lags behind that of actin, but Choi et al. have made progress in isolating an active γ-tubulin ring complex and assaying its nucleation of tubulin *in vitro*. Other proteins act as polymerases that add subunits to the filament end, and Mizuno and Watanabe (formin) and Al-Bassam (TOG domain proteins) describe elegant means of assaying these activities by microscopy. Other enzymes posttranslationally modify the filament after it has been assembled, and Vemu et al. describe how tubulin-modifying enzymes can be purified and their activities measured.

As described earlier, molecular motor proteins have been a focus for single-molecule investigations. A recent trend has been to develop motility assays that reflect greater and more physiological complexity. For example, many cargos contain multiple, rather than single, motor proteins and even can bind different types of motors that pull the cargo in opposite directions (e.g., kinesin and dynein). What are the properties of such complex multimotor systems? To answer such questions, one approach is to use scaffolds where the numbers and types of bound motors can be precisely controlled. To achieve this, Rogers et al. as well as Goodman and Reck-Peterson attach motors to DNA through precisely controlled linkages. To mimic other scenarios in which motors are bound to the cell cortex and pull on microtubules (causing displacement of the centrosome), Roth et al. attached dynein motors to the walls of microfabricated chambers. It is also important to study how ensembles of motors work on natural cargos. Both Hendricks et al. and Barak et al. describe methods for isolating membranous organelles and assaying their movement along microtubules *in vitro*.

Another challenge is to reconstitute complex cytoskeletal processes using networks of interacting purified proteins. Both Murrell et al. and Boujemaa-Paterski et al. describe amazing self-organization patterns, motility, and contractile properties of actin networks, assembled with different geometric constrains and protein mixtures. Substructures and reactions in the mitotic spindle are also proving amenable to reconstitution. Driver et al. describe methods for isolating kinetochores and studying their interactions with microtubules, and Fourniol et al. have reconstituted the basic ingredients of the microtubule overlap zone that forms during anaphase. And why not start to bridge the worlds of actin and microtubules? By adding proteins that interact with both actin and microtubules, Preciado López et al. have developed assays where the interactions and self-organization of actin and microtubules can be studied *in vitro*.

For some activities, it is not yet possible (and perhaps not even desirable, since regulatory mechanism might be lost) to achieve a complete reconstitution with purified proteins. In such a case, reconstitution can be performed in a crude extract, which still enables types of experiments that cannot be easily achieved with intact cells (e.g., immunoprecipitations, certain drug treatments, kinetic experiments, etc). The Xenopus egg extract, because of its very high protein concentration, has been a remarkable source for reconstitution experiments. In this volume, Groen et al. and Field et al. describe very useful tricks for preparing different types of extracts and ways in which these extracts can be used to study the self-organization of microtubules or actin.

From examining these chapters, it is apparent that new trends are emerging in reconstitution work. First, the reconstitution of cytoskeletal functions is attracting and benefitting from the interdisciplinary research of biologists, chemists, physicists, and engineers. For example, the microfabrication of chambers, lithography, and microfluidics are involved in many of the methods described in this volume. Furthermore, the self-organization and mechanical properties of the cytoskeleton *in vitro* are attractive experimental and theoretical problems for physicists (many of the authors in this series were originally trained as physicists). A second trend is the use of light microscopy for the vast majority of the assays in this book. Microscopy-based assays provide much more detailed information and require less material than bulk population measurements using fluorimeters or spectrophotometers. For such microscopy assays, one has to prepare specific types of chambers, clean and passivate the glass, and attach specific proteins of interest to the surface. Most of the chapters describe such methods, which are often derived from painstaking development work. There is a wealth of information in browsing and comparing the methods for surface preparation in these different chapters.

For students and postdocs reading this volume, reconstitution, while very powerful, requires attention to detail; protein handling, buffers, and concentrations matter a great deal. However, in this era of "black-box," commercial "kits," it is particularly rewarding to work out, step by step, a reconstitution method, with the reward being seeing a biochemical process "come to life" or witnessing the self-organization of molecules and polymers into larger structures. Exciting challenges also lie ahead for young scientists in this field. Might it be possible to build something as complex as a mitotic spindle? An axoneme? A microvillus? A centrosome? Perhaps, one day, these far-fetched dreams may not be so far-fetched after all.

ACKNOWLEDGMENT

I would like to thank Peter Bieling for his helpful suggestions in the planning of this volume.

REFERENCES

Heald, R., Tournebize, R., Blank, T., Sandalzopoulos, R., Becker, P., Hyman, A., & Karsenti, E. (1996). Self-organization of microtubules into bipolar spindles around artificial chromosomes in Xenopus egg extracts. *Nature, 382*, 420–425.

Inoue, S. (1953). Polarization optical studies of the mitotic spindle. 1. The demonstration of spindle fibers in living cells. *Chromosoma, 5*, 487–500.

Kasai, M., Asakura, S., & Oosawa, F. (1960). The cooperative nature of the G-F transformation of actin. *Biochimica et Biophysica Acta, 57*, 22–30.

Loisel, T. P., Boujemaa, R., Pantaloni, D., & Carlier, M. F. (1999). Reconstitution of actin-based motility of Listeria and Shigella using pure proteins. *Nature, 401*, 613–616.

Mitchison, T., & Kirschner, M. (1984). Dynamic instability of microtubule growth. *Nature, 312*, 237–242.

Shamu, C. E., & Murray, A. W. (1992). Sister chromatid separation in frog egg extracts requires DNA topoisomerase II activity during anaphase. *The Journal of Cell Biology, 117*, 921–934.

Sheetz, M. P., & Spudich, J. A. (1983). Movement of myosin-coated fluorescent beads on actin cables in vitro. *Nature, 303*, 31–35.

Summers, K. E., & Gibbons, I. R. (1971). Adenosine triphosphate-induced sliding of tubules in trypsin-treated flagella of sea-urchin sperm. *Proceedings of the National Academy of Sciences of the United States of America, 12*, 3092–3096.

Szent-Györgyi, A. (1942). The contraction of myosin threads. *Studies from the Institute of Medical Chemistry University Szeged, 1*, 17–26.

Theriot, J. A., Rosenblatt, J., Portnoy, D. A., Goldschmidt-Clermont, P. J., & Mitchison, T. J. (1994). Involvment of profilin in the actin-based motility of L. monocytogenes in cells and in cell-free extracts. *Cell, 76*, 505–517.

Vale, R. D., Schnapp, B. J., Reese, T. S., & Sheetz, M. P. (1985). Organelle, bead, and microtubule translocations promoted by soluble factors from the squid giant axon. *Cell, 40*, 559–569.

Weisenberg, R. C. (1972). Microtubule formation *in vitro* in solutions containing low calcium concentrations. *Science, 177*, 1104–1105.

Woodrum, D. T., Rich, S. A., & Pollard, T. D. (1975). Evidence for a biased directional polymerization of actin filaments using heavy meromyosin prepared by an improved method. *The Journal of Cell Biology, 67*, 231–237.

Ronald D. Vale
Department of Cellular and Molecular Pharmacology
Howard Hughes Medical Institute
University of California, San Francisco
California, USA

SECTION I

Polymer Dynamics

CHAPTER ONE

Actin Filament Dynamics Using Microfluidics

Marie-France Carlier[1], Guillaume Romet-Lemonne[1], Antoine Jégou[1]
Laboratoire d'Enzymologie et Biochimie Structurales, CNRS, Gif-sur-Yvette, France
[1]Corresponding authors: e-mail addresses: carlier@lebs.cnrs-gif.fr; romet@lebs.cnrs-gif.fr; jegou@lebs.cnrs-gif.fr

Contents

Abstract

We describe how combining microfluidics with TIRF and epifluorescence microscopy can greatly facilitate the quantitative analysis of actin assembly dynamics and its regulation, as well as the exploration of issues that were often out of reach with standard single-filament microscopy, such as the kinetics of processes linked to actin self-assembly or the kinetics of interaction with regulators. We also show how the viscous drag force exerted by fluid flowing on the filaments can be calibrated in order to assess the mechanosensitivity of end-binding protein machineries such as formins or adhesion proteins. We also discuss how microfluidics, in conjunction with other techniques, could be used to address the mechanism of coordination between heterogeneous populations of filaments, or the behavior of individual filaments during regulated treadmilling. These techniques also can be applied to study the assembly and regulation of other cytoskeletal polymers such as microtubules, septins, intermediate filaments, as well as the transport of cargoes by molecular motors under a flow-produced load.

Methods in Enzymology, Volume 540
ISSN 0076-6879
http://dx.doi.org/10.1016/B978-0-12-397924-7.00001-7

1. INTRODUCTION

Actin self-assembly into filaments has been extensively studied *in vitro*, following Fumio Oosawa's pioneering works, using bulk solution kinetic and thermodynamic measurements. However, in cellular motile processes, actin filaments cannot be simply considered as free polymers in solution in dynamic equilibrium with monomers and soluble regulators. Instead, they are assembled into higher order, modular, structurally, and functionally distinct meshworks, by protein machineries that operate like "factories" (Small, Stradal, Vignal, & Rottner, 2002) in a cell signaling controlled fashion. The resulting motile processes that produce force and movement by actin assembly at the cell scale are understood through *in vitro* reconstitution studies, which bridge a wide gap from molecular to macroscopic scales (Loisel, Boujemaa, Pantaloni, & Carlier, 1999). However, the intermediate stages of understanding how force and movement are generated by single filaments assembled in a site-directed fashion, or how a collective behavior arises from specific interactions between a small number of polymers and soluble regulators, are not well understood. In particular, in live cells, filament nucleation and polarized growth are mediated by membrane-bound regulators of barbed end dynamics. A complete understanding of actin-based motile processes, therefore, requires the observation and quantitative analysis of actin assembly dynamics and its control by immobilized end-binding proteins at the scale of individual filaments. At a higher scale, thermodynamics systems consisting of a few filaments will be built as an intermediate step preceding the reconstitution of larger self-organized systems mimicking the *in vivo* context.

Total Internal Reflection Fluorescence (TIRF) microscopy emerged as a significant advance for visualizing single molecules and single filaments (Funatsu, Harada, Tokunaga, Saito, & Yanagida, 1995; Vale et al., 1996). Visualizing single filaments using actin labeled with an extrinsic fluorophore provided a vivid illustration of the polymerization time course (Kuhn & Pollard, 2005), kinetics of filament barbed end capping (Kuhn & Pollard, 2007), and processive growth mediated by formins (Breitsprecher, Kiesewetter, Linkner, & Faix, 2009; Kovar & Pollard, 2004; Romero et al., 2004). *In vivo*, TIRF microscopy monitored actin filaments and microtubules localized at the basal plasma membrane of adherent cells (Manneville, 2006). TIRF and epifluorescence single-molecule microscopy also were used as a more accurate alternative to FRAP and photoactivatable fluorophores to monitor actin filament dynamics in motile structures such as lamellipodia and focal adhesions (Adams et al., 2004; Watanabe, 2012).

In vitro analysis of single-filament assembly by TIRF faces technical limitations, which can introduce artifacts and impose conditions that differ from the cellular context. To be constrained within the TIRF field (<200 nm), actin filaments are usually bound to the cover slip by anchored actin side-binding proteins, thus impairing the 2D and 3D mobility of the filaments. Moreover, the lack of control in the position of the tips of growing filaments turns the filament contour analysis into a cumbersome process. The inability to observe filaments during tens of seconds while changing the biochemical conditions prevents the accurate measurement of rapid kinetics.

The recent implementation of microfluidics in an epifluorescence/TIRF microscope setup abolishes these concerns and opens new and exciting experimental possibilities (Jégou, Carlier, & Romet-Lemonne, 2011, 2013; Jégou, Niedermayer, et al., 2011; Niedermayer et al., 2012). Here, we describe the design of a microfluidic method for analyzing actin and briefly outline its advantages for examining actin assembly, most specifically in conditions in which filaments are anchored only at one or the other end.

Future applications of the method include (1) the functional characterization of a large number of actin regulators, individually and in combination; (2) the application of mechanical constraints in order to investigate potential conformation changes and their impact on the affinity for regulatory proteins (Galkin, Orlova, & Egelman, 2012; Yogurtcu, Kim, & Sun, 2012); and (3) the use of multiple nanopatterned solid or soft surfaces to analyze heterogeneous populations of filaments or the coordinated behavior of various actin meshworks. Further advances are expected from the use of sophisticated microfluidics designs involving rapid premixing of reactants or gradient formation, and combining these methods with single-molecule imaging techniques, such as single-molecule FRET (Holden et al., 2010; Sugawa, Arai, Iwane, Ishii, & Yanagida, 2007) or PhADE (Photo-Activation, Diffusion, Excitation) (Loveland, Habuchi, Walter, & van Oijen, 2012).

2. KEY TECHNICAL ASPECTS OF THE MICROFLUIDIC METHOD

We describe the basic microfluidic setup on an inverted microscope, which we have used for several studies on individual actin filaments. More sophisticated systems also can be used, and possible developments of this setup are discussed in Section 4.

2.1. Overview

The flow cell consists of a block of PDMS (polydimethylsiloxane) in which the microchamber is molded, assembled to a standard glass cover slip (other types of chambers can also be used) (Brewer & Bianco, 2008; Tabeling, 2005). Polyetheretherketone (PEEK) tubing connects the microchamber to solution reservoirs in which the pressure is controlled (Fig. 1.1A). Pressure differences drive solutions from the different reservoirs through the microchamber, where they flow side by side. The solution with the dominant flow rate occupies most of the microchamber.

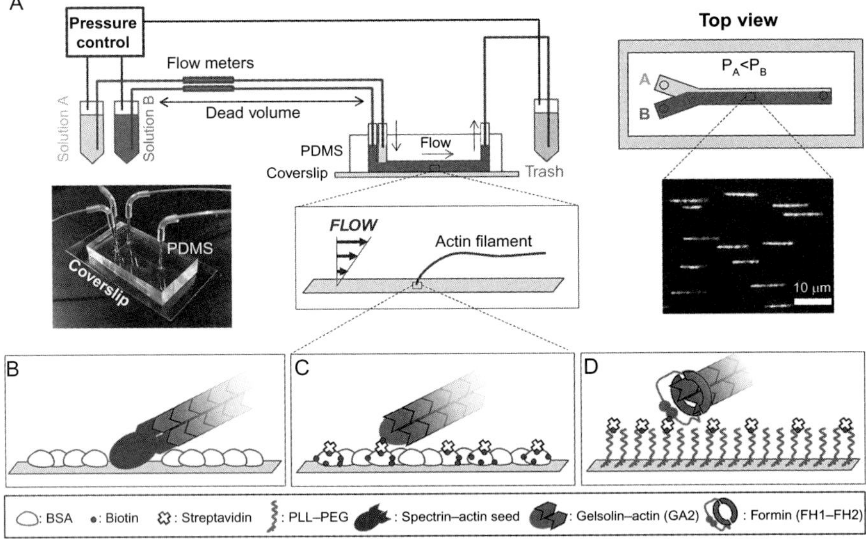

Figure 1.1 General sketch of the microfluidic system. (A) Setup. Left: photo of a cross-shaped microfluidics chamber with three entry channels and schematic side view of the setup; right: top view emphasizing how the dominant flow imposes the solution conditions on anchored filaments and epifluorescence image of labeled actin filaments. (B) Anchoring of filament pointed ends by spectrin–actin seeds immobilized on a BSA-passivated surface. (C) Anchoring of filament barbed ends on biotinylated BSA attached to biotinylated gelsolin–actin (GA2) complexes via streptavidin. (D) Anchoring of filament barbed ends via biotinylated PLL–PEG coupled to an antibody against a tag fused to a formin. (For color version of this figure, the reader is referred to the online version of this chapter.)

The accessible flow velocities are determined by the microchamber cross section, the tubing length, the viscosity of the fluid, and the applied pressures. We typically use chambers that are a few tens of micrometers high, a few hundreds of micrometers wide, connected with roughly a meter of tubing, with a pressure difference of a few millibars to a few tens of millibars, and a fluid with the same viscosity as water (0.9 cP at 25 °C). The resulting flow rates are in the microliter per minute range, corresponding to flow velocities of tens of micrometer per second just above (~100 nm) the cover slip surface.

Individual actin filaments are anchored at one end, in a biochemically controlled fashion (Fig. 1.1B–D), to the bottom of a flow chamber comprising multiple entry channels. Filaments align with the flow and can be exposed to different incoming protein solutions (see Section 3). The flowing fluid exerts a viscous drag force on the filaments, which can be calibrated.

Practically, a typical experiment consists in the following steps: (1) Place the microchamber on the microscope stage and connect it to the microfluidic system; (2) load the whole system with buffer; (3) successively flow in, incubate, and rinse the different solutions required to passivate and functionalize the surface; (4) load the desired protein solutions; (5) perform the experiment by acquiring images while exposing the field of view to the different solutions from the reservoirs; (6) as long as the surface is in a satisfactory state, consecutive experiments can be performed in the same chamber and new stock solutions can be loaded in the system; and (7) disconnect the flow chamber, and rinse and dry the tubing of the microfluidic system. Reservoirs are usually disposable. The tubing can usually be cleaned with surfactants or bases. As the PDMS chamber is not covalently bound to the glass surface, it can be peeled off, rinsed, dried, and reused with a new, clean cover slip.

2.2. Controlling the flow

Controlling the flow rate in each channel is crucial for a microfluidic experiment. The flow rates must be stable over time and respond rapidly to user-imposed changes. Rapidly changing chemical conditions requires the incoming flow from different channels to be precisely controlled (see Section 3). Such control has been greatly improved recently with the advent of microfluidic devices that efficiently control the pressure in the vapor phase of the reservoirs (e.g., from Fluigent or Elvesys), instead of using classical syringe pumps. Flow meters allow the constant and accurate monitoring of each incoming/outgoing flow rate during the experiment.

2.3. Controlling surface properties

As in most microscopy and micromanipulation experiments, it is essential to use a clean surface where the desired proteins are anchored and other proteins will not bind (passivation). A variety of surface passivation techniques can be found throughout the literature. The choice may be based on available time and equipment and on a specific protein's tendency to adsorb on the surface. For some experiments, for example, with only actin and profilin (Jégou, Niedermayer, et al., 2011; Niedermayer et al., 2012), a simple protocol relying on adsorption and passivation with proteins can prove satisfactory (see example 1). Others require a more stringent procedure, such as using a polymer brush to passivate the surface and biotin–streptavidin bonds for the specific anchoring of proteins (example 2).

Example 1—Clean cover slip with detergent solution (e.g., Hellmanex), rinse with water and then with ethanol, blow dry, and assemble to PDMS block. Adsorb spectrin–actin seeds (a few picomolar for a few minutes) and then rinse with buffer. Passivate the surface by incubating with a solution of 5–10% BSA (bovine serum albumin) for a few minutes and then rinse with buffer. Seeds will remain bound to the surface as fluorescent filaments are grown from them and maintained close to the surface by the microfluidic flow, without sticking, and with a low fluorescent background from the surface.

Example 2—Clean cover slip with sequential steps of sonic vibration in pure water, ethanol, acetone, and 1 *M* NaOH, each for 10 min. Rinse extensively with pure water, blow dry, and expose to oxygen (air) plasma for 2 min. Assemble to PDMS block and incubate with a solution of 1 mg mL^{-1} PLL–PEG (poly-L-lysine–polyethyleneglycol), 1% biotinylated, for a few minutes. Rinse thoroughly. Incubate with streptavidin, rinse, and incubate with the biotinylated protein of interest, for example, formins in complex with biotinylated antibodies.

2.4. Calibrating the applied force

The fluid exerts a friction force (viscous drag) on the anchored filament. In order to calibrate the resulting pulling force exerted on the anchoring molecules, one must know the filament's friction (drag) coefficient and determine the local flow rate to which the filament is exposed.

The longitudinal friction coefficient can be measured using an actin filament anchored to a bead held in an optical trap, a few micrometers above the surface. We have thus measured a longitudinal friction coefficient (per unit length) of 6×10^{-4} pN s μm^{-2} for a viscosity of 0.9 cP.

The filaments are anchored to the bottom of the flow cell (Fig. 1.1) where they fluctuate within a local flow velocity gradient. In principle, this local gradient is linear and the flow velocity drops to zero on the cover slip surface, but these features can be altered if the surface is hydrophobic or if it contains structures. The flow profile can be measured by monitoring the velocity of microbeads at different heights, as well as fragments detaching from anchored filaments. The different positions of the fluctuating filament above the surface can be determined by measuring its fluorescence intensity profile using a well-calibrated TIRF illumination. In our experimental conditions, we found that filaments fluctuate rapidly around a plane located 250 nm above the surface (Jégou, Carlier, & Romet-Lemonne, 2013). Their average contour can nonetheless be monitored in TIRF using a sufficiently long image acquisition time, typically 50–100 ms.

The friction force can thus be integrated over the whole filament contour to determine the resulting instantaneous force on the anchoring point. Due to filament fluctuations, this force fluctuates within 30% of its average value, in our conditions.

2.5. Caveats and control experiments

Microscopic air bubbles that reside, unseen, on the walls of the microfluidic system will eventually coalesce into larger bubbles that can block the flow and/or dramatically damage the protein-coated surfaces as they flow through the chamber. Bubbles are efficiently avoided by degasing buffer solutions beforehand and bringing them to the temperature of the setup before loading. At the beginning of the experiment, the whole system should be flushed with buffer at maximum pressure. When disconnecting a reservoir (for refilling or changing its stock solution), a gentle retrograde flow in the corresponding channel will avoid introducing air in the system.

A newly introduced solution requires some time to flow through the tubing connecting the reservoir to the flow cell and fully replace the previous solution. This delay (typically a few minutes) can be measured, for example, by monitoring the fluorescence intensity increase in the microchamber as a fluorescent solution is introduced. These experiments, therefore, require a minimum volume (typically 200 μL) of each solution. To spare chemicals, a smaller volume of the initial surface passivation solution may be directly injected using a micropipette before connecting the microchamber to the microfluidic system.

It is important to control that the fluid flow does not alter the molecular reactions that are being studied (this can be checked by repeating the experiment at different flow rates). Between image acquisitions, if desired, the flow rate can be decreased to a value where filaments are not aligned close to the surface.

3. ASSETS FOR THE STUDY OF ACTIN DYNAMICS

3.1. Improving the observation of single filaments

Compared to standard single-filament studies, the microfluidic setup provides a number of improvements. First, the filaments are maintained close to the surface, enabling TIRFM observation, but without being pinned down by numerous anchoring points or adding methylcellulose, hence avoiding possible artifacts. Second, a large number of filaments are aligned, quasi-straight, and parallel, thereby facilitating position or length measurements. Third, the incoming solution is constantly renewed, and thus filaments are exposed to a stable protein or nucleotide concentration. Fourth, the solution to which the filaments are exposed can be changed very rapidly (<1 s), suppressing the dead time one encounters when flowing a new solution in classical open flow cell and facilitating the monitoring of the same filaments throughout the process. Buffer exchange can be done repeatedly which offers a wealth of new experimental possibilities, as discussed in Section 3.2.

3.2. New experimental capabilities

Here, we highlight some of the key possibilities offered by microfluidics for studies on individual actin filaments and give examples from published work. Many other useful situations can be imagined along these lines for future experiments.

3.2.1 Kinetic measurements and more

Having a tight control over the chemical medium flowing in the microchamber opens numerous avenues for new experiments. Being able to add or remove a protein from the medium, while continuously monitoring the filaments, allows one to perform accurate kinetic measurements of their association to/dissociation from filament ends or sides. For instance, using filaments growing from anchored spectrin–actin seeds in the presence of profilin–actin and formin (Fig. 1.1B), formin processivity can be readily assessed by observing the sustained rapid elongation of a formin-bound barbed end following removal of free formins from the medium. In a variety of experimental designs, it is useful, often essential, to image filaments by

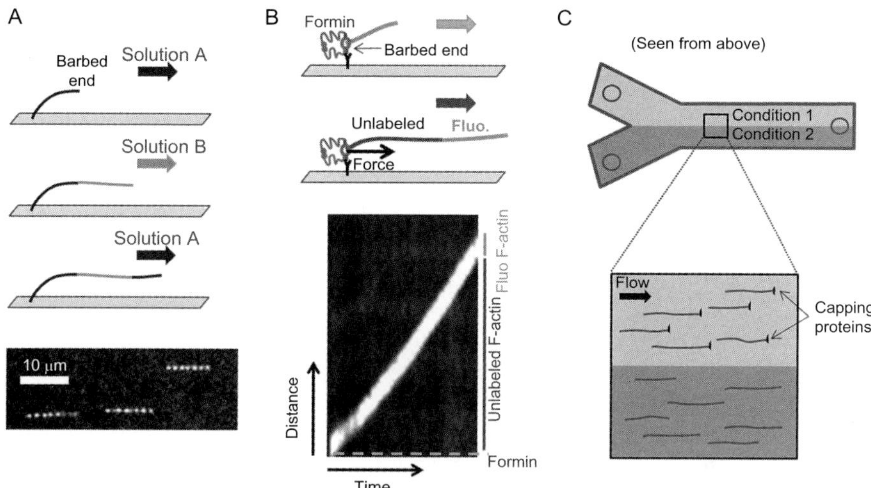

Figure 1.2 Some experimental designs offered by microfluidics. (A) Sequential exposure allows the assembly of hybrid filaments made of actin domains of different nature (label, nucleotide state, or any chemical modification). Bottom: epifluorescence image of "striped filaments" composed of alternating dark and labeled segments. (B) The elongation of filaments from unlabeled actin at surface-bound formins can be monitored tracking a fluorescent segment near the pointed end. Bottom: corresponding kymograph. (C) Two different populations of filaments can coexist in two distinct regions exposed to different incoming solutions. (See the color plate.)

transiently exposing them to a fully unlabeled actin/protein solution, thus enabling epifluorescence monitoring with a very high signal-to-noise ratio, and working at concentrations beyond the range that is attainable with labeled actin/protein and TIRF (Fig. 1.2A). Consecutive exposure to various proteins allows one to monitor how filaments with specific structures/functions are built from various sets of proteins binding cooperatively.

3.2.2 Building hybrid filaments

Being able to flow different solutions in the microchamber allows one to build filaments comprising segments of different nature (Fig. 1.2). This feature has been instrumental in recent studies. For example, elongating filaments from MgATP-actin followed by CrATP-actin results in filaments that have a strict CrADP-Pi cap near their barbed ends, mimicking the situation that would result from a vectorial Pi release mechanism (Jégou, Niedermayer, et al., 2011). Filaments can also be built with an unlabeled segment, in order to study the disassembly dynamics in the absence of fluorophore (Niedermayer et al., 2012). Polymerization from unlabeled actin can be directly monitored when filaments are processively elongated by surface-anchored formins (Fig. 1.2B; Jégou et al., 2013). Building filaments with segments of different natures (nucleotide state, isoform, etc.) allows one to readily study the affinity of side-binding proteins for these different filamentous actin "species." Monitoring the disassembly of such hybrid filaments in the presence of barbed-end-binding proteins (Spire, formin, profilin, capping protein, etc.) allows one to derive their affinities for the various actin species consecutively exposed the depolymerizing barbed end.

3.2.3 Observing subpopulations of filaments

Understanding molecular mechanisms can benefit from being able to follow the individual histories of each filament throughout the experiment, thus providing information on subpopulations or stochastic behaviors that cannot be measured in bulk populations. For example, if only a subset of filaments are capped using a low concentration of capping protein, subsequent exposure to another protein targeting the barbed ends will reveal potential differences in the affinity for capped versus free barbed ends (Montaville et al., 2014). One also can simultaneously monitor filament barbed and pointed end dynamics by growing filaments from anchored spectrin–actin seeds (free barbed ends) and from anchored gelsolin–actin complexes (free pointed ends) in the same chamber, and observe their dynamic behaviors in the presence of controlled concentrations of G-actin and treadmilling regulators. The two filament subpopulations can be distinguished based on differences in their assembly dynamics or labeling; one also can build segregated subpopulations by exposing different regions of the microchamber to different solutions (Fig. 1.2C). These assays provide insight into the treadmilling mechanism at the scale of a population of filaments.

3.2.4 Applying mechanical constraints to many filaments in parallel

Due to filament fluctuations, microfluidics is less accurate than optical trapping as a tool to apply forces to individual filaments. In the optical trap, the force barely fluctuates and the trapped bead position can be measured with nanometric resolution. However, while optical traps typically manipulate one filament at a time, a microfluidic flow can apply tension to tens of filaments in parallel. This feature was essential in collecting sufficient data for statistics on the mechanosensitivity of formins (Jégou et al., 2013).

In addition, experiments are made easier by the fact that microfluidics allows one to apply tension to filaments by anchoring only one end (to the glass surface), as opposed to having the two ends attached to trapped beads.

When filaments grow from spectrin–actin seeds, which are short filament segments anchored to the surface with a random orientation, they bend in order to align with the flow. Microfluidics hence appears as a tool to reversibly apply curvature to tens of filaments simultaneously. As we discuss in Section 4, the ability to orient and curve filaments can be exploited further.

4. PERSPECTIVES

4.1. Expanding the setup

4.1.1 Orienting filaments toward micropatterns or structures

The local orientation of the flow can be modulated in order to orient filaments in different directions (Jégou, Carlier, et al., 2011). Introducing micropatterns or three-dimensional structures in the microchamber provides interesting possibilities for the manipulation of actin filaments. For example, filaments can be oriented toward patterns coated with specific regulatory proteins, or toward series of micropillars that will deviate the flow lines and locally impose desired curvatures to the filaments, thus mimicking cellular obstacles and allowing one to monitor how curvature affects the binding of regulators (Fig. 1.3A).

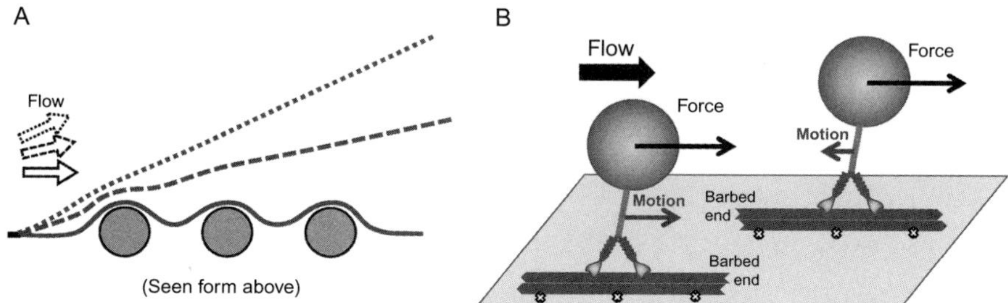

Figure 1.3 Examples of further developments of the microfluidics approach. (A) The curvature of dynamic filaments can be controlled using micropillars and by orienting the flow in different directions. (B) The flow can be used to apply a load to the cargo transported by a barbed-end-directed myosin motor (the same could be done with pointed end motors, or with microtubules and dyneins or kinesins). Motors moving upstream, against an opposing force from the flow, and motors moving downstream, assisted by flow, can be monitored simultaneously in the same assay. (For color version of this figure, the reader is referred to the online version of this chapter.)

4.1.2 Lipid layers

Anchoring proteins to a layer of lipids covering the cover slip surface can be used to mimic physiological conditions of membranes and also enable proteins to diffuse and reorganize on the fluid lipid surface. Filaments growing from the surface move with the flow, and solid barriers can be used to immobilize them. This technique has been developed for the study of DNA filaments and leads to the formation of "filament curtains" along the barriers (Gibb, Silverstein, Finkelstein, & Greene, 2012). The lipid-bound anchor is pressed against the barrier, and this contact may impair the function of the anchor if it is an active protein. The impact of the barrier on an active anchor was recently illustrated when filaments were grown from formins attached to a lipid layer, and formin activity was shown to depend on the height of the barrier (Courtemanche, Lee, Pollard, & Greene, 2013). Future developments, perhaps using barriers of different compositions, may solve this problem and provide a means to exploit the promising possibilities offered by a fluid surface.

4.1.3 Coupling to other techniques

The microfluidics approach can be readily coupled to virtually any optical microscope technique. For example, single-molecule fluorescence, the use of photoactivatable probes, and PhADE technique (Loveland et al., 2012) could be used in a microfluidic setup, in order to provide further insight into molecular reactions. In addition, molecular force sensors could be used to provide an *in situ* measurement of the applied forces (Grashoff et al., 2010).

4.2. Beyond single-filament actin dynamics

Microfluidics is emerging as a powerful method to manipulate cellular and subcellular processes at physiologically relevant scales, and study mechanochemical coupling in reactions that take place at or across membranes in gliding of microorganisms (Ducret, Theodoly, & Mignot, 2013), intracellular uptake of materials (Wang, Zhan, Bao, & Lu, 2012), or cell–cell and cell–ECM adhesion. We will not discuss these aspects here. Beyond actin, micro- and nano-multiple patterning of the cover slip (for reviews, see Vignaud, Blanchoin, & Thery, 2012 and Chapter 16) also will be instrumental to solve more complex issues in the cytoskeleton field, including the dynamics of microtubules, intermediate filaments, and molecular motors. A few, but not exhaustive, possibilities are discussed below.

4.2.1 Other cytoskeletal filaments

We have shown here how microfluidics can be used to manipulate and observe actin filaments; most of these features apply to virtually any semi-flexible filaments. For instance, many of the issues addressed for actin have analogies in microtubules, including studying the dynamics of the GTP cap and the binding of end-tracking proteins. Microfluidics also could be used to curve microtubules, although with a gentler curvature due to their larger persistence length than actin. Microfluidics should also be very useful to study the assembly dynamics of less-characterized systems, such as intermediate filaments or bacterial cytoskeletal filaments.

4.2.2 Motor proteins

Depolymerizing motors that disassemble the plus ends of microtubules can be tracked with an approach inspired by formins tracking along depolymerizing actin filaments (Jégou et al., 2013). Microfluidics also can be used to apply forces to microbeads or vesicles of various sizes attached to motors walking along the sides of actin filaments or microtubules. While less precise than optical trapping, this method enables simultaneous measurements on a large population of individual motors (Fig. 1.3B). The effect of a force opposing or assisting motor movement can be monitored in the same assay using filaments with either one or the other end pointing downstream.

4.2.3 Filament networks

Networks of cross-linked and/or branched actin filaments (comet tails), growing from anchored beads or micropatterns, also could be sequentially exposed to different protein solutions mimicking signaling events and the response of the actin network monitored. In these situations, microfluidics would be used simply as a rapid method of protein exchange. One may want to avoid exerting a force on the filament network, and this could be achieved by using a structure (e.g., a porous membrane or an array of micropillars) that would mechanically isolate the filaments from the flowing fluid, while allowing the rapid diffusion of proteins. Such integrated systems represent a new avenue for microfluidics-assisted microscopy, bridging the gap between single-molecule and cell-scale studies of the cytoskeleton.

REFERENCES

Adams, M. C., Matov, A., Yarar, D., Gupton, S. L., Danuser, G., & Waterman-Storer, C. M. (2004). Signal analysis of total internal reflection fluorescent speckle microscopy (TIR-FSM) and wide-field epi-fluorescence FSM of the actin cytoskeleton and focal adhesions in living cells. *Journal of Microscopy*, *216*, 138–152.

Breitsprecher, D., Kiesewetter, A. K., Linkner, J., & Faix, J. (2009). Analysis of actin assembly by in vitro TIRF microscopy. *Methods in Molecular Biology, 571*, 401–415.

Brewer, L. R., & Bianco, P. R. (2008). Laminar flow cells for single-molecule studies of DNA-protein interactions. *Nature Methods, 5*, 517–525.

Courtemanche, N., Lee, J. Y., Pollard, T. D., & Greene, E. C. (2013). Tension modulates actin filament polymerization mediated by formin and profilin. *Proceedings of the National Academy of Sciences of the United States of America, 110*, 9752–9757.

Ducret, A., Theodoly, O., & Mignot, T. (2013). Single cell microfluidic studies of bacterial motility. *Methods in Molecular Biology, 966*, 97–107.

Funatsu, T., Harada, Y., Tokunaga, M., Saito, K., & Yanagida, T. (1995). Imaging of single fluorescent molecules and individual ATP turnovers by single myosin molecules in aqueous solution. *Nature, 374*, 555–559.

Galkin, V. E., Orlova, A., & Egelman, E. H. (2012). Actin filaments as tension sensors. *Current Biology, 22*, R96–R101.

Gibb, B., Silverstein, T. D., Finkelstein, I. J., & Greene, E. C. (2012). Single-stranded DNA curtains for real-time single-molecule visualization of protein-nucleic acid interactions. *Analytical Chemistry, 84*, 7607–7612.

Grashoff, C., Hoffman, B. D., Brenner, M. D., Zhou, R., Parsons, M., Yang, M. T., et al. (2010). Measuring mechanical tension across vinculin reveals regulation of focal adhesion dynamics. *Nature, 466*, 263–266.

Holden, S. J., Uphoff, S., Hohlbein, J., Yadin, D., Le Reste, L., Britton, O. J., et al. (2010). Defining the limits of single-molecule FRET resolution in TIRF microscopy. *Biophysical Journal, 99*, 3102–3111.

Jégou, A., Carlier, M. F., & Romet-Lemonne, G. (2011). Microfluidics pushes forward microscopy analysis of actin dynamics. *BioArchitecture, 1*, 271–276.

Jégou, A., Carlier, M. F., & Romet-Lemonne, G. (2013). Formin mDia1 senses and generates mechanical forces on actin filaments. *Nature Communications, 4*, 1883.

Jégou, A., Niedermayer, T., Orban, J., Didry, D., Lipowsky, R., Carlier, M. F., et al. (2011). Individual actin filaments in a microfluidic flow reveal the mechanism of ATP hydrolysis and give insight into the properties of profilin. *PLoS Biology, 9*, e1001161.

Kovar, D. R., & Pollard, T. D. (2004). Insertional assembly of actin filament barbed ends in association with formins produces piconewton forces. *Proceedings of the National Academy of Sciences of the United States of America, 101*, 14725–14730.

Kuhn, J. R., & Pollard, T. D. (2005). Real-time measurements of actin filament polymerization by total internal reflection fluorescence microscopy. *Biophysical Journal, 88*, 1387–1402.

Kuhn, J. R., & Pollard, T. D. (2007). Single molecule kinetic analysis of actin filament capping. Polyphosphoinositides do not dissociate capping proteins. *Journal of Biological Chemistry, 282*, 28014–28024.

Loisel, T. P., Boujemaa, R., Pantaloni, D., & Carlier, M.-F. (1999). Reconstitution of actin-based motility of Listeria and Shigella using pure proteins. *Nature, 401*, 613–616.

Loveland, A. B., Habuchi, S., Walter, J. C., & van Oijen, A. M. (2012). A general approach to break the concentration barrier in single-molecule imaging. *Nature Methods, 9*, 987–992.

Manneville, J. B. (2006). Use of TIRF microscopy to visualize actin and microtubules in migrating cells. *Methods in Enzymology, 406*, 520–532.

Montaville, P., Jégou, A., Pernier, J., Compper, C., Guichard, B., Mogessie, B., et al. (2014). Spire and Formin 2 synergize and antagonize in regulating actin assembly in meiosis by a ping-pong mechanism. *PLoS Biology*, in press.

Niedermayer, T., Jégou, A., Chieze, L., Guichard, B., Helfer, E., Romet-Lemonne, G., et al. (2012). Intermittent depolymerization of actin filaments is caused by photo-induced dimerization of actin protomers. *Proceedings of the National Academy of Sciences of the United States of America, 109*, 10769–10774.

Romero, S., Le Clainche, C., Didry, D., Egile, C., Pantaloni, D., & Carlier, M.-F. (2004). Formin is a processive motor that requires profilin to accelerate actin assembly and associated ATP hydrolysis. *Cell, 119*, 419–429.

Small, J. V., Stradal, T., Vignal, E., & Rottner, K. (2002). The lamellipodium: where motility begins. *Trends in Cell Biology, 12*, 112–120.

Sugawa, M., Arai, Y., Iwane, A. H., Ishii, Y., & Yanagida, T. (2007). Single molecule FRET for the study on structural dynamics of biomolecules. *Biosystems, 88*, 243–250.

Tabeling, P. (2005). *Introduction to microfluidics.* Oxford: Oxford University Press.

Vale, R. D., Funatsu, T., Pierce, D. W., Romberg, L., Harada, Y., & Yanagida, T. (1996). Direct observation of single kinesin molecules moving along microtubules. *Nature, 380*, 451–453.

Vignaud, T., Blanchoin, L., & Thery, M. (2012). Directed cytoskeleton self-organization. *Trends in Cell Biology, 22*, 671–682.

Wang, J., Zhan, Y., Bao, N., & Lu, C. (2012). Quantitative measurement of quantum dot uptake at the cell population level using microfluidic evanescent-wave-based flow cytometry. *Lab on a Chip, 12*, 1441–1445.

Watanabe, N. (2012). Fluorescence single-molecule imaging of actin turnover and regulatory mechanisms. *Methods in Enzymology, 505*, 219–232.

Yogurtcu, O. N., Kim, J. S., & Sun, S. X. (2012). A mechanochemical model of actin filaments. *Biophysical Journal, 103*, 719–727.

CHAPTER TWO

Bacterial Actin-Like Proteins: Purification and Characterization of Self-Assembly Properties

Natalie A. Petek[*], **R. Dyche Mullins**[*,†,1]

[*]Department of Cellular and Molecular Pharmacology, University of California, San Francisco, California, USA

[†]Howard Hughes Medical Institute, Chevy Chase, Maryland, USA

[1]Corresponding author: e-mail address: dyche@mullinslab.ucsf.edu

Contents

Abstract

The sequence, structure, and assembly dynamics of eukaryotic actins are conserved across phyla. In contrast, actin-like proteins (ALPs) from eubacteria share little sequence homology, form polymers with different architectures, and assemble with different kinetics. The structural and functional diversity of the bacterial ALPs appears to arise from their (i) high degree of functional specialization and (ii) small number of regulatory factors. To understand the molecular mechanism by which a given ALP carries out its biological function, we must, therefore, understand its unique architecture and assembly dynamics. In this chapter, we provide a basic toolbox for studying the self-assembly of uncharacterized ALPs, including methods for characterizing the architecture and stability of polymers, specifying the mechanism of their nucleation, and measuring their rate of growth. Determining these basic properties provides a stable base for more complex reconstitutions of biological function.

Methods in Enzymology, Volume 540
ISSN 0076-6879
http://dx.doi.org/10.1016/B978-0-12-397924-7.00002-9

1. INTRODUCTION

Actin filaments are everywhere. They (and their relatives) are found in every domain of life. They exist in every known eukaryotic cell as well as in many eubacterial and archaeal cells. Muscle actin was first discovered in the 1940s (Straub, 1943); nonmuscle actins were discovered in the 1960s (Hatano & Oosawa, 1966); and bacterial actin-like proteins (ALPs) were discovered in the early 2000s (Jones, Carballido-López, & Errington, 2001). As a consequence of this order of discovery, we know much more about the biochemical properties and cellular functions of eukaryotic actins than prokaryotic ALPs. Cell biological functions of these ALPs include coordinating cell division machinery (FtsA) (Pichoff & Lutkenhaus, 2005), controlling cell wall synthesis (MreB) (Dominguez-Escobar et al., 2011; Garner et al., 2011; van Teeffelen et al., 2011), segregating DNA (ParM, AlfA, Alp7A) (Becker et al., 2006; Campbell & Mullins, 2007; Derman et al., 2009), and aligning organelles (MamK) (Komeili, Li, Newman, & Jensen, 2006). Unlike the actin cytoskeleton in eukaryotes, whose architecture and function are determined by a multitude of regulatory proteins, each bacterial actin appears to be adapted to a specific function and to possess specific biochemical properties that reduce its need for accessory factors.

A current frontier in prokaryotic cell biology is to understand the assembly dynamics and regulation of these ALPs and the mechanisms by which they help to establish order in eubacterial and archael cytoplasm. Although they participate in analogous processes, prokaryotic and eukaryotic actins differ in many ways. For example, they (1) share low sequence homology (Bork, Sander, & Valencia, 1992; Derman et al., 2009), (2) assemble with different dynamics (Garner, Campbell, & Mullins, 2004; Polka, Kollman, Agard, & Mullins, 2009; Rivera, Kollman, Polka, Agard, & Mullins, 2011), (3) form filaments with different architectures (Polka et al., 2009; van den Ent, Møller-Jensen, Amos, Gerdes, & Löwe, 2002), and (4) are controlled by different regulatory factors (Draper et al., 2011). Sequence analysis (Derman et al., 2009) also reveals a low level of sequence conservation among members of the 40 families of ALPs in eubacteria and archaea (ALP1–ALP41; 30% sequence similarity). As might be expected from this low sequence similarity, the architecture and dynamics of ALPs *in vitro* differ significantly from one other as well as from conventional actin.

The most well-studied bacterial actin, ParM, promotes inheritance of plasmids in Gram-negative enteric pathogens. ParM filaments are simple, polymerization-based motors that push plasmids to the poles of rod-shaped cells (Campbell & Mullins, 2007). Four basic biochemical properties of ParM, none of which are true of eukaryotic actin, appear to contribute to this cellular function: (1) dynamic instability (Garner et al., 2004), (2) rapid spontaneous nucleation (Campbell & Mullins, 2007), (3) symmetrical, bidirectional elongation (Garner, Campbell, Weibel, & Mullins, 2007), and (iv) weak antiparallel self-association (Gayathri et al., 2012). Arguably, the most surprising property of ParM is its dynamic instability, an energy-dependent switch from filament growth to catastrophic shortening. This inherent instability of ParM filaments produces a high concentration of monomeric ParM that is available to do polymerization-coupled work. Interestingly, another DNA-segregating ALP, AlfA, has a remarkably different set of assembly properties from ParM: (1) it is not inherently dynamically unstable (Polka et al., 2009), (2) its nucleation is controlled by accessory factors (Polka, Kollman, Mullins, 2014), (3) its elongation is polarized (Polka, Kollman, Mullins, 2014), and (4) it strongly self-associates and forms large filament bundles (Polka et al., 2009).

The goal of this chapter is to provide a set of techniques for characterizing the basic assembly properties of a prokaryotic actin-like protein. We will begin by discussing general strategies for expressing and purifying bacterial ALPs. We will then describe how to use three basic techniques—right-angle light scattering, electron microscopy, and total internal reflection fluorescence microscopy—to characterize polymer assembly. We will describe how to use right-angle light scattering to (1) measure the critical concentration for polymerization; (2) determine whether the polymer is dynamically unstable; (3) estimate the size of the filament nucleus; and (4) determine the rate of spontaneous nucleation. We will explain how to use electron microscopy to characterize polymer architecture and determine whether filament assembly is symmetrical or polarized. Finally, we will describe how to use TIRF microscopy to directly observe filament elongation. Determining the assembly properties of ALPs by these techniques is an essential step in understanding the molecular mechanisms by which they carry out their cellular functions.

1.1. Protein expression and purification

1.1.1 Expression

Unlike eukaryotic actins, which do not fold properly in bacteria, high expression of properly folded ALPs is relatively easy to achieve by conventional bacterial expression methods. We typically use expression strains of *E. coli* and induce protein expression overnight at 18 °C with 0.2 m*M* IPTG. For proteins not native to *E. coli*, say an ALP normally expressed in Gram-positive organisms, efficient expression may require the gene to be codon optimized (Polka et al., 2009). We generally avoid affinity tags and purify wild-type (or nearly wild-type) proteins using a combination of selective precipitation and conventional chromatography. Polyhistidine affinity tags, in particular, have been shown to alter the solubility and assembly properties of actin-like polymers (unpublished). For labeling the protein with maleimide derivatives (e.g., fluorescent dyes or biotin), we sometimes append a tripeptide sequence (lysine-cysteine-lysine) to the C-terminus with a short linker. In the cases we have tested, (R1 ParM, pB171 ParM, AlfA), this does not alter filament architecture or assembly dynamics.

1.1.2 Protein purification

To purify native, bacterial ALPs, we rely on three general properties: (1) high expression level; (2) relatively low solubility in ammonium sulfate; and (3) ATP-dependent polymerization. A 2 L culture of *E. coli* (~40 g packed cells) often yields 50–150 mg of purified protein. Briefly (Fig. 2.1), we lyse cells (emulsiflex or microfluidizer) and then clarify the lysate with a high-speed spin (60 min at 140,000 × *g*) to remove cellular debris and insoluble protein. We then fractionate the extract using ammonium sulfate. Optimum ammonium sulfate concentrations for selective precipitation should be determined for each protein (30–50% for AlfA, 10% for ParM). After resuspending the protein, we dialyze it overnight into polymerization buffer without ATP. We then warm the protein to room

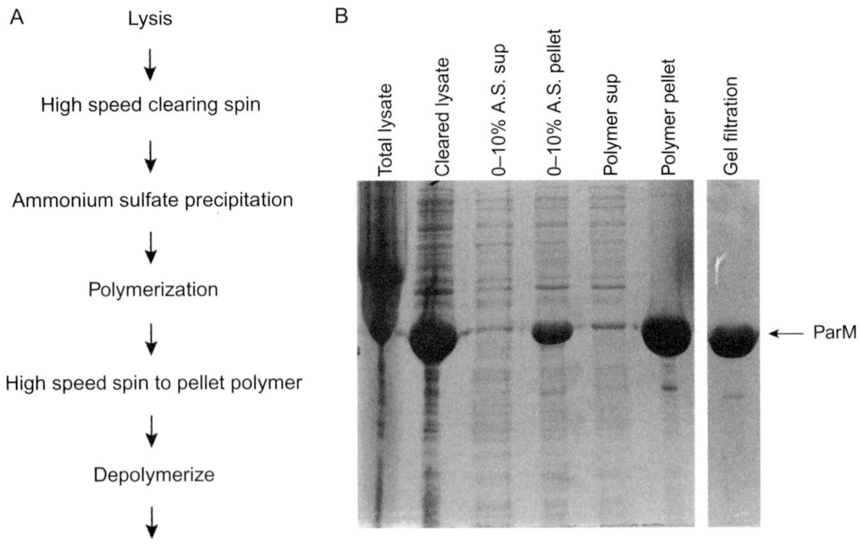

Figure 2.1 (A) Schematic of a generalized purification scheme. (B) Gel of purification scheme for R1 ParM.

temperature and induce polymerization by adding ATP and $MgCl_2$ to a final concentration of 5–10 m*M* each. We pellet the polymer by high-speed centrifugation (30 min at 140,000 × *g*) and then resuspend the pellet in 1/10 volume of depolymerization buffer. We typically dialyze the resuspended protein overnight to remove residual ATP and ensure complete depolymerization. Finally, we gel filter the protein on Superdex S-200 resin, pool the purest protein fractions and dialyze them into storage buffer containing a cryoprotectant, typically 20% glycerol. We divide the protein into small aliquots and snap freeze them in liquid nitrogen. The ALPs that we have studied are all stable at −80 °C and, unlike eukaryotic actin, they do not require the presence of nucleotide in the storage buffer.

1.1.3 Buffers

Lysis buffer

25 m*M* Tris (pH 7.5–8), 100 m*M* KCl, 1 m*M* EDTA, 1 m*M* DTT, 1 m*M* PMSF

Depolymerization Buffer

25 m*M* Tris (pH 7.5–8), 300 m*M* KCl, 5 m*M* EDTA, 1 m*M* DTT

Note: EDTA concentration should be equivalent to $MgCl_2$ added for polymerization.

Storage buffer

25 m*M* Tris (pH 7.5–8), 100 m*M* KCl, 1 m*M* $MgCl_2$, 1 m*M* DTT, 20% glycerol

Polymerization buffer

25 m*M* Tris (pH 7.5–8), 100 m*M* KCl, 1 m*M* $MgCl_2$, 1 m*M* DTT

TIRF buffer

25 m*M* Tris–HCl (pH 7.5), 100 m*M* KCl, 1 m*M* $MgCl_2$, 1 m*M* DTT, 0.8% methylcellulose, 0.5% BSA

Cross-linking buffer

15 m*M* Imidazole (pH 7) (no primary amines, i.e., Tris), 100 m*M* KCl, 1 m*M* $MgCl_2$

Note: No reducing agent in this buffer.

1.2. Bulk measurements

1.2.1 Light scattering versus FRET

We use bulk assays, such as FRET or light scattering, to determine various polymerizations, such as the critical concentration versus the steady-state monomer concentration, kinetics of polymerization, and nucleation rates. Light scattering is a powerful technique as it enables observation of kinetic properties in the absence of protein dyes, which have been shown to affect the kinetics of polymerization for actins and ALPs. Light scattering, however, can be quite noisy, making it sometimes difficult to obtain data with an interpretable signal-to-noise. In such cases, we turn to FRET as an alternative. FRET is also linear over a greater range of protein concentrations and filament lengths. Protein should always be buffer exchanged to remove glycerol and hard spun to remove any aggregate formed as a result of freeze/thaw.

For optimal results, the fluorimeter settings and assay conditions should be optimized for a particular system. Of particular concern is making sure that there is a sufficiently high intensity of incident light to minimize signal-to-noise, while still remaining in the linear range of the detector across the range of conditions for a particular experiment. The intensity of the incident beam can be modulated either with slits that physically obstruct light or dialing the diaphragm appropriately. If necessary, the gain on the PMT can be tuned to obtain a proper signal. For comparison of reactions, it is crucial that the fluorimeter settings remain identical for all conditions tested. For light scattering, particular care must be taken with respect to cleanliness of the quartz cuvette. Dirt or protein aggregate on the inside of the cuvette can scatter light or nucleate small bubbles, both of which will severely diminish the quality of data. Likewise, dirt in or on the window can contribute to scattered light. For this reason, we recommend soaking cuvettes in nitric acid for 30 min prior to use.

1.2.2 Light scattering

1. Set up fluorimeter. Optimize settings, including excitation intensity and detector sensitivity.
2. The excitation wavelength should be set at 320 nm (wavelength should be sufficiently high to avoid buffer scattering and absorbance by aromatic residues, but low enough to observe the smallest polymers possible).
3. Make a 2× solution of ALP protein in polymerization buffer and a 2× solution of equimolar ATP/Mg^{++}. ATP concentration should be saturating and will need to be experimentally determined.
4. To begin the polymerization reaction, place quartz cuvette containing 50 μL of 2× protein solution in the fluorimeter and start the program (should set collection time for 60–2000+s depending on the kinetics of polymerization).
5. Once the program is running, manually close the shutter and add saturating amount of ATP/Mg^{++} to initiate polymerization. Mix rapidly by carefully pipetting up and down three times. Be extremely careful to not

to generate bubbles. Reopen the shutter and observe polymerization. *Note*: This step should be done as quickly as possible.

6. Polymerization is observed as an increase in light scattered at 320 nm.

1.2.3 FRET

The basic strategy is to polymerize a combination of unlabeled protein with protein labeled with two different fluorophores (a FRET pair). Copolymerization of these constructs will result in a FRET signal proportional to the amount of polymer in solution.

1. Fluorimeter settings should be such that the excitation monochrometer is set to an appropriate wavelength for your donor FRET dye. The emission monochrometer should be set to as near peak emission of your acceptor dye as possible (avoid overlap with donor dye emission spectrum).
2. Polymerization reactions should proceed as with light scattering except that your 2× protein mix will be composed of a mixture of unlabeled monomer, ~15% donor-labeled monomer, and ~15% acceptor-labeled monomer (actual % labeled monomer will need to be experimentally determined). Polymerization is then observed as an increase in intensity at the appropriate acceptor wavelength.

1.2.3.1 Critical concentration versus steady-state monomer concentration

Using these techniques, valuable parameters of polymer assembly can be extracted, such as the steady-state monomer concentration and the critical concentration. The maximum intensity of a polymerization curve scales with protein concentration linearly. Therefore, the steady-state monomer concentration can be extracted from these data by plotting the steady-state intensities (maximum intensity) from a protein titration versus protein concentration as shown in Fig. 2.2A. A line fit to the data will intersect the x-axis at the steady-state monomer concentration. The critical concentration can be determined by the same type of experiment in the presence of a nonhydrolyzable ATP analog or with a mutant deficient in hydrolyzing ATP (i.e., ParM_E148A; Fig. 2.2A) (Garner et al., 2004).

1.2.3.2 Nucleation

Spontaneous assembly of actin and actin-like filaments is a nucleation condensation reaction; it occurs in two distinct steps (Oosawa & Kasai, 1962). Nucleation is typically a slow process; once a critical nucleus size is exceeded, rapid elongation is favored. Bacterial ALPs have variable rates of nucleation and a variety of critical nucleus sizes. It is therefore necessary

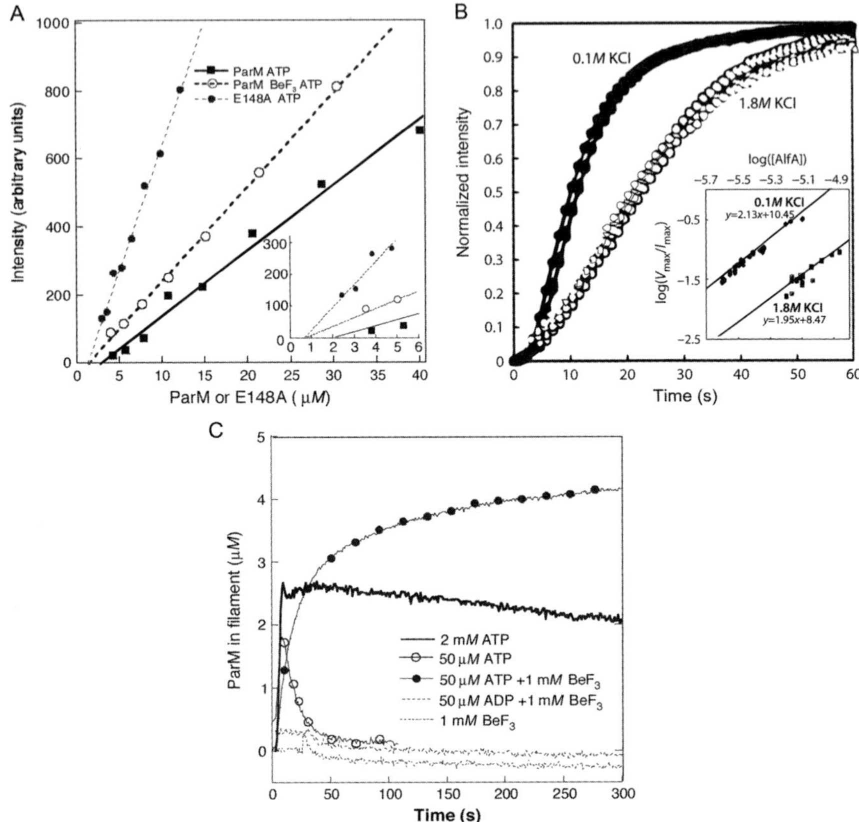

Figure 2.2 Assembly dynamics as determined by bulk measurements. Steady-state monomer concentration of ParM (closed squares), ATP critical concentration (closed circles), and BeF_3 critical concentration (open circles) indicating that ATP hydrolysis is required for filament turnover and that filaments are stabilized in the ATP or ADP-Pi (Garner et al., 2004). (B) Nucleation kinetics of AlfA in the presence (closed circles) and absence (open circles) of DNA-binding protein AlfB. Inset shows that nucleation occurs by a different mechanism in the presence of AlfB as indicated by the slope of the log/log plot. (C) Characteristics of dynamic instability. In saturating ATP (solid line) or ATP supplemented with BeF_3 (solid circles), filaments are stabilized over long periods. In limiting ATP concentrations (open circles) and in ADP supplemented with BeF_3 (dashed line), polymerization is either highly ephemeral or nonexistent, respectively (Garner et al., 2004). *(B) Adapted from Polka et al. (2009).*

to experimentally determine the nucleus size and nucleation kinetics for uncharacterized bacterial actins.

Using a rapid mixer attachment for the fluorimeter with either FRET or light scattering, it is possible to measure the lag phase for nucleation, as well as to determine the number of kinetically resolvable nucleation steps (and therefore an estimate of nucleus size). This analysis is described in detail by Nishida and Sakai (1983) and involves a log/log plot of the ratio of maximum polymerization velocity to maximum intensity versus protein concentration (Fig. 2.2B). Twice the slope ($2n$) of this line represents the number of kinetically resolvable steps involved in nucleating the particular filament. Therefore, $2n+1$ estimates the size of the nucleus, which is defined as the number of monomers that need to assemble for elongation to be favored over disassembly (Fig. 2.2B). Additionally, putative regulatory factors can be tested as potential nucleators using this technique.

1.2.3.3 Dynamic instability

Dynamic instability can be inferred from light scattering or FRET data by the following approaches. In limiting ATP, a characteristic polymerization curve is generated by dynamically unstable proteins. The initial elongation phase is followed by a rapid decrease in intensity, then a slow increase to equilibrium or a return to baseline (Fig. 2.2C). This overshoot and rapid decline does not occur in saturating ATP, or in mutants incapable of ATP hydrolysis, indicating that the rapid decrease is the result of dynamic instability-driven depolymerization. Likewise, the dynamic instability can be repressed by addition of excess phosphate analog (BeF_3), which stabilizes filaments by maintaining monomers within the filament in an ADP-Pi state (Fig. 2.2C).

Note: It is critical to verify that the increase in light scattering is due to polymerization and not protein aggregation. Light scattering and FRET experiments should be accompanied by negative stain EM using identical conditions for polymerization. Light scattering without EM is like driving in the dark without your headlights on.

1.2.4 Negative stain electron microscopy

Interestingly, the bacterial ALPs form a variety of filamentous structures. ParM forms two stranded filaments with a left-handed helical twist (Fig. 2.3A, left), while AlfA forms bundles of filaments (Fig. 2.3A, center). Under high-salt conditions, AlfA bundles can be broken into two stranded filaments (Fig. 2.3A, right). Filament reconstructions of ParM (van den Ent et al., 2002) and AlfA (Polka et al., 2009) show that both proteins form two stranded, helical filaments, but with distinct architectures. Both form helical filaments opposite in handedness to eukaryotic actin; additionally, AlfA filaments are highly twisted, which enhances lateral filament interactions and bundling. Explicit staining procedures are described by Huxley (1963).

Briefly, grids are stained by

1. Applying 5 μL of polymerized ALPs to glow discharged grids and incubating for 30 s.

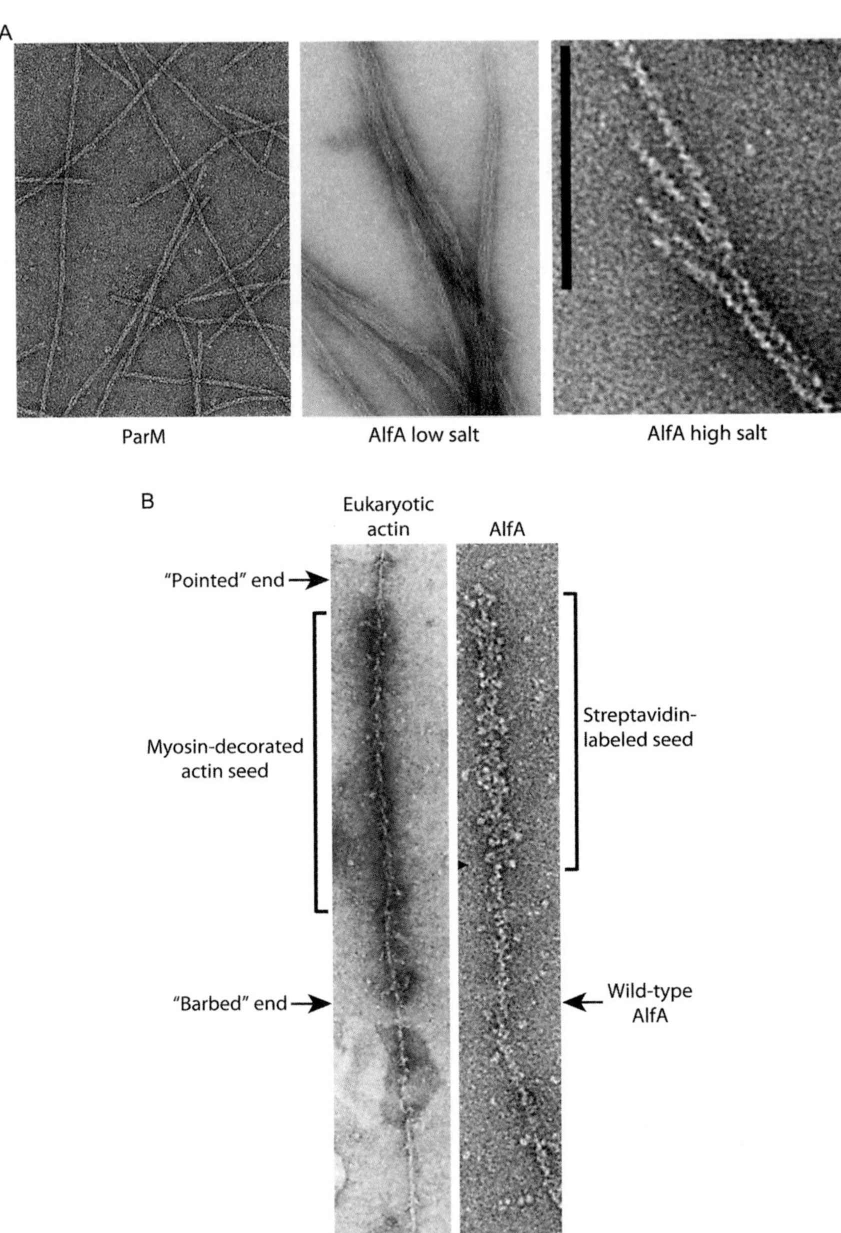

Figure 2.3 Filament architecture and polarity. (A) Filament architecture of ParM and AlfA. ParM forms two stranded helical filaments (left). AlfA forms bundles of filaments (center), which can be broken up by high salt (right). (B) Polymerization of wild-type polymer from a stabilized, labeled seed to determine filament polarity. As an example, wild-type actin elongates bidirectionally from meromyosin-labeled stable seeds (left) (Woodrum, Rich, & Pollard, 1975). AlfA (right) seeds with wild-type protein growing from a streptavidin-labeled seed (Polka, Kollman, Mullins, 2014). *(A) AlfA images from Polka et al. (2009).*

2. Washing grids 3 × in polymerization buffer, blotting between washes.
3. Staining 3 × with 0.75% uranyl formate, third stain step is done for 30 s.

1.3. TIRF: Visualizing filament dynamics

Filament dynamics are best characterized by direct visualization using TIRF microscopy (Fig. 2.4). Using this approach, one can directly measure the rate at which single filaments assemble, whether catastrophic disassembly occurs (dynamic instability), whether filaments undergo treadmilling, and how any accessory factors may modulate these dynamic behaviors.

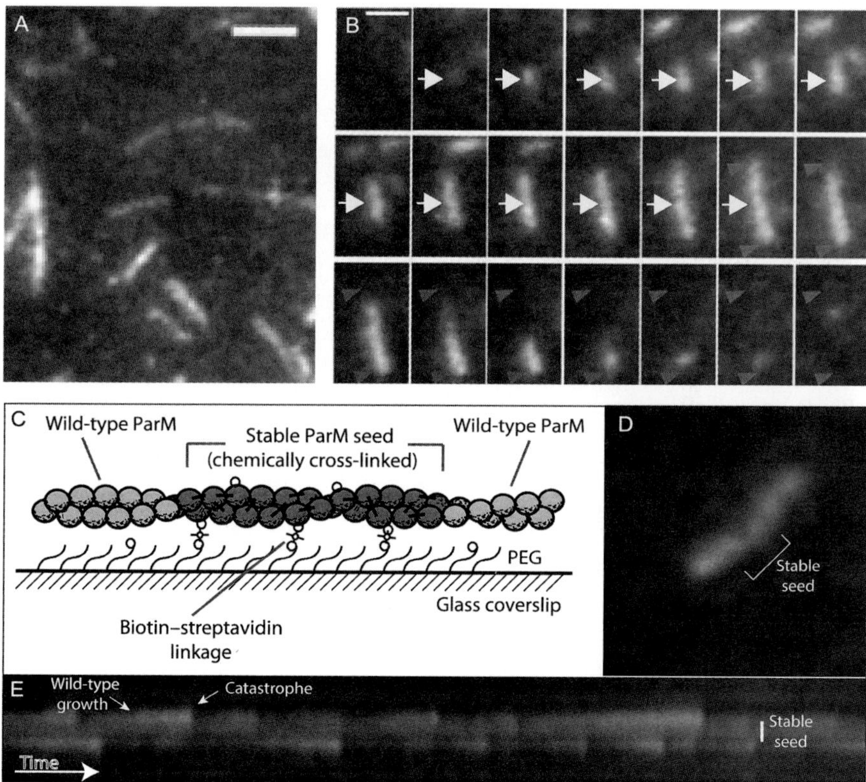

Figure 2.4 Single-filament dynamics by TIRF. (A) A typical TIRF field with GFP-ParM filaments polymerized in AMP-PNP. Scale bar 5 μm. (B) A montage of a single filament over time shows bidirectional growth (white arrow) and catastrophic disassembly (red arrows). Scale bar 2 μm (Garner et al., 2004). (C) Schematic of two-color seeded filament assembly by TIRF microscopy. (D) A representative filament shows bidirectional growth. (E) Kymograph analysis shows bidirectional growth and catastrophic disassembly over many cycles. (See the color plate.)

1.3.1 Labeling strategies

Labeling of bacterial actins can be done by purifying the bacterial actin of choice with a c-terminal GSKCK amino acid sequence added to the construct for covalent cross-linking to maleimide dyes or biotin. The two positively charged lysines cause the flanked cysteine to be hyperreactive, enabling efficient and rapid labeling. If possible, endogenous cysteines should be mutated to alanines to create a cys-lite mutant. This will ensure that maleimide labeling will yield a protein with only a single dye attached. If this is not possible, stoichiometry of dye:protein can be adjusted to favor a single-labeling event of the highly reactive c-terminal cysteine.

1.3.2 Surface chemistry: APTES-PEG-biotin glass

1.3.2.1 Silanized glass

1. Place coverslips (mark with diamond pen) in a 500–1000 mL glass beaker.
2. Sonicate in 200 mL of 3 *M* NaOH for 30 min. Swirl every 10 min.
3. Wash thoroughly with MilliQ water.
4. Wash with 100% isopropanol.
5. Make 200–300 mL of 5% APTES (3-aminopropyltriethoxysilane—Sigma Aldrich) solution in isopropanol.
6. Next, remove coverglass individually, rinse with isopropanol, and then drop into 5% APTES solution made in isopropanol.
7. Sonicate for 20 min.
8. Rinse all glass 3 × with 100% isopropanol.
9. Remove coverslips one-by-one. Dip into a series of 2–3 beakers containing isopropanol. Rack coverslips in ceramic holders.
10. Bake for ≥2 h at 90 °C.
11. Place coverslips in beaker containing 100% EtOH, sonicate for 15 min.
12. Wash 2 × in 100% EtOH.

1.3.2.2 Pegylation

1. Rinse coverslips in MilliQ water. Oven or spin dry.
2. Weigh ~27 mg of CH_3O-PEG-NHS and 2–3 mg Biotin-PEG-NHS (JenKem Technology, USA). Add 1 mL dimethylformamide. 30 mg/mL final concentration.
3. Place coverslip in glass weigh jars.
4. Spot 75 μL of solution on coverslip.
5. Place second coverslip on top of first to make sandwich.
6. Incubate at 75 °C for 1–2 h.
7. Remove glass and place in a large beaker of MilliQ water. Wash multiple times with water.

8. Remove coverslips one-by-one. Dip into a series of 2–3 beakers containing MilliQ water.
9. Rack and dry glass. Store in covered container at 4–25 °C.

1.3.3 TIRF microscopy

1. Clean the counterglass with ethanol.
2. Assemble functionalize glass with double-sided tape.
3. Wash with TIRF buffer.
4. Incubate for 1 min.
5. Flow 40 μL of 50 n*M* streptavidin in TIRF buffer. Incubate for 30 s.
6. Wash with 40 μL TIRF buffer.
7. Begin polymerization reaction by adding 20 μL of 2 × ATP/Mg^{++} in TIRF buffer to 20 μL of 2 × protein in TIRF buffer, mix well and flow in all 40 μL. Final concentrations of protein in this reaction should be composed of 0.1–0.5% biotinylated monomer and 10–25% fluorescently labeled monomer.

   Note: Oxygen scavengers should be added fresh to each polymerization reaction to a final concentration of

   - 200–320 μg/mL glucose oxidase (Biophoretics/Serva Electrophoresis #22778.01)
   - 30–60 μg/mL catalase (Sigma C40-100MG; from Bovine Liver)
   - 20 m*M* glucose
   - 20 m*M* β-mercaptoethanol

1.3.4 Polarity

1.3.4.1 Making cross-linked stable seeds

1. Preincubate your KCK-hydrolysis dead mutant with 5 m*M* TCEP. *Note*: Mutants capable of binding ATP, but hydrolysis deficient will form stable polymers that disassemble extremely slowly (i.e., ParM_E148A). If unable to obtain mutant, polymerize wild-type protein with a nonhydrolyzable nucleotide analog (i.e., AMP-PNP).
2. Buffer exchange your KCK-hydrolysis dead into labeling buffer.
3. Resuspend NHS-PEG-biotin (JenKem Technology, USA) reagent in DMSO to 20 m*M*.
4. Add biotin in a 5:1 molar ratio to your protein (if endogenous cysteines present, use 1:1 ratio).
5. Incubate at RT for 30 min.
6. Quench by adding DTT to 10 m*M* final concentration.
7. Buffer exchange into cross-linking buffer (no primary amines) + DTT to remove excess biotin.

8. Hard spin to remove any aggregate (20 min, 175,000 × *g*, 4 °C).
   Note: Very important step, this is the only opportunity to remove any aggregate that may have formed during biotin labeling. Aggregate = bad for EM and your assays.
9. Polymerize protein with excess nucleotide in the presence of equimolar (or slightly excess) cross-linking reagent for 30 min (e.g., DSS or BS3).
   Note: This polymerization mixture will be a mix of about 15% biotinylated and 15% dye-labeled protein if being used for TIRF, or 100% biotinylated hydrolysis dead mutant if for EM. This will need to be optimized.
10. Quench reaction by adding excess 1 *M* Tris (pH 8).
    Note: *Do not pipette* up and down to mix, this will sheer filaments. Flick gently.
11. Pellet seeds at 175,000 × *g*, 20 min, 4 °C.
12. Resuspend in reaction buffer *very gently* with a p1000 pipetteman with the pipette tip cut at end.
13. Aliquot and freeze if using for TIRF, store at 4 °C and use within a few days if for EM.
14. Check seeds by SDS-PAGE gel to verify cross-linking and visualize by negative stain to verify that most are cross-linked single filaments.
    Note: Too much cross-linker can result in cross-linked bundles which is bad as it will convolute interpretation of experiments. Optimization of cross-linker:protein may be necessary to yield mostly single cross-linked filaments.

For EM

1. Incubate seeds with equimolar streptavidin for 10 min.
2. Start polymerization reaction by mixing 2× protein solution (seeds + monomeric protein + any additional factors of interest) with 2× nucleotide/Mg^{++}.
3. Make grids and image.
   Analysis: Measure the filament length distribution on either side of the seed in the presence and absence of any factor that may affect bidirectional growth.

   For TIRF

1. Seeds can be attached to surface-functionalized glass (described above) using streptavidin–biotin linkage.
2. Polymerization mixture can then be added to the seeds. Importantly, the assembling monomeric protein must be of a different color so that seeds can be distinguished from assembling protein.
3. Observe dynamics by time-lapse TIRF microscopy.

1.4. Other considerations

1.4.1 Nucleotide affinity and preference

Bacterial actins are unique among cytoskeletal proteins in that they are general NTPases as opposed to strict ATPases (actin) or GTPases (eukaryotic and bacterial tubulin). While polymerization kinetics can vary for each nucleotide, the affinity of a bacterial actin, taken together with the relative abundance of nucleotides within the cell, is the best metric for nucleotide preference. An approach for directly measuring nucleotide affinity is described by Rivera et al. (2011).

2. CONCLUSION

The study of bacterial ALPs is in its infancy but it is rapidly expanding and promises to reveal new biology and new principles of self-organization. The ALPs share very little sequence identity (as low as 11%) (Derman et al., 2009) and their filament structures and assembly properties can vary significantly. Therefore, to understand the molecular mechanism by which any ALP carries out its biological function, it is essential to characterize its basic structure and assembly dynamics. On a wider scale, such studies may help us to better understand the functions and the family history of one of the most conserved and abundant eukaryotic proteins: actin.

ACKNOWLEDGMENTS

We thank J. Polka and C. Rivera for data, and S. Hansen for protocols for TIRF and glass surface chemistry, which were adapted for bacterial ALPs. We also thank B. Gesserit for assistance in developing polarity assays.

REFERENCES

Becker, E., Herrera, N. C., Gunderson, F. Q., Derman, A. I., Dance, A. L., Sims, J., et al. (2006). DNA segregation by the bacterial actin AlfA during Bacillus subtilis growth and development. *EMBO Journal*, *25*, 5919–5931.

Bork, P., Sander, C., & Valencia, A. (1992). An ATPase domain common to prokaryotic cell cycle proteins, sugar kinases, actin, and hsp70 heat shock proteins. *Proceedings of the National Academy of Sciences of the United States of America*, *89*, 7290–7294.

Campbell, C. S., & Mullins, R. D. (2007). In vivo visualization of type II plasmid segregation: Bacterial actin filaments pushing plasmids. *Journal of Cell Biology*, *179*, 1059–1066.

Derman, A. I., Becker, E. C., Truong, B. D., Fujioka, A., Tucey, T. M., Erb, M. L., et al. (2009). Phylogenetic analysis identifies many uncharacterized actin-like proteins (Alps) in bacteria: Regulated polymerization, dynamic instability and treadmilling in Alp7A. *Molecular Microbiology*, *73*, 534–552.

Dominguez-Escobar, J., Chastanet, A., Crevenna, A. H., Fromion, V., Wedlich-Soldner, R., & Carballido-López, R. (2011). Processive movement of MreB-associated cell wall biosynthetic complexes in bacteria. *Science*, *333*, 225–228.

Draper, O., Byrne, M. E., Li, Z., Keyhani, S., Barrozo, J. C., Jensen, G., et al. (2011). MamK, a bacterial actin, forms dynamic filaments in vivo that are regulated by the acidic proteins MamJ and LimJ. *Molecular Microbiology, 82*, 342–354.

Garner, E. C., Bernard, R., Wang, W., Zhuang, X., Rudner, D. Z., & Mitchison, T. (2011). Coupled circumferential motions of the cell wall synthesis machinery and MreB filaments in B. subtilis. *Science, 333*, 222–225.

Garner, E. C., Campbell, C. S., & Mullins, R. D. (2004). Dynamic instability in a DNA-segregating prokaryotic actin homolog. *Science, 306*, 1021–1025.

Garner, E. C., Campbell, C. S., Weibel, D. B., & Mullins, R. D. (2007). Reconstitution of DNA segregation driven by assembly of a prokaryotic actin homolog. *Science, 315*, 1270–1274.

Gayathri, P., Fujii, T., Møller-Jensen, J., van den Ent, F., Namba, K., & Löwe, J. (2012). A bipolar spindle of antiparallel ParM filaments drives bacterial plasmid segregation. *Science, 338*, 1334–1337.

Hatano, S., & Oosawa, F. (1966). Extraction of an actin-like protein from the plasmodium of a myxomycete and its interaction with myosin a from rabbit striated muscle. *Journal of Cellular Physiology, 68*, 197–202.

Huxley, H. E. (1963). Electron microscope studies on the structure of natural and synthetic protein filaments from striated muscle. *Journal of Molecular Biology, 7*, 281–308.

Jones, L. J., Carballido-López, R., & Errington, J. (2001). Control of cell shape in bacteria: Helical, actin-like filaments in Bacillus subtilis. *Cell, 104*, 913–922.

Komeili, A., Li, Z., Newman, D. K., & Jensen, G. J. (2006). Magnetosomes are cell membrane invaginations organized by the actin-like protein MamK. *Science, 311*, 242–245.

Nishida, E., & Sakai, H. (1983). Kinetic analysis of actin polymerization. *Journal of Biochemistry, 93*, 1011–1020.

Oosawa, F., & Kasai, M. (1962). A theory of linear and helical aggregations of macromolecules. *Journal of Molecular Biology, 4*, 10–21.

Pichoff, S., & Lutkenhaus, J. (2005). Tethering the Z ring to the membrane through a conserved membrane targeting sequence in FtsA. *Molecular Microbiology, 55*, 1722–1734.

Polka, J. K., Kollman, J. M., Agard, D. A., & Mullins, R. D. (2009). The structure and assembly dynamics of plasmid actin AlfA imply a novel mechanism of DNA segregation. *Journal of Bacteriology, 191*, 6219–6230.

Polka, J. K., Kollman, J. M., & Mullins, R. D. (2014). Accessory factors promote AlfA-dependent plasmid segregation by regulating filament nucleation, disassembly, and bundling. *Proceedings of the National Academy of Sciences of the United States of America*, In Press.

Rivera, C. R., Kollman, J. M., Polka, J. K., Agard, D. A., & Mullins, R. D. (2011). Architecture and assembly of a divergent member of the ParM family of bacterial actin-like proteins. *Journal of Biological Chemistry, 286*, 14282–14290.

Straub, F. B. (1943). Actin, II. *Studies of the Institute of Medical Chemistry University of Szeged, 3*, 23–37.

van den Ent, F., Møller-Jensen, J., Amos, L. A., Gerdes, K., & Löwe, J. (2002). F-actin-like filaments formed by plasmid segregation protein ParM. *EMBO Journal, 21*, 6935–6943.

van Teeffelen, S., Wang, S., Furchtgott, L., Huang, K. C., Wingreen, N. S., Shaevitz, J. W., et al. (2011). The bacterial actin MreB rotates, and rotation depends on cell-wall assembly. *Proceedings of the National Academy of Sciences of the United States of America, 108*, 15822–15827.

Woodrum, D. T., Rich, S. A., & Pollard, T. D. (1975). Evidence for biased bidirectional polymerization of actin filaments using heavy meromyosin prepared by an improved method. *Journal of Cell Biology, 67*, 231–237.

CHAPTER THREE

Quantitative Analysis of Microtubule Self-assembly Kinetics and Tip Structure

Louis S. Prahl*, **Brian T. Castle***, **Melissa K. Gardner**†, **David J. Odde***,[1]
*Department of Biomedical Engineering, University of Minnesota, Minneapolis, Minnesota, USA
†Department of Genetics, Cell Biology & Development, University of Minnesota, Minneapolis, Minnesota, USA
[1]Corresponding author: e-mail address: oddex002@umn.edu

Contents

Abstract

Microtubules are dynamic polymers of the cytoskeleton, which play important roles in cell division, polarization, and intracellular transport. Self-assembly of microtubule polymer from αβ-tubulin heterodimers is highly variable, with stochastic switching between alternate states of net growth and net shortening, a phenomenon known as dynamic instability. Microtubule tip structures are also variable and directly influence the kinetics of assembly and vice versa. TipTracker, a semiautomated, image processing-based tool, permits high spatial and temporal resolution measurements from fluorescence microscopy images (~10–40 nm, or 1–5 dimer lengths, at 1–10 Hz) with simultaneous tip structure estimation. We provide a walkthrough of the TipTracker code to demonstrate methods used to (1) fit the coordinates of the microtubule backbone; (2) track microtubule tip position; and (3) estimate tip structure from the spatial decay of the tip fluorescence distribution, discuss possible sources of error, and include an example protocol for nanometer-scale tip tracking in living cells. Additionally, we evaluate TipTracker's accuracy on simulated digital images and fixed microtubules to estimate

Methods in Enzymology, Volume 540
ISSN 0076-6879
http://dx.doi.org/10.1016/B978-0-12-397924-7.00003-0

accuracy under realistic imaging conditions. In summary, this chapter demonstrates the use of TipTracker in making robust, high-resolution measurements of microtubule tip dynamics and structures, facilitating quantitative investigations into nanoscale/molecular control of microtubule assembly. Although our primary focus is on microtubules, these methods are, in principle, suitable for other polymer structures, such as F-actin.

1. INTRODUCTION

Microtubules are cytoskeletal polymers that mediate diverse cellular functions, from mitosis, to cell polarization and intracellular transport. The microtubule lattice typically consists of 13 protofilaments (linear strands of αβ-tubulin heterodimers, each 8 nm in length, stacked head to tail), which are in turn arranged laterally in a circle to form a cylinder with a closed lumen with outer diameter 25 nm. The interactions between protofilaments are stabilized via lateral bonds between adjacent dimers, with offsets between neighboring protofilaments that give rise to a helical pitch within the lattice (Mandelkow, Mandelkow, & Milligan, 1991; Nogales, Wolf, Khan, Ludueña, & Downing, 1995; Wade, Chrétien, & Job, 1990). Assembly and disassembly are driven by addition and loss of dimers from the plus end, where the β-tubulin subunit is distal. GTP hydrolysis within the lattice-bound exchangeable site associated with β-tubulin fuels stochastic switches between growth and rapid shortening phases at the plus end, a process known as "dynamic instability" (Horio & Hotani, 1986; Mitchison & Kirschner, 1984; Walker et al., 1988).

Light microscopy has been the primary tool for observing single microtubule dynamics (Cassimeris, Pryer, & Salmon, 1988; Chrétien, Fuller, & Karsenti, 1995; Gildersleeve, Cross, Cullen, Fagen, & Williams, 1992; Horio & Hotani, 1986; Rusan, Fagerstrom, Yvon, & Wadsworth, 2001; Rusan, Tulu, Fagerstrom, & Wadsworth, 2002; Sammak & Borisy, 1988; Schulze & Kirschner, 1988; Walker et al., 1988). The earliest studies typically used transmitted light microscopy (dark field and differential interference contrast) because it obviated the need to synthesize and purify fluorescently tagged tubulin. However, with the advent of engineered fluorescent fusion proteins and more sensitive digital cameras, fluorescence microscopy is currently the dominant approach for imaging single microtubules (example image shown in Fig. 3.1A). In addition, dual-color imaging allows for simultaneous tracking of microtubules and microtubule-associated proteins, many of which regulate microtubule assembly dynamics.

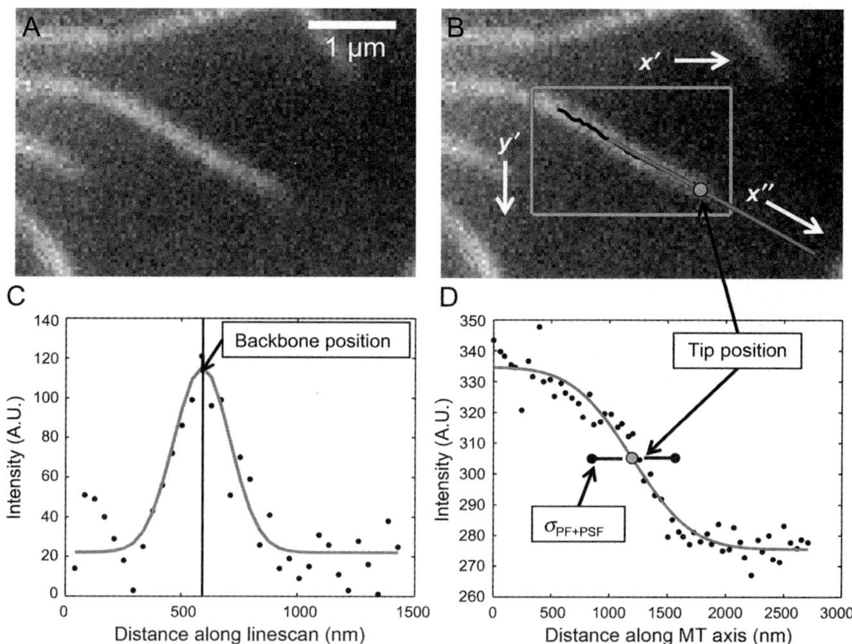

Figure 3.1 Functions used to fit microtubule backbone and tip. (A) An example fluorescence image of an individual microtubule is selected for tip-tracking analysis. (B) Two user-supplied clicks serve as "initial guesses" and define the boundaries of the green box. Within this region, microtubule backbone coordinates (red) are estimated using a Gaussian fit along line scans perpendicular to the major microtubule axis (in this case, the *y*-axis). (C) Plot of the Gaussian function (purple curve) used to fit microtubule backbone coordinates, with black dots representing single pixel intensity values (A.U.) and red vertical line representing the Gaussian mean (backbone coordinate). (D) Plot of the Gaussian survival function (blue curve) used to fit the microtubule tip in (B), with black dots representing intensity values (A.U.) along the x''-axis. Note that TipTracker subtracts the minimum from intensity values for the Gaussian line scan fit (C), but uses the original pixel values to fit the survival function in (D). Black bars mark the standard deviation of the error function (σ_{PF+PSF}), which is used to estimate tip taper. (See the color plate.)

In earlier work, microtubule assembly was tracked by manual clicking with a cursor, where the operator recorded points corresponding to the microtubule tip position as judged by eye with a time interval of ~1–3 s (Cassimeris et al., 1988; Gildersleeve et al., 1992; Rusan et al., 2001, 2002; Walker et al., 1988). In one study, comparison of two independent operators led to an estimate of ~160 nm single time point precision for differential interference contrast imaging (Gildersleeve et al., 1992). Other contemporary studies reported accuracies of ~150 nm *in vitro* (Odde, Buettner, & Cassimeris, 1996) and <500 nm in newt lung epithelial cells (Cassimeris et al., 1988). Hadjidemetriou, Toomre, and Duncan (2008) developed an automated tip-tracking algorithm that follows microtubule contours, although the estimated accuracy (260 nm) does not confer improvement over point-click tracking. Wan et al. (2009) fit an error function to kinetochore microtubule bundles to estimate bundle tip position with <100 nm precision. Fully automated tip-tracking algorithms can track multiple plus-end tips simultaneously through the use of fluorescently labeled end-binding proteins (Applegate et al., 2011; Garrison et al., 2012; Matov et al., 2010), but the spatial resolution of these methods has not been described.

High-resolution *in vitro* studies of microtubule assembly using optical tweezers in microfabricated chambers show that growth phases are dominated by rapid addition and loss of single subunits (the αβ-heterodimer) and punctuated by brief shortening episodes (Kerssemakers et al., 2006; Schek, Gardner, Cheng, Odde, & Hunt, 2007). While these measurements provide unprecedented spatial and temporal resolution (~3–10 nm at 10–25 Hz), the optical tweezers methodology is far more technically challenging and lower throughput than the fluorescence microscopy-based method that we describe in this chapter. Technical descriptions of the optical tweezers methodology have been published previously (Charlebois, Schek, & Hunt, 2010; Kerssemakers et al., 2006; Schek et al., 2007; Schek & Hunt, 2005). A unique feature of the optical trapping method is that it presumably tracks the longest protofilament, rather than mean protofilament length, providing a complementary statistic to fluorescence microscopy-based approaches. However, a limitation is that it does not provide information about tip structure in terms of protofilament length variability.

Protofilament length variability at the microtubule tip is an important determinant of assembly kinetics, with more variable tips correlating with a less stable microtubule (Coombes, Yamamoto, Kenzie, Odde, & Gardner, 2013; Gardner et al., 2011). Monte Carlo studies showed that the strengths of lateral and longitudinal interactions between tubulin subunits within the lattice are expected to determine the kinetics of assembly and disassembly and give rise to variable tip structures (VanBuren, Cassimeris, & Odde, 2005; VanBuren, Odde, & Cassimeris, 2002). Plus-end tip structures observed using cryoelectron microscopy (the classic method for observing tip structures) are highly variable; growing microtubules tend to range from blunt to tapered (mean taper ~100–200 nm, max up to ~2 μm), while shortening microtubules tend to have blunt or slightly fraying ends (Chrétien et al., 1995; Mandelkow et al., 1991). More recently, it was shown that microtubule protofilament length variability could be quantified using fluorescence microscopy (Demchouk, Gardner, & Odde, 2011; Gardner et al., 2011) and that the observed variability correlated with observations by electron microscopy (Coombes et al., 2013).

Here, we provide a detailed description of an algorithm, implemented as a MATLAB (The Mathworks Inc., Natick, MA) script, designed to track microtubule tip position on the ~10–40 nm length scale and ~1–10 Hz timescale while simultaneously estimating the protofilament length variability. Using this method of tracking microtubule tips *in vitro* revealed much greater variability of assembly than expected based on previous lower resolution measurements (Gardner et al., 2011). Synthetic digital images, generated via "model convolution," provide a means to validate the precision of these measurements by including imaging system blurring and noise (Gardner et al., 2010) while imaging microtubules in fixed cells provide an experimental estimate of measurement error in the absence of assembly or disassembly (Demchouk et al., 2011). The method we describe here is based on the method described in Demchouk et al., which is similar to that described by Ruhnow, Zwicker, and Diez (2011), who used a 2D Gaussian fit to track microtubule tips *in vitro* with an estimated accuracy of 9 nm, except that the Demchouk et al. method also estimates tip structure. This chapter gives a walkthrough of the TipTracker script, provides example protocols as guidelines, and discusses the features and practical limitations of this method, as well as considerations for optimization. The most current version of TipTracker and associated documentation on compatibility with image file-types are available for download from http://oddelab.umn.edu.

2. TIP-TRACKING ALGORITHM WALKTHROUGH

TipTracker uses a semi-automated method to determine (1) the coordinates of the microtubule backbone, (2) tip position, and (3) protofilament length standard deviation. For each tracking session, two user-supplied "initial guesses" define a rectangular region used to identify a single microtubule tip. Sections 2.1–2.3 describe the functions used to obtain the microtubule backbone and tip coordinates, as well as the tip structure, and provide a "pseudo-code" walkthrough of the algorithm. Information on the specific input variables and code features, as well as output variables, can be found in supporting documentation on the download website.

2.1. Fitting the microtubule backbone

An example microtubule image for tracking is shown in Fig. 3.1A. Two user-supplied "guesses" (mouse clicks) define a rectangular region (x'- and y'-axes along the sides of the rectangle) near a microtubule tip, as shown by the green box in Fig. 3.1B. The first (left) click lies along the microtubule backbone, away from the tip, whereas the second (right) click lies proximal to the microtubule tip, but still along the backbone. To avoid scenarios where the two clicks lie along a vertical or horizontal line (which would give the rectangular region zero area), an artificial "padding" of several pixels is added. This value is equal to a user-specified number multiplied by the rounded estimate of the imaging system point-spread function (PSF) standard deviation, σ_{PSF}, which is calculated based on user-specified variables for the imaging system pixel size, wavelength, and numerical aperture (NA).

Center of mass coordinates (R_{CM}) for the rectangular region are calculated using

$$R_{\mathrm{CM}} = \frac{\sum m_i r_i}{\sum m_i} \tag{3.1}$$

where m_i and r_i correspond to the brightness and position of the ith pixel within the defined region. The microtubule image is then rotated about R_{CM} in 90° intervals to find the orientation where the microtubule lies best along the x-axis. Rotations of 90° multiples avoid distorting pixels, which might occur at other rotation angles (Bicek et al., 2009; Demchouk et al., 2011). New coordinates (x', y') are then defined for the transposed region. Vertical (y'-axis) line scans along the x'-axis estimate microtubule backbone positions (red points, Fig. 3.1B) using a Gaussian fit to pixel intensities:

$$I(y') = I_{\mathrm{BG}} + I_{\mathrm{MT}} \mathrm{e}^{-(y'-y'_{\mathrm{m}})^2/2\sigma^2} \tag{3.2}$$

where I_{BG} is the background intensity, I_{MT} is the microtubule signal, y'_{m} is the mean value of the Gaussian line scan fit taken to be the microtubule backbone position, and σ is the standard deviation of the Gaussian function (Fig. 3.1C). The value of σ is constrained to the interval $[\sigma_{\mathrm{PSF}}, 3\sigma_{\mathrm{PSF}}]$ to minimize fitting to noise (Bicek et al., 2009; Bicek, Tüzel, Kroll, & Odde, 2007). Coordinates obtained from vertical line scans are then used to fit the microtubule axis (x'', blue line in Fig. 3.1B), permitting estimation of the microtubule tip position and structure.

2.2. Measuring tip position

Positional and structural information is contained within the fluorescence distribution near the microtubule tip. Once the microtubule backbone coordinates are defined, the algorithm extends a line along the microtubule axis (x''-axis) and across the tip (blue line, Fig. 3.1B and D). The line is centered near the second user click, and its length is defined by a user-specified variable multiplied by σ_{PSF}. Pixel intensity values along the x''-axis are fit to a Gaussian survival function

$$I(x'') = \frac{1}{2} I_{\mathrm{MT}} \mathrm{erfc}\left(\frac{x'' - \mu_{\mathrm{PF}}}{\sqrt{2}\sigma_{\mathrm{PF+PSF}}}\right) + I_{\mathrm{BG}} \tag{3.3}$$

where I_{MT} and I_{BG} are signal and background intensities, μ_{PF} is the mean tip position (cyan dot, Fig. 3.1B and D), $\sigma_{\mathrm{PF+PSF}}$ is the combined standard deviation of the microtubule tip taper and microscope PSF, and erfc is the error function complement.

2.3. Measuring tip structure

Fluorescence intensities along the microtubule axis drop from higher signal intensities (I_{MT}) to the background intensity (I_{BG}), as shown in Fig. 3.1D. How steeply or gradually this intensity profile drops off depends on the protofilament length distribution for different tip conformations, as well as the imaging system PSF (Demchouk et al., 2011). For example, blunt tips, where all protofilaments are approximately the same length, will have a smaller length standard deviation (σ_{PF}) than tapered tips with a gradual loss of protofilaments out to the microtubule end. These variances are related by

$$\sigma^2_{\mathrm{PF+PSF}} = \sigma^2_{\mathrm{PF}} + \sigma^2_{\mathrm{PSF}} \tag{3.4}$$

where $\sigma^2_{\mathrm{PF+PSF}}$ is the combined variance of the intensity profile and σ^2_{PF} and σ^2_{PSF} are the variances of protofilament length and microscope PSF, respectively.

3. *In Vivo* Tip-Tracking Protocol

LLC-PK1 cells are a reliable system to study microtubule assembly dynamics and tip structures *in vivo* because individual microtubules can be easily identified within adherent cells' broad lamella (Fig. 3.2A and B). Additionally, these cells' lamellae are thin (~500 nm in height, approximately equivalent to the axial depth of field for a 1.4-NA lens; Bicek et al.,

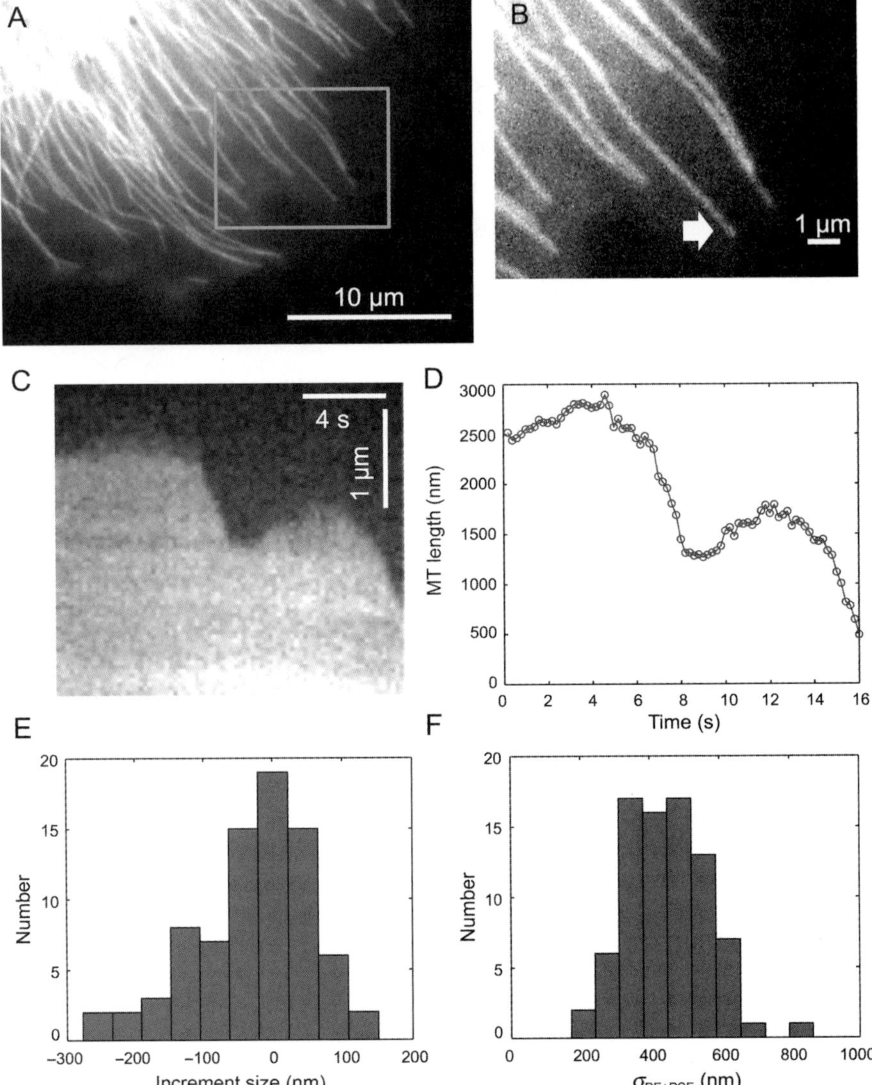

Figure 3.2 TipTracker algorithm in LLC-PK1 cells. (A) An example frame from a time-lapse image of an LLC-PK1 cell expressing EGFP-α-tubulin. Frames were collected using MetaMorph's "Stream" function (80 frames, 200 ms exposure, 100% PhotoFluor brightness). (B) Inset (red box in Fig. 3.1A) shows that individual microtubules (white arrow) are clearly identifiable in the cell periphery. A kymograph (C) and length versus time output (D) of the selected microtubule from (B) show alternating phases of net assembly and disassembly. Blue circles mark microtubule length at each point along the growth contour, in red. (E) Histograms of length increments from the tracking experiment show a distribution of negative and positive incremental length changes between frames. (F) Histograms of σ_{PF+PSF} show variability in tip conformation over the imaging duration. For this tip, the mean σ_{PF+PSF} was 451 ± 119 nm, corresponding to a tip taper of ~1000 nm. (For interpretation of the references to color in this figure legend, the reader is referred to the online version of this chapter.)

2007), which confines microtubules to a single focal plane and thereby minimizes errors from out-of-focus tip structures (Demchouk et al., 2011). We provide a general protocol for tracking single microtubule tips and tip structures in LLC-PK1 cells and discuss experimental considerations for tracking in other cell lines.

3.1. Cell culture and imaging

Porcine kidney epithelial cells stably expressing EGFP-α-tubulin (LLC-PK1α) (Rusan et al., 2001) are cultured in Gibco® Opti-MEM® media (Invitrogen Corporation, Carlsbad, CA) containing 10% fetal bovine serum. Cells are plated at ~20,000 cells/dish on MatTek 35 mm No. 1.5 dishes (MatTek Corporation, Ashland, MA) and incubated at 37 °C in 5% CO_2 for approximately 24 h prior to imaging, to allow sufficient time to adhere. To evenly distribute cells throughout the dish, cells can be suspended in 2 mL of media in a 15-mL tube before plating.

LLC-PK1 cells are imaged at 37 °C using a Nikon TE200 epifluorescence microscope (Nikon Instruments Inc., Melville, NY) equipped with a Ludl BioPrecision stage (Ludl Electronic Products, Ltd., Hawthorne, NY) under control of MetaMorph version 7.4 imaging software (Molecular Devices LLC, Sunnyvale, CA). Images are acquired through an ET-EGFP filter set (49002; Chroma Technology Corporation, Bellows Falls, VT), using a Photometrics CoolSnap HQ^2 CCD camera (Photometrics, Tuscon, AZ), with illumination provided by a PhotoFluor® II metal halide light source (89 North, Burlington, VT). A 60 × Plan Apo total internal reflection fluorescence (TIRF) objective (1.49 NA) with a 2.5 × projection lens (150 × total magnification) gives this configuration spatial sampling of 42 nm/pixel (Bicek et al., 2009; Demchouk et al., 2011).

Images for TipTracker analysis are typically acquired using MetaMorph's "Stream Acquisition" or "Acquire Timelapse" function. Typical stream acquisition rates vary from ~100 to 300 ms per frame, depending on the desired temporal resolution. For longer time-lapse experiments (~1–2 min at ~1 frame per second), using a similar exposure time will minimize photobleaching.

Before initializing TipTracker, select input settings in the initialization script window. When the program is run, a file window will prompt the user for an image stack. A single image from an example stack is shown in Fig. 3.2A, while a microtubule of interest is shown in the inset (red box, and white arrow in Fig. 3.2B). To make it easier to click on individual microtubules, larger movies (~500 × 500 pixels or greater) should be cropped to smaller regions for tracking. The user then supplies two input guesses (mouse clicks, as described previously) for each frame included in the analysis (unless options for fixed cell tracking are enabled, in which case user only inputs on the first frame). Using these inputs to define a region of interest, as shown in Fig. 3.1B, TipTracker will then fit backbone coordinates and Gaussian survival function as described previously. Once the program has run through all frames, it will generate outputs as MATLAB variables and windows (Fig. 3.2D–F).

For the example microtubule-tracking experiment (identified in Fig. 3.2A and B), a kymograph (Fig. 3.2C) and plot of tip position versus time (Fig. 3.2D) show variability in assembly and disassembly. A histogram of length changes between frames (Fig. 3.2E) reveals individual length changes that range from approximately −200 to 150 nm between frames. Negative length changes correspond to shortening, and positive length changes to growth (Demchouk et al., 2011). A histogram of σ_{PF+PSF} for all frames as shown in Fig. 3.2F shows tip structure variability over the duration of the imaging experiment.

3.2. Experimental setup considerations

Noise, background, and signal intensity are all important considerations when obtaining quantitative measurements from digital fluorescence images, as they affect the accuracy and precision of these measurements (Waters, 2009). In general, one would like to maximize signal while minimizing background and noise. This means that imaging parameters, such as exposure time, fluorophore brightness, and spatial sampling, must be considered for each application. The optimization of some parameters requires a trade-off with others. For example, finer spatial sampling and/or increased temporal resolution reduce the signal intensity per pixel. To guide users in optimal code performance, TipTracker accuracy has been evaluated on simulated images obtained by model convolution (Demchouk et al., 2011; Gardner et al., 2010) with an eye toward (1) signal intensity above background and (2) spatial sampling.

3.2.1 Signal above background

Fluorescence signal intensity is an important factor in determining TipTracker accuracy. Pixel intensity values along the microtubule are determined in part by the quantity of labeled tubulin subunits (fraction labeled). Figure 3.3A shows that, for spatial sampling of 40 or 100 nm/pixel, as the

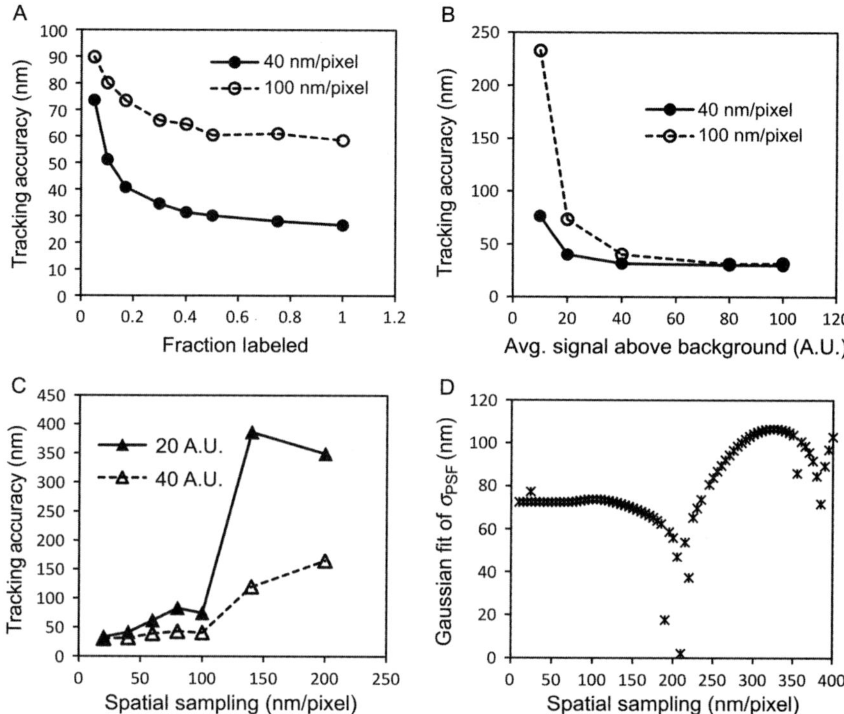

Figure 3.3 Accuracy dependence on fluorescence signal and spatial sampling from analysis of model-convolved microtubule images. (A) TipTracker accuracy (nm) plotted as a function of fraction labeled (percentage of labeled tubulin) and (B) as a function of average signal counts above background (A.U.). Both fraction labeled and accuracy were evaluated at spatial sampling of 40 and 100 nm/pixel. All points in (A) assume 20 A.U. above background per pixel. (C) Accuracy plotted as a function of spatial sampling, for signal values of 20 and 40 A.U. above background. All points in (B) and (C) assume 20% tubulin labeling. (D) Gaussian estimates of the point-spread function standard deviation σ_{PSF} (nm) as a function of spatial sampling (at 40 A.U. above background).

fraction of labeled tubulin in the microtubule lattice increases, resolution reaches a maximum (~50% labeling). In LLC-PK1α cells, labeling fraction is fixed at ~17% (Rusan et al., 2001), but *in vitro*, the labeling fraction is controllable. At low average signal above background, the noise is dominated by photon-counting noise, yielding a less accurate measurement, while at higher values (>40 A.U. above background), the measurement reaches a maximal accuracy (Fig. 3.3B). This maximal accuracy is independent of spatial sampling for either pixel size evaluated (40 or 100 nm/pixel). These simulations were run on tapered tips comparable to those estimated *in vivo*, so the accuracy converges to ~36 nm, but blunt tips have higher accuracy, to ~15 nm (Demchouk et al., 2011). On average, tips are more tapered *in vivo* compared to *in vitro* (Demchouk et al., 2011; Gardner et al., 2011); therefore, accuracy is expected to be closer to 15 nm *in vitro*. For very low fraction labeled (less than 0.1), microtubules are in the so-called "speckle" regime (Waterman-Storer & Salmon, 1998), which, while optimal for imaging microtubule transport, yields poor estimation of tip position. Figure 3.3 plots were generated by running TipTracker on simulated microtubule images obtained via model convolution, constructed using a 2 nm × 2 nm "fine grid" and binned to appropriate pixel sizes (Demchouk et al., 2011; Gardner et al., 2010).

3.2.2 Spatial sampling

Spatial sampling refers to the size of individual pixels that comprise a digital image and is determined by the pixel size of the CCD camera and the magnification power of the optics. For example, a Nikon TE200 microscope, equipped with a Photometrics CoolSnap HQ^2 CCD camera (6.45 μm/pixel) and 60 × lens, has a spatial sampling of 105 nm/pixel, or 42 nm/pixel with an intermediate 2.5 × projection lens (150 × total magnification). Larger pixels may collect more photons and give higher signal, but smaller pixels provide finer positional information and improve function fitting. Evaluating tip-tracking accuracy on model-convolved microtubule images over a range of spatial sampling (Fig. 3.3C), we show that TipTracker accuracy plateaus at ~36 nm for tapered tips for pixel sizes below ~100 nm/pixel. Tracking accuracy below this threshold is relatively insensitive to signal above background (20 or 40 A.U.). However, accuracy precipitously declines for spatial sampling greater than this value. The explanation lies in using a Gaussian to fit the imaging system PSF. With increasing pixel size, fewer pixels are available to fit σ_{PSF}, resulting in information loss about the microtubule tip position. In particular, around the so-called optical resolution of the system (defined by the Abbe diffraction limit, ~200–250 nm), the fit becomes unreliable because essentially the entire PSF is contained in 1–2 pixels used to fit the 1D Gaussian (Fig. 3.3D). In general, when measuring features such as microtubule tips from digital images, larger pixels (at or near the diffraction limit) may contribute errors to function fits. Pixel binning (either preprocessing or via a built-in function) is a viable strategy to increase signal, so long as binned pixels stay below ~100 nm/pixel. Otherwise, using a projection lens (or optovar) to reduce pixel size is recommended.

4. TIPTRACKER OPTIMIZATION USING FIXED CELLS

While model convolution (as employed above) provides a convenient way to test algorithm accuracy, it may be easiest to optimize the experimental setup for tracking accuracy on fixed cells using established protocols for tubulin immunostaining (Demchouk et al., 2011; Seetapun & Odde, 2010; Witte, Neukirchen, & Bradke, 2008). As the tip position does not change between frames, deviations from the mean microtubule length (ΔLE) are due to experimental noise. Thus, the standard deviation of ΔLE (σ_{LE}) is an estimate of the tracking accuracy. To demonstrate the effects of frame integration, we took 200 ms streaming videos of microtubules in the periphery of fixed LLC-PK1α cells ($N=54$ microtubules) using our previously described microscope setup, integrated frames (1 ×, 2 ×, and 3 × frame integration, effective exposure times of 200, 400, and 600 ms), and measured deviations from the mean length (Fig. 3.4A). Increasing exposure time increases TipTracker accuracy (reduces the standard deviation of length elements, σ_{LE}). An important consideration in using fixed cells for TipTracker optimization is that GFP brightness is reduced ~2 × by cell fixation (users should measure the loss for each specific application) (Demchouk et al., 2011).

Pixel binning is commonly employed in digital image processing to boost signal by summing intensities for a region of several pixels into a single pixel, both during acquisition and as a postprocessing technique. To estimate the effects of binning on TipTracker accuracy, we took the same fixed cell images and recorded ΔLE values for binned pixel sizes (1 × 1, 2 × 2, and 4 × 4 binning). Tip-tracking accuracy decreased with higher binning (Fig. 3.4B), especially in the case of 4 × 4 binning for the previously described system (168 nm/pixel). Note that this spatial sampling is near the Rayleigh criterion for the system, where it was determined that the fit to σ_{PSF} breaks down (Fig. 3.3D), resulting in a less accurate estimate of tip position. As previously suggested, binning should be employed with caution, keeping spatial sampling under ~100 nm/pixel to avoid contributing function fitting errors.

5. CONSIDERATIONS FOR *IN VITRO* ASSAYS AND OTHER FILAMENT SYSTEMS

Many *in vitro* microtubule assembly assays use fluorescence microscopy-based methods to measure growth and shortening rates under

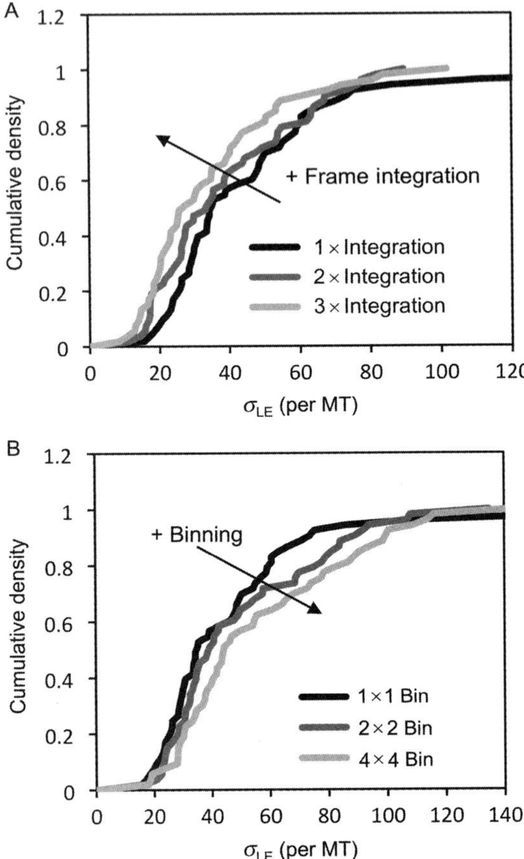

Figure 3.4 Effects of binning and frame integration on TipTracker analysis on fixed cells. Image stacks of fixed microtubules in the periphery of LLC-PK1α cells ($n=54$ microtubules) were analyzed using TipTracker. Images were acquired at 150× magnification (42 nm/pixel) with 200 ms streaming acquisition. (A) Effects of integrating multiple frames (1×, 2×, and 3× integration, effective exposure times of 200, 400, and 600 ms) on standard deviation of length changes (σ_{LE}). (B) Effects of increasing bin size (1×1, 2×2, and 4×4 binning, effective pixel sizes of 42, 84, and 168 nm/pixel) on σ_{LE}. (For color version of this figure, the reader is referred to the online version of this chapter.)

various experimental conditions. Recent studies have used the tip-tracking methods described here in combination with TIRF microscopy to examine tip structures in microtubules grown on coverslips from GMPCPP seeds (Coombes et al., 2013; Gardner et al., 2011). In these assays, thermal fluctuations may cause tips of longer microtubules to drift upward and away from the coverslip. This reduces the signal from TIRF illumination, decreasing TipTracker's performance and increasing apparent tip taper. These studies compared microtubules of similar length in different free tubulin concentrations; increasing free tubulin increases tip taper, allowing them to determine these artifacts' contribution to tip structure (Coombes et al., 2013; Gardner et al., 2011).

An additional consideration for these systems is that they offer significantly more control over tubulin fluorophore labeling density than *in vivo*. Typical concentrations for GTP-tubulin in these assays are 7–12 μ*M* at 28 °C or 0.7–1.5 μ*M* for GMPCPP-tubulin under the same conditions (Coombes et al., 2013; Gardner et al., 2011). By increasing the pool of labeled monomers, one can increase signal-to-noise ratio (SNR) and improve both backbone fit and tip measurement resolution (Fig. 3.3A). However, as the labeled fraction of tubulin is increased, background will also increase, reducing the dynamic range of the imaging system. These studies employed ~10–15% labeling, although in this *in vitro* tubulin preparation, up to 25% labeling ensures visualization without inhibiting interaction of microtubule-associated proteins (Gell et al., 2010). For the lower free tubulin concentrations used in GMPCPP experiments, the labeling fraction may be increased up to ~55% (Gardner et al., 2011).

Although the methods described in this chapter focus primarily on tubulin, in principle, they can be applied to other filament systems, such as actin. Actin filaments are less rigid than microtubules (Felgner, Frank, & Schliwa, 1996; Gitties, Mickey, Nettleton, & Howard, 1993), with a measured persistence length of 8–17 μm *in vitro* compared to several millimeters for microtubules (Brangwynne et al., 2007; Isambert et al., 1995; Janson & Dogterom, 2004; Ott, Magnasco, Simon, & Libchaber, 1993). Because TipTracker relies upon a straight line fit along the filament backbone to estimate tip position and structure, curved filaments may introduce additional error in tracking if the best-fit axis is skewed from the tip position. The highly cross-linked structure of many actin networks may present additional challenges to *in vivo* tracking of individual filaments or bundles. However, individual actin filaments contain only two protofilaments, meaning that signal per unit length will be lower even at constant labeling fraction. At the same time, F-actin tip structures will be much less variable, similar to blunt microtubule tips so that σ_{PF+PSF} will likely be equal to σ_{PSF}, which by itself will improve accuracy in estimating tip position. In general, the method is relevant to estimating tip position and structure of any self-assembled linear polymer.

ACKNOWLEDGMENTS

We would like to thank Dr. Patricia Wadsworth (University of Massachusetts) for sharing the LLC-PK1α cells used to generate data for figures. L. S. P. is supported by a 3M Science and Technology Doctoral Fellowship and B. T. C. was supported by NIH fellowship T32 EB008389. This work was supported by NIH Grants R01 GM71522 and R01 GM76177 to D. J. O. and R01 GM103833 to M. K. G.

REFERENCES

Applegate, K. T., Besson, S., Matov, A., Bagonis, M. H., Jaqaman, K., & Danuser, G. (2011). PlusTipTracker: Quantitative image analysis software for the measurement of microtubule dynamics. *Journal of Structural Biology*, *176*(2), 168–184.

Bicek, A. D., Tüzel, E., Demtchouk, A., Uppalapati, M., Hancock, W. O., Kroll, D. M., et al. (2009). Anterograde microtubule transport drives microtubule bending in LLC-PK1 epithelial cells. *Molecular Biology of the Cell*, *20*(12), 2943–2953.

Bicek, A. D., Tüzel, E., Kroll, D. M., & Odde, D. J. (2007). Analysis of microtubule curvature. *Methods in Cell Biology*, *83*, 237–268.

Brangwynne, C. P., Koenderink, G. H., Barry, E., Dogic, Z., MacKintosh, F. C., & Weitz, D. A. (2007). Bending dynamics of fluctuating biopolymers probed by automated high-resolution filament tracking. *Biophysical Journal*, *93*(1), 346–359.

Cassimeris, L., Pryer, N. K., & Salmon, E. D. (1988). Real-time observations of microtubule dynamic instability in living cells. *The Journal of Cell Biology*, *107*(6 Pt 1), 2223–2231.

Charlebois, B. D., Schek, H. T., & Hunt, A. J. (2010). Nanometer-resolution microtubule polymerization assays using optical tweezers and microfabricated barriers. *Methods in Cell Biology*, *95*, 207–219.

Chrétien, D., Fuller, S. D., & Karsenti, E. (1995). Structure of growing microtubule ends: Two-dimensional sheets close into tubes at variable rates. *The Journal of Cell Biology*, *129*(5), 1311–1328.

Coombes, C. E., Yamamoto, A., Kenzie, M. R., Odde, D. J., & Gardner, M. K. (2013). Evolving tip structures can explain age-dependent microtubule catastrophe. *Current Biology*, *23*(14), 1342–1348.

Demchouk, A. O., Gardner, M. K., & Odde, D. J. (2011). Microtubule tip tracking and tip structures at the nanometer scale using digital fluorescence microscopy. *Cellular and Molecular Bioengineering*, *4*(2), 192–204.

Felgner, H., Frank, R., & Schliwa, M. (1996). Flexural rigidity of microtubules measured with the use of optical tweezers. *Journal of Cell Science*, *109*(Pt 2), 509–516.

Gardner, M. K., Charlebois, B. D., Jánosi, I. M., Howard, J., Hunt, A. J., & Odde, D. J. (2011). Rapid microtubule self-assembly kinetics. *Cell*, *146*(4), 582–592.

Gardner, M. K., Sprague, B. L., Pearson, C. G., Cosgrove, B. D., Bicek, A. D., Bloom, K., et al. (2010). Model convolution: A computational approach to digital image interpretation. *Cellular and Molecular Bioengineering*, *3*(2), 163–170.

Garrison, A. K., Shanmugam, M., Leung, H. C., Xia, C., Wang, Z., & Ma, L. (2012). Visualization and analysis of microtubule dynamics using dual color-coded display of plus-end labels. *PLoS One*, 7(11), e50421.

Gell, C., Bormuth, V., Brouhard, G. J., Cohen, D. N., Diez, S., Friel, C. T., et al. (2010). Microtubule dynamics reconstituted in vitro and imaged by single-molecule fluorescence microscopy. *Methods in Cell Biology*, *95*, 221–245.

Gildersleeve, R. F., Cross, A. R., Cullen, K. E., Fagen, A. P., & Williams, R. C. (1992). Microtubules grow and shorten at intrinsically variable rates. *The Journal of Biological Chemistry*, *267*(12), 7995–8006.

Gitties, F., Mickey, B., Nettleton, J., & Howard, J. (1993). Flexural rigidity of microtubules and actin filaments measured from thermal fluctuations in shape. *The Journal of Cell Biology*, *120*(4), 923–934.

Hadjidemetriou, S., Toomre, D., & Duncan, J. (2008). Motion tracking of the outer tips of microtubules. *Medical Image Analysis*, *12*(6), 689–702.

Horio, T., & Hotani, H. (1986). Visualization of the dynamic instability of individual microtubules by dark-field microscopy. *Nature*, *321*(6070), 605–607.

Isambert, H., Venier, P., Maggs, A. C., Fattoum, A., Kassab, R., Pantaloni, D., et al. (1995). Flexibility of actin filaments derived from thermal fluctuations. *The Journal of Biological Chemistry*, *270*(19), 11437–11444.

Janson, M. E., & Dogterom, M. (2004). A bending mode analysis for growing microtubules: Evidence for a velocity-dependent rigidity. *Biophysical Journal*, *87*(4), 2723–2736.

Kerssemakers, J. W., Munteanu, E. L., Laan, L., Noetzel, T. L., Janson, M. E., & Dogterom, M. (2006). Assembly dynamics of microtubules at molecular resolution. *Nature*, *442*(7103), 709–712.

Mandelkow, E. M., Mandelkow, E., & Milligan, R. A. (1991). Microtubule dynamics and microtubule caps: A time-resolved cryo-electron microscopy study. *The Journal of Cell Biology*, *114*(5), 977–991.

Matov, A., Applegate, K., Kumar, P., Thoma, C., Krek, W., Danuser, G., et al. (2010). Analysis of microtubule dynamic instability using a plus-end growth marker. *Nature Methods*, 7(9), 761–768.

Mitchison, T., & Kirschner, M. (1984). Dynamic instability of microtubule growth. *Nature*, *312*(5991), 237–242.

Nogales, E., Wolf, S. G., Khan, I. A., Ludueña, R. F., & Downing, K. H. (1995). Structure of tubulin at 6.5 A and location of the taxol-binding site. *Nature*, *375*(6530), 424–427.

Odde, D. J., Buettner, H. M., & Cassimeris, L. (1996). Spectral analysis of microtubule assembly dynamics. *AIChE Journal*, *42*(5), 1434–1442.

Ott, A. M., Magnasco, M., Simon, A., & Libchaber, A. (1993). Measurement of the persistence length of polymerized actin using fluorescence microscopy. *Physical Review. E, Statistical Physics, Plasmas, Fluids, and Related Interdisciplinary Topics*, *48*(3), R1642–R1645.

Ruhnow, F., Zwicker, D., & Diez, S. (2011). Tracking single particles and elongated filaments with nanometer precision. *Biophysical Journal*, *100*(11), 2820–2828.

Rusan, N. M., Fagerstrom, C. J., Yvon, A. M., & Wadsworth, P. (2001). Cell cycle-dependent changes in microtubule dynamics in living cells expressing green fluorescent protein-alpha tubulin. *Molecular Biology of the Cell*, *12*(4), 971–980.

Rusan, N. M., Tulu, U. S., Fagerstrom, C., & Wadsworth, P. (2002). Reorganization of the microtubule array in prophase/prometaphase requires cytoplasmic dynein-dependent microtubule transport. *The Journal of Cell Biology*, *158*(6), 997–1003.

Sammak, P. J., & Borisy, G. G. (1988). Direct observation of microtubule dynamics in living cells. *Nature*, *332*(6166), 724–726.

Schek, H. T., Gardner, M. K., Cheng, J., Odde, D. J., & Hunt, A. J. (2007). Microtubule assembly dynamics at the nanoscale. *Current Biology*, *17*(17), 1445–1455.

Schek, H. T., & Hunt, A. J. (2005). Micropatterned structures for studying the mechanics of biological polymers. *Biomedical Microdevices*, 7(1), 41–46.

Schulze, E., & Kirschner, M. (1988). New features of microtubule behaviour observed in vivo. *Nature*, *334*(6180), 356–359.

Seetapun, D., & Odde, D. J. (2010). Cell-length-dependent microtubule accumulation during polarization. *Current Biology*, *20*(11), 979–988.

VanBuren, V., Cassimeris, L., & Odde, D. J. (2005). Mechanochemical model of microtubule structure and self-assembly kinetics. *Biophysical Journal*, *89*(5), 2911–2926.

VanBuren, V., Odde, D. J., & Cassimeris, L. (2002). Estimates of lateral and longitudinal bond energies within the microtubule lattice. *Proceedings of the National Academy of Sciences of the United States of America, 99*(9), 6035–6040.

Wade, R. H., Chrétien, D., & Job, D. (1990). Characterization of microtubule protofilament numbers. How does the surface lattice accommodate? *Journal of Molecular Biology, 212*(4), 775–786.

Walker, R. A., O'Brien, E. T., Pryer, N. K., Soboeiro, M. F., Voter, W. A., Erickson, H. P., et al. (1988). Dynamic instability of individual microtubules analyzed by video light microscopy: Rate constants and transition frequencies. *The Journal of Cell Biology, 107*(4), 1437–1448.

Wan, X., O'Quinn, R. P., Pierce, H. L., Joglekar, A. P., Gall, W. E., DeLuca, J. G., et al. (2009). Protein architecture of the human kinetochore microtubule attachment site. *Cell, 137*(4), 672–684.

Waterman-Storer, C. M., & Salmon, E. D. (1998). How microtubules get fluorescent speckles. *Biophysical Journal, 75*(4), 2059–2069.

Waters, J. C. (2009). Accuracy and precision in quantitative fluorescence microscopy. *The Journal of Cell Biology, 185*(7), 1135–1148.

Witte, H., Neukirchen, D., & Bradke, F. (2008). Microtubule stabilization specifies initial neuronal polarization. *The Journal of Cell Biology, 180*(3), 619–632.

SECTION II

Polymer Nucleation and Regulation

CHAPTER FOUR

Biochemical Reconstitution of the WAVE Regulatory Complex

Baoyu Chen, Shae B. Padrick, Lisa Henry, Michael K. Rosen[1]

Department of Biophysics and Howard Hughes Medical Institute, University of Texas Southwestern Medical Center at Dallas, Dallas, Texas, USA

[1]Corresponding author: e-mail address: Michael.Rosen@utsouthwestern.edu

Contents

Abstract

The WAVE regulatory complex (WRC) is a 400-kDa heteropentameric protein assembly that plays a central role in controlling actin cytoskeletal dynamics in many cellular processes. The WRC acts by integrating diverse cellular cues and stimulating the actin nucleating activity of the Arp2/3 complex at membranes. Biochemical and biophysical studies of the underlying mechanisms of these processes require large amounts of purified WRC. Recent success in recombinant expression, reconstitution, purification, and crystallization of the WRC has greatly advanced our understanding of the inhibition, activation, and membrane recruitment mechanisms of this complex. But many important questions remain to be answered. Here, we summarize and update the methods developed in our laboratory, which allow reliable and flexible production of tens of milligrams of recombinant WRC of crystallographic quality, sufficient for many biochemical and structural studies.

Methods in Enzymology, Volume 540
ISSN 0076-6879
http://dx.doi.org/10.1016/B978-0-12-397924-7.00004-2

1. INTRODUCTION

Actin nucleation, the formation of a new actin filament from actin monomers, is a crucial and rate-limiting step in actin cytoskeletal dynamics (Pollard & Cooper, 2009). The Arp2/3 complex is a central actin nucleator that binds to the side of existing filaments to promote new filament growth as a branch, creating a complex cortical actin network beneath membranes. The Arp2/3 complex has low basal activity in actin nucleation and requires stimulation by so-called nucleation promoting factors (NPFs). Members of the ubiquitous WASP (Wiskott–Aldrich syndrome protein) family constitute the largest and best-understood group of NPFs (Campellone & Welch, 2010; Padrick & Rosen, 2010; Rotty, Wu, & Bear, 2013). WASP-family proteins are defined by a conserved C-terminal VCA (Verprolin-homology, Central, Acidic) sequence that is sufficient to stimulate the Arp2/3 complex. Members of the WASP family include WASP/N-WASP, WAVE1/2/3 (WASP family verprolin homologous protein 1, 2, or 3), and the more recently discovered WASH (WASP and SCAR homolog), WHAMM (WASP homolog associated with actin, membranes, and microtubules), and JMY (junction-mediating and regulatory protein) (Pollitt & Insall, 2009; Takenawa & Suetsugu, 2007).

Many WASP family members are basally inhibited and, in response to activation signals, release the VCA to stimulate actin assembly. WASP and N-WASP are inhibited *in cis* through intramolecular contacts between the VCA and an N-terminal GTPase-binding domain (Kim, Kakalis, Abdul-Manan, Liu, & Rosen, 2000; Miki, Sasaki, Takai, & Takenawa, 1998; Prehoda, Scott, Mullins, & Lim, 2000; Rohatgi et al., 1999). In contrast, the WAVE proteins are inhibited *in trans* by incorporation into a~400-kDa heteropentameric protein assembly, referred to as the WAVE regulatory complex (WRC). The WRC consists of five proteins (Fig. 4.1A): Sra1/Cyfip1 (or the ortholog PIR121/Cyfip2), Nap1/Hem2/Kette (or the ortholog Hem1), Abi2 (or the orthologs Abi1 and Abi3), HSPC300/Brick1, and WAVE1/SCAR (or the orthologs WAVE2 and WAVE3) (Eden, Rohatgi, Podtelejnikov, Mann, & Kirschner, 2002). Different orthologs of each component seem exchangeable, allowing assembly of different WRC isoforms (Stovold, Millard, & Machesky, 2005). Within the WRC, the VCA is sequestered *in trans* through intracomplex interactions (Chen et al., 2010) (Fig. 4.1A).

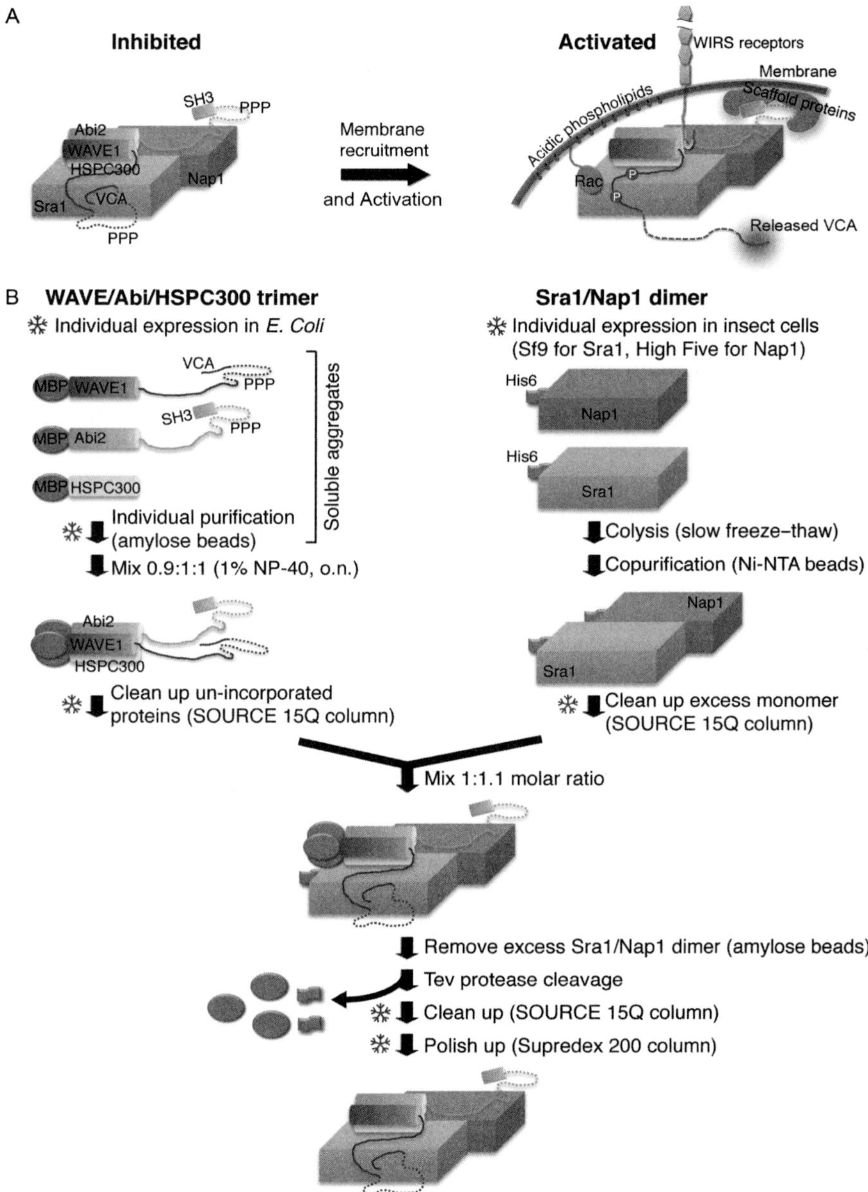

Figure 4.1 Activation mechanism and purification strategy of the WRC. (A) Schematic of WRC inhibition, activation, and membrane recruitment. Dotted lines indicate unstructured sequences. (B) Schematic of WRC *in vitro* reconstitution. Snowflake symbols indicate steps after which samples can be frozen and stored for future use. (See the color plate.)

To function, the inhibited WRC needs to be both recruited to and activated at the membrane by diverse signaling molecules as illustrated in Fig. 4.1A. These include small GTPases (Rac and Arf), acidic phospholipids (phosphatidylinositol (3,4,5)-trisphosphate, PIP_3), kinases (Abl, Cdk5, and ERK2), scaffolding proteins (IRSp53, Toca1, and WRP) (Chen et al., 2010; Fricke et al., 2009; Koronakis et al., 2011; Mendoza, 2013; Miki, Yamaguchi, Suetsugu, & Takenawa, 2000; Oikawa et al., 2004; Soderling et al., 2007; Takenawa & Suetsugu, 2007; Westphal, Soderling, Alto, Langeberg, & Scott, 2000), and the recently identified WIRS (WRC-interacting receptor sequence)-containing family consisting of a large number of membrane receptors (Chen et al., 2014). These ligands link the WRC to many cellular processes (adhesion, migration, division, fusion, etc.) across diverse biological systems, including embryogenesis, neuron morphogenesis and plasticity, immune cell activation and chemotaxis, and cancer invasion and metastasis (Pollitt & Insall, 2009; Takenawa & Suetsugu, 2007).

Mechanistic biochemical and biophysical studies of WRC/ligand interactions require access to purified WRC. Over the last decade, three major strategies have been developed to generate such material. The first involves purification from natural sources, including animal brains, blood, or cultured cells (Eden et al., 2002; Gautreau et al., 2004; Kim et al., 2006; Lebensohn & Kirschner, 2009; Weiner et al., 2006). This method allowed the discovery of the WRC and produces materials preserving native posttranslational modifications. As described in chapter 20 of the same issue (Hume, Humphreys, & Koronakis, 2014), Koronakis and colleagues recently further developed a new strategy to purify the native WRC from porcine brain extract by using phospholipid bilayer-coated silica microbeads, which led to identification of a new WRC activator, Arf (Koronakis et al., 2011). The above purifications cannot be readily scaled up and do not allow genetic modification of the WRC components for structure/function studies. The second method is *in vivo* reconstitution, involving (co-)expression of one or multiple affinity-tagged WRC subunits in cultured mammalian or insect cells (Derivery, Lombard, Loew, & Gautreau, 2009; Ismail, Padrick, Chen, Umetani, & Rosen, 2009; Mendoza et al., 2011). The recombinant WRC is assembled while expressed in cells and is purified using the affinity tags. This method had produced the WRC of sufficient quantity and purity for rigorous biochemical assays, which led to the final reconciliation of debates about whether the WRC is intrinsically inhibited. Here, we focus on the third method, *in vitro* reconstitution, developed and optimized in our laboratory over the last 10 years (Chen et al., 2010, 2013; Ismail et al., 2009) (Fig. 4.1B). This method improves the yield (up to tens of milligrams), the purity (yielding crystal structures of the WRC), and readily allows engineering of the complex to answer mechanistic questions. Through this method, we have been able to accomplish multiple structure–function studies of the WRC (Chen et al., 2010, 2014; Ismail et al., 2009; Padrick et al., 2008).

2. OVERVIEW OF THE *IN VITRO* RECONSTITUTION METHOD

Generation of recombinant WRC presents substantial challenges. First, expression of the WRC subunits in bacteria either gives poor yield or produces aggregated or truncated proteins. Second, coexpression and purification from eukaryotic hosts suffers from high cost and low yield. Third, inhibition of WAVE depends on a fully assembled WRC. Partially assembled subcomplexes containing WAVE, but missing other components, have high basal activity that obscures biochemical assays. This was actually a problem during our early preparations of the WRC generated by coexpressing all five components in insect cells. Such partial assemblies coincidentally copurified with the WRC pentamer, giving variable activities to the final material. We were finally able to consistently obtain inhibited WRC preparations by removing the contaminants through an added cation exchange step (Ismail et al., 2009). Similarly, aggregation of the WRC also gives anomalous activation (Lebensohn & Kirschner, 2009). Thus, purifications must rigorously exclude all subcomplexes and aggregates. These complications likely account for previous conflicting observations regarding the basal activity of the WRC (Derivery et al., 2009; Eden et al., 2002; Innocenti et al., 2004; Ismail et al., 2009; Kim et al., 2006; Lebensohn & Kirschner, 2009). Below we describe our strategy, which addresses these various difficulties.

The WRC can be viewed as two subcomplexes: a trimer of WAVE, Abi, and HSPC300 held together by an N-terminal four-helix bundle, and a dimer formed by the two large structurally homologous proteins, Sra1 and Nap1 (Chen et al., 2010; Ismail et al., 2009) (Fig. 4.1). The *in vitro* reconstitution follows these logistics: WAVE/Abi/HSPC300 trimer + Sra1/Nap1 dimer = WRC (Fig. 4.1B).

- *WAVE/Abi/HSPC300 trimer.* Initially, soluble, but aggregated maltose-binding protein (MBP)-fusions of the proteins are separately expressed and purified from *Escherichia coli*. The three proteins are then mixed in the presence of detergent and incubated for several hours. This allows aggregates to spontaneously disassemble and reassemble into a WAVE/Abi/HSPC300 trimer. Further purification removes the detergent and unincorporated protein or subcomplexes.

- *Sra1/Nap1 dimer.* His_6-tagged Sra1 and Nap1 are separately expressed in insect cells. The cells are colysed allowing formation of the Sra1/Nap1 heterodimer. This heterodimer is then purified and frozen.
- *WRC pentamer.* The heteropentamer is assembled from the purified WAVE/Abi/HSPC300 trimer and the Sra1/Nap1 dimer, followed by purification to remove the excess dimer and affinity tags after protease cleavage.

It is noteworthy that all the affinity tags were fused to the N-termini of the WRC subunits; this produces the desired proteins and preserves the activity while not disturbing the inhibited state of the WRC. We have not explored placing tags at alternative positions for most subunits. However, for the full-length Abi and WAVE, bacterial expression sometimes gives significant amounts of truncated products, which can be difficult to separate using conventional chromatography (note that this was not a problem in our previous approach, where all subunits were coexpressed in insect cells, but yields were much lower in that case) (Ismail et al., 2009). To remove these products, we have recently added an additional C-terminal or internal tag to be used in a second-affinity purification step. For the full-length Abi, we directly add a His_6 tag to the C-terminus. For the full-length WAVE, since a C-terminal tag next to the VCA substantially decreases the activity of the WRC in *in vitro* assays, we insert a His_6 or StrepII tag to an internal region between the proline-rich sequence and the VCA (Chen et al., 2014).

The modular nature of this method provides excellent versatility. First, through this method, we have been able to produce a variety of WRC isoforms by incorporating different constructs (different truncations or tags), orthologs (Abi1, 2, WAVE1, 2, 3), and homologs (human, *Drosophila*). Second, we have found that individual subunits or subcomplexes can be frozen, allowing quick combinatorial assembly of different WRC mutants. Third, this method allows for modification of individual subunits prior to reconstitution, suitable for metabolic labeling or chemical modification of specific subunits. For example, the Sra1/Nap1 dimer was selenomethionine labeled to improve X-ray diffraction quality and to solve the phase problem for crystallographic structure determination of the WRC (Chen et al., 2010).

3. KEY REAGENTS

- *Lysis buffer*: 20 m*M* Tris–HCl pH 8.0, 200 m*M* NaCl, 1 m*M* EDTA, and 5 m*M* β-mercaptoethanol (BME)
- *MBP-Elution buffer*: 20 m*M* Tris–HCl pH 8.0, 1 m*M* EDTA, 1% (w/v) maltose, 10% (w/v) glycerol, 2 m*M* DTT
- *QA buffer*: 20 m*M* Tris–HCl pH 8.0, 1 m*M* EDTA, and 5 m*M* BME
- *QB buffer*: QA buffer + 1 *M* NaCl
- *WRC-Lysis buffer*: 20 m*M* Tris–HCl pH 8.0, 20 m*M* imidazole pH 8.0, 200 m*M* NaCl, 20% (w/v) glycerol, and 5 m*M* BME (no EDTA, as this will block subsequent binding to Ni-NTA affinity resin)
- *WRC-Ni-Elution buffer*: 20 m*M* Tris–HCl pH 8.0, 300 m*M* imidazole pH 8.0, 100 m*M* NaCl, 20% (w/v) glycerol, and 5 m*M* BME (no EDTA)
- *WRC-MBP-Elution buffer*: 20 m*M* Tris–HCl pH 8.0, 50 m*M* NaCl, 1% (w/v) maltose, 20% (w/v) glycerol, 1 m*M* EDTA, and 5 m*M* BME
- *WRC-QA buffer*: 20 m*M* Tris–HCl pH 8.0, 20% (w/v) glycerol, 1 m*M* EDTA, and 5 m*M* BME
- *WRC-QB buffer*: WRC-QA buffer + 1 *M* NaCl
- *WRC-GF buffer*: 20 m*M* Tris–HCl pH 8.0, 150 m*M* NaCl, 12% (w/v) glycerol, 1 m*M* DTT
- *WRC-100KMEI20G buffer*: 10 m*M* imidazole pH 7.0, 100 m*M* KCl, 20% (w/v) glycerol, 1 m*M* $MgCl_2$, 1 m*M* EGTA, 1 m*M* DTT
- *Protease inhibitors (PI)*: 2 μg/mL leupeptin, 2 μg/mL antipain, and 2 m*M* benzamidine, diluted from 500 × individual stock solutions in water
- *Ampicillin (Amp)*: 100 μg/mL, diluted from 1000 × stock solution in water
- *Gentamycin (Gen)*: 20 μg/mL, diluted from 1000 × stock solution in water

We note that the inclusion of glycerol (10–20%, w/v) in many of the buffers is essential to maintain stability of the WRC. Without glycerol, the complex precipitates over time and shows highly variable activity.

4. RECONSTITUTION OF THE WAVE/Abi/HSPC300 TRIMER

4.1. Expression and purification of individual trimer components

WAVE, Abi, and HSPC300 are individually expressed and purified from *E. coli*. We routinely use N-terminal MBP fusion proteins to improve solubility and to facilitate affinity purification. Although MBP has an intermediate affinity for amylose beads, this is substantially improved by multiple MBP tags on the proteins (e.g., multiple copies on the soluble aggregates of individual proteins, or three copies on the reconstituted trimer or WRC).

Below we give a detailed protocol (Fig. 4.1B) that has been used to purify many different WAVE/Abi/HSPC300 trimers. In the case of full-length WAVE1, 2, 3 or Abi 1, 2, we fuse a His_6-tag to or near the protein C-terminus and add a nickel affinity purification step to remove the truncated proteins (Chen et al., 2014). In some cases, we use di-MBP-HSPC300 (with two tandem MBP proteins at the N-terminus) to facilitate pull-down assays (Chen et al., 2014).

4.1.1 Expression constructs for trimer proteins

All constructs were cloned into a vector derived from pMAL-c2X (New England Biolabs). This vector has a Tev protease cleavage site introduced between the MBP and the protein of interest as shown in Fig. 4.2A. Here, we use the following constructs as examples: WAVE217 for human WAVE1 (1–217)-$(GGS)_6$-(485–559), in which the proline-rich region (PPP in Fig. 4.1) is replaced by a $(Gly\text{-}Gly\text{-}Ser)_6$ linker; Abi158 for human Abi2 (1–158), in which the proline-rich region and the C-terminal SH3 domain are removed; and full-length human HSPC300.

4.1.2 Expression of individual proteins in E. coli

We generally use two host strains. We use BL21 (DE3) $T1^R$ cells (New England Biolabs) for MBP-tagged Abi158, HSPC300, or shorter constructs of WAVE. We use ArcticExpress™ (DE3) RIL cells (ArcticExpress cells for short, Stratagene) for MBP-tagged WAVE217 or longer constructs of WAVE and Abi. Protocols are different for the two strains.

Day 1: Transform cells using standard procedures. Incubate plates at 37 °C overnight.

Day 2: Use a single colony to inoculate 6 mL of LB/Amp medium in a 15-mL culture tube. Grow this starter culture at 37 °C, 250 rpm overnight.

- For ArcticExpress cells, separately inoculate 2–3 starter cultures (LB/Amp/Gen, Gen selects for the chaperonin-expressing plasmid).

Day 3: Spin down starter culture (2500 g, 10 min, 4 °C), resuspend the cell pellets in 1–2 L of fresh medium in 4 L flasks supplemented with Amp and 0.2% (w/v) glucose (from sterile 40%, w/v, stock).

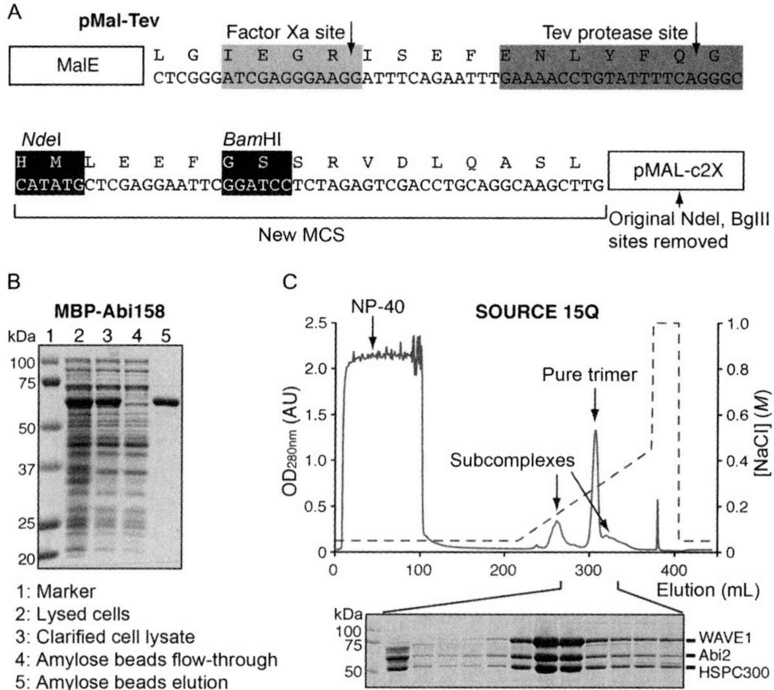

Figure 4.2 Production of the WAVE/Abi/HSPC300 trimer. (A) Sequence of the pMal-Tev vector used for all trimer constructs. (B) SDS-PAGE gel showing affinity purification of MBP-Abi158 by amylose beads. Other trimer proteins are purified similarly. (C) Anion-exchange chromatography of the reconstituted WAVE217/Abi158/HSPC300 trimer. Coomassie brilliant blue stained SDS-PAGE analysis of the indicated fractions is shown below. (For color version of this figure, the reader is referred to the online version of this chapter.)

- For BL21 (DE3) cells, use LB medium. Grow the culture at 37 °C, 250 rpm until the OD_{600nm} reaches 0.6–0.8. Add 1 m*M* IPTG to induce expression at 18 °C overnight (16 h).
- For ArcticExpress cells, use TB (terrific broth) and baffled flasks. Grow the culture at 37 °C, 250 rpm until the OD_{600nm} reaches 0.4–0.6. Then change the temperature to 30 °C until the OD_{600nm} reaches 1.5–2. Transfer the flasks to ice water. After 30 min, add 1 m*M* IPTG to induce expression at 10 °C, 250 rpm for 24 h.

Day 4: Pellet the cells by centrifugation in 1-L bottles at 1600 g for 20 min. Resuspend in 30 mL of Lysis buffer per liter of culture (supplemented with PI). Cells may be stored at −80 °C.

4.1.3 Purification of individual proteins by amylose beads

Perform all purification steps at 4 °C. Thaw the frozen cell suspension in tap water for ~10 min. Add PMSF to 1 m*M* and immediately lyse the cells by sonication (on ice water) or extrusion (we use an Avestin EmulsiFlex-C5 microfluidizer). Clarify the lysed cells by centrifugation at 48,000 g for 30 min at 4 °C. Add the supernatant to 10 mL of amylose beads (New England Biolabs) in an Econo-Column® glass column (2.5 × 10 cm, Bio-Rad) and mix by nutation for 30 min. Drain the lysate through the column and wash the beads three times with 30 mL of Lysis buffer. Elute bound proteins by resuspending the beads in 20 mL of MBP-Elution buffer, waiting 5 min, and collecting the eluted proteins by draining through the column. Repeat with an additional 30 mL of MBP-Elution buffer and pool the two eluates. Examine protein purity by SDS-PAGE (Fig. 4.2B). Measure protein concentration by OD_{280nm} using extinction coefficients calculated by ExPASy (Gasteiger et al., 2003). Proteins can be flash frozen and stored at −80 °C for at least 2 years.

4.2. Reconstitution of the WAVE/Abi/HSPC300 trimer

Most individually purified WAVE, Abi, and HSPC300 proteins are soluble aggregates. When mixed, trimer assembly from these aggregates is very slow, but inclusion of 1–2% (w/v) NP-40 substantially accelerates assembly. Once the trimer has formed, high-resolution anion exchange (in some cases, both cation and anion) is used to remove the detergent and unincorporated proteins or subcomplexes. Below is a representative protocol, but modifications can be made to adapt to more difficult situations. For instance, choice of detergents or chaotropic agents may vary: although all human WAVE/Abi/HSPC300 trimers can be efficiently assembled with NP-40, a *Drosophila* WAVE/Abi/HSPC300 trimer could only be assembled with 1% (w/v) CHAPS or OG (Octyl-beta-glucopyranoside), but not NP-40. Thus, this protocol provides a useful template for trimer assembly, especially for proteins from other species.

4.2.1 Reconstitution of trimer

Mix purified WAVE217, Abi158, and HSPC300 in a 0.9:1:1 molar ratio. Add 2 m*M* DTT, PI, and 1% (w/v) NP-40 (now available as Igepal® CA-630, Sigma). Incubate the mixture at 4 °C for 12–48 h.

4.2.2 Purification of trimer

Dilute the trimer mixture with an equal volume of QA buffer. Centrifuge at 2500 g for 10 min. Apply supernatant to an 8-mL, 10-mm wide, SOURCE

15Q (GE Healthcare) column using a chromatography system capable of linear gradients at back pressures of 2–3 MPa (e.g., ÄKTA FPLC or similar) (Fig. 4.2C). After loading, wash the column with 8 column volumes (CV) of QA buffer and elute the bound proteins using a gradient of 5–35% QB buffer developed over 20 CV. The WAVE/Abi/HSPC300 trimer elutes as the major peak, flanked by minor peaks containing partial assemblies (Fig. 4.2C). In some difficult situations (i.e., WRCs containing full-length WAVE or Abi), separation of trimer from partial assemblies is improved at pH 7.0 and/or requires an additional cation exchange step (SOURCE 15S, GE Healthcare). Examine trimer purity by SDS-PAGE (Fig. 4.2C). Measure protein concentration by OD_{280nm}. Add glycerol to the pooled fractions to 10% (w/v). If not immediately used, the timer can be flash frozen in liquid nitrogen and kept at −80 °C for at least 2 years.

5. RECONSTITUTION OF THE Sra1/Nap1 DIMER

Sra1 and Nap1 are expressed individually in insect cells, and the Sra1/Nap1 dimer is readily formed upon mixing and colysing the cells. A similar strategy can be used to produce complexes with PIR121, or *Drosophila* Sra and Nap. Variations to consider include separate versus coexpression, and choice of insect cells.

One precaution to take when handling the Sra1/Nap1 dimer (or individual proteins, and to a lesser extent, the WRC) is that the dimer tends to denature on air/water interfaces, forming a sticky protein film. Therefore, it is important to avoid rough pipetting and foaming (i.e., that created by continued mixing with beads in the presence of air). We also routinely include 20% (w/v) glycerol and 100–200 m*M* NaCl during purification to lessen these problems.

5.1. Constructs for dimer proteins

Constructs were cloned into the pAV5a vector, which is derived from pFastBacHTa (Invitrogen) modified by insertion of an L21 promoter sequence from lobster tropomyosin before the start codon (Sano, Maeda, Oki, & Maeda, 2002) (Fig. 4.3A).

5.2. Insect cell maintenance and baculovirus production

We refer readers to the 2010 Invitrogen *Bac-to-Bac* and 2002 Invitrogen *Growth and Maintenance of Insect Cell Lines* manuals for details of baculovirus

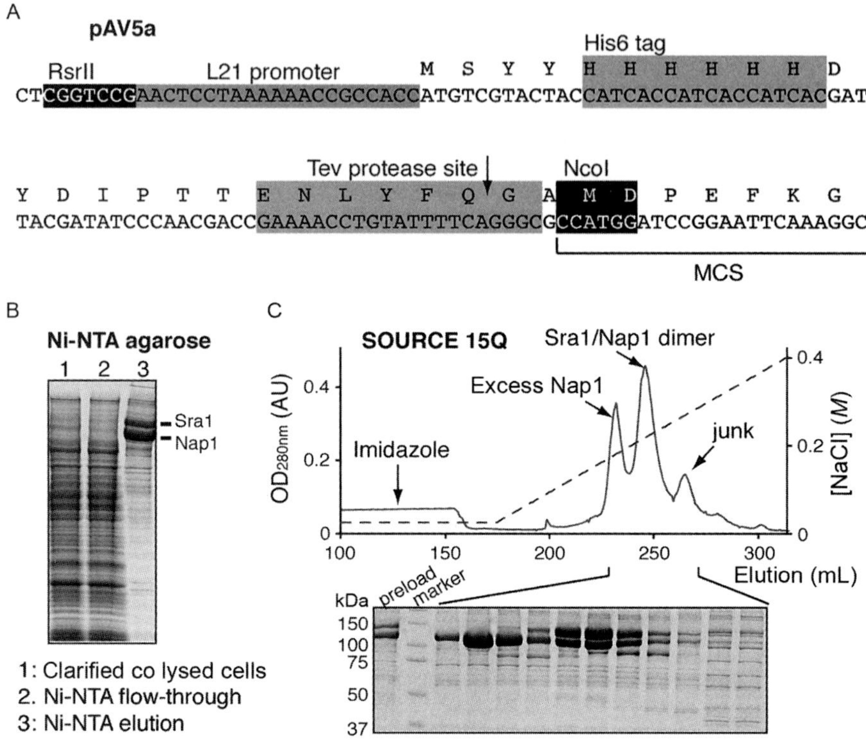

Figure 4.3 Production of the Sra1/Nap1 dimer. (A) Modification of the pFastBacHTa vector to enhance expression of dimer proteins by inserting an L21 sequence. (B) Coomassie brilliant blue stained SDS-PAGE analysis following the copurification of His_6-tagged Sra1 and His_6-tagged Nap1 by Ni-NTA beads from colysed cells. In this case, Nap1 is in excess. (C) Chromatogram of SOURCE 15Q anion-exchange purification of the Sra1/Nap1 dimer. Coomassie brilliant blue stained SDS-PAGE analysis of indicated fractions is shown below. (For color version of this figure, the reader is referred to the online version of this chapter.)

production and insect cell maintenance, respectively. Below we describe parameters specific to Sra1 and Nap1.

We use Sf9 cells to produce baculoviruses for both Sra1 and Nap1. To produce P1 virus, production and transfection of viral bacmids are performed according to supplier protocols. We examine the cells 72–96 h post-transfection and collect the virus when cells begin to dissociate from the culture dish into the medium. P2 and P3 viruses are prepared as described in the *Invitrogen* manual. All viruses are stored at 4 °C before use. Instead of determining virus titers, we usually estimate the potency of virus by evaluating protein expression level and percentage of dead cells (normally aiming at ~20%) after 72 h of small-scale infection. As protein expression levels decline with virus age, we make new P1 viruses every 6 months to 1 year. We use Sf-900™ II SFM medium (Invitrogen) to maintain and expand the Sf9 cells as suspension culture at 27 °C, shaking at 125 rpm. We use ESF 921 medium (Expression Systems) for High Five™ cells, which are maintained as stationary culture in T75 flasks at 27 °C in a humidified incubator but are grown similar to the Sf9 cells as suspension culture upon expansion for large-scale protein expression. All cultures are supplemented with the Antibiotic–Antimycotic solution (Invitrogen) at 5 ml per L.

5.3. Infection and protein expression

For the human Nap1/Sra1 heterodimer, we have found that separate expression of Sra1 in Sf9 cells and Nap1 in High Five™ cells gives the best overall yield by a significant margin over other combinations (e.g., both proteins together in either Sf9 or High Five™ cells). To express Sra1 or Nap1 proteins, culture 900 mL of desired insect cells at 27 °C, 125 rpm in a 2-L Delong culture flask (Bellco) until they reach 2×10^6 cells/ml. Infect the cells with 1.8 mL of P2 or P3 baculovirus suspensions (500-fold dilution). After infection for 65–70 h, harvest the cells by centrifugation at 1600 g for 30 min. Resuspend the cells in 30 mL of WRC-Lysis buffer per flask of culture (supplemented with PI), transfer to freezer, and store at −80 °C until needed.

5.4. Purification of Sra1/Nap1 dimer

Estimate protein expression level for each batch of infection by Coomassie brilliant blue stained SDS-PAGE. Mix cells to give roughly a 1:1 Sra1:Nap1 ratio. If the viruses are of similar age, this typically requires three parts Sra1-expressing cells with one part Nap1-expressing cells. The protocol below is written for cells from 6×0.9 L Sra1 culture mixed with 2×0.9 L Nap1 culture.

Lyse the cells by slow thawing in wet ice for 1 hr. Pool all cells and add PMSF to 1 m*M*. Centrifuge the colysed cells at 48,000 g for 45 min. Load the supernatant to 20 mL of Ni-NTA agarose beads (Qiagen) in an Econo-Column® glass column (5×10 cm, Bio-Rad) by gravity flow. Wash the beads with 3×60 mL of the WRC-Lysis buffer. Elute the bound proteins with 100 mL of the WRC-Ni-Elution buffer (Fig. 4.3B). The Ni-NTA agarose beads may be regenerated and reused for subsequent preparations.

Dilute the eluate with an equal volume of the WRC-QA buffer. Centrifuge at 2500 g for 10 min and apply the supernatant to 8-mL of

SOURCE 15Q packed in a 10-mm wide column using a system capable of performing linear gradients at >3 MPa (e.g., an ÄKTA FPLC system or equivalent) (Fig. 4.3C). After loading, wash with 2 CV of 95% WRC-QA/5% WRC-QB buffer and elute the bound proteins by a gradient of 5–35% WRC-QB buffer developed over 20 CV. The first peak is usually excess monomers (Sra1 or Nap1), and the second peak is the Sra1/Nap1 dimer (Fig. 4.3C). Assess dimer purity by SDS-PAGE (Fig. 4.3C). Measure protein concentration by OD_{280nm}. If not immediately used, the dimer can be flash frozen and kept at −80 °C for at least 2 years.

6. RECONSTITUTION OF THE WRC PENTAMER

The pentameric WRC spontaneously assembles from the WAVE/Abi/HSPC300 trimer and the Sra1/Nap1 dimer subcomplexes (Fig. 4.1B). The WAVE, Abi, and HSPC300 proteins still bear MBP fusions, allowing capture of the WRC on amylose beads and removal of unincorporated Nap1/Sra1 dimer (which is added in excess) (Fig. 4.4A). The WRC is further purified by one anion-exchange column and a gel filtration step.

6.1. Reconstitution of the WRC

Mix the WAVE/Abi/HSPC300 trimer and the Sra1/Nap1 dimer subcomplexes at a 1:1.1 molar ratio. Note that it is critical to use excess dimer to eliminate unincorporated, free trimer; as described above, the trimer has high activity and can be difficult to separate from the full WRC, and its presence even in small amounts can impart anomalous activity to WRC preparations. After incubation for 1 hr, absorb the WRC to 5 mL of amylose beads in an Econo-Column® glass column (2.5 × 10 cm, Bio-Rad) by rotary mixing for 30 min. Use of gentle agitation to keep the amylose beads suspended is acceptable at this point as the WRC is less prone to denaturation on air–water interfaces than is the Sra1/Nap1 dimer. The bound WRC is purified by gravity flow. After three washes with 15 mL of WRC-Lysis buffer, the bound WRC is eluted with 25 mL of the WRC-MBP-Elution buffer (Fig. 4.4A).

6.2. Tev cleavage and final purification of the WRC

The MBP and His_6 tags are removed using Tev protease cleavage after the amylose bead purification. Add 2 m*M* DTT and 200 μL of 3 mg/mL-purified

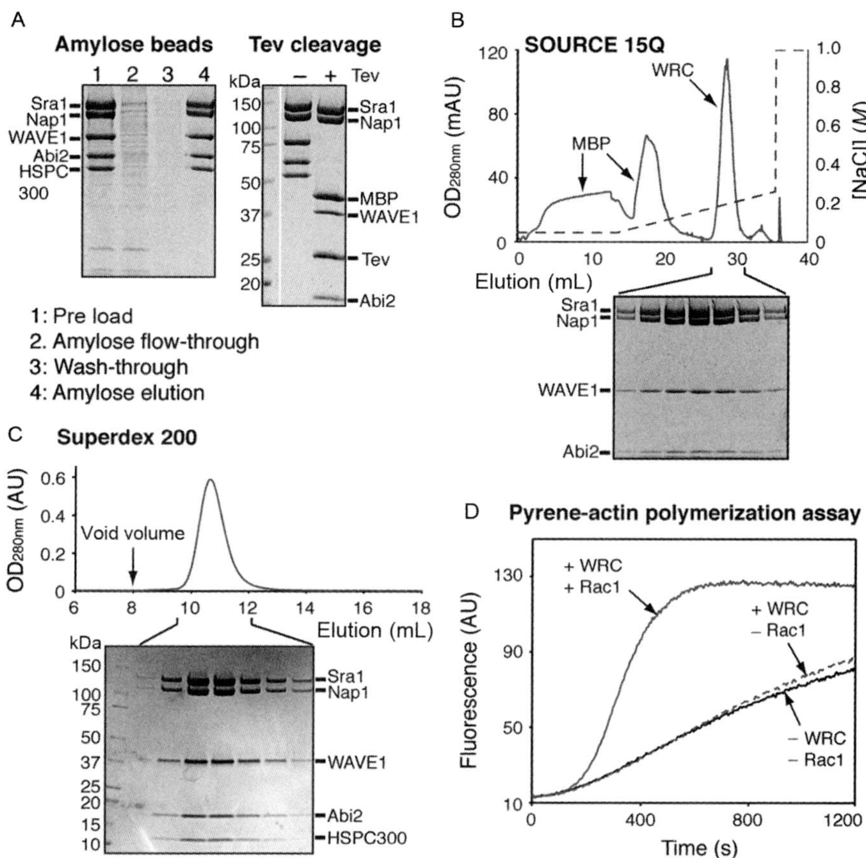

Figure 4.4 Reconstitution of the WRC pentamer. (A) Purification of the reconstituted WRC (WAVE217/Abi158/HSPC300/Sra1/Nap1) by amylose beads, followed by Tev protease cleavage. (B–C) UV traces of anion exchange and size exclusion chromatography steps of the WRC purification. Coomassie brilliant blue stained SDS-PAGE analysis of indicated fractions is shown for each chromatogram. In (A, B), untagged HSPC300 (which is smaller than 10 kDa) has been run off of the bottom of both gels. (D) Results from a representative pyrene-actin polymerization assay. All reactions contain 4 μ*M* actin (5% pyrene labeled) and 10 n*M* Arp2/3 complex. Reconstituted WRC (WAVE217/Abi158/HSPC300/Sra1/Nap1) was added at 100 n*M* concentration, along with 15 μ*M* GMPNPN-loaded Rac1 as indicated. (For color version of this figure, the reader is referred to the online version of this chapter.)

Tev protease to the WRC eluted from the amylose beads and incubate at 22 °C for 8 h. Confirm complete cleavage by SDS-PAGE before continuing (Fig. 4.4A).

After Tev treatment, the untagged WRC is purified using the SOURCE 15Q anion-exchange protocol described in Section 5.4 (Fig. 4.4B). Pool fractions containing pure WRC and concentrate the complex using an Amicon Ultra-15 centrifugal filter unit, MWCO 30 kDa (Millipore) at 2500 g, 4 °C to reduce the sample volume to 10 mL (or 1 mL if purifying less than 10 mg of protein). Remove any aggregates by centrifugation. Apply the supernatant to a HiLoad 26/600 Superdex 200 column (or a 24-mL Superdex 200 10/300 GL column if loading less than 10 mg of protein) (GE Healthcare), equilibrated with the WRC-GF buffer (for general purposes) or the WRC-100KMEI20G buffer (for actin polymerization assays) (Fig. 4.4C). Measure protein concentration by OD_{280nm}. Concentrate the purified WRC if desired; at least 10 mg/mL is generally achievable without producing protein precipitation. Aliquot, flash freeze, and store the WRC at −80 °C. WRC can be stored in this fashion for at least 1 year. We routinely use the Arp2/3-mediated pyrene-actin assembly assay to evaluate the quality of the reconstituted WRC (Chen et al., 2010, 2013; Ismail et al., 2009). As shown in Fig. 4.4D, the purified WRC should exhibit marginal basal activity and be readily activated by GMPPNP-bound Rac1 (Fig. 4.4D).

ACKNOWLEDGMENTS

We thank Ayman Ismail, Zhucheng Chen, Da Jia, and Junko Umetani for their contributions to this project. B. C. was supported by a fellowship from the American Heart Association. M. K. R. was supported by the NIH (R01-GM056322), the Welch Foundation (I-1544), and the Howard Hughes Medical Institute.

REFERENCES

Campellone, K. G., & Welch, M. D. (2010). A nucleator arms race: Cellular control of actin assembly. *Nature Reviews. Molecular Cell Biology*, *11*(4), 237–251.

Chen, Z., Borek, D., Padrick, S. B., Gomez, T. S., Metlagel, Z., Ismail, A. M., et al. (2010). Structure and control of the actin regulatory WAVE complex. *Nature*, *468*(7323), 533–538.

Chen, B., Brinkmann, K., Chen, Z., Pak, C. W., Liao, Y., Shi, S., et al. (2014). The WAVE regulatory complex links diverse receptors to the actin cytoskeleton. *Cell*, *156*(1–2), 195–207.

Derivery, E., Lombard, B., Loew, D., & Gautreau, A. (2009). The Wave complex is intrinsically inactive. *Cell Motility and the Cytoskeleton*, *66*(10), 777–790.

Eden, S., Rohatgi, R., Podtelejnikov, A. V., Mann, M., & Kirschner, M. W. (2002). Mechanism of regulation of WAVE1-induced actin nucleation by Rac1 and Nck. *Nature*, *418*(6899), 790–793.

Fricke, R., Gohl, C., Dharmalingam, E., Grevelhorster, A., Zahedi, B., Harden, N., et al. (2009). Drosophila Cip4/Toca-1 integrates membrane trafficking and actin dynamics through WASP and SCAR/WAVE. *Current Biology*, *19*(17), 1429–1437.

Gasteiger, E., Gattiker, A., Hoogland, C., Ivanyi, I., Appel, R. D., & Bairoch, A. (2003). ExPASy: The proteomics server for in-depth protein knowledge and analysis. *Nucleic Acids Research*, *31*(13), 3784–3788.

Gautreau, A., Ho, H. Y., Li, J., Steen, H., Gygi, S. P., & Kirschner, M. W. (2004). Purification and architecture of the ubiquitous Wave complex. *Proceedings of the National Academy of Sciences of the United States of America, 101*(13), 4379–4383.

Hume, P. J., Humphreys, D., & Koronakis, V. (2014). WAVE Regulatory Complex Activation. *Methods in Enzymology, 540*, 55–72.

Innocenti, M., Zucconi, A., Disanza, A., Frittoli, E., Areces, L. B., Steffen, A., et al. (2004). Abi1 is essential for the formation and activation of a WAVE2 signalling complex. *Nature Cell Biology, 6*(4), 319–327.

Ismail, A. M., Padrick, S. B., Chen, B., Umetani, J., & Rosen, M. K. (2009). The WAVE regulatory complex is inhibited. *Nature Structural & Molecular Biology, 16*(5), 561–563.

Kim, A. S., Kakalis, L. T., Abdul-Manan, N., Liu, G. A., & Rosen, M. K. (2000). Autoinhibition and activation mechanisms of the Wiskott-Aldrich syndrome protein. *Nature, 404*(6774), 151–158.

Kim, Y., Sung, J. Y., Ceglia, I., Lee, K. W., Ahn, J. H., Halford, J. M., et al. (2006). Phosphorylation of WAVE1 regulates actin polymerization and dendritic spine morphology. *Nature, 442*(7104), 814–817.

Koronakis, V., Hume, P. J., Humphreys, D., Liu, T., Horning, O., Jensen, O. N., et al. (2011). WAVE regulatory complex activation by cooperating GTPases Arf and Rac1. *Proceedings of the National Academy of Sciences of the United States of America, 108*(35), 14449–14454.

Lebensohn, A. M., & Kirschner, M. W. (2009). Activation of the WAVE complex by coincident signals controls actin assembly. *Molecular Cell, 36*(3), 512–524.

Mendoza, M. C. (2013). Phosphoregulation of the WAVE regulatory complex and signal integration. *Seminars in Cell & Developmental Biology, 24*(4), 272–279.

Mendoza, M. C., Er, E. E., Zhang, W., Ballif, B. A., Elliott, H. L., Danuser, G., et al. (2011). ERK-MAPK drives lamellipodia protrusion by activating the WAVE2 regulatory complex. *Molecular Cell, 41*(6), 661–671.

Miki, H., Sasaki, T., Takai, Y., & Takenawa, T. (1998). Induction of filopodium formation by a WASP-related actin-depolymerizing protein N-WASP. *Nature, 391*(6662), 93–96.

Miki, H., Yamaguchi, H., Suetsugu, S., & Takenawa, T. (2000). IRSp53 is an essential intermediate between Rac and WAVE in the regulation of membrane ruffling. *Nature, 408*(6813), 732–735.

Oikawa, T., Yamaguchi, H., Itoh, T., Kato, M., Ijuin, T., Yamazaki, D., et al. (2004). PtdIns(3,4,5)P3 binding is necessary for WAVE2-induced formation of lamellipodia. *Nature Cell Biology, 6*(5), 420–426.

Padrick, S. B., Cheng, H. C., Ismail, A. M., Panchal, S. C., Doolittle, L. K., Kim, S., et al. (2008). Hierarchical regulation of WASP/WAVE proteins. *Molecular Cell, 32*(3), 426–438.

Padrick, S. B., & Rosen, M. K. (2010). Physical mechanisms of signal integration by WASP family proteins. *Annual Review of Biochemistry, 79*, 707–735.

Pollard, T. D., & Cooper, J. A. (2009). Actin, a central player in cell shape and movement. *Science, 326*(5957), 1208–1212.

Pollitt, A. Y., & Insall, R. H. (2009). WASP and SCAR/WAVE proteins: The drivers of actin assembly. *Journal of Cell Science, 122*(Pt. 15), 2575–2578.

Prehoda, K. E., Scott, J. A., Mullins, R. D., & Lim, W. A. (2000). Integration of multiple signals through cooperative regulation of the N-WASP-Arp2/3 complex. *Science, 290*(5492), 801–806.

Rohatgi, R., Ma, L., Miki, H., Lopez, M., Kirchhausen, T., Takenawa, T., et al. (1999). The interaction between N-WASP and the Arp2/3 complex links Cdc42-dependent signals to actin assembly. *Cell, 97*(2), 221–231.

Rotty, J. D., Wu, C., & Bear, J. E. (2013). New insights into the regulation and cellular functions of the ARP2/3 complex. *Nature Reviews. Molecular Cell Biology, 14*(1), 7–12.

Sano, K., Maeda, K., Oki, M., & Maeda, Y. (2002). Enhancement of protein expression in insect cells by a lobster tropomyosin cDNA leader sequence. *FEBS Letters*, *532*(1–2), 143–146.

Soderling, S. H., Guire, E. S., Kaech, S., White, J., Zhang, F., Schutz, K., et al. (2007). A WAVE-1 and WRP signaling complex regulates spine density, synaptic plasticity, and memory. *The Journal of Neuroscience*, *27*(2), 355–365.

Stovold, C. F., Millard, T. H., & Machesky, L. M. (2005). Inclusion of Scar/WAVE3 in a similar complex to Scar/WAVE1 and 2. *BMC Cell Biology*, *6*(1), 11.

Takenawa, T., & Suetsugu, S. (2007). The WASP-WAVE protein network: Connecting the membrane to the cytoskeleton. *Nature Reviews. Molecular Cell Biology*, *8*(1), 37–48.

Weiner, O. D., Rentel, M. C., Ott, A., Brown, G. E., Jedrychowski, M., Yaffe, M. B., et al. (2006). Hem-1 complexes are essential for Rac activation, actin polymerization, and myosin regulation during neutrophil chemotaxis. *PLoS Biology*, *4*(2), e38.

Westphal, R. S., Soderling, S. H., Alto, N. M., Langeberg, L. K., & Scott, J. D. (2000). Scar/WAVE-1, a Wiskott-Aldrich syndrome protein, assembles an actin-associated multi-kinase scaffold. *The EMBO Journal*, *19*(17), 4589–4600.

CHAPTER FIVE

Rotational Movement of Formins Evaluated by Using Single-Molecule Fluorescence Polarization

Hiroaki Mizuno, Naoki Watanabe[1]

Laboratory of Single-Molecule Cell Biology, Tohoku University Graduate School of Life Sciences, Sendai, Miyagi, Japan

[1]Corresponding author: e-mail address: nwatanabe@m.tohoku.ac.jp

Contents

Abstract

Formin homology proteins (formins) are responsible for the formation of actin structures such as actin stress fibers, actin cables, and cytokinetic contractile rings. Formins are the major actin filament (F-actin) nucleators in the cell. Because formins remain bound to the barbed end after nucleating an actin filament, it was expected that formins might rotate along the double-helical structure of F-actin during processive actin elongation

Methods in Enzymology, Volume 540
ISSN 0076-6879
http://dx.doi.org/10.1016/B978-0-12-397924-7.00005-4

(helical rotation). Here, we describe a method to detect the rotational movement of F-actin elongating from immobilized formins using single-molecule fluorescence polarization (FL_P). Tetramethylrhodamine (TMR) attached to Cys-374 of actin emits polarized fluorescence at ≈45° with respect to the filament axis. When the TMR-labeled F-actin laying at 45° in the visual field rotates, the vertical- and horizontal-polarized fluorescence (FL_V and FL_H, respectively) of TMR alternately become bright. This technique allowed us to demonstrate the helical rotation of mDia1, a mammalian formin. Adenosine triphosphate (ATP) hydrolysis in actin subunits is not required for helical rotation; however, ATP appears to contribute to accelerating actin elongation by mDia1. When helical rotation is limited by trapping both mDia1 and the pointed-end side, the processive filament elongation is blocked. Thus, mDia1 faithfully rotates along the long-pitch helix of F-actin. In this chapter, we introduce the theoretical concept of single-molecule FL_P, the optical setup, the preparation of adenosine diphosphate-bound actin, and the procedure to observe the rotational movement of F-actin elongating from immobilized formins.

1. INTRODUCTION

Formin homology proteins (formins) are the major actin nucleator (Evangelista, Zigmond, & Boone, 2003; Watanabe, 2010; Watanabe, Kato, Fujita, Ishizaki, & Narumiya, 1999). Formins have two conserved domains, formin homology domains 1 and 2 (FH1 and FH2), in the C-terminal half (Goode & Eck, 2007). Remarkably, formins FH2 processively elongate filamentous actin (F-actin) (Higashida et al., 2004; Pring, Evangelista, Boone, Yang, & Zigmond, 2003). We speculated that formins might rotate along the double-helical structure of F-actin during processive elongation (herein referred to as helical rotation).

Several studies have used fluorescence polarization (FL_P) for monitoring the directional transition of cytoskeletal structures, such as actin, myosin, and septin. For example, Corrie et al. (1999) revealed the change in orientation of myosin light chain during contraction, relaxation, and rigor state. Vrabioiu and Mitchison (2006) found that septin filaments align along the bud neck and then rotate by 90° before cytokinesis in budding yeast. Thus, the FL_P of bulk filamentous structures could provide information about their orientation. However, in order to monitor the rotation of the filament, the observation of single-molecule FL_P is required.

The method for monitoring the rotational movements of F-actin was originally introduced by Sase, Miyata, Ishiwata, and Kinosita (1997). Using single-molecule FL_P of tetramethylrhodamine (TMR)-labeled F-actin, it was revealed that, during its sliding movement on heavy meromyosins immobilized to a glass surface, F-actin rotates slowly (~1 μm per rotation), but not along the long-pitch actin helix.

By combining the single-molecule FL_P technique with our method to visualize processive actin elongation (Higashida et al., 2004), we discovered the helical rotation of mDia1, a mammalian formin, during actin polymerization and depolymerization (Mizuno et al., 2011).

2. THEORETICAL CONCEPTS AND OPTICS FOR SINGLE-MOLECULE FLUORESCENCE POLARIZATION

In this section, we describe the theoretical concepts of single-molecule FL_P and our optics.

2.1. Theoretical concepts

Fluorophores generally have a dipole moment, and the excitation dipole moment of a fluorophore can be different from its emission dipole moment. The dipole moment denotes the strength of the dipole and its direction. Fluorophores strongly absorb light when the direction of the excitation dipole moment corresponds with the oscillating direction of the electric field of the polarized light. In general, the direction of the dipole moment of a fluorophore changes quickly because the fluorophore moves randomly by Brownian motion. Thus, freely moving fluorophores emit polarized fluorescence at various orientations. It is therefore impracticable to estimate the direction of a protein from the angle of the dipole moment of the attached fluorescent dye. However, if the dye attaches to a protein within a rigid structure and the motion of the dye is restricted within a narrow angle with respect to the protein, it becomes possible to monitor the change in the direction of the protein based on the polarization of the emitted fluorescence.

By observing FL_P of F-actin labeled at Cys-374 with TMR (30% labeled), Sase et al. (1997) estimated that the dipole moment of TMR lay at ~45° with respect to the filament axis. When TMR-labeled F-actin was laid parallel to the polarizer axis, it showed an FL_P of 0.18 ± 0.04 through an objective with a numerical aperture (NA) of 1.2. This value was close to the predicted value for the case in which the dipole moments of TMR bound to F-actin are distributed on the surface of a corn of semi-angle 46° around the F-actin axis (Kinosita et al., 1991). Based on the measurement of the fluorescence anisotropy of TMR-labeled F-actin (3%), the rate of the rotary diffusion of TMR on actin was found to be confined within a cone of semiangle <25°. From these results, they concluded that TMR is not freely mobile on actin and that its transition moment lies at 45° to the filament axis.

When F-actin that is labeled sparsely with TMR is laid at an angle of ~45° in the visual field, individual TMR-actin molecules emit different levels of vertical- and horizontal-polarized fluorescence (FL_V and FL_H, respectively) depending on the position of the molecule (Fig. 5.1). If F-actin rotates, the ratio between FL_V and FL_H periodically alters. Using this technique, we observed the rotational movement of F-actin elongating from immobilized mDia1.

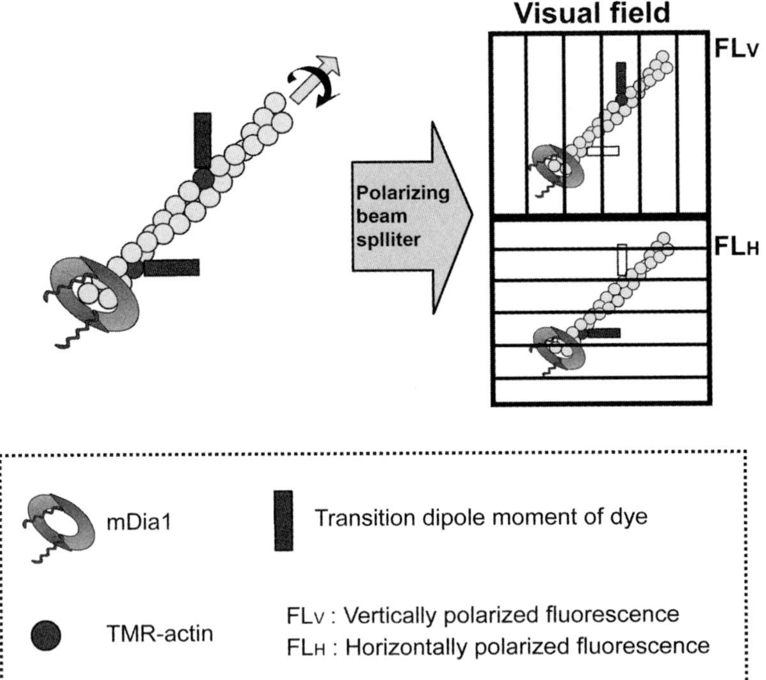

Figure 5.1 A schematic of the detection of the polarized fluorescence from single-molecule tetramethylrhodamine (TMR) on growing actin filaments. F-actin is nucleated by immobilized mDia1 and TMR-G-actin (1.5%). TMR emits polarized fluorescence at an angle of 45° to the filament axis (Sase et al., 1997). When the filament lies diagonally, the orientation of the TMR-actin subunits can be monitored by the vertically and horizontally polarized fluorescence using a polarizing beamsplitter. *The figure is modified from our previous study (Mizuno et al., 2011).*

2.2. Optical setup

2.2.1 Setup of a microscope for single-molecule fluorescence polarization

TMR-labeled F-actin elongating from immobilized mDia1 was visualized using an epifluorescence IX81 microscope (Olympus) equipped with an Olympus PlanApo 100× oil objective lens (NA=1.40). The large NA of the objective decreases the polarization of the emitted fluorescence (Axelrod, 1979; Kinosita et al., 1991). Nonetheless, we could detect the FL_P from TMR-F-actin elongating from immobilized mDia1 through that objective. We excited TMR with a mercury lamp (OSRAM GmbH, #69182-1). The filter set was purchased from Semrock (TRITC-A; http://www.semrock.com/SetDetails.aspx?id=2681). The fluorescence of TMR was separated into FL_V and FL_H through a polarizing cube beamsplitter (PBS; Edmund Optics, #49-002) attached to the G-ang base (G-Angstrom) (Fig. 5.2).

The G-ang base is a small box in which optics, such as the polarizing beamsplitter, lenses, and mirrors, are easily arranged for capturing two visual fields on a single camera. Figure 5.2B shows the light path of the fluorescence emitted from the samples through the G-ang base. In the IX81 microscope, an intermediary image plane is formed outside of the C-mount. When the G-ang base is connected to the microscope through the C-mount, a slit (S in Fig. 5.2B) of G-ang base is placed on the intermediary image plane. The slit is adjustable, allowing the image size to be limited so that the two separate images can be aligned on the image sensor of the camera. The beam through the slit is separated into FL_V and FL_H by a PBS (Fig. 5.2B). The separated beams are converted to parallel beams by two lenses (L_1 and L_2 in Fig. 5.2B), respectively. The two beams are aligned by using three mirrors (M_{1-3} in Fig. 5.2B) side by side before projection to the camera. The two-aligned beams are imaged on the camera sensor by a lens (L_3 in Fig. 5.2B). Different setups for FL_P microscopy are described in the literature (Kinosita et al., 1991).

We recorded the separated FL_V and FL_H on a charge-coupled device (CCD) camera, CoolSNAP HQ (Roper Scientific), simultaneously (Fig. 5.3A). The images were acquired using MetaMorph software (Molecular Device).

2.2.2 Preparation of the flow cell

In this section, we describe the method to prepare the flow cell (uncoated), which is used to detect FL_P without trapping the pointed-end side of F-actin (see Sections 4.1 and 4.2).

A

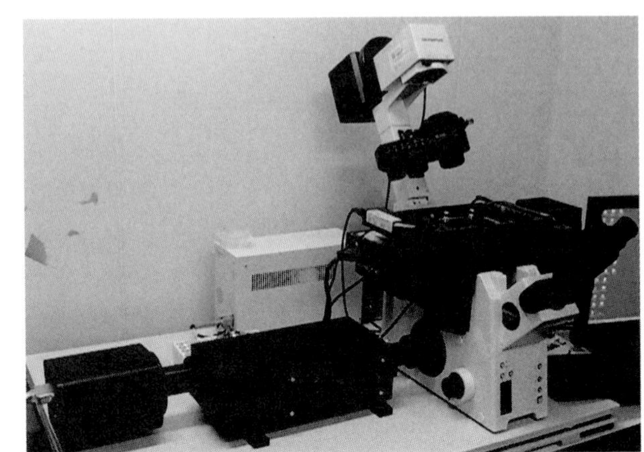

B

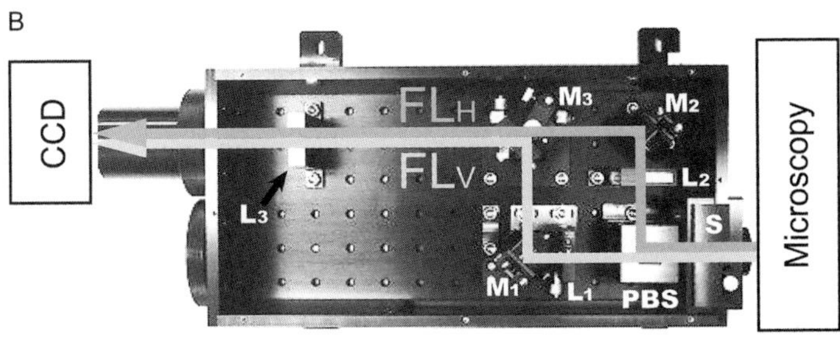

Figure 5.2 The setup of the microscope for acquiring the polarized fluorescence from a single fluorophore. (A) The G-ang base connected between an IX81 microscope and a CCD camera. (B) Interior of the G-ang base. The G-ang base relays the images from the C-mount port of the microscope to a camera. In the G-ang base, a polarizing cube beamsplitter (PBS), three lenses (L_{1-3}), and three mirrors (M_{1-3}) are set at predetermined positions. A slit (S) is located at the entry port of the G-ang base. The polarized fluorescence emitted from the microscope is separated into vertical- and horizontal-polarized fluorescence (FL_V and FL_H) through a PBS. The separated FL_V and FL_H are aligned parallel to each other by using three mirrors (M_{1-3}) before being projected to the camera. *Mizuno et al. (2011).*

Required materials

- Large coverslip (Matsunami Glass Ind., #C024501; $24 \times 50 \times 0.12$–0.17 mm, lwh)
- Top cover slip (Matsunami Glass Ind., #C218181; $18 \times 18 \times 0.12$–0.17 mm, lwh)
- Scotch double-stick tape (3 *M*, 665-3-12; 35 m × 12 mm × 0.1 mm, lwh)
- 1 *M* potassium hydroxide
- 1 × SCAT: SCAT 20 × phosphate-free (Dai-ichi Clean Chemical, Inc., #0145020) is diluted with deionized H_2O

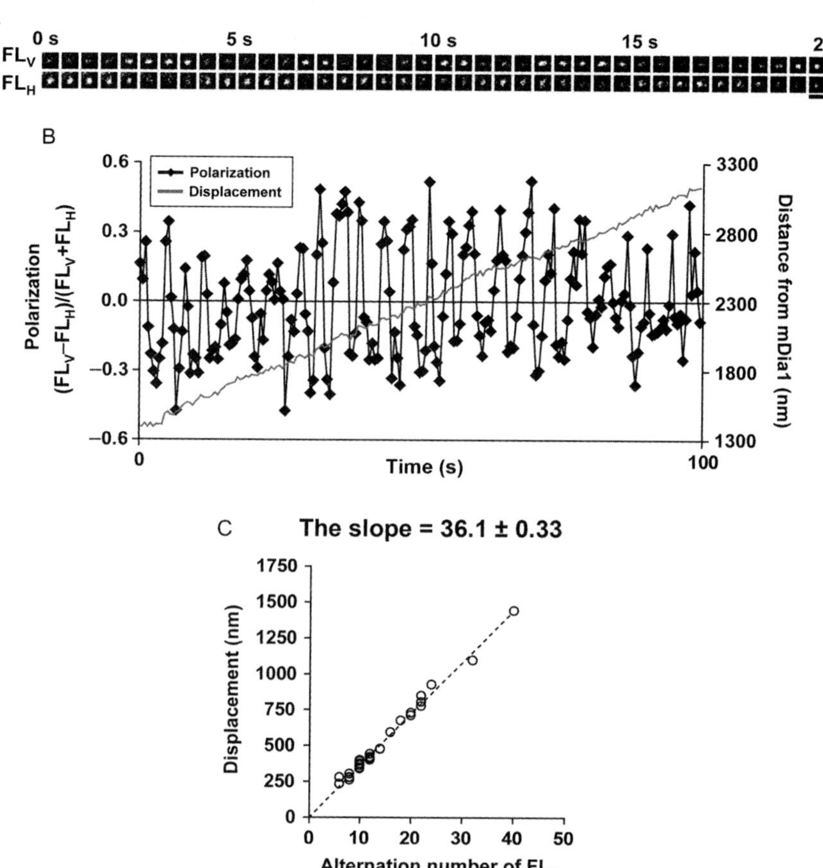

Figure 5.3 Fluorescence polarization and spot displacement during processive ATP-actin elongation by mDia1. (A) Time-lapse images of FL_V and FL_H at 0.5-s intervals of the single molecule of TMR on the F-actin elongating from immobilized mDia1. Scale bar is 1 μm. (B) Fluorescence polarization (FL_P) and distance from mDia1 aggregates of the spot shown in (A). (C) Distance per half rotation of ATP-F-actin (mean ± SEM; 19 independent experiments, $n = 28$ spots, total 428 alternations). *These figures are modified from our previous study (Mizuno et al., 2011).*

Procedure for cleaning the glass surface

1. The large cover slip is soaked in 10 mL of 1 *M* KOH in a 500-mL beaker.
2. The large cover slip is sonicated in the 500-mL beaker using the sonication bath for 1 h. The large cover slip is then washed 10 times with deionized H_2O. Keep the surface of the cover slip clean. The washed large cover slip can be stored in deionized H_2O for 1 week.
3. The top cover slip is used for covering the top of the flow cell. The top cover slip is cut to half size ($18 \times 9 \times 0.12$ mm, lwh) and is soaked in

1 × SCAT for 1 h. Then, the top cover slip is washed 10 times with deionized H_2O.

4. The top cover slip is dried and stored in a box for weeks.

Procedure for making the flow cell

The flow cell should be prepared before use.

1. Before use, the washed large cover slip is dried by spraying nitrogen gas.
2. The flow cell is prepared by sandwiching two pieces of double-layer double-stick tape between the top cover slip and the large cover slip. The volume of our prepared flow cell was ~9 μL (5 × 9 × 0.2 mm, lwh). During observation, the flow cell is placed on the stage of the microscope so that the large cover slip is at the bottom. Perfusion of solutions is carried out by placing ~20 μL of perfusion solution at the edge of the top cover slip and sucking it out from the other side using filter paper.

3. PROTEINS

3.1. Purification of proteins

The expression and purification of glutathione *S*-transferase (GST)-fused mDia1ΔN3, which contains the FH1 and FH2 domains (amino acids 543–1192), using *Escherichia coli* BL21DE3 (BioLabs, #C25271) were previously described in detail (Higashida et al., 2004). Globular actin (G-actin) is purified from the rabbit skeletal muscle according to the method of Spudich and Watt (1971). Then, G-actin is further purified by gel filtration using the Superdex 200 pg HiLoad 16/60 (16 × 600 mm; GE Healthcare). Purification of human profilin I was previously described (Higashida et al., 2008).

3.2. Labeling of actin

In this section, we describe the procedure for the labeling of actin. To detect single-molecule FL_P, F-actin was labeled with TMR (TMR-actin) at Cys-374 according to the Sase method (Sase, Miyata, Corrie, Craik, & Kinosita, 1995) with modification. To visualize the shape of F-actin (see Section 5), F-actin was labeled with Alexa Fluor 488 carboxylic acid succinimidyl ester (Alexa 488-actin).

3.2.1 Labeling of actin at Cys-374 with TMR

Required materials for TMR-actin

- G-actin (see Section 3.1)

- G-buffer: 2 m*M* Tris–HCl, 0.2 m*M* $CaCl_2$, 0.1 m*M* adenosine triphosphate (ATP), 1 m*M* dithiothreitol (DTT), pH 8.0
- DTT(−)-G-buffer: 2 m*M* Tris–HCl, 0.2 m*M* $CaCl_2$, 0.1 m*M* ATP, pH 8.0
- TMR-labeling buffer: 0.1 *M* KCl, 2 m*M* 3-(*N*-morpholino)propanesulfonic acid, 1 m*M* $MgCl_2$, 0.1 m*M* ATP, pH 7.0
- 10 × polymerizing solution: 1 *M* KCl, 20 m*M* $MgCl_2$
- TMR-5-iodoacetamide (Invitrogen, #T6006)
- Protein Assay CBB solution (Nacalai Tesque, #29449-15)

Procedure for TMR labeling of actin

1. To remove DTT from the G-actin solution, 6 mL of 20 μ*M* G-actin is mixed with a 1/9 volume of 10 × polymerizing solution and polymerized for 1 h at 37 °C.
2. F-actin is pelleted by centrifugation at 100,000 × *g* for 1 h at 4 °C.
3. The pellet is dissolved in 6 mL of cold DTT(−)-G-buffer and dialyzed against 500 mL of DTT(−)-G-buffer overnight at 4 °C. During dialysis, DTT(−)-G-buffer is changed at least once.
4. After dialysis, the concentration of G-actin is determined using the Protein Assay CBB solution.
5. Dialyzed G-actin (10 μ*M*) is polymerized in 12 mL of TMR-labeling buffer for 1 h at 37 °C.
6. Before use, TMR is dissolved at 30 m*M* in dimethyl sulfoxide (DMSO).
7. TMR solution is added to 10 μ*M* F-actin at a final concentration of 50 μ*M*. It is recommended that DMSO is kept at a concentration below 0.5% in the actin solution.
8. The mixture is gently stirred using a magnetic stirrer for 15 h at 4 °C in the dark.
9. The reaction is stopped by adding 10 m*M* DTT, and then labeled F-actin is pelleted by centrifugation at 100,000 × *g* for 1 h at 4 °C.
10. The pellet of labeled F-actin is gently rinsed with cold G-buffer twice. Then, the pellet is dissolved in 2 mL of cold G-buffer and dialyzed against 500 mL of G-buffer overnight at 4 °C in the dark. During dialysis, the G-buffer is changed at least once.
11. Unsolved aggregates in labeled G-actin solution are removed by centrifugation at 100,000 × *g* for 1 h at 4 °C.
12. The supernatant is mixed with a 1/9 volume of 10 × polymerizing solution and polymerized for 1 h at 37 °C. F-actin is pelleted by centrifugation at 100,000 × *g* for 1 h at 4 °C.

13. To remove the unbound fluorescent dye, steps 10–12 should be repeated at least four times.
14. The concentration of G-actin is determined using the Protein Assay CBB solution with bovine serum albumin (BSA) as the standard. The concentration of TMR bound to actin is estimated by using $\varepsilon_{543} = 87{,}000\ M^{-1}\ \mathrm{cm}^{-1}$ according to the manufacturer's protocol.
15. Labeled G-actin is stored on ice in the dark. If long-term storage is required, labeled G-actin is quickly frozen in liquid nitrogen and stored at −80 °C.

3.2.2 Labeling of actin with Alexa Fluor 488

We here describe the procedure for labeling actin with Alexa Fluor 488. Alexa 488-actin is used in the procedures described in Section 5.

Required materials for Alexa 488-actin

- Purified G-actin (see Section 3.1)
- G-buffer (see Section 3.2.1)
- Piperazine-*N*,*N*-bis(ethanesulfonic acid) (PIPES)-G-buffer: 5 m*M* PIPES, 0.2 m*M* $CaCl_2$, 0.1 m*M* ATP, pH 6.8
- Alexa-F-buffer: 50 m*M* KCl, 50 m*M* PIPES, 0.2 m*M* $CaCl_2$, 0.2 m*M* ATP, pH 6.8
- 10× polymerizing solution (see Section 3.2.1)
- Alexa Fluor 488 carboxylic acid succinimidyl ester (Alexa 488; Invitrogen, #A20000)
- Protein Assay CBB solution

Procedure for Alexa 488-actin

1. To remove Tris from the G-actin solution, 3 mg/mL G-actin (4.5 mL) is mixed with a 1/9 volume of 10× polymerizing solution and polymerized for 1 h at 37 °C.
2. F-actin is pelleted by centrifugation at 100,000 × *g* for 1 h at 4 °C.
3. The pellet is dissolved in 4.5 mL of cold PIPES-G-buffer and dialyzed against 500 mL of PIPES-G-buffer overnight at 4 °C. During dialysis, the PIPES-G-buffer is changed at least once.
4. After dialysis, the concentration of G-actin is determined by using the Protein Assay CBB solution with BSA as the standard.
5. Dialyzed G-actin (2 mg/mL) is polymerized in 6.5 mL of Alexa-F-buffer for 1 h at 37 °C.
6. Before use, Alexa 488 is dissolved at 50 m*M* in DMSO.

7. Alexa 488 solution is added to 2 mg/mL of F-actin at a final concentration of 240 μ*M*. It is recommended that DMSO be kept below 0.5% in the actin solution.
8. The mixture is stirred gently using a magnetic stirrer for 15 h at 4 °C in the dark.
9. The reaction is stopped by adding 100 m*M* Tris–HCl (pH 8.0), and then labeled F-actin is pelleted by centrifugation at 100,000 × *g* for 1 h at 4 °C.
10. Steps 10–13 are performed according to the procedure described for TMR-actin preparation in Section 3.2.1.
11. After several cycles, the concentration of G-actin is determined using the Protein Assay CBB solution with BSA as the standard. The concentration of Alexa 488 bound to actin is estimated from $\varepsilon_{495} = 73{,}000\ M^{-1}\ \mathrm{cm}^{-1}$ according to the manufacturer's protocol.
12. Storage of the labeled G-actins is performed according to the procedure described for TMR-actin.

3.3. Preparation of adenosine diphosphate-bound actin

In this section, we describe the procedure for preparing G-actin bound to adenosine diphosphate (ADP) (ADP-actin). ADP-actin enables testing whether ATP hydrolysis is required for helical rotation.

In our previous work, we used ADP-actin to address an additional issue. Although profilin accelerates the elongation of ATP-actin at the formin-bound barbed ends, its effect on ADP-actin had remained controversial (Kovar, Harris, Mahaffy, Higgs, & Pollard, 2006; Romero et al., 2007). Based on the observation of defective ADP-actin elongation from mDia1 in the presence of profilin, Romero et al. (2007) concluded that ATP hydrolysis is required for the formin-mediated actin elongation in the presence of profilin. However, Kovar et al. (2006) reported that profilin accelerates the rate of formin-mediated ADP-actin elongation, albeit less efficiently than ATP-actin elongation.

To address the above issues, special care is required. ADP-actin must be prepared within a short time because part of ADP-G-actin is irreversibly denatured within days (Drewes & Faulstich, 1991). ADP-G-actin prepared with this method has successfully been used to visualize ADP-actin elongation as well as processive depolymerization by mDia1 (see Sections 4.1.2 and 4.2). The latter was the first example of the direct demonstration of processive actin depolymerization (Mizuno et al., 2011).

Required materials

- Purified G-actin (see Section 3.1)
- TMR-actin (see Section 3.2.1)
- 2× ADP-polymerizing buffer: 0.2 *M* KCl, 20 m*M* imidazole–HCl, 2 m*M* $MgCl_2$, 0.4 m*M* ethylene glycol tetraacetic acid (EGTA), 0.2 m*M* ADP, 2 m*M* DTT, 10 m*M* glucose, pH 7.0
- ADP-G-buffer: 2 m*M* Tris–HCl, 0.1 m*M* $CaCl_2$, 0.2 m*M* ADP, 1 m*M* DTT, 5 m*M* glucose, pH 8.0
- Hexokinase (Sigma)

Procedure

1. ADP-actin should be prepared before use. To deplete ATP in solutions, 15 U/mL hexokinase is added to 2× ADP-polymerizing buffer and ADP-G-buffer. Then, these buffers are incubated for 5 min at room temperature (r.t.). After 5 min, these buffers are incubated on ice.
2. G-actin (20 μ*M*; unlabeled) or 20 μ*M* TMR-actin in 100 μL G-buffer is mixed with 100 μL of 2× ADP-polymerizing buffer containing 15 U/mL hexokinase on ice and polymerized for 30 min at r.t.
3. F-actin is precipitated by centrifugation at 340,000 × *g* for 20 min at 4 °C.
4. Pellets of F-actins are immediately dissolved in 200 μL of cold ADP-G-buffer by extensive pipetting on ice.
5. These suspensions are sonicated for 3 s and then incubated for 1 h on ice. Before use, the concentration of ADP-G-actin is estimated by using the Protein Assay CBB solution with BSA as the standard. ADP-G-actin must be used within 3 h after preparation.

4. OBSERVATION OF ROTATIONAL MOVEMENT OF TMR-F-ACTIN

4.1. Observation of rotational polymerization

In this section, we describe the procedure to observe the rotational movement of F-actin elongating from immobilized mDia1 from ATP-actin or ADP-actin (Mizuno et al., 2011).

Several factors should be considered for the observation of the rotational movement and mDia1-mediated F-actin elongation. To fix mDia1 on the glass surface tightly, mDia1 is trapped in protein aggregates composed of anti-GST and secondary antibodies. G-actin is centrifuged at high speed to eliminate actin aggregates before use and is kept on ice; centrifuged G-actin should be used within 4 h after centrifugation. Profilin should be excluded from the solution during the F-actin nucleation step by formins because it prevents the nucleation of F-actin (Paul & Pollard, 2008). Because profilin accelerates the formin-mediated ATP-actin polymerization (Kovar et al., 2006), the concentrations of actin and profilin must be adjusted to limit the filament elongation rate in keeping with the temporal resolution of single-molecule imaging.

4.1.1 Observation of ATP-TMR-F-actin elongating from immobilized mDia1

First, we describe the procedure to observe the rotational movement of ATP-F-actin elongating from immobilized mDia1 in the presence or absence of profilin. This section includes the basic procedures for all observation conditions.

Required materials

- Anti-GST antibody (GE Healthcare, #27-4577-01).
- Alexa Fluor 488-conjugated secondary anti-goat IgG antibodies (Invitrogen, A11055).
- Purified GST-mDia1ΔN3 (see Section 3.1)
- TMR-ATP-G-actin (see Section 3.2)
- Unlabeled-ATP-actin (see Section 3.1)
- Human profilin I (see Section 3.1)
- Flow cell (see Section 2.2.2)
- mDia1-aggregation (MA) buffer: 50 m*M* KCl, 20 m*M* 4-(2-hydroxyethyl) piperazine-1-ethanesulfonic acid (HEPES), and 1 m*M* ethylenediaminetetraacetic acid, pH 7.0
- Basic buffer: 50 m*M* KCl, 10 m*M* imidazole–HCl, 1 m*M* $MgCl_2$, 1 m*M* EGTA, 50 μ*M* $CaCl_2$, 0.5% 2-mercaptoethanol, 0.5% methylcellulose, 100 μg/mL glucose oxidase, 20 μg/mL catalase, 4.5 mg/mL glucose, 2 m*M* ATP, pH 7.0
- 10% BSA in Basic buffer

Procedure

1. Purified GST-mDia1ΔN3 (3.7 μg) is mixed with 25 μg of anti-GST antibodies in 10 μL of MA buffer and then incubated for 30 min on ice.
2. The mixture (10 μL) is mixed with 5 μL of Alexa Fluor 488-conjugated secondary anti-goat IgG antibodies (12.5 μg) and then incubated for 30 min on ice. This mixture (herein referred to as mDia1 aggregates) must be used on the same day.
3. TMR-G-actin (200 μL) and 200 μL of unlabeled actin are centrifuged at 340,000 × *g* for 1 h at 4 °C.

4. During centrifugation, the collection tube should be cooled on ice to avoid actin polymerization. One half of the supernatants should be carefully collected to the precooled tube.
5. The concentration of actin in the supernatant is determined using the Protein Assay CBB solution with BSA as the standard.
6. mDia1 aggregates are diluted to 3% in Basic buffer.
7. Diluted mDia1 aggregates are adsorbed on the glass surface of the flow cell for 1 min.
8. Free aggregates are washed out twice by perfusion with 20 μL of 10% BSA. Then, BSA is incubated in the flow cell to block the binding of other proteins for 5 min.
9. The flow cell is washed out five times by perfusion with 20 μL of Basic buffer.
10. For F-actin nucleation, 1 μ*M* TMR-ATP-G-actin (1.5% labeled) in 20 μL of Basic buffer is added to the flow cell by perfusion and incubated with immobilized mDia1 aggregates on the glass surface for 1 min.
11. The flow cell is then perfused three times with 20 μL of Basic buffer containing 1 μ*M* unlabeled G-actin. In case observation is made in the presence of profilin, 1 μ*M* G-actin and 15 μ*M* profilin in 20 μL of Basic buffer are perfused to the flow cell instead. After perfusion, a small amount of TMR-actin remains in the flow cell. Any remaining TMR-actin is often incorporated to the barbed ends of processively elongating filaments. During processive actin elongation, incorporated TMR-actin moves unidirectionally away from the protein aggregate.
12. TMR-actin is excited with a mercury lamp without a neutral density filter. TMR fluorescence is imaged at 0.3- to 0.5-s intervals to obtain a sufficient signal-to-noise ratio.

4.1.2 Observation of ADP-TMR-F-actin elongating from immobilized mDia1

In this section, we describe the procedure to observe the rotational movement of ADP-F-actin elongating from immobilized mDia1.

As described in Section 3.3, the effect of profilin on formin-mediated ADP-F-actin elongation had been controversial. We found that profilin decreases the elongation rate of ADP-F-actin by mDia1 during both polymerization and depolymerization. Remarkably, the addition of inorganic phosphate abolished the profilin effect. Inorganic phosphate is known to abrogate the actin off-rate at the native actin barbed end (Fujiwara, Vavylonis, & Pollard, 2007). We thus found that profilin decelerates mDia1-mediated ADP-F-actin elongation by enhancing the off-rate at the barbed end (Mizuno et al., 2011).

Required materials

- mDia1 aggregates (see Section 4.1.1)
- TMR-ADP-G-actin and unlabeled-ADP-actin (see Section 3.3)
- Human profilin I (see Section 3.1)
- Flow cell (see Section 2.2.2)
- ADP-Basic buffer: 50 m*M* KCl, 10 m*M* imidazole–HCl, 1 m*M* $MgCl_2$, 1 m*M* EGTA, 50 μ*M* $CaCl_2$, 0.5% 2-mercaptoethanol, 0.5% methylcellulose, 100 μg/mL glucose oxidase, 20 μg/mL catalase, 15 U/mL hexokinase, 9.5 mg/mL glucose, 2 m*M* ADP, pH 7.0.
- 10% BSA in ADP-Basic buffer

Procedure

1. mDia1 aggregates are prepared according to the procedure described in Section 4.1.1; mDia1 aggregates are diluted to 30% in ADP-Basic buffer.
2. Adsorption of mDia1 aggregates and blocking by 10% BSA on the glass surface are carried out according to the procedure described in Section 4.1.1.
3. ADP-TMR-actin and ADP-actin are prepared according to the procedure described in Section 3.3.
4. For F-actin nucleation, 5 μ*M* ADP-TMR-G-actin (1.5% labeled) in 20 μL of ADP-Basic buffer is added to the flow cell by perfusion and incubated for 5 min.
5. The flow cell is then washed out three times by perfusion with 20 μL of ADP-Basic buffer containing 5 μ*M* unlabeled-ADP-G-actin. In case observation is made in the presence of profilin, 5 μ*M* unlabeled-ADP-G-actin and 6 μ*M* profilin in 20 μL of ADP-Basic buffer is perfused to the flow cell instead.
6. Excitation of sample and acquisition of images are performed according to the procedure described in Section 4.1.1.

4.2. Observation of the processive depolymerization at immobilized mDia1

In this section, we describe the procedure to observe the depolymerization of ADP-TMR-F-actin bound to mDia1 immobilized onto the glass surface. Processive depolymerization and helical rotation were demonstrated for the first time elsewhere (Mizuno et al., 2011).

Required materials

- Human profilin I (see Section 3.1)
- ADP-Basic buffer (see Section 4.1.2)

Procedure

1. Observation of ADP-F-actin elongation by immobilized mDia1 is performed according to the procedure described in Section 4.1.2.
2. Free G-actin in the flow cell is washed out by perfusion with ADP-Basic buffer including 0–20 μ*M* profilin.
3. TMR spots on F-actin move unidirectionally toward mDia1 aggregates. Excitation of TMR and acquisition of TMR fluorescence images are performed according to the procedure described in Section 4.1.1.

5. LIMITING AXIAL ROTATION OF F-ACTIN USING BIOTINYLATED ACTIN

In this section, we describe the procedure to perturb the axial rotation of F-actin elongating from immobilized mDia1. In our previous work (Mizuno et al., 2011), we employed this method to address the questions raised by a prior related study (Kovar & Pollard, 2004). They reported that when F-actin elongating from the formin protein Bni1p was trapped on the glass, it continued elongating and formed a bent loop. The authors suggested that Bni1p might rotate around the actin barbed end like a bearing. However, this observation could also arise from the slippage between Bni1p and the glass surface. To test these possibilities more precisely, (i) we directly observed the helical rotation after trapping the pointed-end of F-actin and (ii) compared the actin elongation activity of differently immobilized mDia1 on the glass surface: one was tightly trapped in protein aggregates and the other was loosely attached on the glass by nonspecific adsorption. To address issue (ii), the direct observation of the processive F-actin elongation can be carried out by total internal reflection fluorescence (TIRF) microscopy.

Concerning issue (i), we found that the helical rotation stops when the pointed-end side of F-actin is trapped. Concerning issue (ii), F-actin elongating from mDia1 aggregates often prematurely stopped elongation after bending slightly, whereas nonspecifically adsorbed mDia1 frequently continued elongation of the bending F-actin (Mizuno et al., 2011). We interpreted that the nonspecific adsorption method of mDia1 gave rise to an increase in slippage between mDia1 FH2 and the glass surface. These observations provided evidence that formins rotate faithfully along the actin long-pitch helix during processive elongation.

To limit axial rotation, a small amount of biotinylated actin is included in F-actin. The streptavidin-coated glass was used to observe F-actin processively elongating from mDia1. During processive actin elongation, part of F-actin is occasionally trapped on the glass surface through the biotin–avidin interaction.

The preparation of the streptavidin-coated flow cell is described below along with the preparation of three actin mixtures (ARB-, and AR-actin; A, R, and B stand for Alexa 488, TMR, and biotin, respectively).

5.1. Preparation of the streptavidin-coated flow cell

The streptavidin-coated flow cell is made by coating the large cover slip sequentially with aminosilane, biotin, and streptavidin.

Required materials

- Large cover slip and top cover slip (see Section 2.2.2)
- 30% hydrogen peroxide (H_2O_2) aqueous solution (Santoku Chemical Industries Co., Ltd.; CAS No. 7722-84-1)
- 3-Aminopropyltrimethoxysilane (Sigma Aldrich, #281778)
- 6-[(Biotinoyl)amino]hexanoic acid succinimidyl ester (0.1 m*M*; Biotin-X SE, #90051) in 50 m*M* HEPES–NaOH (pH 7.5)
- 0.1 *M* Tris–HCl (pH 8.0)
- Streptavidin (Nacalai Tesque, #32243-11)
- Streptavidin-coating (SA) buffer: 50 m*M* KCl, 10 m*M* imidazole–HCl (pH 7.0), 1 m*M* $MgCl_2$, 1 m*M* EGTA, 50 μ*M* $CaCl_2$, and 0.5% 2-mercaptoethanol

Procedure

The streptavidin-coated flow cell should be prepared before use.

1. The large cover slip and the top cover slip are prepared according to the procedure for cleaning the glass surface described in Section 2.2.2.
2. The washed large cover slip is soaked in a 1:1 mixture of 1 *M* HCl and 30% H_2O_2 for 5 min at r.t.
3. Then, the large cover slip is washed five times with deionized H_2O and is soaked in acetone containing 3-aminopropyltrimethoxysilane (4%, v/v) for 30 min at r.t.
4. The aminosilane-coated large cover slip is washed five times with deionized H_2O and dried by spraying nitrogen gas before use.

5. The flow cell is made up according to the procedure for making the flow cell described in Section 2.2.2.
6. The flow cell is filled with 0.1 m*M* Biotin-X SE in 50 m*M* HEPES–NaOH (pH 7.5) and incubated for 30 min at r.t. To prevent evaporation, the flow cell is kept in a humidified chamber.
7. The reaction of Biotin-X SE is stopped by perfusing 20 μL of 0.1 *M* Tris–HCl (pH 8.0) three times. Then, the flow cell is washed two times with 20 μL of SA buffer.
8. Streptavidin (0.1 μ*M*) in 20 μL of SA buffer is added to the flow cell by perfusion and incubated for 5 min. Then, the flow cell is washed three times with 20 μL of SA buffer. The streptavidin-coated flow cell is kept in a humidified chamber and must be used within 5 h after preparation.

5.2. Arrest of the helical rotation of F-actin by trapping the pointed-end side

In this section, we describe the procedure to observe the rotational movement of F-actin elongating from immobilized mDia1 after trapping the pointed-end side. mDia1 is nonspecifically adsorbed to the glass surface without protein aggregates. Alexa 488-actin is used to observe the shape of F-actin and TMR-actin is used to detect FL_P. For detecting the helical rotation, TMR-actin molecules must be included in a diagonally orientated portion of F-actin forming a bent loop.

Required materials

- Purified GST-mDia1ΔN3 (see Section 3.1)
- ATP-G-actin (unlabeled actin)
- TMR-ATP-G-actin and Alexa 488-actin (see Section 3.2)
- Biotinylated actin (Cytoskeleton, Inc., #AB07)
- Streptavidin-coated flow cell (see Section 5.1)
- Basic buffer (see Section 4.1.1)

Procedure

1. Two actin mixtures are prepared by mixing Alexa 488-actin, biotinylated actin, and TMR-actin in 200 μL of G-buffer on ice as below.

   AR-actin: 1.4 μ*M* Alexa 488-actin and 4.5 μ*M* TMR-actin.

   ARB-actin: AR-actin and 30 n*M* biotinylated actin.
2. Centrifugation of actin mixtures and estimation of the concentration of actin in the supernatant are performed according to the procedure described in Section 4.1.1.
3. The streptavidin-coated flow cell is prepared according to the procedure described in Section 5.1.

4. GST-mDia1ΔN3 (50–100 μ*M*) in 20 μL of Basic buffer is added to the streptavidin-coated flow cell by perfusion and incubated for 1 min.
5. Blocking by BSA on the glass surface is performed according to the procedure described in Section 4.1.1.
6. The nucleation of F-actin is performed using 1 μ*M* ARB-actin according to the procedure described in Section 4.1.1.
7. Basic buffer (20 μL) containing 80 n*M* AR-actin and 420 n*M* unlabeled actin is perfused twice.
8. In our previous work, we first recorded the images of Alexa 488-fluorescence of F-actin at 2-s intervals for 2 min and then recorded the images of TMR fluorescence of the same F-actin at 0.4-s intervals.

5.3. Elongation of F-actin assembled by loosely or rigidly immobilized mDia1 after limiting the axial rotation of the filament

We here describe the procedure to observe F-actin elongating from immobilized mDia1 after trapping the pointed-end side. The observation is carried out with TIRF microscopy. In our previous work, we compared the actin elongation activity of differently immobilized mDia1 (i.e., mDia1 aggregates and nonspecifically adsorbed mDia1) (see Section 5).

The streptavidin-coated flow cell is used (see Section 5.1). The immobilization of mDia1 is described in Section 4.1.1 (mDia1 aggregates) or 5.2 (nonspecifically adsorbed mDia1). Blocking of the glass surface by BSA is described in Section 4.1.1. The actin mixture is prepared by mixing 1.4 μ*M* Alexa 488-actin, 60 n*M* biotinylated actin, and 8.5 μ*M* unlabeled actin in 200 μL of G-buffer on ice. The actin mixture is centrifuged before use. The observation is carried out after perfusing 1 μ*M* actin mixture as described in Section 4.1.1. Excitation of samples is performed by TIRF microscopy. Alexa 488 fluorescence images are recorded at 2-s intervals.

5.3.1 The setup of TIRF microscope

TIRF microscopy enables the direct observation of the elongation of F-actin forming a bent loop between immobilized mDia1 and the trapped pointed-end side. An IX71 microscope (Olympus) equipped with an Olympus PlanApo 100× TIRFM objective lens (NA = 1.45), BCD1 Blue DDD laser excitation (488 nm, 20 mW; Melles Griot), and a TIRF condenser (IX2-RFAEVA-2; Olympus) was used.

6. DATA ANALYSIS

FL_P of the moving TMR molecule is measured from the intensities of FL_V and FL_H. The fluorescence intensities of the spot and the background intensity in each visual field can be measured using analysis software such as MetaMorph and ImageJ. The background values are subtracted from the measured intensities, yielding the fluorescence intensities in each field, FL_{V0} and FL_{H0}.

In our optical settings, FL_V tended to be brighter than FL_H due to differences in the transmission efficiency through the polarizing beamsplitter. To correct for the difference, we measured the intensity of mDia1 aggregates containing multiple Alexa 594-conjugated secondary antibodies in each visual field. The intensity of FL_{V0} of mDia1 aggregates was ~1.2-fold brighter than the intensity of FL_{H0}. We used this value as the correction factor and normalized the intensity of the horizontal fluorescence as $FL_H = 1.2 \times FL_{H0}$; moreover, $FL_V = FL_{V0}$. FL_P is then calculated by the following formula: $FL_P = (FL_V - FL_H)/(FL_V + FL_H)$. Figure 5.3B shows FL_P calculated from the intensities of FL_V and FL_H during the processive ATP-actin elongation reported in Fig. 5.3A.

The displacement of the fluorescent spots can be measured from the merged images of FL_V and FL_H. The position of the fluorescent spots was measured using the 2D-Gaussian fitting algorithm in the G-track software (G-Angstrom). Recently, our colleagues developed an open source ImageJ plugin that includes a 2D-Gaussian fitting algorithm optimized for centroid detection of fluorescent speckles (Smith et al., 2011; Speckle TrackerJ, http://athena.physics.lehigh.edu/speckletrackerj/). This plugin is user-friendly and provides multiple optional algorithms for automatic and semi-automatic tracking of particle images. Currently, this plugin is extensively used in our laboratory.

To correct for the drift of the visual field, we measured the drift of ~10 glass-bound fluorescent spots. The average drift of these spots was subtracted from the position of the TMR spots.

In our previous work, the barbed end of F-actin elongating from mDia1 was buried in the protein aggregate. During the observation, the filaments often move laterally around the barbed end at the center. Because our analysis was only concerned with the displacement caused by the filament elongation, we empirically chose the putative position of the barbed end among several places in the aggregates so that the fluctuation of the spot displacement due to the filament lateral movement was minimized. The displacement of the spots was calculated based on its distance from the putative barbed end (Fig. 5.3B).

The distance per half rotation of F-actin can be calculated from the FL_P data and the displacement data (Fig. 5.3C) (Mizuno et al., 2011).

7. PERSPECTIVES

Our FL_P detection method is a useful tool to visualize the twist of "live" actin filaments elongating from formins without fixation. Further improvement of the spatiotemporal precision of this method may enable the detection of the change in the local twist on F-actin. The structural change in the proximity of the barbed end may provide important clues for understanding the role of ATP in actin treadmilling as well as for elucidating the acceleration mechanism of formin/profilin-catalyzed fast actin elongation. A detailed discussion of the possible applications of this technique is provided in our recent review article (Mizuno & Watanabe, 2012).

ACKNOWLEDGMENTS

This work was supported by JSPS KAKENHI Grant Number 24770175 (H. M.), by the Cabinet Office, Government of Japan through the Funding Program for Next Generation World-Leading Researchers (LS013) (N. W.), and by Takeda Science Foundation (N. W.).

REFERENCES

Axelrod, D. (1979). Carbocyanine dye orientation in red cell membrane studied by microscopic fluorescence polarization. *Biophysical Journal, 26*, 557–573.

Corrie, J. E., Brandmeier, B. D., Ferguson, R. E., Trentham, D. R., Kendrick-Jones, J., Hopkins, S. C., et al. (1999). Dynamic measurement of myosin light-chain-domain tilt and twist in muscle contraction. *Nature, 400*, 425–430.

Drewes, G., & Faulstich, H. (1991). A reversible conformational transition in muscle actin is caused by nucleotide exchange and uncovers cysteine in position 10. *The Journal of Biological Chemistry, 266*, 5508–5513.

Evangelista, M., Zigmond, S., & Boone, C. (2003). Formins: Signaling effectors for assembly and polarization of actin filaments. *Journal of Cell Science, 116*, 2603–2611.

Fujiwara, I., Vavylonis, D., & Pollard, T. D. (2007). Polymerization kinetics of ADP- and ADP-Pi-actin determined by fluorescence microscopy. *Proceedings of the National Academy of Sciences of the United States of America, 104*, 8827–8832.

Goode, B. L., & Eck, M. J. (2007). Mechanism and function of formins in the control of actin assembly. *Annual Review of Biochemistry, 76*, 593–627.

Higashida, C., Miyoshi, T., Fujita, A., Oceguera-Yanez, F., Monypenny, J., Andou, Y., et al. (2004). Actin polymerization-driven molecular movement of mDia1 in living cells. *Science, 303*, 2007–2010.

Higashida, C., Suetsugu, S., Tsuji, T., Monypenny, J., Narumiya, S., & Watanabe, N. (2008). G-actin regulates rapid induction of actin nucleation by mDia1 to restore cellular actin polymers. *Journal of Cell Science, 121*, 3403–3412.

Kinosita, K., Jr., Itoh, H., Ishiwata, S., Hirano, K., Nishizaka, T., & Hayakawa, T. (1991). Dual-view microscopy with a single camera: Real-time imaging of molecular orientations and calcium. *The Journal of Cell Biology, 115*, 67–73.

Kovar, D. R., Harris, E. S., Mahaffy, R., Higgs, H. N., & Pollard, T. D. (2006). Control of the assembly of ATP- and ADP-actin by formins and profilin. *Cell, 124*, 423–435.

Kovar, D. R., & Pollard, T. D. (2004). Insertional assembly of actin filament barbed ends in association with formins produces piconewton forces. *Proceedings of the National Academy of Sciences of the United States of America, 101*, 14725–14730.

Mizuno, H., Higashida, C., Yuan, Y., Ishizaki, T., Narumiya, S., & Watanabe, N. (2011). Rotational movement of the formin mDia1 along the double helical strand of an actin filament. *Science, 331*, 80–83.

Mizuno, H., & Watanabe, N. (2012). mDia1 and formins: Screw cap of the actin filament. *Biophysics, 8*, 95–102.

Paul, A. S., & Pollard, T. D. (2008). The role of the FH1 domain and profilin in formin-mediated actin-filament elongation and nucleation. *Current Biology, 18*, 9–19.

Pring, M., Evangelista, M., Boone, C., Yang, C., & Zigmond, S. H. (2003). Mechanism of formin-induced nucleation of actin filaments. *Biochemistry, 42*, 486–496.

Romero, S., Didry, D., Larquet, E., Boisset, N., Pantaloni, D., & Carlier, M. F. (2007). How ATP hydrolysis controls filament assembly from profilin-actin: Implication for formin processivity. *The Journal of Biological Chemistry, 282*, 8435–8445.

Sase, I., Miyata, H., Corrie, J. E., Craik, J. S., & Kinosita, K., Jr. (1995). Real time imaging of single fluorophores on moving actin with an epifluorescence microscope. *Biophysical Journal, 69*, 323–328.

Sase, I., Miyata, H., Ishiwata, S., & Kinosita, K., Jr. (1997). Axial rotation of sliding actin filaments revealed by single-fluorophore imaging. *Proceedings of the National Academy of Sciences of the United States of America, 94*, 5646–5650.

Smith, M. B., Karatekin, E., Gohlke, A., Mizuno, H., Watanabe, N., & Vavylonis, D. (2011). Interactive, computer-assisted tracking of speckle trajectories in fluorescence microscopy: Application to actin polymerization and membrane fusion. *Biophysical Journal, 101*, 1794–1804.

Spudich, J. A., & Watt, S. (1971). The regulation of rabbit skeletal muscle contraction. I. Biochemical studies of the interaction of the tropomyosin-troponin complex with actin and the proteolytic fragments of myosin. *The Journal of Biological Chemistry, 246*, 4866–4871.

Vrabioiu, A. M., & Mitchison, T. J. (2006). Structural insights into yeast septin organization from polarized fluorescence microscopy. *Nature, 443*, 466–469.

Watanabe, N. (2010). Inside view of cell locomotion through single-molecule: Fast F-/G-actin cycle and G-actin regulation of polymer restoration. *Proceedings of the Japan Academy. Series B, Physical and Biological Sciences, 86*, 62–83.

Watanabe, N., Kato, T., Fujita, A., Ishizaki, T., & Narumiya, S. (1999). Cooperation between mDia1 and ROCK in Rho-induced actin reorganization. *Nature Cell Biology, 1*, 136–143.

CHAPTER SIX

Single-Molecule Studies of Actin Assembly and Disassembly Factors

Benjamin A. Smith[*,1], **Jeff Gelles**[*,2], **Bruce L. Goode**[†,2]
[*]Department of Biochemistry, Brandeis University, Waltham, Massachusetts, USA
[†]Department of Biology and Rosenstiel Basic Medical Sciences Research Center, Brandeis University, Waltham, Massachusetts, USA
[1]Present address: Biogen Idec, Cambridge, Massachusetts, USA
[2]Corresponding authors: e-mail address: gelles@brandeis.edu; goode@brandeis.edu

Contents

Abstract

The actin cytoskeleton is very dynamic and highly regulated by multiple associated proteins *in vivo*. Understanding how this system of proteins functions in the processes of actin network assembly and disassembly requires methods to dissect the mechanisms of activity of individual factors and of multiple factors acting in concert. The advent of single-filament and single-molecule fluorescence imaging methods has provided a powerful new approach to discovering actin-regulatory activities and obtaining direct, quantitative insights into the pathways of molecular interactions that regulate actin network architecture and dynamics. Here we describe techniques for acquisition and analysis of single-molecule data, applied to the novel challenges of studying the filament assembly and disassembly activities of actin-associated proteins *in vitro*. We discuss

Methods in Enzymology, Volume 540
ISSN 0076-6879
http://dx.doi.org/10.1016/B978-0-12-397924-7.00006-6

the advantages of single-molecule analysis in directly visualizing the order of molecular events, measuring the kinetic rates of filament binding and dissociation, and studying the coordination among multiple factors. The methods described here complement traditional biochemical approaches in elucidating actin-regulatory mechanisms in reconstituted filamentous networks.

1. INTRODUCTION

The actin cytoskeleton plays fundamental roles in a diverse set of cellular processes, rapidly responding to cellular or environmental cues, including directed cell migration, cell and tissue morphogenesis, endocytosis, cell division, and pathogen infection (Haglund & Welch, 2011; Pollard & Cooper, 2009; Ridley, 2011). Spatial and temporal control of these functions is achieved by an arsenal of actin-associated proteins, which regulate the nucleation, elongation, cross-linking, and disassembly of filaments (Siripala & Welch, 2007). For example, at the sites of endocytosis in yeast and mammals, more than 50 different proteins function to regulate the rapid assembly and turnover of a distinctive, architecturally complex actin network and drive endocytic vesicle internalization within 15–20 s (Boettner, Friesen, Andrews, & Lemmon, 2011; Weinberg & Drubin, 2012). New actin-regulatory factors are being discovered at an accelerated pace (Firat-Karalar & Welch, 2011), and it is now clear that a single animal cell can express hundreds of different proteins that directly bind actin. While conventional biochemical studies provide excellent tools for identifying interactions among actin-associated proteins, quantifying their equilibrium thermodynamic properties, and determining the efficacy of their control over actin dynamics in ensemble reactions, the question facing many researchers is: *How do I better understand the activity and mechanism of my protein*?

Direct observation of single-filament dynamics in real time (Amann & Pollard, 2001; Yanagida, Nakase, Nishiyama, & Oosawa, 1984) (see also the preceding chapters of this volume), using methods such as total internal reflection fluorescence (TIRF) microscopy, has led to major advances in our understanding of the regulation of actin filament nucleation, network architecture, and disassembly. Bulk assays (e.g., pyrene–actin) measure polymer mass and are not well suited to detecting changes in the interconnection of filaments that occurs during network remodeling. However, direct observation of the individual filaments, for example, within dendritic networks formed by Arp2/3 complex, has led to important discoveries about how branches are initially assembled and subsequently pruned or debranched (Blanchoin, Pollard, & Mullins, 2000; Cai, Makhov, Schafer, & Bear, 2008; Chan, Beltzner, & Pollard, 2009; Gandhi et al., 2010; Martin, Welch, & Drubin, 2006). Yet even with this advance, we are left to infer the molecular mechanisms, indirectly, from observations of filament dynamics. Similarly, mathematical modeling can provide valuable insights into whether a hypothesized mechanism can explain macroscopic behavior (Beltzner & Pollard, 2008; Mogilner, Wollman, & Marshall, 2006; Schaus, Taylor, & Borisy, 2007). However, the available data are often insufficient to accurately constrain all of the unknown and interdependent parameters in such models. What is often lacking is a means to directly validate the underlying mechanistic assumptions and kinetic pathways of molecular interactions.

Single-molecule imaging has the exciting potential to circumvent the above limitations and advance our understanding of actin-regulatory mechanisms (Breitsprecher et al., 2012; Fujiwara, Remmert, & Hammer, 2010; Hansen & Mullins, 2010; Mizuno et al., 2011; Smith, Daugherty-Clarke, Goode, & Gelles, 2013; Smith, Padrick, et al., 2013). By monitoring the molecular sequence of events in filament formation or disassembly in real time, and quantifying the rate constants for each step, these studies can elucidate the precise functional mechanisms of actin-regulatory proteins. Dual observation of individual molecules and filaments provides a means to (1) measure filament binding and dissociation kinetics, (2) define the order of steps and the intermediates in a mechanism, (3) identify alternate pathways, and (4) detect rare events that may be critically important for function among a majority that are not. Furthermore, the roles of multiple factors in jointly regulating actin dynamics can be examined by either testing the effects of a secondary (unlabeled) factor on the behavior of the primary (labeled) factor or labeling both proteins and performing multicolor analysis. Recently reported examples that we highlight in this chapter include the coordination of Arp2/3 complex and its activator (the VCA domain of WASp family proteins) in actin filament branch formation (Smith, Daugherty-Clarke, Goode, & Gelles, 2013; Smith, Padrick, et al., 2013), the dual activity of the formin mDia1 and adenomatous polyposis coli (APC) in actin filament nucleation and elongation (Breitsprecher et al., 2012), and the cooperative severing and disassembly of filaments by cofilin and Srv2/cyclase-associated protein (Srv2/CAP) (Chaudhry et al., 2013). These studies have helped define critical pieces of the puzzle of cytoskeletal dynamics, but more importantly, they show that this approach can help researchers define the mechanisms of numerous other actin regulators.

The methods described here are intended for researchers who wish to perform single-molecule imaging and analysis of actin-associated proteins, but could be applied more generally to study any macromolecule–polymer interactions in the case where the polymer is elongated on optical length scales (filamentous polymers).

2. PROTEIN TAGS, LABELING, AND TETHERING

The goal of protein labeling for single-molecule fluorescence imaging is to achieve high efficiency, covalent attachment of a fluorophore with good photophysical properties onto the protein of interest without disrupting its function. Desirable properties include photostability (high photon yield prior to photobleaching), brightness (product of extinction coefficient and quantum yield), minimal blinking (transient interruptions in emission), and suitable spectral shape (narrow excitation and emission spectra, particularly for observation of multiple labels). Certain small-molecule fluorophores, including the cyanine dyes (GE Healthcare) (Levitus & Ranjit, 2011) and AlexaFluor dyes (Life Technologies), are often preferable to fluorescent proteins (e.g., GFP) because of their superior photophysical properties.

The photophysical properties of single-molecule fluorescence can often be improved by inclusion of certain agents in the imaging buffer, for example, an oxygen scavenging system such as glucose oxidase/catalase or protocatechuic acid/protocatechuate-3,4-dioxygenase (Crawford, Hoskins, Friedman, Gelles, & Moore, 2008) to improve photon yield, and reducing agents or triplet-state quenchers such as Trolox to reduce blinking and improve photostability (Ha & Tinnefeld, 2012). However, these agents can acidify samples (Shi, Lim, & Ha, 2010) and have adverse effects on protein function, so optimization is critical. In addition, each fluorophore should be tested for nonspecific accumulation on the microscope slide surface, which can be either spontaneous or photo-induced.

The most frequently used methods for fluorescent labeling of proteins employ maleimide- or succinimidyl ester derivatives of small-molecule dyes to target surface-exposed sulfhydryls (cysteine residues) or primary amines (lysine residues, protein N-terminus), respectively. Single-residue labeling techniques often require some protein engineering (such as substitution of endogenous cysteines) and additional purification steps to remove unreacted dye, and yet they can still result in nonspecific or inefficient labeling (Means & Feeney, 1971).

Chemical biology techniques for protein labeling that use tags such as SNAP (New England Biolabs), CLIP (New England Biolabs), Halo (Promega), DHFR (Jing & Cornish, 2013), sortagging (Popp, Antos, Grotenbreg, Spooner, & Ploegh, 2007), or ybbR (Yin et al., 2005) can combine the benefits of genetically encoded tags with those of small-molecule fluorophores. First, the labeling tag is engineered to be at one end of the target protein, along with any additionally desired purification tags such as 6His, GST, and MBP. Then, the labeling tag is reacted with any of a wide array of available small-molecule dyes conjugated to the small-molecule target of the labeling tag. The labeling reactions are highly specific and efficient and can be performed during purification so as to achieve effective clearance of unreacted dye conjugates. Recent development of hetero-bifunctional substrates for the SNAP tag has been particularly useful in designing novel TIRF experiments. These substrates link both a small-molecule fluorophore and a biotin moiety to the SNAP tag and can be used to tether a protein of interest to a microscope slide and also visualize its location (Smith, Padrick, et al., 2013).

Regardless of the labeling method used, investigators should take care to verify that the labeled protein displays the same activity as the unlabeled, wild-type protein. For ensemble activity assays, it is usually necessary to achieve high fractional labeling of the protein to ensure that activity is not confined to the unlabeled subpopulation.

3. LABELING AND SURFACE CONFINEMENT OF ACTIN FILAMENTS

For studying actin regulation by TIRF, filaments are fluorescently labeled and tethered or confined to a microscope slide surface. To label filaments, a mixture of unlabeled and fluorescently labeled (~10%) actin monomers is used. Higher percentage mixtures of labeled actin (>50%) can have undesirable effects on polymerization or actin-associated protein binding (Amann & Pollard, 2001; Kuhn & Pollard, 2005). A short wavelength fluorophore (excitable by a blue 488-nm laser), such as AlexaFluor488 or Oregon Green, is the best choice for labeling actin in multiwavelength experiments for a number of reasons. Despite the suboptimal photophysical properties typical of short wavelength fluorophores, filaments are readily visible at 10% labeling because they contain multiple subunits per optically resolvable length (~100 subunits). Furthermore, since the spectra of most dyes are broader on the short wavelength side than the long wavelength side of the excitation peak, filaments labeled with the shortest wavelength dye will be less excited by the longer wavelength lasers, used at higher power to image single molecules of labeled actin-associated proteins. For TIRF, cysteine-reactive probes (such as Oregon Green-maleimide) are usually preferable, because actin has a single reactive cysteine, and therefore labeling is specific. Cysteine-labeled actin yields polymerization rates nearly identical to those of unlabeled actin (Amann & Pollard, 2001; Kuhn & Pollard, 2005). An alternative is to use amine-reactive probes (such as AlexaFluor488 carboxylic acid, tetrafluorophenyl ester; AF488-TFPE), which offers the advantage of not interfering with binding of some actin-associated proteins that bind near the cysteine (Isambert et al., 1995; Kovar, Harris, Mahaffy, Higgs, & Pollard, 2006).

To confine filaments near the surface, enabling observation by TIRF, two recommended approaches are (a) inclusion of a crowding agent or (b) tethering of biotin–actin to a streptavidin surface. Crowding agent polymers such as methyl cellulose or dextran will force filaments against the slide surface due to entropic effects (Uyeda, Kron, & Spudich, 1990). These agents, used without filament tethering, are particularly recommended to visualize filament severing or debranching, since they allow the severed products to diffuse apart, increasing the chance of detecting such events (Gandhi et al., 2010). The amount of agent included needs to be optimized, as too much can induce filament bundling.

In some experiments, it is desirable to physically tether filaments to the surface, for example, so that they remain in the field of view under flow, allowing new ingredients to be introduced during reactions. A small amount of biotinylated actin (~1%) can be included, often in combination with a polymer crowding agent, to tether filaments to a streptavidin-coated surface. Firmly tethering filaments can also simplify quantitative image analysis, but potentially interferes with detection of some regulatory activities (e.g., bundling, severing, or debranching).

4. SLIDE SURFACE TREATMENT

In experiments where one is observing the interactions of labeled actin-binding proteins with surface-confined filaments, it is critical to minimize nonspecific binding of the actin-binding proteins to the surface. This can be achieved in two complementary ways. First, the slide surfaces can be covalently modified with polymer chains such as polyethylene glycol (PEG), which interferes with the approach of proteins to the surface. Second, the surface can be passivated by absorbing high concentrations (1 mg/ml) of a carrier protein such as bovine serum albumin to block nonspecific binding sites. A protocol for microscope slide surface preparation is included in Fig. 6.1. Once slides have been coated with PEG, they should be used within ~24 h or frozen at −80 °C under nitrogen gas for long-term storage. Streptavidin should be applied immediately before the biotinylated protein to be tethered.

5. REQUIREMENTS FOR SINGLE-MOLECULE DETECTION

To detect the weak emission from single molecules, two conditions must be met. First, in order to observe fluorescently labeled single molecules interacting with a target (e.g., actin filament or other actin regulator), the target should be immobilized for at least the duration of a single-image frame acquisition. Immobilization is required so that the emitted photons from the single molecule are imaged as a discrete diffraction-limited spot and can be accomplished by tethering or confining the target filament or actin regulator as described above. The second requirement is that both the solution concentration of the fluorescent protein and the microscope optics must be optimized to ensure that the fluorescent spot is visible over the diffuse background of labeled molecules diffusing freely in the surrounding solution. The evanescent field created by a totally internally reflected excitation beam (in TIRF microscopy) at the surface of a microscope slide provides a means of exciting only the molecules within ~50–200 nm of the slide surface. This minimizes the background fluorescence from most of the molecules in solution. As a rule of thumb, the background will be low enough for good detection if the concentration of fluorescent molecules in the reaction is <50 n*M*; however, the precise limit depends on many factors, including the desired time resolution. Importantly, even proteins with weak binding (in the micromolar K_d range) can be studied very effectively at low nanomolar concentrations by TIRF; this is because even at low occupancy, the binding and dissociation kinetics (k_{on} and k_{off}) can be directly measured.

6. MICROSCOPE CONSIDERATIONS

Many single-molecule experiments can be performed using commercially available TIRF microscopes, which use through-objective laser excitation. The newest of these commercial instruments have all of the necessary features for the most common single-molecule applications. However,

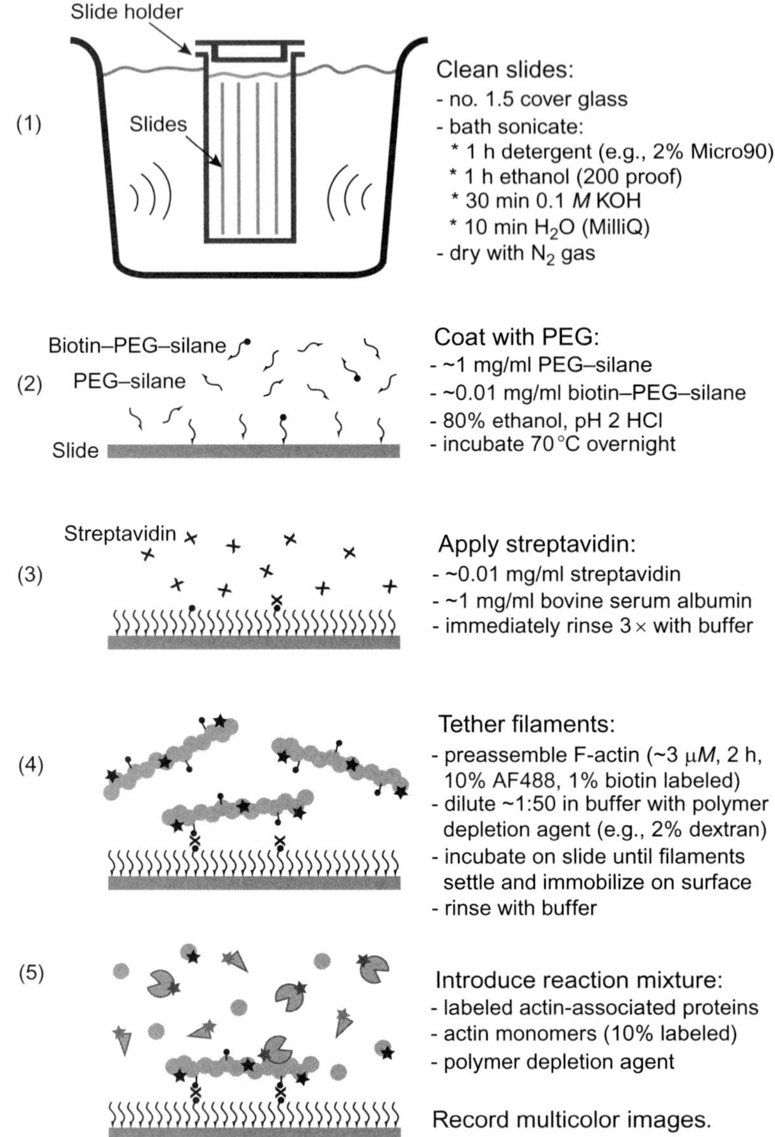

Figure 6.1 Protocol for preparation of microscope slides and observing protein binding to surface-confined actin filaments. Thorough cleaning and passivation of slides, using PEG–silane (Laysan Bio, Arab, AL), is essential to reduce unwanted fluorescent signals and minimize nonspecific binding of proteins. Immobilization of filaments, using biotin–PEG–silane (Laysan Bio), streptavidin, and biotinylated actin, facilitates prolonged analysis of protein–filament interactions. (For color version of this figure, the reader is referred to the online version of this chapter.)

home-built instruments (Friedman, Chung, & Gelles, 2006; Roy, Hohng, & Ha, 2008) can provide improved performance, for example, in experiments requiring rapid imaging of multiple fluors and/or the highest image quality. First-time microscope builders will find that a prism-based TIRF microscope (Fig. 6.2A) is easier to assemble and use than a through-objective TIRF microscope (Fig. 6.2B). However, prism-based TIRF involves imaging across the specimen chamber, which is not as sensitive as through-objective TIRF that images the chamber surface proximal to the objective (Axelrod, 2003).

Microscope features to consider in any TIRF single-molecule imaging experiment are (1) a camera with high sensitivity (high quantum efficiency and low noise), for example, an electron multiplying charged coupled device (EMCCD) camera, (2) high-quality optical filter sets that maximize recovery of the emitted photons, and (3) a high numerical aperture ($NA \geq 1.4$) oil-immersion objective (if through-objective TIRF is used). Microscope optics should be configured so that a diffraction-limited spot spans more than one pixel, allowing spots to be distinguished from noise transients, but a spot size that includes too many pixels can reduce the accuracy of spot detection.

A variation on through-objective TIRF, designed to optimize multi-color single-molecule fluorescence imaging, uses broadband micromirrors, rather than a dichroic mirror, to spatially instead of spectrally separate excitation from emission light (Friedman et al., 2006) (Fig. 6.2C). Multi-wavelength imaging most commonly uses multiple colors of excitation lasers, for example, 488, 532, and 633 nm (blue, green, and red, respectively). Solid-state lasers are available at these and other wavelengths at powers in the range of 10–300 mW. The power needed depends on the microscope design since different optical systems vary in the efficiency with which they transfer laser power into the sample.

7. IMAGING ACTIN FILAMENTS AND SINGLE MOLECULES

Successful imaging depends on proper selection of laser power and image acquisition time. These two factors in turn depend on the timescale of the events to be monitored, the photostability of the fluors, and the degree to which the molecules under observation are immobilized or confined at the surface. If laser powers and acquisition times are too high, rapid photobleaching will occur, and if they are too low, the images will have low signal-to-noise ratios (i.e., poor quality).

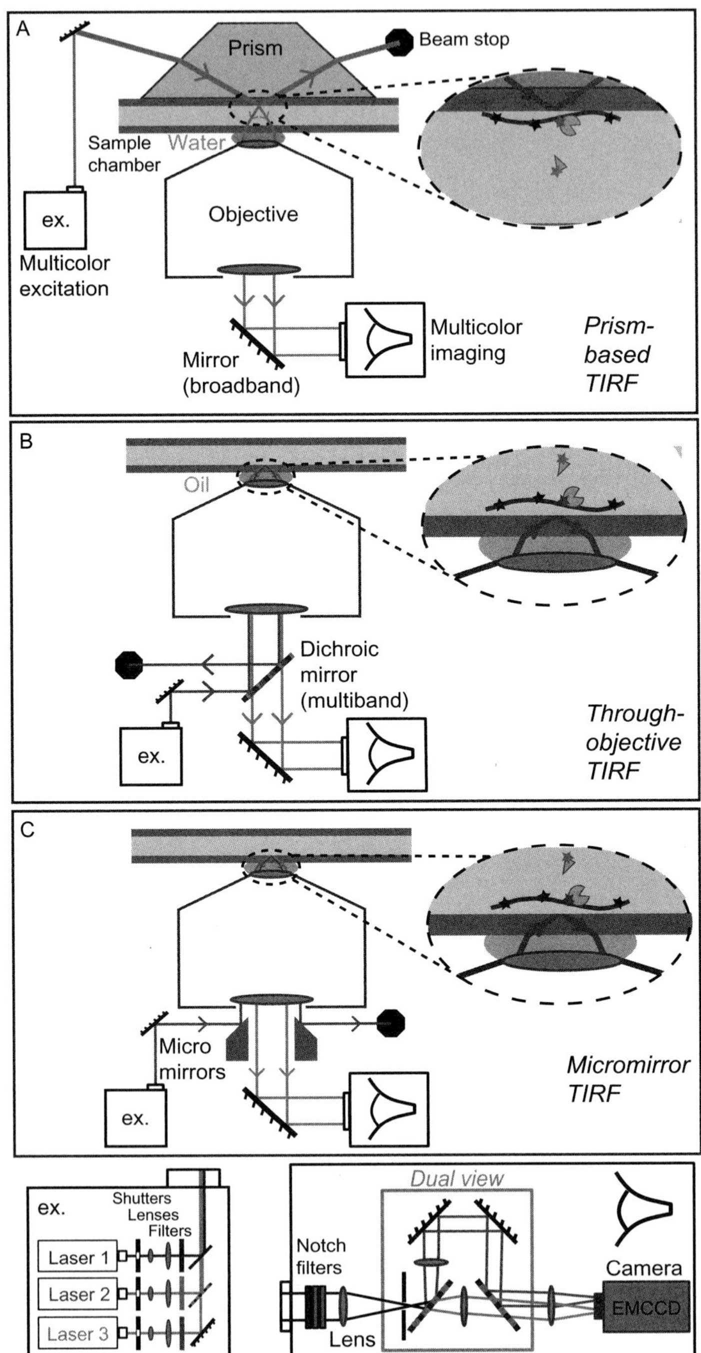

Figure 6.2 Schematics of various multicolor TIRF microscope designs. Prism-based TIRF (A) is in general simpler to set up than through-objective TIRF (B), but is less sensitive since imaging across the full thickness of sample chamber is required. Custom

TIRF reactions used to monitor actin filament growth typically contain 1 μ*M* G-actin, which means that the barbed ends grow at ~10 subunits/s (~27 nm/s). At these rates, recording filament growth does not require image acquisition more than about once every 5–10 s, which helps to minimize photobleaching. A separate consideration from the time interval between image acquisitions is the duration of single-image acquisitions. If filaments are not rigidly attached, their ends can undergo rapid Brownian motion and will be blurred unless the individual frame acquisition time is reduced to <0.3 s. At this acquisition time, filaments grown from 10% labeled actin (incorporating ~38 dyes/μm) can still be imaged at low excitation powers, <1 mW for illumination of a ~50 μm diameter field of view. In contrast, much higher laser powers may be required to monitor labeled proteins transiently interacting with the filaments. For example, Arp2/3 complex binds filament sides with lifetimes as short as ~0.2 s, such that observation required 0.05 s/frame acquisition with ~5 mW excitation laser power (Smith, Padrick, et al., 2013). Higher laser powers can also be required for quantitative analysis of protein complex stoichiometry by stepwise photobleaching (Leake et al., 2006), and for the accurate measurements of dye photostability required for some kinds of kinetics analysis.

Another important consideration in designing multiwavelength single-molecule experiments is to ask whether truly simultaneous acquisition at multiple wavelengths is required. If the reaction dynamics are slow, it is usually sufficient to alternate between image records of the dye labels on filaments and those on associated proteins. However, faster reaction dynamics can make it desirable to capture simultaneous multichannel fluorescence image sequences, particularly if more than one dye-labeled actin-associated protein is being visualized (Smith, Padrick, et al., 2013).

micromirror TIRF (C) further improves the broad spectrum sensitivity by spatially separating excitation (green) and emission (red) optical paths. In each case, excitation (ex.) can be generated by a combination of different color laser beams (bottom left module), and fluorescence emissions can be split into two (or more) color channels, focused on an EMCCD camera, for multicolor image sequence acquisition (bottom right module). (See the color plate.)

8. DUAL-COLOR TIRF IMAGING OF ACTIN-REGULATORY MECHANISMS

A dual-color experiment that monitors labeled actin-regulatory molecules interacting with labeled filaments provides a real-time record of filament association and dissociation events and the order of events in a mechanism. Analysis of these records can define critical aspects of a mechanism, for example, the time delays between association of an actin-regulatory protein with a filament and the event in which the filament state is altered (e.g., severing or branched nucleation). Furthermore, by counting the number of filament-binding events in a window of time and the number of those events that lead to the activity being monitored, one can quantify the efficiency of the actin-regulatory protein. We now discuss examples of such analyses.

8.1. Actin branch formation by the Arp2/3 complex

In a study that examined the mechanism of branched actin nucleation by Arp2/3 complex (Smith, Daugherty-Clarke, Goode, & Gelles, 2013), the delay between time of Arp2/3 complex association with the side of a pre-existing (mother) filament and the nucleation of a new (daughter) filament was directly observed (Fig. 6.3A). For these experiments, *S. cerevisiae* Arp2/3 complex was purified from a strain carrying an integrated SNAP-TEV-3HA tag at the C-terminus of the Arc18/ARPC3 subunit, and labeled with a benzyl guanine-derivatized Dyomics-549 dye (SNAP Surface 549; New England Biolabs). Actin was labeled with AF488-TFPE (10%) and biotin (1%), and unlabeled VCA was included to activate Arp2/3 complex. Using micromirror TIRF microscopy with alternating 488/532 nm laser excitation, Arp2/3-filament-binding events were detected by the appearance of an Arp2/3-SNAP549 fluorescence spot at locations where AF488-filament fluorescence was also observed. That the spots were single molecules was confirmed by single-step photobleaching of the SNAP549 dye. The time at which branched nucleation occurred was determined by tracking the elongation of the daughter filaments, measuring filament lengths, and extrapolating to zero filament length. The delay between filament side binding of Arp2/3 complex and daughter nucleation was found to be short ($<\sim 5$ s), and the efficiency of nucleation from Arp2/3-filament complexes was very low ($<2\%$). These results provided valuable new insights into the kinetic mechanism of filament branch formation (Smith, Daugherty-Clarke, Goode, & Gelles, 2013).

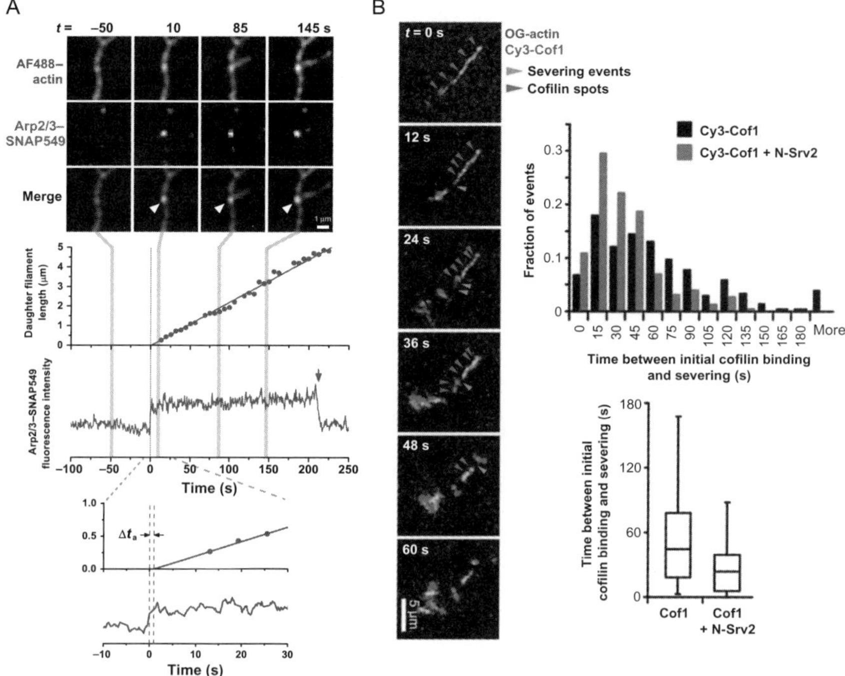

Figure 6.3 Dual-color TIRF studies of actin filaments and actin-associated proteins. (A) Two-color imaging of actin and individual Arp2/3 complexes showed a short activation time delay (Δt_a ~1 s) between binding of Arp2/3 complex to the side of a filament (at $t=0$) and nucleation of a new filament branch (Smith, Daugherty-Clarke, Goode, & Gelles, 2013). Single-molecule observation was confirmed by single-step photobleaching (red arrow) of the dye label on Arp2/3 complex incorporated into a stable branch junction. (B) Real-time observation of actin and cofilin during cooperative filament severing by Srv2/CAP and cofilin. Srv2 shortened the delay between cofilin binding and severing compared to controls without Srv2 (see histogram and whisker plot). *Reproduced from Chaudhry et al. (2013).* (See the color plate.)

8.2. Filament severing by cofilin and Srv2/CAP

In a different study, regulated actin filament disassembly by cofilin and Srv2/CAP was studied using dual-color imaging (Chaudhry et al., 2013). Here, cofilin (with amino acid substitutions T46C/C62A) was labeled with Cy3-maleimide, actin was labeled with Oregon Green 488 iodoacetamide (10%) and biotin (0.5%), and the effect of unlabeled Srv2/CAP on cofilin-mediated severing was tested. Cooperative binding of labeled cofilin on the surface-tethered filaments was visualized as spots of fluorescence using a commercial through-objective TIRF microscope (Nikon-Ti200) and interleaved two-color fluorescence image acquisition (Fig. 6.3B).

The time delay between appearance of labeled cofilin and local severing events was recorded, which revealed that Srv2/CAP had no effect on rate of cofilin accumulation but substantially decreased the delay between cofilin binding and severing (Chaudhry et al., 2013). Since each severing event is likely induced by the cooperative binding of many cofilin molecules in a single spot (Suarez et al., 2011), more advanced single-molecule analysis (e.g., multistep photobleaching; Leake et al., 2006) could be used in the future to investigate the stoichiometry of protein complexes around sites of filament severing.

9. MEASURING KINETICS OF SINGLE-MOLECULE INTERACTIONS WITH ACTIN FILAMENTS

An analysis of single actin-regulatory molecules interacting with filaments can provide a deep quantitative understanding of the molecular mechanism. This analysis can help resolve the step-by-step rate constants and intermediate states along a pathway of activity, and identify steps at which regulation occurs. The unique challenge presented by optical studies of actin-associated proteins is that there are many potential binding sites per optically resolvable length of filament (~0.5 μm, or ~180 actin subunits). Therefore, to obtain reliable single actin-binding protein kinetics, experiments should be conducted under conditions of low occupancy (i.e., much less than one protein per 0.5 μm filament length). It is also important to accurately measure the fraction of protein that is labeled. This is required to estimate the true occupancy from the number of fluorescence spots per filament length. At low occupancy, the following method can be used to evaluate rate constants for association and dissociation of actin-associated proteins on the sides of filaments. The protocol is based on the assumption that protein molecules bind independently to sites on the filament; if there is evidence of cooperative binding, for example, as seen with cofilin (Fig. 6.3B), modifications to the method would be required.

Step 1. Record dual-color fluorescence image sequences of actin-associated proteins interacting with actin filaments immobilized on a microscope slide (Fig. 6.1). Movies should be recorded for long enough time to observe at least one binding event in each resolvable segment of filament, and at a frame rate and laser power that allows observation of most binding events for at least three consecutive frames.

Step 2. Track the positions of all individual actin-associated protein fluorescent spots in the image sequence. Although this can be done manually, we instead use an automated method to identify spots (Crocker & Grier,

1996), which makes the analysis far less laborious. Use of the algorithm requires adjustment of several parameters to optimize detection and tracking, including an intensity threshold, a nominal spot size, and a maximum displacement for tracking a spot between consecutive frames. The displacement tolerance is particularly important when the spots are moving, for example, for motor proteins, for proteins that diffuse on filaments, or when filaments are weakly tethered to the slide. For particularly noisy recordings, it is advantageous to use a dual threshold analysis, where analysis with a high-intensity threshold identifies binding events and analysis with a low-intensity threshold identifies dissociation events (Friedman, Mumm, & Gelles, 2013), minimizing false positives and negatives, respectively.

Step 3. In the filament fluorescence image sequence, trace the contour of a filament. In Matlab (MathWorks, Natick, MA), this can be accomplished using the *improfile* function. Then create a mask of fixed width (w) along the contour of the filament (Fig. 6.4A). Adjust the width so that filament movements are enclosed by the mask boundary throughout the course of the observation.

Step 4. Select the subset of individual actin-associated protein "tracks" identified in Step 2 that appear within the filament mask constructed in Step 3 (Fig. 6.4B). Manually inspect tracks to eliminate spots that are a single pixel or are large and asymmetrical (presumed to not represent *bona fide* single molecules). To measure association rates, divide the filament mask into segments of equal length, and tabulate the times at which a protein is first observed in each segment. The smaller the segment length, the more data are obtained from a single filament, but the segment length should not be smaller than the optical resolution. To measure dissociation rates, record the duration of all tracks that fall within a filament mask. Steps 3 and 4 should be repeated as needed with additional filaments in order to obtain sufficient data. Approximately 100 filament segments or more are needed to obtain precise association rates, but at least 1000 bound protein observations may be required to provide sufficient data to identify rare bound states within the population.

Step 5. As a control, measure the kinetics of interactions of the actin-associated protein with the microscope slide. To accomplish this, repeat Steps 3 and 4 for similarly shaped and sized regions of the recorded field of view that do not contain filaments (Fig. 6.4B).

Step 6. Generate a binding curve by plotting the fraction of filament segments that received at least one binding event (Step 4) versus time (Fig. 6.4C), and fit to an exponential to obtain the apparent first-order

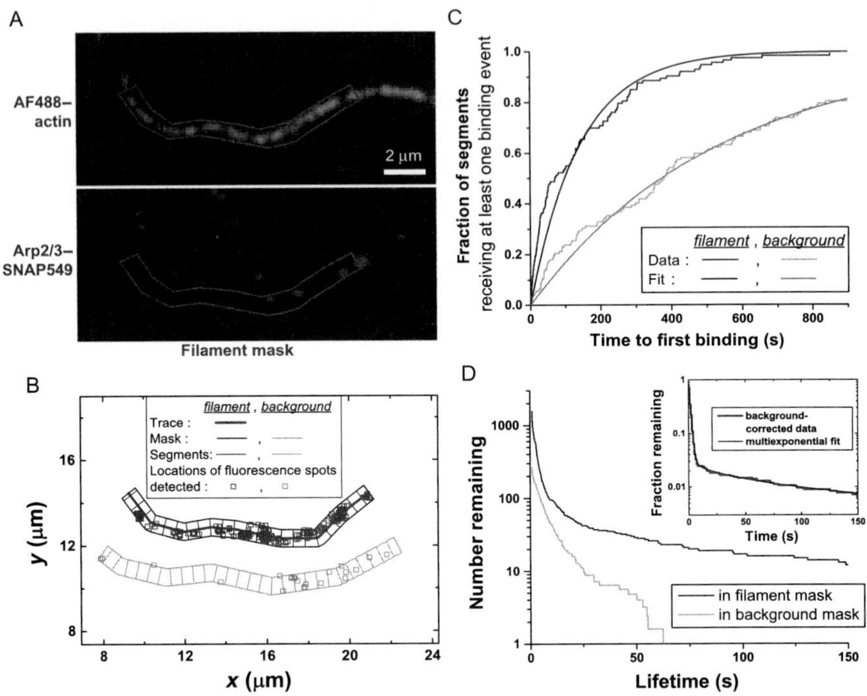

Figure 6.4 Analysis of the kinetics of a protein binding to and dissociating from filaments. (A) Dual-color observations of surface-tethered actin filaments and Arp2/3 complex were analyzed by first tracing a section of filament and generating a mask of uniform width (blue). (B) Fluorescence spots corresponding to single Arp2/3 complexes were tracked over a ~15-min recording. Spots that appeared in the filament mask (red squares) and in control masks covering regions of the field of view that did not contain filament (orange squares in cyan background mask) were retained for analysis. (C) Cumulative distribution of times of first appearance of Arp2/3 complex in each segment of filament and background masks in B. Single-exponential fits are shown (smooth curves), yielding the binding rate constants. (D) Dissociation rate constants were analyzed by generating survival plots of Arp2/3 complexes in filament and background masks. A specific Arp2/3-filament lifetime distribution was generated (inset: red line) by subtracting the background survival plot from the filament survival plot. Multi-exponential fit (black curve) to the background-corrected data reveals that multiple bound states of the Arp2/3-filament complex exist. *Adapted from Smith, Daugherty-Clarke, Goode, and Gelles (2013).* (See the color plate.)

association rate constant. Apply the same analysis to slide surface binding data (Step 5) and calculate the corrected rate constant by subtracting the rate of binding to regions of the slide that do not contain filament from regions that do (assuming homogeneous binding in each case). If binding is consistently inhomogeneous across filament or background masks, the binding curves will not fit to a single exponential function with an amplitude equal to one, and background correction is more complicated (Smith, Daugherty-Clarke, Goode, & Gelles, 2013). Finally, divide the binding rate by the number of potential binding sites in each segment of filament (even if sites overlap; typically one per actin subunit) and by the labeled protein concentration (i.e., the product of the total protein concentration and the labeling fraction) to calculate the second-order binding rate constant per site. Measuring the first binding event in each segment is more accurate than measuring the overall frequency of binding to the filament since it (1) avoids overcounting of binding events due to dye blinking, (2) minimizes errors caused by occupation of sites by photobleached molecules, (3) circumvents the problem of detecting multiple proteins binding to multiple sites within the spatial resolution of the microscope, and (4) is sensitive to heterogeneities in the filament, at least on length scales longer than the microscope resolution.

Step 7. To analyze the kinetics of dissociation of actin-associated proteins from filaments, make a semi-logarithmic plot of the number of proteins remaining as a function of time since each first appeared (i.e., fraction with lifetimes greater than time t as a function of t; Fig. 6.4D). Correct these survival curves for interactions with the microscope slide at each time by subtracting the number remaining in background masks from the number remaining in filament masks. The result is a specific protein–filament complex survival plot (Fig. 6.4D, inset), the logarithm of which decreases linearly with time when there is a single-filament-bound state of the protein. Nonlinear log-survival curves (as in the example shown in Fig. 6.4D) are evidence for the presence of multiple bound states. The points on the survival curve are not independent measurements and thus the data should be fit using maximum likelihood methods (Colquhoun & Sigworth, 1983) instead of by conventional least-squares methods. In measurements of protein dissociation kinetics by single-molecule fluorescence, it is essential that the disappearance of the observed fluorescence spots be from protein dissociation and not photobleaching of the attached dye. Verify this by repeating experiments over a ~3-fold or greater range of laser powers and checking that the dissociation rates measured at different powers do not differ significantly. Alternatively, the photobleaching effect can be corrected by fitting measured rates to a linear function of laser power. The intercept of this fit at zero laser power identifies the true dissociation rate.

The results of the above stepwise analysis should yield accurate association and dissociation rate constants for interactions between actin-associated proteins and filaments, although some mathematical modeling may be required to relate observed rates to rate constants of individual reaction steps. These data, in combination with results from single-molecule activity analysis (e.g., Fig. 6.3) and bulk assays, can then be used to test hypothesized reaction mechanisms (e.g., Smith, Daugherty-Clarke, Goode, & Gelles, 2013).

10. COORDINATION OF MULTIPLE ACTIN-REGULATORY FACTORS FROM MULTICOLOR IMAGING

Many cellular processes are controlled by multiple actin-associated proteins working in concert. Thus, it is often desirable to directly observe multiple factors acting simultaneously to regulate actin filament dynamics. This was accomplished in two recent studies that used triple-color fluorescence analysis to visualize how different pairs of actin nucleation factors cooperate in assembling either unbranched or branched actin filaments (Breitsprecher et al., 2012; Smith, Padrick, et al., 2013) (Fig. 6.5). Both studies addressed key questions relating to the stoichiometry and order of assembly of transient prenucleation complexes, as well as the fate of the assembly factors following nucleation, all of which could be directly visualized in real time using single-molecule TIRF microscopy.

10.1. Cooperative filament assembly by APC and mDia1

In the first example (Breitsprecher et al., 2012), APC and the formin mDia1 were visualized by generating SNAP-tag fusions and labeling them with SNAP647 and SNAP549, respectively. Actin was labeled with Oregon Green-maleimide on 10% of monomers, and reactions included unlabeled capping protein (CapZ) and profilin (Fig. 6.5A). The combination of APC and mDia1 produced rapid formation of actin filaments despite the presence of inhibitory effects of profilin on nucleation and of CapZ on elongation. Using single-molecule analysis, it was shown that the mechanism of nucleation involves the interaction of an APC dimer with an mDia1 dimer and actin monomers. Upon formation of the filament, the APC–mDia1 complex dissociated, leaving the APC dimer stably associated with the filament near the nucleation site and the formin dimer moving away on the growing barbed end of the filament (Fig. 6.5A). This "rocket launcher" mechanism of unbranched filament assembly may also be used by other pairs of assembly factors, for example, Spire-formin and Bud6-formin.

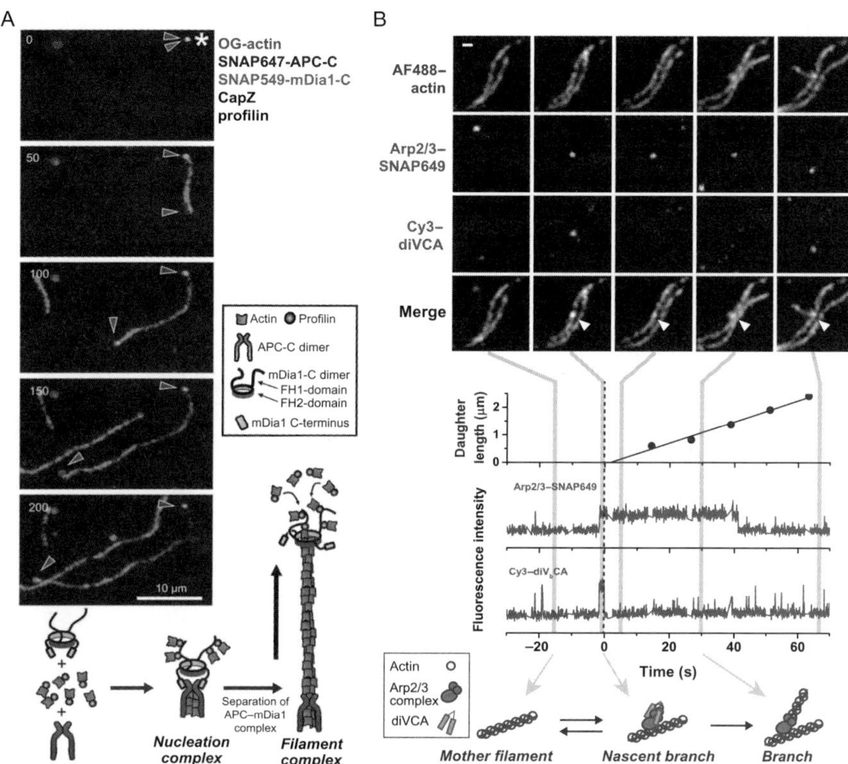

Figure 6.5 Directly observing the coordination of multiple factors regulating actin filament dynamics. (A) Three-color image sequence (left) showing the coordinated action of APC (blue) and the formin mDia1 (red) in nucleation and elongation of an actin filament (green). Prior to filament elongation, APC and formin colocalized (asterisk), but during elongation APC remained with the filament pointed end whereas formin processively tracked the growing barbed end, defining the "rocket launcher" mechanism shown (cartoon below). (B) Three-color imaging (top) of single molecules of dimeric VCA (green) and Arp2/3 complex (red) binding the side of a preexisting mother filament and generating a new daughter filament (forming an actin branch; blue) (Smith, Padrick, et al., 2013). Fluorescence intensity traces and daughter length records (middle) reveal that Arp2/3 complex and VCA appear on filament sides as a complex, but VCA departs rapidly prior to initiation of daughter filament elongation, indicating that VCA release from the nascent branch may trigger branch formation (simplified kinetic scheme, below). *Panel (A): Reproduced from Breitsprecher et al. (2012). Panel (B): Portions modified from Smith, et al. (2013).* (See the color plate.)

10.2. Coordination of Arp2/3 complex and its activator in actin branch formation

In the second example (Smith, Padrick, et al., 2013), branched actin filament nucleation was visualized using Arp2/3 complex (SNAP tagged and dye labeled), a dimeric VCA activator (N-terminally linked with bis-maleimide-Cy3), and actin (10% AlexaFluor488 labeled), and by surface tethering either the Arp2/3 complex or preassembled filaments. Simultaneous high-speed recording of individual Arp2/3 complex and VCA dimers (50 ms/frame; dual view), interleaved with time-lapse actin imaging (every ~10 s), was essential for this study. VCA dimers were observed to bind tethered Arp2/3 complex with high affinity when not associated with an actin filament, but did not bind Arp2/3 complex following branch formation. Correspondingly, when filaments were tethered, Arp2/3 complex was most often associated with VCA dimers upon binding to filament and Arp2/3–VCA complexes rapidly dissociated from filament as a unit without initiating filament formation. However, in cases where branch formation was observed, VCA dimers rapidly released from Arp2/3–filament complexes prior to initiation of new filament elongation (Fig. 6.5B). This suggested that WASp family proteins serve a dual role in both promoting the assembly of an actin nucleus for branch formation and in limiting new filament growth, such that VCA release from the nascent branch is required to trigger elongation.

Both of the studies described above (Fig. 6.5) contributed significant advances to our understanding of the mechanisms of reconstituted collaborative actin assembly. The quantitative kinetic and stoichiometric data obtained should prove invaluable in building systems-level models of cytoskeletal function. We expect that future studies will be similarly successful in elucidating the processes of actin disassembly orchestrated by multiple factors.

ACKNOWLEDGMENTS

Preparation of this article was funded in part by National Institutes of Health grants R01GM098143 (to J. G.), GM063691 (to B. L. G.), and GM098143 (to J. G. and B. L. G.), as well as National Science Foundation grant DMR-MRSEC-0820492 (to J. G. and B. L. G.). The authors thank Dennis Breitsprecher, Julian Eskin, Larry Friedman, Silvia Jansen, and Casey Ydenberg for comments on the manuscript.

REFERENCES

Amann, K. J., & Pollard, T. D. (2001). Direct real-time observation of actin filament branching mediated by Arp2/3 complex using total internal reflection fluorescence microscopy. *Proceedings of the National Academy of Sciences of the United States of America, 98*(26), 15009–15013. http://dx.doi.org/10.1073/pnas.211556398.

Axelrod, D. (2003). Total internal reflection fluorescence microscopy in cell biology. *Methods in Enzymology, 361*, 1–33.

Beltzner, C. C., & Pollard, T. D. (2008). Pathway of actin filament branch formation by Arp2/3 complex. *The Journal of Biological Chemistry, 283*(11), 7135–7144. http://dx.doi.org/10.1074/jbc.M705894200.

Blanchoin, L., Pollard, T. D., & Mullins, R. D. (2000). Interactions of ADF/cofilin, Arp2/3 complex, capping protein and profilin in remodeling of branched actin filament networks. *Current Biology, 10*(20), 1273–1282.

Boettner, D. R., Friesen, H., Andrews, B., & Lemmon, S. K. (2011). Clathrin light chain directs endocytosis by influencing the binding of the yeast Hip1R homologue, Sla2, to F-actin. *Molecular Biology of the Cell, 22*(19), 3699–3714. http://dx.doi.org/10.1091/mbc.E11-07-0628.

Breitsprecher, D., Jaiswal, R., Bombardier, J. P., Gould, C. J., Gelles, J., & Goode, B. L. (2012). Rocket launcher mechanism of collaborative actin assembly defined by single-molecule imaging. *Science, 336*(6085), 1164–1168. http://dx.doi.org/10.1126/science.1218062.

Cai, L., Makhov, A. M., Schafer, D. A., & Bear, J. E. (2008). Coronin 1B antagonizes cortactin and remodels Arp2/3-containing actin branches in lamellipodia. *Cell, 134*(5), 828–842. http://dx.doi.org/10.1016/j.cell.2008.06.054.

Chan, C., Beltzner, C. C., & Pollard, T. D. (2009). Cofilin dissociates Arp2/3 complex and branches from actin filaments. *Current Biology, 19*(7), 537–545. http://dx.doi.org/10.1016/j.cub.2009.02.060.

Chaudhry, F., Breitsprecher, D., Little, K., Sharov, G., Sokolova, O., & Goode, B. L. (2013). Srv2/cyclase-associated protein forms hexameric shurikens that directly catalyze actin filament severing by cofilin. *Molecular Biology of the Cell, 24*(1), 31–41. http://dx.doi.org/10.1091/mbc.E12-08-0589.

Colquhoun, D., & Sigworth, F. J. (1983). Fitting and statistical analysis of single-channel records. In *Single-channel recording* (pp. 191–263). New York: Plenum.

Crawford, D. J., Hoskins, A. A., Friedman, L. J., Gelles, J., & Moore, M. J. (2008). Visualizing the splicing of single pre-mRNA molecules in whole cell extract. *RNA, 14*(1), 170–179. http://dx.doi.org/10.1261/rna.794808.

Crocker, J. C., & Grier, D. G. (1996). Methods of digital video microscopy for colloidal studies. *Journal of Colloid and Interface Science, 179*(1), 298–310. http://dx.doi.org/10.1006/jcis.1996.0217.

Firat-Karalar, E. N., & Welch, M. D. (2011). New mechanisms and functions of actin nucleation. *Current Opinion in Cell Biology, 23*(1), 4–13. http://dx.doi.org/10.1016/j.ceb.2010.10.007.

Friedman, L. J., Chung, J., & Gelles, J. (2006). Viewing dynamic assembly of molecular complexes by multi-wavelength single-molecule fluorescence. *Biophysical Journal, 91*(3), 1023–1031. http://dx.doi.org/10.1529/biophysj.106.084004.

Friedman, L. J., Mumm, J. P., & Gelles, J. (2013). RNA polymerase approaches its promoter without long-range sliding along DNA. *Proceedings of the National Academy of Sciences of the United States of America, 110*(24), 9740–9745. http://dx.doi.org/10.1073/pnas.1300221110.

Fujiwara, I., Remmert, K., & Hammer, J. A., 3rd (2010). Direct observation of the uncapping of capping protein-capped actin filaments by CARMIL homology domain 3. *The Journal of Biological Chemistry, 285*(4), 2707–2720. http://dx.doi.org/10.1074/jbc.M109.031203.

Gandhi, M., Smith, B. A., Bovellan, M., Paavilainen, V., Daugherty-Clarke, K., Gelles, J., et al. (2010). GMF is a cofilin homolog that binds Arp2/3 complex to stimulate filament debranching and inhibit actin nucleation. *Current Biology, 20*(9), 861–867. http://dx.doi.org/10.1016/j.cub.2010.03.026.

Ha, T., & Tinnefeld, P. (2012). Photophysics of fluorescent probes for single-molecule biophysics and super-resolution imaging. *Annual Review of Physical Chemistry, 63*, 595–617. http://dx.doi.org/10.1146/annurev-physchem-032210-103340.

Haglund, C. M., & Welch, M. D. (2011). Pathogens and polymers: Microbe-host interactions illuminate the cytoskeleton. *The Journal of Cell Biology, 195*(1), 7–17. http://dx.doi.org/10.1083/jcb.201103148.

Hansen, S. D., & Mullins, R. D. (2010). VASP is a processive actin polymerase that requires monomeric actin for barbed end association. *The Journal of Cell Biology, 191*(3), 571–584. http://dx.doi.org/10.1083/jcb.201003014.

Isambert, H., Venier, P., Maggs, A. C., Fattoum, A., Kassab, R., Pantaloni, D., et al. (1995). Flexibility of actin filaments derived from thermal fluctuations. Effect of bound nucleotide, phalloidin, and muscle regulatory proteins. *The Journal of Biological Chemistry*, *270*(19), 11437–11444.

Jing, C., & Cornish, V. W. (2013). A fluorogenic TMP-tag for high signal-to-background intracellular live cell imaging. *ACS Chemical Biology*, *8*, 1704–1712. http://dx.doi.org/10.1021/cb300657r.

Kovar, D. R., Harris, E. S., Mahaffy, R., Higgs, H. N., & Pollard, T. D. (2006). Control of the assembly of ATP- and ADP-actin by formins and profilin. *Cell*, *124*(2), 423–435. http://dx.doi.org/10.1016/j.cell.2005.11.038.

Kuhn, J. R., & Pollard, T. D. (2005). Real-time measurements of actin filament polymerization by total internal reflection fluorescence microscopy. *Biophysical Journal*, *88*(2), 1387–1402. http://dx.doi.org/10.1529/biophysj.104.047399.

Leake, M. C., Chandler, J. H., Wadhams, G. H., Bai, F., Berry, R. M., & Armitage, J. P. (2006). Stoichiometry and turnover in single, functioning membrane protein complexes. *Nature*, *443*(7109), 355–358. http://dx.doi.org/10.1038/nature05135.

Levitus, M., & Ranjit, S. (2011). Cyanine dyes in biophysical research: The photophysics of polymethine fluorescent dyes in biomolecular environments. *Quarterly Reviews of Biophysics*, *44*(1), 123–151. http://dx.doi.org/10.1017/S0033583510000247.

Martin, A. C., Welch, M. D., & Drubin, D. G. (2006). Arp2/3 ATP hydrolysis-catalysed branch dissociation is critical for endocytic force generation. *Nature Cell Biology*, *8*(8), 826–833. http://dx.doi.org/10.1038/ncb1443.

Means, G. E., & Feeney, R. E. (1971). *Chemical modification of proteins*. San Francisco: Holden-Day, Inc.

Mizuno, H., Higashida, C., Yuan, Y., Ishizaki, T., Narumiya, S., & Watanabe, N. (2011). Rotational movement of the formin mDia1 along the double helical strand of an actin filament. *Science (New York, NY)*, *331*(6013), 80–83. http://dx.doi.org/10.1126/science.1197692.

Mogilner, A., Wollman, R., & Marshall, W. F. (2006). Quantitative modeling in cell biology: What is it good for? *Developmental Cell*, *11*(3), 279–287. http://dx.doi.org/10.1016/j.devcel.2006.08.004.

Pollard, T. D., & Cooper, J. A. (2009). Actin, a central player in cell shape and movement. *Science (New York, NY)*, *326*(5957), 1208–1212. http://dx.doi.org/10.1126/science.1175862.

Popp, M. W., Antos, J. M., Grotenbreg, G. M., Spooner, E., & Ploegh, H. L. (2007). Sortagging: A versatile method for protein labeling. *Nature Chemical Biology*, *3*(11), 707–708. http://dx.doi.org/10.1038/nchembio.2007.31.

Ridley, A. J. (2011). Life at the leading edge. *Cell*, *145*(7), 1012–1022. http://dx.doi.org/10.1016/j.cell.2011.06.010.

Roy, R., Hohng, S., & Ha, T. (2008). A practical guide to single-molecule FRET. *Nature Methods*, *5*(6), 507–516. http://dx.doi.org/10.1038/nmeth.1208.

Schaus, T. E., Taylor, E. W., & Borisy, G. G. (2007). Self-organization of actin filament orientation in the dendritic-nucleation/array-treadmilling model. *Proceedings of the National Academy of Sciences of the United States of America*, *104*(17), 7086–7091. http://dx.doi.org/10.1073/pnas.0701943104.

Shi, X., Lim, J., & Ha, T. (2010). Acidification of the oxygen scavenging system in single-molecule fluorescence studies: In situ sensing with a ratiometric dual-emission probe. *Analytical Chemistry*, *82*(14), 6132–6138. http://dx.doi.org/10.1021/ac1008749.

Siripala, A. D., & Welch, M. D. (2007). SnapShot: Actin regulators I. *Cell*, *128*(3), 626. http://dx.doi.org/10.1016/j.cell.2007.02.001.

Smith, B. A., Daugherty-Clarke, K., Goode, B. L., & Gelles, J. (2013). Pathway of actin filament branch formation by Arp2/3 complex revealed by single-molecule imaging.

Proceedings of the National Academy of Sciences of the United States of America, *110*(4), 1285–1290. http://dx.doi.org/10.1073/pnas.1211164110.

Smith, B. A., Padrick, S. B., Doolittle, L. K., Daugherty-Clarke, K., Corrêa, I. R., Jr., Xu, M. Q., et al. (2013). Three-color single molecule imaging shows WASP detachment from Arp2/3 complex triggers actin filament branch formation. *eLife*, *2*, e01008. http://dx.doi.org/10.7554/eLife.01008.

Suarez, C., Roland, J., Boujemaa-Paterski, R., Kang, H., McCullough, B. R., Reymann, A. C., et al. (2011). Cofilin tunes the nucleotide state of actin filaments and severs at bare and decorated segment boundaries. *Current Biology*, *21*(10), 862–868. http://dx.doi.org/10.1016/j.cub.2011.03.064.

Uyeda, T. Q., Kron, S. J., & Spudich, J. A. (1990). Myosin step size. Estimation from slow sliding movement of actin over low densities of heavy meromyosin. *Journal of Molecular Biology*, *214*(3), 699–710. http://dx.doi.org/10.1016/0022-2836(90)90287-V.

Weinberg, J., & Drubin, D. G. (2012). Clathrin-mediated endocytosis in budding yeast. *Trends in Cell Biology*, *22*(1), 1–13. http://dx.doi.org/10.1016/j.tcb.2011.09.001.

Yanagida, T., Nakase, M., Nishiyama, K., & Oosawa, F. (1984). Direct observation of motion of single F-actin filaments in the presence of myosin. *Nature*, *307*(5946), 58–60.

Yin, J., Straight, P. D., McLoughlin, S. M., Zhou, Z., Lin, A. J., Golan, D. E., et al. (2005). Genetically encoded short peptide tag for versatile protein labeling by Sfp phosphopantetheinyl transferase. *Proceedings of the National Academy of Sciences of the United States of America*, *102*(44), 15815–15820.

CHAPTER SEVEN

Assaying Microtubule Nucleation by the γ-Tubulin Ring Complex

Yuk-Kwan Choi, Robert Z. Qi[1]

Division of Life Science, The Hong Kong University of Science and Technology, Hong Kong, China

[1]Corresponding author: e-mail address: qirz@ust.hk

Contents

Abstract

Microtubule organization by microtubule-organizing centers such as the centrosome requires γ-tubulin, which exists in the γ-tubulin ring complex (γTuRC) that nucleates microtubules. The γTuRC is a ring-shaped, macromolecular complex whose core components are γ-tubulin and the γ-tubulin complex proteins. Despite the recent identification of additional γTuRC components, the molecular composition and regulatory properties of the complex remain poorly understood. The ability to purify the γTuRC at a large scale for characterization may hold a key to understanding the mechanism by which the γTuRC nucleates microtubules. In this chapter, we describe methods to isolate the γTuRC from human cell cultures and to perform assays on the purified γTuRC.

Methods in Enzymology, Volume 540
ISSN 0076-6879
http://dx.doi.org/10.1016/B978-0-12-397924-7.00007-8

1. INTRODUCTION

The spatial and temporal organization of microtubule arrays requires γ-tubulin, a highly conserved protein that plays a principal role in initiating microtubule assembly. Cells contain γ-tubulin complexes (γTuCs) of two distinct sizes: the γ-tubulin small complex (γTuSC) and the γ-tubulin ring complex (γTuRC) (Kollman, Merdes, Mourey, & Agard, 2011; Luders & Stearns, 2007; Raynaud-Messina & Merdes, 2007; Wiese & Zheng, 2006). The γTuSC is a tetramer composed of two γ-tubulins and two other γ-tubulin complex proteins (GCPs), GCP2 and GCP3. In the γTuRC, several copies of the γTuSC are arranged into a ring-shaped, macromolecular structure with additional proteins such as GCP4, GCP5, and GCP6. Despite recent advances in delineating the composition of the γTuRC (Choi, Liu, Sze, Dai, & Qi, 2010; Hutchins et al., 2010; Teixido-Travesa et al., 2010), large gaps exist in our knowledge of the molecular assembly of the complex.

Among γTuCs, the γTuRC acts as a major nucleator of microtubules in animal cells. In an *in vitro* assay, the γTuSC showed only minimal nucleating activity, whereas the γTuRC exhibited substantially higher microtubule-nucleating activity (~150-fold higher) (Oegema et al., 1999). More recently, the budding yeast γTuSC has been shown to assemble into a γTuRC-like ring structure with 13-fold symmetry that nucleates microtubules (Kollman, Polka, Zelter, Davis, & Agard, 2010). In addition to nucleating microtubules, the γTuRC forms a minus-end cap of microtubules (Keating & Borisy, 2000; Moritz, Braunfeld, Guenebaut, Heuser, & Agard, 2000; Wiese & Zheng, 2000). At centrosomes, the γTuRC mediates the nucleation of microtubules and the anchoring of a radial array of microtubules. During mitosis, at least two other mechanisms are used for microtubule nucleation and hence spindle assembly: one of these mechanisms is chromatin-based nucleation and the other is microtubule-dependent amplification of spindle microtubules. Both of these acentrosomal mechanisms depend on the γTuRC (Goshima, Mayer, Zhang, Stuurman, & Vale, 2008; Luders, Patel, & Stearns, 2006; Mishra, Chakraborty, Arnaoutov, Fontoura, & Dasso, 2010; Zhu, Coppinger, Jang, Yates, & Fang, 2008).

Biochemical isolation of γTuCs by using conventional chromatography methods was unable to achieve high purity at least partly because the complexes are present in low quantities in cells (Zheng, Wong, Alberts, & Mitchison, 1998). Furthermore, the fragile nature of the γTuRC renders its isolation extremely challenging. Therefore, the isolation of γTuCs has relied on immunoaffinity-based methods in which specific γ-tubulin or GCP2 antibodies are used (Detraves et al., 1997; Murphy et al., 2001; Oegema et al., 1999; Zheng, Wong, Alberts, & Mitchison, 1995; Zheng et al., 1998). Although the γTuSC can be reconstituted by coexpressing its three components in insect cells (Gunawardane et al., 2000; Vinh, Kern, Hancock, Howard, & Davis, 2002), such methods have not been reported for the preparation of the γTuRC.

CDK5RAP2 is a centrosomal protein whose mutations cause defects in neurogenic mitosis, leading to primary microcephaly (Bond et al., 2005). Molecular characterization of CDK5RAP2 has revealed that this protein interacts with the γTuRC through a short, conserved segment located near its amino terminus (Fong, Choi, Rattner, & Qi, 2008). Furthermore, the γTuRC-binding domain of CDK5RAP2 is present in several γTuC-binding proteins found in lower organisms, including *Drosophila* centrosomin and *Schizosaccharomyces pombe* Mto1p and Pcp1p (Fong, Sato, & Toda, 2010; Fong et al., 2008; Sawin, Lourenco, & Snaith, 2004; Terada, Uetake, & Kuriyama, 2003; Venkatram et al., 2004). Remarkably, this domain of CDK5RAP2 stimulates the microtubule-nucleating activity of the γTuRC and is therefore referred to as the γTuRC-mediated nucleation activator (γTuNA) (Choi et al., 2010). Exploiting the specific interaction between the γTuRC and the γTuNA, we devised a protocol for rapidly purifying the γTuRC from cell cultures, and the isolated γTuRC was highly pure and effectively nucleated microtubules (Choi et al., 2010). In this chapter, we describe methods for isolating the γTuRC from human cells and subsequently characterizing the isolated complex.

2. γTuRC ISOLATION

The γTuRC is isolated based on its ability to bind to the conserved CDK5RAP2 region 51–100, referred to as the γTuNA (Choi et al., 2010; Fong et al., 2008). The isolation procedure involves preparing cell cultures that ectopically express the CDK5RAP2 fragment and isolating the γTuNA-bound γTuRC. The human embryonic kidney (HEK) 293-derived cells HEK293A or HEK293T can be used for the isolation, although HEK293A cells yield more consistent results in the isolation.

2.1. Reagents

- Polyethyleimine (PEI) stock solution (1 mg/mL): To 10 mL of water prewarmed to 70 °C, 10 mg of PEI powder is added. The solution is stirred until PEI dissolves completely and then the solution is filtered using 0.22-μm syringe filters. Aliquots of the PEI stock can be stored long term at −20 °C or at 4 °C for short-term usage (up to 1 month).

- Lysis buffer: 50 m*M* HEPES–KOH, pH 7.2, 150 m*M* NaCl, 1 m*M* EGTA, 1 m*M* $MgCl_2$, 1 m*M* dithiothreitol, 0.5% IGEPAL, 0.1 m*M* GTP, and Complete Protease Inhibitor Cocktail (Roche).
- Washing buffer: 50 m*M* HEPES–KOH, pH 7.2, 150 m*M* NaCl, 1 m*M* EGTA, 1 m*M* $MgCl_2$, 1 m*M* DTT, 0.01% IGEPAL, and 0.1 m*M* GTP.
- Elution buffer: 0.2 mg/mL FLAG peptide (DYKDDDDK; purity >95%) in washing buffer.
- Anti-FLAG M2-coupled agarose (ANTI-FLAG® M2 Affinity Gel, Sigma-Aldrich).

2.2. Cell culture

The γTuRC is isolated by immunoprecipitating the complex from a 1:1 mixture of FLAG-γTuNA-transfected and -untransfected HEK293A cells. A total of 3×10^6 cells are seeded in 40–150-mm plates and maintained in DMEM medium supplemented with 10% fetal bovine serum, 100 U/mL of penicillin, 100 μg/mL of streptomycin, and 2 m*M* L-glutamine (Invitrogen) in a humidified incubator at 37 °C with 5% CO_2; the culture medium is replaced with fresh medium every 2–3 days. Twenty plates are grown to confluence for collecting untransfected cells. Plasmid transfection is performed on the other 20 plates of cells at ~90% confluence.

2.3. PEI-mediated transfection

Transfecting cells by using PEI is a simple and cost-effective method for large-scale transfection of mammalian cell cultures (Boussif et al., 1995). Before transfection, the culture medium in each 150-mm plate is replaced with 9 mL of fresh medium. To transfect one 150-mm plate of cells, plasmid DNA (10 μg) and PEI (25 μg) from the stock solutions are diluted to 500 μL in separate tubes with serum-free Opti-MEM (Invitrogen). After incubating at room temperature for 5 min, the two solutions are combined, mixed by inverting, and then incubated at room temperature for 20 min. To transfect 20 culture plates, a master mix of DNA/PEI is prepared for 20 transfections in a 50-mL conical tube. Transfection is initiated by adding the DNA/PEI mixture (1 mL) to the medium in each culture plate. The medium is mixed by gently tilting or shaking the culture plates, and the cells are then cultured in a tissue culture incubator. At 24 h posttransfection, 10 mL of fresh culture medium is added to each culture plate.

Transfection conditions are optimized by transfecting a GFP-expressing construct such as pEGFP-C1 (Invitrogen) into one 150-mm dish of cells before large-scale experiments. GFP expression is monitored using a fluorescence microscope. The transfection efficiency achieved is generally in the 40–50% range under the conditions described above.

2.4. Cell harvest

Transfected and untransfected cells are scraped from the culture plates, transferred to 1-L centrifuge bottles (Nalgene centrifuge bottles, Style 3120), and pelleted by centrifugation at 3000 × *g* for 5 min at room temperature. Cell pellets are rinsed twice with PBS (137 m*M* NaCl, 2.7 m*M* KCl, 4.3 m*M* Na_2HPO_4, and 1.47 m*M* KH_2PO_4, pH 7.4) and transferred to 50-mL conical tubes (BD Falcon). Typically, ~7 g of cells can be obtained from a culture of 40–150-mm plates. Cell pellets are immediately frozen and stored at −80 °C.

2.5. Immunoaffinity isolation

The entire procedure is performed at 4 °C unless stated otherwise.

1. The cell pellet of ~7 g is resuspended in 2.5 volumes (~17.5 mL) of ice-chilled lysis buffer by disrupting the pellet with a 21G × 1½″ needle and then pipetting repeatedly; the mixture is placed on ice for 10 min.
2. Cell extracts are clarified by centrifugation at 100,000 × *g* (TLA100.4 rotor, Beckman Coulter) for 30 min at 4 °C.
3. In parallel, 300 μL of anti-FLAG M2-coupled agarose (50% slurry) is washed with and equilibrated in lysis buffer devoid of the protease inhibitors.
4. The cell extracts are filtered using 0.22-μm syringe filters (Millipore).
5. FLAG-γTuNA is immunoprecipitated by incubating the cell extracts with the preequilibrated anti-FLAG M2-coupled agarose for 2 h with rotation.
6. Anti-FLAG M2 beads are collected by centrifugation (400 × *g* for 1 min) and washed with lysis buffer (at least twice) and then with washing buffer (at least twice).
7. FLAG-γTuNA and proteins bound to it are eluted by incubating the beads with elution buffer for 30 min in a spin column (Micro Bio-Spin™ Column, Bio-Rad) capped at the bottom. The eluate is collected by spinning the column briefly. The elution procedure is repeated once and the eluates are combined.

2.6. Sucrose gradient sedimentation

The eluate containing FLAG-γTuNA and bound proteins is loaded onto a 2.2-mL gradient of 5–40% sucrose in washing buffer prepared in advance in a TLS-55 centrifuge tube (Beckman Coulter). The samples are centrifuged at ~200,000 × *g* (TLS-55 rotor, Beckman Coulter) for 3 h at 4 °C. After centrifugation, the gradient is collected from top to bottom in 14 fractions (~160 μL/fraction). The γTuRC appears in Fractions 9–11 and peaks in Fraction 10.

2.7. SDS-PAGE and mass spectrometry

Fractions of the sucrose gradient are resolved using Tris–Glycine SDS-PAGE, and the gels are stained with SYPRO Ruby protein stain. Gel images are acquired on an imager (Typhoon Trio; GE Healthcare). The stoichiometry of γTuRC components is determined by quantifying proteins in the γTuRC peak fraction (Fraction 10) with ImageQuant TL software (GE Healthcare).

To identify proteins, gel bands are excised and subjected to reduction, alkylation, and in-gel tryptic digestion (Shevchenko, Tomas, Havlis, Olsen, & Mann, 2006). Extracted peptides are introduced into a tandem mass spectrometer (LTQ Velos linear ion trap, Thermo Fisher Scientific) equipped with an online liquid chromatography system (C18 reversed-phase column of 100-μm internal diameter). Protein identities are determined by searching sequence databases with generated mass spectra.

3. ELECTRON MICROSCOPY

To assess the quality of the isolated γTuRC, its morphology is examined using electron microscopy after negative staining. In a high-quality preparation, most of the γTuRC should display an intact, ring-shaped structure with a diameter of ~20 nm (Zheng et al., 1995):

1. The γTuRC samples (2.5 μL) are pipetted onto carbon grids, which are placed at room temperature for 5 min.
2. Residual solution is aspirated from grid edges by using filter papers.
3. The grids are rinsed briefly by dipping into a drop of water (30 μL) placed on parafilm and then the solution on the grids is aspirated with filter papers.
4. To stain the samples, grid is dipped briefly into 1% uranyl acetate (freshly prepared and filtered through 0.22-μm syringe filters) and then the

solution from the edge of the grids is removed using filter papers. The procedure is repeated thrice, and then the grids are air dried.

5. The grids are examined using transmission electron microscopy.

4. MICROTUBULE NUCLEATION

4.1. Reagents

- BRB80: 80 m*M* PIPES–KOH, pH 6.8, 1 m*M* $MgCl_2$, and 1 m*M* EGTA.
- HB: 50 m*M* HEPES–KOH, pH 7.2, 150 m*M* NaCl, 1 m*M* EGTA, and 1 m*M* $MgCl_2$.
- α/β-Tubulin is prepared from porcine brain by using two cycles of temperature-dependent polymerization and depolymerization followed by phosphocellulose chromatography; samples are stored in aliquots at −80 °C (Gell et al., 2011).

4.2. Microtubule nucleation

1. The γTuRC gradient fractions are pooled and then desalted using Zeba™ Spin Desalting columns (40 K MWCO, Thermo Scientific) preequilibrated with washing buffer.
2. Rhodamine-labeled α/β-tubulin (Cytoskeleton, Inc.) and -unlabeled α/β-tubulin are mixed at a 1:12 ratio to obtain a final concentration of 2 mg/mL in BRB80 supplemented with 2 m*M* GTP. The solution is clarified by centrifugation at 287,000 × *g* (TLA100.1 rotor; Beckman Coulter) for 8 min at 2 °C.
3. To generate 5 μL reaction mixtures, 2.5 μL of the α/β-tubulin solution prepared above is added to 2.5 μL of the desalted γTuRC or HB buffer. To test the effect of the γTuNA, recombinant γTuNA is preincubated with purified γTuRC for 30 min on ice.
4. The tubes are transferred to a water bath at 37 °C to allow microtubule polymerization for 3 min.
5. Reactions are terminated by adding 50 μL of prewarmed 1% glutaraldehyde/BRB80. The solutions are mixed gently and allowed to sit at room temperature for 3 min. Each sample is diluted to 1 mL with chilled BRB80. Further dilution may be required to obtain a concentration of nucleated microtubules that is optimal for microscopic examination.
6. Samples (0.1 mL) are loaded onto a 0.75-mL cushion of 15% glycerol/BRB80 in TLS-55 tubes (Beckman Coulter) that house a 5-mm cover

slip supported by a Teflon platform at the bottom. Microtubules are sedimented onto the cover slips by centrifuging at 173,000 × g for 8 min at 4 °C.

7. Solutions are aspirated carefully, and the cover slips are withdrawn and mounted on glass slides.
8. Microtubules are examined under an inverted fluorescence microscope (Axio Observer ZI, Carl Zeiss) and counted in 20 random fields to determine the average number of microtubules per field.

5. SUMMARY

The methods described above are designed for isolating the functional γTuRC from mammalian cells, examining γTuRC morphology, and assaying for microtubule nucleation by the γTuRC. Using these methods, we have isolated intact γTuRC to high purity (Figs. 7.1 and 7.2) and

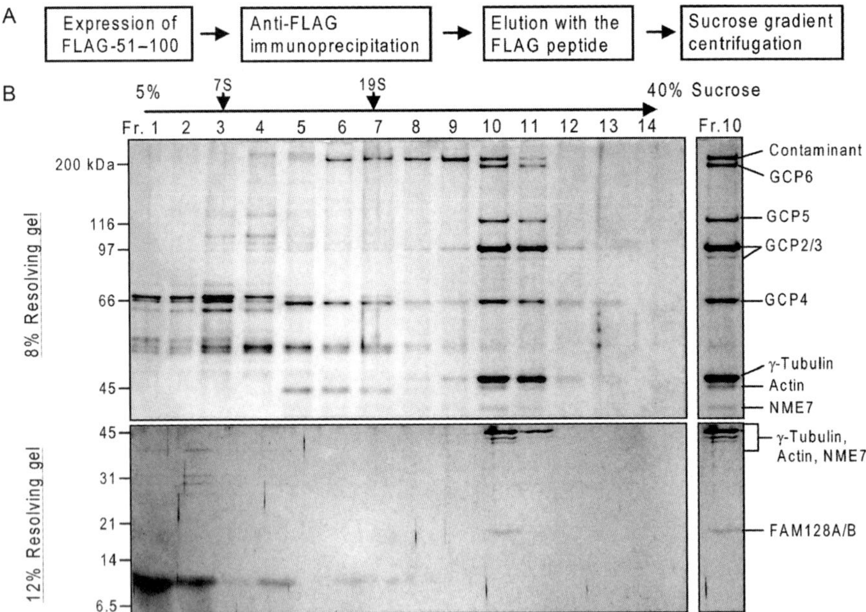

Figure 7.1 Isolation of the γTuRC bound to CDK5RAP2 (51–200). (A) Schematic outline of the isolation procedure. (B) After gradient centrifugation, aliquots of each fraction were resolved by SDS-PAGE, and the gels were silver stained. Proteins resolved from the peak fraction of the γTuRC (Fr. 10) were identified by mass spectrometry. The contaminant protein above GCP6 also appeared in the precipitates of blank beads. *Reproduced with permission from Choi et al. (2010). Copyright ©2010 The Journal of Cell Biology.*

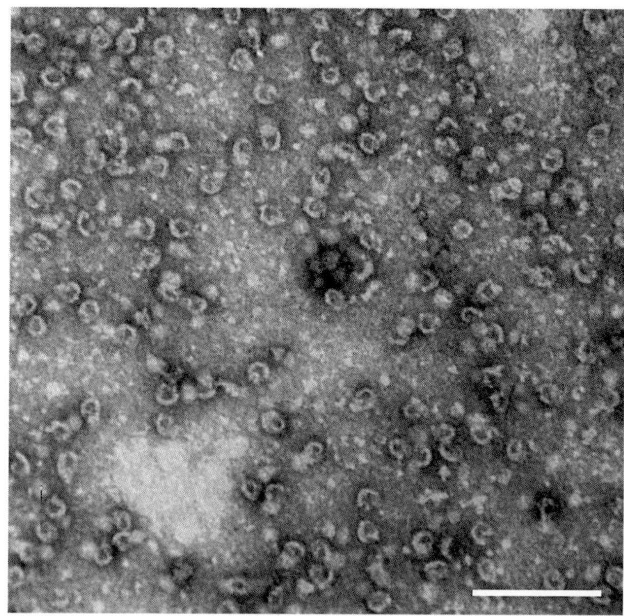

Figure 7.2 Examination of purified γTuRC by electron microscopy. Micrographs were obtained using a transmission electron microscope equipped with a tungsten electron source (JEM 100 CXII, JEOL) at 29K magnification (1024 × 1024 pixels at 9.30 A/pixel). Scale bar, 200 nm.

demonstrated that CDK5RAP2 stimulates the microtubule-nucleating activity of the γTuRC (Fig. 7.3). Compared with other procedures used to isolate γTuCs, this method enables the γTuRC to be purified specifically and quickly without using antibodies against γ-tubulin or GCPs. This isolation method can be improved in the future by generating a stable cell line expressing the γTuRC-binding domain of CDK5RAP2 under an inducible condition and by using suspension cell cultures to facilitate batch harvesting.

In conclusion, the γTuRC-binding domain of CDK5RAP2 serves as a useful tool for specifically isolating the γTuRC from mammalian cells. This isolation strategy can likely be applied using other γTuRC-binding sequences such as the one found in GCP-WD/Nedd1 (Haren et al., 2006; Luders et al., 2006). Biochemical isolation of the γTuRC may in several ways facilitate enhanced understanding of its microtubule-nucleating activity. First, the isolation could lead to the identification of new γTuRC-associating proteins, and the subsequent characterization of the association of these proteins with the γTuRC may reveal the mechanisms that regulate γTuRC functions. Moreover, the γTuRC preparation enables the potential

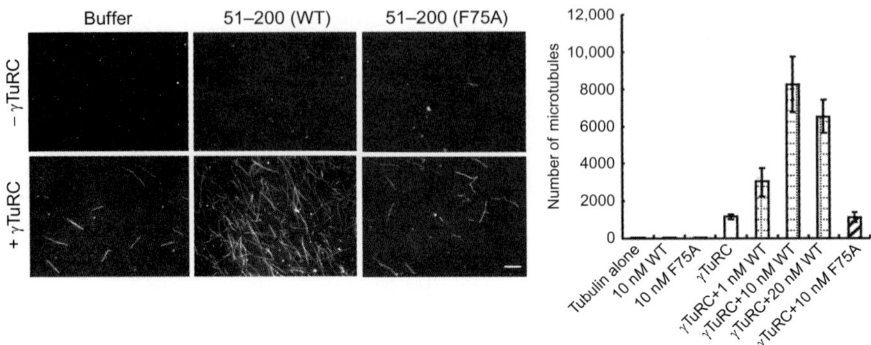

Figure 7.3 Stimulation of the γTuRC for microtubule nucleation. Microtubules were polymerized with or without the isolated γTuRC and CDK5RAP2 (51–200); representative microscopic fields of polymerized microtubules are shown. Microtubules were counted from 20 random fields to determine the average numbers of microtubules. Data are shown as mean ± SD of three independent experiments. Scale bar, 10 μm. *Reproduced with permission from Choi et al. (2010). Copyright ©2010 The Journal of Cell Biology.*

regulation of γTuRC-dependent nucleation by proteins of interest to be tested directly, for example, by using the assay presented herein (Fig. 7.3). Finally, the isolation may yield the γTuRC at a quantity and quality required for performing structural analysis and gaining insights into γTuRC-mediated microtubule nucleation.

ACKNOWLEDGMENTS

This work was supported by grants from the Research Grants Council (General Research Fund and Theme-based Research Scheme) of Hong Kong, the National Key Basic Research Program of China (2013CB530900), the University Grants Committee (Area of Excellence Scheme and Special Equipment Grant) of Hong Kong, and the TUYF Charitable Trust.

REFERENCES

Bond, J., Roberts, E., Springell, K., Lizarraga, S., Scott, S., Higgins, J., et al. (2005). A centrosomal mechanism involving CDK5RAP2 and CENPJ controls brain size. *Nature Genetics*, *37*, 353–355.

Boussif, O., Lezoualc'h, F., Zanta, M. A., Mergny, M. D., Scherman, D., Demeneix, B., et al. (1995). A versatile vector for gene and oligonucleotide transfer into cells in culture and in vivo: Polyethylenimine. *Proceedings of the National Academy of Sciences of the United States of America*, *92*, 7297–7301.

Choi, Y. K., Liu, P., Sze, S. K., Dai, C., & Qi, R. Z. (2010). CDK5RAP2 stimulates microtubule nucleation by the γ-tubulin ring complex. *The Journal of Cell Biology*, *191*, 1089–1095.

Detraves, C., Mazarguil, H., Lajoie-Mazenc, I., Julian, M., Raynaud-Messina, B., & Wright, M. (1997). Protein complexes containing γ-tubulin are present in mammalian brain microtubule protein preparations. *Cell Motility and the Cytoskeleton*, *36*, 179–189.

Fong, K. W., Choi, Y. K., Rattner, J. B., & Qi, R. Z. (2008). CDK5RAP2 is a pericentriolar protein that functions in centrosomal attachment of the γ-tubulin ring complex. *Molecular Biology of the Cell*, *19*, 115–125.

Fong, C. S., Sato, M., & Toda, T. (2010). Fission yeast Pcp1 links polo kinase-mediated mitotic entry to γ-tubulin-dependent spindle formation. *The EMBO Journal*, *29*, 120–130.

Gell, C., Friel, C. T., Borgonovo, B., Drechsel, D. N., Hyman, A. A., & Howard, J. (2011). Purification of tubulin from porcine brain. *Methods in Molecular Biology*, *777*, 15–28.

Goshima, G., Mayer, M., Zhang, N., Stuurman, N., & Vale, R. D. (2008). Augmin: A protein complex required for centrosome-independent microtubule generation within the spindle. *The Journal of Cell Biology*, *181*, 421–429.

Gunawardane, R. N., Martin, O. C., Cao, K., Zhang, L., Dej, K., Iwamatsu, A., et al. (2000). Characterization and reconstitution of Drosophila γ-tubulin ring complex subunits. *The Journal of Cell Biology*, *151*, 1513–1524.

Haren, L., Remy, M. H., Bazin, I., Callebaut, I., Wright, M., & Merdes, A. (2006). NEDD1-dependent recruitment of the γ-tubulin ring complex to the centrosome is necessary for centriole duplication and spindle assembly. *The Journal of Cell Biology*, *172*, 505–515.

Hutchins, J. R., Toyoda, Y., Hegemann, B., Poser, I., Heriche, J. K., Sykora, M. M., et al. (2010). Systematic analysis of human protein complexes identifies chromosome segregation proteins. *Science*, *328*, 593–599.

Keating, T. J., & Borisy, G. G. (2000). Immunostructural evidence for the template mechanism of microtubule nucleation. *Nature Cell Biology*, *2*, 352–357.

Kollman, J. M., Merdes, A., Mourey, L., & Agard, D. A. (2011). Microtubule nucleation by γ-tubulin complexes. *Nature Reviews. Molecular Cell Biology*, *12*, 709–721.

Kollman, J. M., Polka, J. K., Zelter, A., Davis, T. N., & Agard, D. A. (2010). Microtubule nucleating γ-TuSC assembles structures with 13-fold microtubule-like symmetry. *Nature*, *466*, 879–882.

Luders, J., Patel, U. K., & Stearns, T. (2006). GCP-WD is a γ-tubulin targeting factor required for centrosomal and chromatin-mediated microtubule nucleation. *Nature Cell Biology*, *8*, 137–147.

Luders, J., & Stearns, T. (2007). Microtubule-organizing centres: A re-evaluation. *Nature Reviews. Molecular Cell Biology*, *8*, 161–167.

Mishra, R. K., Chakraborty, P., Arnaoutov, A., Fontoura, B. M., & Dasso, M. (2010). The Nup107-160 complex and γ-TuRC regulate microtubule polymerization at kinetochores. *Nature Cell Biology*, *12*, 164–169.

Moritz, M., Braunfeld, M. B., Guenebaut, V., Heuser, J., & Agard, D. A. (2000). Structure of the γ-tubulin ring complex: A template for microtubule nucleation. *Nature Cell Biology*, *2*, 365–370.

Murphy, S. M., Preble, A. M., Patel, U. K., O'Connell, K. L., Dias, D. P., Moritz, M., et al. (2001). GCP5 and GCP6: Two new members of the human γ-tubulin complex. *Molecular Biology of the Cell*, *12*, 3340–3352.

Oegema, K., Wiese, C., Martin, O. C., Milligan, R. A., Iwamatsu, A., Mitchison, T. J., et al. (1999). Characterization of two related Drosophila γ-tubulin complexes that differ in their ability to nucleate microtubules. *The Journal of Cell Biology*, *144*, 721–733.

Raynaud-Messina, B., & Merdes, A. (2007). γ-Tubulin complexes and microtubule organization. *Current Opinion in Cell Biology*, *19*, 24–30.

Sawin, K. E., Lourenco, P. C., & Snaith, H. A. (2004). Microtubule nucleation at non-spindle pole body microtubule-organizing centers requires fission yeast centrosomin-related protein mod20p. *Current Biology*, *14*, 763–775.

Shevchenko, A., Tomas, H., Havlis, J., Olsen, J. V., & Mann, M. (2006). In-gel digestion for mass spectrometric characterization of proteins and proteomes. *Nature Protocols*, *1*, 2856–2860.

Teixido-Travesa, N., Villen, J., Lacasa, C., Bertran, M. T., Archinti, M., Gygi, S. P., et al. (2010). The γTuRC revisited: A comparative analysis of interphase and mitotic human γTuRC redefines the set of core components and identifies the novel subunit GCP8. *Molecular Biology of the Cell*, *21*, 3963–3972.

Terada, Y., Uetake, Y., & Kuriyama, R. (2003). Interaction of Aurora-A and centrosomin at the microtubule-nucleating site in Drosophila and mammalian cells. *The Journal of Cell Biology*, *162*, 757–763.

Venkatram, S., Tasto, J. J., Feoktistova, A., Jennings, J. L., Link, A. J., & Gould, K. L. (2004). Identification and characterization of two novel proteins affecting fission yeast γ-tubulin complex function. *Molecular Biology of the Cell*, *15*, 2287–2301.

Vinh, D. B., Kern, J. W., Hancock, W. O., Howard, J., & Davis, T. N. (2002). Reconstitution and characterization of budding yeast γ-tubulin complex. *Molecular Biology of the Cell*, *13*, 1144–1157.

Wiese, C., & Zheng, Y. (2000). A new function for the γ-tubulin ring complex as a microtubule minus-end cap. *Nature Cell Biology*, *2*, 358–364.

Wiese, C., & Zheng, Y. (2006). Microtubule nucleation: γ-Tubulin and beyond. *Journal of Cell Science*, *119*, 4143–4153.

Zheng, Y., Wong, M. L., Alberts, B., & Mitchison, T. (1995). Nucleation of microtubule assembly by a γ-tubulin-containing ring complex. *Nature*, *378*, 578–583.

Zheng, Y., Wong, M. L., Alberts, B., & Mitchison, T. (1998). Purification and assay of γ tubulin ring complex. *Methods in Enzymology*, *298*, 218–228.

Zhu, H., Coppinger, J. A., Jang, C. Y., Yates, J. R., III, & Fang, G. (2008). FAM29A promotes microtubule amplification via recruitment of the NEDD1-γ-tubulin complex to the mitotic spindle. *The Journal of Cell Biology*, *183*, 835–848.

CHAPTER EIGHT

Reconstituting Dynamic Microtubule Polymerization Regulation by TOG Domain Proteins

Jawdat Al-Bassam[1]
Molecular Cellular Biology, University of California Davis, California, USA
[1]Corresponding author: e-mail address: jawdat@ucdavis.edu

Contents

Abstract

Microtubules (MTs) polymerize from soluble αβ-tubulin and undergo rapid dynamic transitions to depolymerization at their ends. Microtubule-associated regulator proteins modulate polymerization dynamics *in vivo* by altering microtubule plus end conformations or influencing αβ-tubulin incorporation rates. Biochemical reconstitution of dynamic MT polymerization can be visualized with total internal reflection fluorescence (TIRF) microscopy using purified MT regulators. This approach has provided extensive details on the regulation of microtubule dynamics. Here, I describe a general approach to reconstitute MT dynamic polymerization with TOG domain microtubule regulators from the XMAP215/Dis1 and CLASP families using TIRF microscopy. TIRF imaging strategies require nucleation of microtubule polymerization from surface-attached, stabilized MTs. The approaches described here can be used to study the mechanism of a wide variety of microtubule regulatory proteins.

Methods in Enzymology, Volume 540
ISSN 0076-6879
http://dx.doi.org/10.1016/B978-0-12-397924-7.00008-X

1. INTRODUCTION

Microtubules (MTs) are polarized dynamic intracellular polymers that are required for producing and organizing forces during cell division and development. MTs serve as tracks for motor proteins and generate pulling and pushing forces by dynamic polymerization and depolymerization activities at their plus ends. MTs polymerize from αβ-tubulin dimers at MT plus ends, and GTP hydrolysis is activated in newly polymerized tubulin dimers near plus ends. Tubulin GTP hydrolysis at plus ends leads to stochastic switch-like transitions, where MT ends become unstable and depolymerize rapidly. These switch-like transitions, termed "dynamic instability," allow dynamic MTs to reorganize cells and produce physical pulling forces while coupled to kinetochores, chromosomes, or other MTs (Akhmanova & Steinmetz, 2008; Desai & Mitchison, 1997). Over the past three decades, diverse and conserved classes of MT-associated proteins (MAPs) emerged with unique functions in regulating MT organization, polymerization, and dynamic transitions (Akhmanova & Steinmetz, 2008; Al-Bassam & Chang, 2011). MT polymerization regulators that bind soluble αβ-tubulin dimer while bound at MT plus ends can influence MT polymerization and dynamic transitions (Al-Bassam & Chang, 2011). The End Binding (EB) protein family, such as EB1, bind polymerizing MT plus ends by recognizing the GTP-like state of newly polymerized tubulin dimers (Maurer, Bieling, Cope, Hoenger, & Surrey, 2011). In contrast, two unique classes of tumor overexpressed gene (TOG) domain proteins regulate MTs by recruiting soluble tubulin dimer to MT plus ends. XMAP215/Dis1 proteins, such as Alp14 and XMAP215, increase MT dynamic polymerization rates by recruiting soluble αβ-tubulin via their conserved TOG domains to polymerizing MT plus ends (Al-Bassam et al., 2012; Brouhard et al., 2008). The related CLASP family proteins decrease MT transitions to depolymerization (termed MT catastrophe) and activate polymerization reinitiating transitions of MT polymerization (termed MT rescue) by recruiting soluble tubulin via their TOGL to depolymerizing MT plus ends (Al-Bassam & Chang, 2011; Al-Bassam et al., 2010). The effects of these regulators on MT dynamics in cells have been extensively studied (Al-Bassam & Chang, 2011). However, it is also critical to observe and understand their mechanism using *in vitro* reconstitution of dynamic MT polymerization using purified recombinant MAPs in combination with soluble tubulin. Here, I describe the approach and methods for purifying, reconstituting, and visualizing mechanisms of recombinant TOG domain-MT regulators during dynamic MT polymerization using total internal reflection fluorescence (TIRF) light microscopy.

1.1. The use of light microscopy methods for the *in vitro* reconstitution of dynamic MTs

Over the past two decades, the interactions of recombinant or purified MAPs with MTs polymerized from purified tubulin have been studied using light microscopy approaches. Although differential interference contrast (termed DIC) microscopy methods were initially used to directly visualize MTs and their dynamic polymerization (Walker et al., 1988), fluorescence-based light microscopy methods rapidly replaced DIC methods to study MTs *in vitro*, due to increased signal to noise and the ability to visualize the localization of fluorescent MAPs along MTs (Bieling, Telley, Hentrich, Piehler, & Surrey, 2010; Gell et al., 2010; Telley, Bieling, & Surrey, 2011). In the past decade, TIRF microscopy methods emerged as an extremely robust method to study MT-based motors and regulator MT dynamic polymerization mechanisms (Bieling et al., 2007; Brouhard et al., 2008). TIRF microscopy visualizes MT polymerization through their proximity to glass surfaces, and fluorescent MAPs, which bind along these MTs or at their ends. TIRF microscopy is particularly effective in visualizing dynamic MTs polymerization from purified tubulin to explore mechanisms of MT regulators in influencing dynamic MT transitions or organization (Bieling et al., 2010; Gell et al., 2010). Here, I describe a general approach to study dynamic MT polymerization regulators. I will use recent studies as examples for how to study TOG-domain-MT regulators from the XMAP215/Dis1 and CLASP families (Al-Bassam et al., 2010, 2012). The approach described here can be adapted to a variety of MT polymerization regulators.

1.2. Advances in reconstituting dynamic MT polymerization using TIRF microscopy

Several technical advances have facilitated reconstitution studies of MT dynamics by TIRF microscopy, including (A) improvements in manufactured light microscopes, objectives lenses, and single-wavelength solid-state lasers that have made multiwavelength TIRF microscopy accessible to many users; these developments are beyond the scope of this chapter and are described previously (Gell et al., 2010). (B) Chemical strategies to clean and neutralize hydrophobic glass surfaces to prevent aggregation of soluble tubulin and MAPs during reconstitution experiments. I will describe a single method to neutralize glass surfaces, but there are other strategies to accomplish this, which can be found in the literature (Bieling et al., 2010; Gell et al., 2010) as well as other chapters in this book. (C) Chemical cross-linking methods that generate fluorescent, polymerization-competent tubulin dimers (Hyman et al., 1991). (D) Extensive advances in recombinant protein expression methods to produce homogenous, well-behaved full-length MT regulators with fluorescent tags (Hitchman, Locanto, Possee, & King, 2011; Machleidt, Robers, & Hanson, 2007). (E) Strategies for attaching short, stable MTs (termed MT seeds) on glass surfaces, which nucleate dynamic MT polymerization at MT plus ends from soluble tubulin (Bieling et al., 2010; Gell et al., 2010). Unlike previous methods that directly attached MTs to glass surfaces, new chemical or protein scaffolds can be used to attach MTs at a defined distance above glass surfaces. This is critical to ensure that MTs are free to undergo dynamic polymerization at their ends. The rigidity of newly formed, dynamic MTs maintains them in focus in the evanescent field near the glass surface to allow imaging by TIRF microscopy.

2. METHODS

2.1. Preparing fluorescently tagged recombinant MT regulator proteins

2.1.1 Rationale

Recombinant macromolecular overexpression strategies advanced dramatically over the past two decades (Hitchman et al., 2011). Full-length MT regulators or complexes, which are relatively large molecular weight proteins, can be produced using overexpression in yeast or insect cells and purified to homogeneity using biochemical purification strategies. MT regulators can be engineered to add fluorescent tags or mutated to inactivate functional domains to study function (Al-Bassam et al., 2010, 2012). A general overexpression and purification strategy is described below, including dual affinity steps as well as ion-exchange and size-exclusion chromatography techniques. Reconstitution studies with MT regulators using TIRF microscopy require well-behaved proteins purified by size-exclusion chromatography. I will describe a strategy for the purification of fission yeast Cls1 or Alp14 (Al-Bassam et al., 2010, 2012) using insect cells baculovirus expression, using his-Maltose-binding protein (MBP) fusion dual affinity strategy (Sun, Tropea, & Waugh, 2011).

2.1.2 Method

1. Baculoviruses are generated from engineered transfer vectors, which can be prepared using a variety of well-described methods. 6×his-MBP sequences are fused at the N-termini of full-length MT regulators such as Alp14 and Cls1. 9×10^5 Sf9 insect cells are transfected in 6-well plates using lipids or calcium phosphate strategies. Three rounds of virus amplification (P1-P3), each taking 72 h by infecting 9×10^5 Sf9 insect cells with the previous round. The final P3 virus is then utilized for protein expression in the following steps.
2. 1–2 L of 1×10^6 Sf9 or Hi5 insect cells are infected with P3 virus and diluted 50–100-fold dilution to initiate protein expression.
3. Cell pellets are collected 60–72 h post infection. Cells are lysed using lysis buffer (50 m*M* Hepes 300 m*M* KCl, pH 7.0 0.2% Triton X 100, 5 m*M* β-mercaptoethanol, 15 m*M* imidazole supplemented with protease inhibitors to prevent protein degradation) using a dounce homogenizer with 1–2 strokes every 20 s for 5–10 min.
4. Lysates are clarified using centrifugation at 60,000 × *g* for 30 min.
5. His-MBP-Alp14 is bound to Ni-NTA agarose (macherry-Nagel Corp.), washed, and eluted with lysis buffer with 250 m*M* imidazole.
6. A second affinity step is performed by binding protein from the nickel elution to Amylose resin (New England Biolabs), washed extensively in lysis buffer to remove insect cell contaminants, and eluted with lysis buffer with 25 m*M* maltose.
7. Purified his-MBP-Alp14 protein is then treated with TEV protease (Sun et al., 2011) to remove his-MBP tags for 24–48 h and then passed onto Ni-NTA agarose beads to remove tags or uncleaved protein. Purified cleaved protein is collected in the flow through.
8. Alp14 is concentrated using Amicon concentrators and injected onto a Superdex 200 or Superose-6 (GE Healthcare) using a FPLC system equilibrated with lysis buffers.
9. The eluted protein is then analyzed using SDS-PAGE to determine the degree of purity and then utilized for TIRF experiments.

2.2. Fluorescent labeling of high-quality soluble tubulin for TIRF microscopy

2.2.1 Rationale

Reconstitution of dynamic MT polymerization requires high-quality, polymerization-competent tubulin. Soluble tubulin is extracted from bulk tissues through GTP-dependent MT polymerization and cold-driven depolymerization cycles, as described more than three decades ago (Kirschner, Williams, Weingarten, & Gerhart, 1974). A recently modified method produces higher quality polymerization-competent soluble tubulin (Castoldi & Popov, 2003). Fluorescent soluble tubulins, produced by amine-reactive labeling reactions and purified through MT polymerization and depolymerization cycles, are used to monitor soluble tubulin and dynamic MTs by TIRF microscopy. In this method, fluorescent dyes are covalently attached to amine residues in polymerized tubulin which can be either purified in the lab or purchased commercially (Cytoskeleton, Inc.), and active tubulin is selected by two additional GTP-dependent polymerization cycles. High-quality soluble tubulin with moderate (10–50%) labeling is extremely critical for studies using dynamic MTs in TIRF with MT regulators, described here. Below, I present a method that was initially described by the Hyman et al. (Hyman et al., 1991). I recommend performing tubulin fluorescent labeling in your laboratory for successful and reproducible TIRF experiments. I also recommend producing multiple fluorescent dye color-labeled soluble tubulin pools and to determine the experimental dye/tubulin-labeling ratio for each pool. The chosen dyes must be compatible with excitation wavelength for the lasers used and the particular emission and dichroic filter properties of your TIRF microscope (Gell et al., 2010). The fluorescent dyes must have very little or no overlap of their excitation or emission spectra to prevent channel cross-signal contamination in these experiments. An extensive discussion of dye choices is previously described (Gell et al., 2010). Common dyes used for tubulin labeling include Cy5, Cy3, and Texas-Red-amine-reactive dyes. A variety of enhanced chemically related dyes with higher fluorescence intensity are termed Alexa-Fluors, which are sold and marketed by Invitrogen Corp.

2.2.2 Method

1. Dilute 10–20 mg of purified tubulin in BRB-80 (80 m*M* pipes, 1 m*M* EGTA pH 6.8) with 3.5 m*M* GTP to 2 mg/mL and transferred to 37 °C and 50% glycerol is added, and this mixture is incubated at 37 °C. This step promotes MT polymerization.
2. Layer the polymerized mix on a cushion, or a dense solution of 50 m*M* Hepes (pH 8.6), 40% glycerol, 1 m*M* $MgCl_2$. Centrifuge at 80 K × *g* for 45 min at 35 °C. The goal of this step is to transfer the polymerized MTs to a high pH buffer in preparation for labeling reaction.
3. Aspirate the supernatant slowly, with 37 °C warm 50 m*M* Hepes (pH 8.6), 5 m*M* $MgCl_2$, 40% glycerol. The polymerized MTs must

remain warm during resuspension process and should be thoroughly resuspended to ensure full labeling.

4. Add 10- to 20-fold molar excess of Succimidyl-ester Alexa-Fluor-dye (such as Alexa-Fluor-488, Invitrogen) dissolved in DMSO to the polymerized MTs. Each Succimidyl-ester dye is added in two portions over 60 min during the labeling reaction at 37 °C. The labeling reaction is then quenched with 1 m*M* amine to the labeling reaction and mixed well.
5. Layer the quenched labeling reaction onto a cushion, or dense buffer solution, of 50 m*M* pipes pH 6.8, 1 m*M* $MgCl_2$, 40% glycerol and then centrifuged at 80K × g for 20 min at 37 °C.
6. Remove the supernatant and the cushion slowly and wash the MT pellet with 37 °C warmed disassembly buffer (50 m*M* K-glutamate $MgCl_2$, 1 m*M* EGTA, pH 7.0). Resuspend the MT pellet using a dounce homogenizer at 0 °C and let sit on ice for 30 min. Glutamate improves MT depolymerization at 0 °C.
7. Clarify the depolymerized tubulin by centrifugation at 100 K × *g* at 0 °C for 20 min.
8. Repolymerize depolymerized tubulin (supernatant from step 8) by diluting in 80 m*M* pipes 4 m*M* $MgCl_2$ and 1 m*M* GTP with 33% glycerol at 37 °C for 30 min.
9. Layer the MT polymerization reaction on a 1 mL of 50 m*M* pipes pH 6.8, 1 m*M* $MgCl_2$, 40% glycerol in a TLA100.3 tube and pellet MTs at 80 K in a TLA100.3 rotor for 20 min at 37 °C.
10. Aspirate the supernatant slowly, with 37 °C warmed disassembly buffer. The MT pellets are depolymerized by careful resuspension at 0 °C using a 2-mL dounce homogenizer.
11. Clarify the depolymerized tubulin at 80 K × g for 10 min at 0 °C. Recover the supernatant, which consists of depolymerized tubulin dimer. Estimate the tubulin dimer concentration and the degree of labeling using protein and dye absorbance at 280 nm and the unique dye emission at that wavelength.
12. The tubulin is aliquoted in 1–2 μL portions, frozen in liquid nitrogen, and stored at −80 °C. Freshly thawed tubulin aliquots are used in all the experiments described below.

2.3. Preparing MT polymerization dynamics flow chambers for TIRF microscopy

2.3.1 Rationale

TIRF microscopy has become a powerful tool to image dynamic MT polymerization. For these studies, short and stable MTs (termed MT seeds) are used to nucleate polymerization, while coupled in close proximity to the glass coverslip. Thus, high-quality glass surfaces and precise functional coupling of MT to glass are both critical for the success of this approach. Glass surfaces must be clean, neutral, with little height variation to prevent soluble αβ-tubulin or MT regulator MAPs from aggregating in flow chambers. The MT seeds are polymerized using nonhydrolysable GTP analog, GMPCPP. The methods described below are modifications of earlier studies (Bieling et al., 2010; Gell et al., 2010). MT seeds are coupled to functionalized glass, which retains them a short distance away from the glass surface, which is important for MT polymerization reconstitution studies. Although many MT attachment strategies are described, I prefer using antibodies against unique chemical moieties incorporated into the MT seeds. Here, I will describe the methods of preparing glass surfaces and treatments to attach microtubules along these surfaces. I will also describe how flow chambers are assembled and treated in preparation for dynamic MT polymerization studies.

2.3.2 Cleaning and neutralizing glass surfaces

1. Load no 1.5 coverslips onto porcelain racks and wash them in a sonic bath for 20–25 min in the following solutions (in order): 10% commercial dishwasher detergent, 1 *M* KOH, acetone, and 100% ethanol. After each sonic wash step, rinse coverslips four times with distilled water.
2. Dry coverslips along with the porcelain rack in a clean 110–120 °C oven for about 1 h. Also dry a glass vessel to be used in step 4.
3. Plasma etch the dry coverslips in an 100% oxygen environment at 100 W/h for 10 min using an oxygen gas enabled plasma etching device.
4. Silanize coverslips in 0.1–0.2% dimethyldichlolorsilane diluted in ultra-dry tricholorethylene (Sigma-Aldrich, with less than 1 ppm H_2O) and react for 2 h.
5. Rinse twice in 100% methanol for 3 min each.
6. Dip and slowly remove coverslip rack in deionized water to observe wicking effect upon slow removal. When properly silanized, the glass coverslips must be hydrophobic and retain no water drops as they are removed out of the deionized water.
7. Glass must be stored between sheets of lens paper under vacuum at all times. Vacuum storage decreases the dust accumulation on the coverslips.

2.3.3 Polymerizing and isolating stabilized MT seeds

1. Mix freshly thawed 4 μ*M* tubulin dimer mixture containing 80% unlabeled tubulin, 10% fluorescently labeled, and 10% biotin-labeled,

freshly thawed tubulin dimer (prepared using the labeling protocol described above) in 10–20 μL of BRB-80 buffer supplemented with 1 m*M* GMPCPP (Jena Biosciences). The mixture is then warmed up to 37 °C for 90–120 min to polymerize MTs with an average length of 3–6 μm (suitable as seeds from which to grow extensions).

2. The mixture is then diluted with 100 μL of warm BRB-80 and then pipetted into 200-μL tube and centrifuged in a TLA100.3 rotor at 18 K rpm using a Beckmann table top-ultracentrifuge, or equivalent for 20 min.
3. The supernatant is removed and mixture is resuspended in 80 μL of BRB-80 buffer.

2.3.4 Assembling flow chambers and preparation for dynamic MT polymerization

1. Flow chambers are assembled from three layers (Fig. 8.1A): (A) top layer is a 2-cm thick, 20 × 20 cm Quartz Glass Adaptor with four holes positioned in the corners. (B) 20 × 20 mm double adhesive sheets (Grace Biolabs, Inc.) with 20 × 4 mm two rectangular channels cut into each side of the square size. (C) 22 × 22 mm salinized-treated glass, prepared as described above (Fig. 8.1A). The channels are placed in line with two holes in the quartz, to generate two rectangular sealed channels along sides of square. The adhesive should be pressed to ensure no air pockets, aside from the entry (Fig. 8.1B).
2. Inject 200 μL of BRB-80 through each flow channel through entry holes, while keeping a weak house-vacuum stream near the exit hole of that flow channel. This effectively cleans and washes the flow cell from any small particles.
3. Using the above approach pipetting approach, inject the following solutions through the flow chamber for the following periods of time. The protocol described below will attach antibodies through hydrophobic surface interaction to the glass, after which the glass is neutralized with a detergent to form a polar layer along the glass surface to prevent protein aggregation and inject the following solutions into the flow channel in this order, which is summarized in Fig. 8.1C:
   (A) 50 μL of BRB-80 and wait for 1 min.
   (B) 50 μL of 50-fold diluted antibiotin mouse monoclonal antibody (Invitrogen) in BRB-80 and wait for 4 min.
   (C) 50 μL of BRB-80 for 1 min.

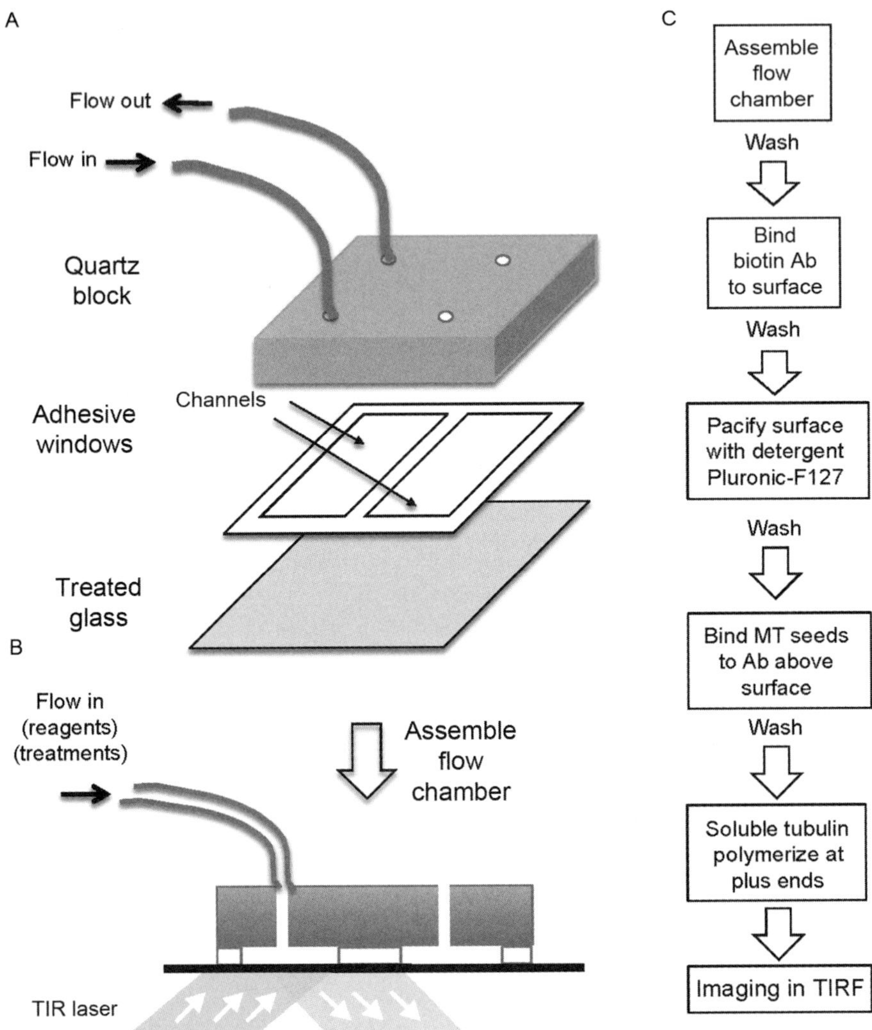

Figure 8.1 Reconstituting MT regulator dynamic MT polymerization using TIRF microscopy. (A) Top panel, composition of Flow chamber components: 20 × 20 mm Quartz block, 20 × 20 mm adhesive window with dual cut 4 × 22 mm channels, and 22 × 22 mm clean, silianized glass, treated as described in Section 2.3.2. Lower panel shows the assembled flow chamber showing the imaging surface. (B) Flow diagram describing a summary of the treatments to reconstitute dynamic MT polymerization along glass-surface; this diagram summarizes Sections 2.3.4, 2.4, and 2.5 in this chapter.

(D) 50 μL of 1% pluronic F127 in BRB-80 (Sigma-Aldrich), pre-filtered through 0.2 μm and wait for 4 min to neutralize the glass surface.

(E) 200 μL of BRB-80 through the flow channel to remove residual Pluronic F127.

(F) 60 μL of 40-fold diluted GMPCPP MT seeds from the original stock described above and wait for 10 min.

(G) 50 μL of BRB-80 to remove excess MT seeds.

(H) 100 μL of imaging buffer (contents of imaging buffer are described below).

2.4. Reconstituting dynamic MT polymerization with MT regulators using TIRF microscopy

2.4.1 Rationale

Surface-attached fluorescent and biotin-containing MT seeds are bound by antibiotin antibodies above the glass surface, within the evanescent field for TIRF microscopy (Fig. 8.2A). Dynamic MT polymerization is nucleated by MT seeds from fluorescently labeled tubulin with a dye color distinct from those in the MT seeds. The newly formed dynamic MTs are observed in a different fluorescent channel from MT seeds, as shown in Fig. 8.2B (Al-Bassam et al., 2010, 2012). I first optimize biochemical conditions, such as MT regulators concentration and solubility, in an appropriate imaging buffer at 37 °C. I recommend that dynamic MTs polymerization assays are performed in a variety of conditions, including (A) varying soluble αβ-tubulin from 6–12 μ*M* in chemical conditions compatible with MT regulator. (B) Increasing MT regulator concentrations and observing MT dynamic parameters at each soluble tubulin concentration (such as 6 or 8 μ*M*). I generally recommend starting at 6 μ*M* αβ-tubulin where MT dynamic polymerization rate is slow and MT catastrophe transitions are frequent, leading the average MT length to be very short. The approach described here was used in prior studies that found that fission yeast Alp14 is a MT polymerase accelerates MT polymerization by threefold, while Cls1 is a MT rescue factor that decreases MT catastrophe frequency and increases MT rescue frequency (Al-Bassam et al., 2010, 2012).

2.4.2 Dynamic MT reconstitution and TIRF microscopy imaging

1. Prepare flow chambers as described above in Section 2.3.4. This method starts at the point after which the MT seeds are added to the coverslip.

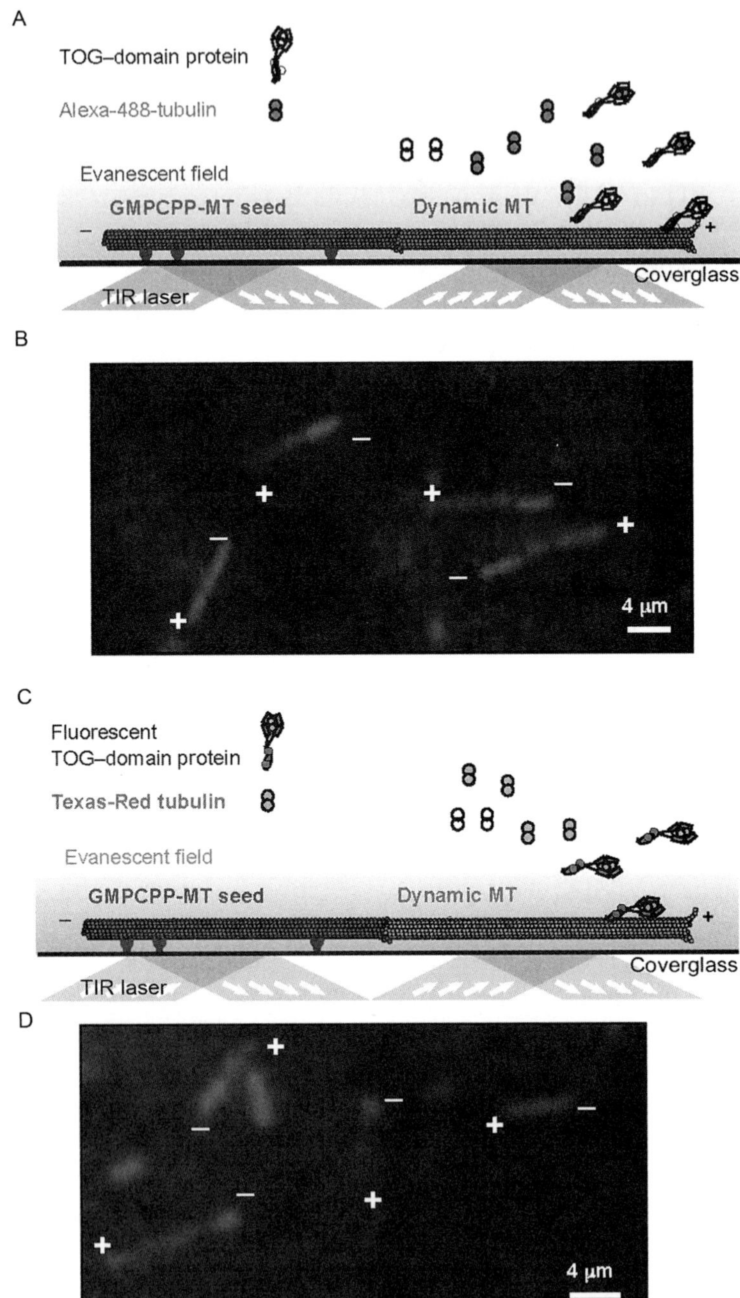

Figure 8.2 Schematic representation of dynamic MT polymerization reconstitution studies. (A) Scheme described in Section 2.4, for dual fluorescent MTs to measure MT

2. Prepare 40 μL of imaging buffer solution containing 50 m*M* buffer such at BRB-80 or others and
   (A) 6–10 μ*M* soluble tubulin with 10% fluorescently labeled of a different dye than the MT seeds attached to the glass surface; for example, Alexa-488-labeled soluble tubulin is prepared in the mix, if MT seeds are labeled with Texas Red (Fig. 8.2A).
   (B) 1–2 m*M* GTP diluted from a 100-m*M* GTP Stock.
   (C) Additives like salt (200 m*M* KCl for Alp14 or 70 m*M* KCl for Cls1)
   (D) Adding MT regulator (such as 0–200 n*M* Alp14 or Cls1 proteins), which should be added last.
3. Filter the mixture through a 0.2-μm spin filter (Amicon) to remove small particles.
4. Inject the imaging mixture in flow channel as described previously.
5. Warm flow chamber to 35–37 °C at the imaging TIRF objective lens.
6. Focus microscope imaging objective TIRF lens to identify the MT seeds, while waiting 5–10 min for tubulin/dynamic MTs to reach 35–37 °C. Apply autofocusing strategy, if available on your microscope. Autofocusing is extremely useful for keeping the sample in focus in the TIRF field.
7. Begin collecting MT polymerization image stacks data for each of the two channels (dynamic MT and static MT seeds). Typically, I acquire every 2–4 s for 10–30 min.

2.4.3 Analysis of dynamic MT image data to determine MT polymerization parameters

The stacks of images (termed movies) collected for each channel can then be corrected and analyzed in parallel using the ImageJ (Rasband, 1997) collection, as previously described (Al-Bassam et al., 2010).

---

polymerization dynamics. Dynamic MTs, shown in green, grow from Alexa-Fluor-488-labeled tubulin, while stabilized GMPCPP MT-seeds, shown in red, are polymerized from Texas-Red-labeled tubulin. MT regulators are nonfluorescent. (B) Example data of reconstituted, dynamic MTs, in the scheme described in Section 2.4. (C) Scheme for dynamic MT polymerization reconstitution using fluorescent MT regulator and fluorescent MTs, as described in Section 2.5. Dynamic MTs are shown in faint red grown 10% Texas-Red tubulin, while stabilized MT seeds are shown in red, polymerized from a higher ratio of Texas-Red-labeled tubulin. MT regulators, shown in green, are labeled with GFP-tags. (D) Example data of reconstituted MT dynamics with fluorescent MT regulators at. All images in this figure are generated with permission from publisher of these references (Al-Bassam et al., 2010, 2012). *(B) Reproduced with permission from Al-Bassam et al. (2010). (D) Reproduced with permission from publisher of reference Al-Bassam et al. (2010).* (See the color plate.)

1. Image stacks are adjusted for photobleaching by applying an average total fluorescence correction.
2. Image linear stage drift is corrected in image stacks by calculating translation parameters in the static MT seed channel and applying these onto each image in the stack.
3. Identify dynamic MT polymerization events and produce kymograph images using imaging tracking algorithms, or manual kymograph functions using the multikymograph plug-in as previously described (Al-Bassam et al., 2010, 2012) to determine the rate and time for MT polymerization and MT depolymerization.
4. Determine average parameters from a collection of kymographs in each data set (Fig. 8.3A), using histogram analysis to determine average numerical values for the dynamic MT polymerization parameters: assembly rate

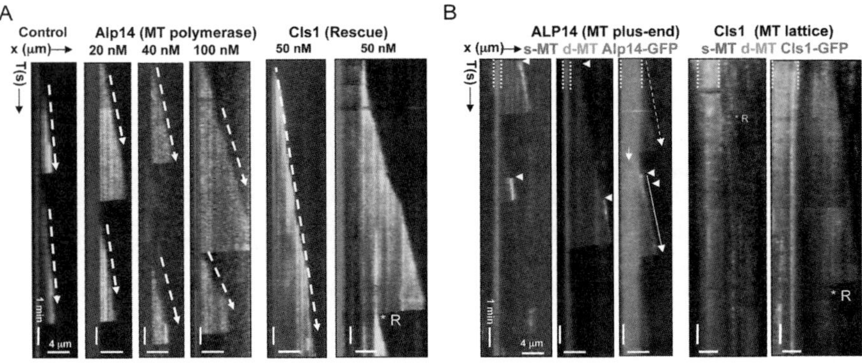

Figure 8.3 Example MT dynamic polymerization kymographs based on studies described. (A) Kymographs of dynamic MTs produced using ImageJ, using methods described in Section 2.4. Dynamic MTs are shown in green, while stabilized MT seeds are shown in red. Left panel, MTs grow slowly at 6 μ*M* tubulin dimer and depolymerize in frequent catastrophe events. Middle panels, increasing ALp14 concentration incrementally increases MT polymerization, without influencing MT catastrophe frequency (Al-Bassam et al., 2012). Right panels, Cls1 decreases in the frequency of MT catastrophe, and occurrence of MT rescues, labeled R^* (Al-Bassam et al., 2010). (B) Kymographs of dynamic MTs produced using ImageJ, using methods described in Section 2.5. Dynamic MTs are shown in light color red, while stabilized MT seeds are shown in intense red. Left panels, Alp14-GFP, shown in green, binds at MT plus ends to increase MT polymerization rate (Al-Bassam et al., 2012). Right panels, Cls1-GFP binds along MT lattices without tracking MT plus ends, where it correlates with absence of MT catastrophes, and its presence coincides with MT rescues, labeled R^* (Al-Bassam et al., 2010).All images in this figure are generated with permission from publisher of these references (Al-Bassam et al., 2010, 2012). (See the color plate.)

(μm/min), disassembly rate (μm/min), catastrophe frequency (event/min assembly time), or rescue frequency (event/min disassembly time).

2.5. Tracking fluorescent MT regulators along dynamic MTs using TIRF microscopy

2.5.1 Rationale

Visualizing dynamic localization of MT regulators along MTs or at their ends during dynamic MT polymerization requires active and fluorescently tagged MT regulators. Many types of fluorescent tags can be genetically or chemically fused to MT regulators for these studies such as fusions of green fluorescent protein (GFP) variants (Ilagan et al., 2010), fusions with domains for specific covalent fluorescent dye attachment, such as CLIP, SNAP, or Halo tags (Gautier et al., 2008), and or short sequences for covalent attachment to biarsenate fluorescent dyes like FlAsH and ReAsH (Machleidt et al., 2007). It is critical to determine that the fluorescently tagged MT regulator behaves similar to native (nonfluorescent) MT regulator using the method described in Section 2.4. Mixtures of fluorescently tagged to native MT regulator may be utilized to decrease average number of labeled molecules without decreasing MT regulatory activity. The fluorescent MT regulator activity must be analyzed the approach described in Section 2.4. These experiments do not necessarily require three channels and can be studied using two channels only, by using two different ratios of a single tubulin color polymerized into the MT seeds (higher ratio) compared to those added in solution and polymerizing into dynamic MTs (low ratio). This leads dynamic MTs to be fainter than the MT seeds in these experiments (Fig. 8.2B). These studies can be performed as described in Section 2.3.4.

2.5.2 Method

1. Prepare flow chambers as described in Section 2.3.4. This method starts at the point after which the MT seeds are added. MT seeds are polymerized with a higher ratio of dye (40%) compared to previously described protocol. This helps distinguish them from the dynamic MTs polymerized from 10% dye-labeled tubulin in the same color.
2. Prepare 40 μL of imaging buffer solution containing
   (A) 6–10 μ*M* freshly thawed soluble tubulin with 10% fluorescently labeled tubulin.
   (B) 1–2 m*M* GTP diluted from a 100-m*M* GTP Stock.
   (C) Additives like salt (200 m*M* for Alp14).

(D) 1–200 n*M* fluorescent MT regulator. The concentration depends on activity range. Single-molecule experiments with MT regulators require low concentration of fluorescent molecules 1–10 n*M*.

(E) Antioxidants and/or oxygen scavengers such as Trolox or the glucose oxidase-catalase system, as previously described (Gell et al., 2010), which maintain fluorescence intensity and prevent photodamage through long exposure periods.

3. Inject the imaging mixture containing items in A–D into flow channel as described.
4. Warm the flow chamber to 35–37 °C at the imaging TIRF objective lens.
5. Focus microscope imaging objective TIRF lens to identify the MT seeds, while waiting 5–10 min for tubulin/dynamic MTs to reach 35–37 °C.
6. Begin collecting MT polymerization image data stacks for each of the two channels (dynamic MT and static MT seeds) every 2–4 s for 10–30 min.

The data in these experiments are processed in the same manner as described above in Section 2.4. The residence time of MT regulators can be determined using particle tracking approaches as previously described. Determining the localization and residence time for a fluorescent MT regulator, such as Alp14 and Cls1, along MT plus ends or MT lattices, respectively, is critical to understand their mechanism in regulating MT polymerization rates or influencing rates of MT dynamic transitions such as activating MT rescues (Fig. 8.3B).

3. CONCLUSION

TIRF microscopy has emerged as a powerful approach to study the mechanisms of MT polymerization regulators or MT organizing proteins determine their unique effects on MT dynamic polymerization and organization. The approach described here can be used to study a variety of MT regulators or complexes of MT regulators with dynamic MTs. The reconstitution of complex MT dynamic regulation or organization activities with multiple MAPs and motors is critical to understand the mechanisms of MT regulation. In the future, I expect MT dynamic polymerization reconstitution and imaging with TIRF microscopy to become the standard approach in studying the biochemistry and mechanisms of MT motor, regulators, and organizer proteins in regulating complex MT assemblies observed in cellular phenomena such as stages of cell division.

ACKNOWLEDGMENTS

I thank Stephen Harrison, Tony Hyman, Jonathon Howard, and Gary Brouhard for support and encouragement to learn and use TIRF microscopy approaches to study microtubule dynamics, Ron Vale for time and help in critiquing and editing this work. I acknowledge support of NIH pathways to independence award (R00-GM08429) and support funds from the University of California Cancer Coordinating Committee.

REFERENCES

Akhmanova, A., & Steinmetz, M. O. (2008). Tracking the ends: A dynamic protein network controls the fate of microtubule tips. *Nature Reviews Molecular Cell Biology*, *9*(4), 309–322.

Al-Bassam, J., & Chang, F. (2011). Regulation of microtubule dynamics by TOG-domain proteins XMAP215/Dis1 and CLASP. *Trends in Cell Biology*, *21*(10), 604–614.

Al-Bassam, J., Kim, H., Brouhard, G., van Oijen, A., Harrison, S. C., & Chang, F. (2010). CLASP promotes microtubule rescue by recruiting tubulin dimers to the microtubule. *Developmental Cell*, *19*(2), 245–258.

Al-Bassam, J., Kim, H., Flor-Parra, I., Lal, N., Velji, H., & Chang, F. (2012). Fission yeast Alp14 is a dose-dependent plus end-tracking microtubule polymerase. *Molecular Biology of the Cell*, *23*(15), 2878–2890.

Bieling, P., Laan, L., Schek, H., Munteanu, E. L., Sandblad, L., Dogterom, M., et al. (2007). Reconstitution of a microtubule plus-end tracking system in vitro. *Nature*, *450*(7172), 1100–1105.

Bieling, P., Telley, I. A., Hentrich, C., Piehler, J., & Surrey, T. (2010). Fluorescence microscopy assays on chemically functionalized surfaces for quantitative imaging of microtubule, motor, and +TIP dynamics. *Methods in Cell Biology*, *95*, 555–580.

Brouhard, G. J., Stear, J. H., Noetzel, T. L., Al-Bassam, J., Kinoshita, K., Harrison, S. C., et al. (2008). XMAP215 is a processive microtubule polymerase. *Cell*, *132*(1), 79–88.

Castoldi, M., & Popov, A. V. (2003). Purification of brain tubulin through two cycles of polymerization-depolymerization in a high-molarity buffer. *Protein Expression and Purification*, *32*(1), 83–88.

Desai, A., & Mitchison, T. J. (1997). Microtubule polymerization dynamics. *Annual Review of Cell and Developmental Biology*, *13*, 83–117.

Gautier, A., Juillerat, A., Heinis, C., Correa, I. R., Jr., Kindermann, M., Beaufils, F., et al. (2008). An engineered protein tag for multiprotein labeling in living cells. *Chemistry & Biology*, *15*(2), 128–136.

Gell, C., Bormuth, V., Brouhard, G. J., Cohen, D. N., Diez, S., Friel, C. T., et al. (2010). Microtubule dynamics reconstituted in vitro and imaged by single-molecule fluorescence microscopy. *Methods in Cell Biology*, *95*, 221–245.

Hitchman, R. B., Locanto, E., Possee, R. D., & King, L. A. (2011). Optimizing the baculovirus expression vector system. *Methods*, *55*(1), 52–57.

Hyman, A., Drechsel, D., Kellogg, D., Salser, S., Sawin, K., Steffen, P., et al. (1991). Preparation of modified tubulins. *Methods in Enzymology*, *196*, 478–485.

Ilagan, R. P., Rhoades, E., Gruber, D. F., Kao, H. T., Pieribone, V. A., & Regan, L. (2010). A new bright green-emitting fluorescent protein—Engineered monomeric and dimeric forms. *The FEBS Journal*, *277*(8), 1967–1978.

Kirschner, M. W., Williams, R. C., Weingarten, M., & Gerhart, J. C. (1974). Microtubules from mammalian brain: Some properties of their depolymerization products and a proposed mechanism of assembly and disassembly. *Proceedings of the National Academy of Sciences of the United States of America*, *71*(4), 1159–1163.

Machleidt, T., Robers, M., & Hanson, G. T. (2007). Protein labeling with FlAsH and ReAsH. *Methods in Molecular Biology*, *356*, 209–220.

Maurer, S. P., Bieling, P., Cope, J., Hoenger, A., & Surrey, T. (2011). GTPgammaS microtubules mimic the growing microtubule end structure recognized by end-binding proteins (EBs). *Proceedings of the National Academy of Sciences of the United States of America*, *108*(10), 3988–3993.

Rasband, W.S., ImageJ, U. S. National Institutes of Health, Bethesda, Maryland, USA, http://imagej.nih.gov/ij/, 1997–2012.

Sun, P., Tropea, J. E., & Waugh, D. S. (2011). Enhancing the solubility of recombinant proteins in Escherichia coli by using hexahistidine-tagged maltose-binding protein as a fusion partner. *Methods in Molecular Biology*, *705*, 259–274.

Telley, I. A., Bieling, P., & Surrey, T. (2011). Reconstitution and quantification of dynamic microtubule end tracking in vitro using TIRF microscopy. *Methods in Molecular Biology*, *777*, 127–145.

Walker, R. A., O'Brien, E. T., Pryer, N. K., Soboeiro, M. F., Voter, W. A., Erickson, H. P., et al. (1988). Dynamic instability of individual microtubules analyzed by video light microscopy: Rate constants and transition frequencies. *The Journal of Cell Biology*, *107*(4), 1437–1448.

CHAPTER NINE

Generation of Differentially Modified Microtubules Using *In Vitro* Enzymatic Approaches

Annapurna Vemu[*,1], **Christopher P. Garnham**[*,1], **Duck-Yeon Lee**[†], **Antonina Roll-Mecak**[*,‡,2]

[*]Cell Biology and Biophysics Unit, National Institute of Neurological Disorders and Stroke, Bethesda, Maryland, USA
[†]Biochemistry Core, National Heart, Lung and Blood Institute, Bethesda, Maryland, USA
[‡]Center for Biophysics, National Heart, Lung and Blood Institute, Bethesda, Maryland, USA
[1]These authors contributed equally
[2]Corresponding author: e-mail address: antonina@mail.nih.gov

Contents

Abstract

Tubulin, the building block of microtubules, is subject to chemically diverse and evolutionarily conserved post-translational modifications that mark microtubules for specific functions in the cell. Here we describe *in vitro* methods for generating homogenous acetylated, glutamylated, or tyrosinated tubulin and microtubules using recombinantly expressed and purified modification enzymes. The generation of differentially modified microtubules now enables a mechanistic dissection of the effects of tubulin post-translational modifications on the dynamics and mechanical properties of microtubules as well as the behavior of motors and microtubule-associated proteins.

Methods in Enzymology, Volume 540
ISSN 0076-6879
http://dx.doi.org/10.1016/B978-0-12-397924-7.00009-1

1. INTRODUCTION

Microtubules are dynamic polymers essential for cell division, intracellular transport, and morphogenesis (Howard & Hyman, 2003; Nogales, 2000). The building block of microtubules is the αβ-tubulin heterodimer. Humans have six α-tubulin (α1A, α1B, α1C, α3A, α4A, and α8) and seven β-tubulin isoforms (βI, βII, βIII, βIVa, βIVb, βV, and βVI) (Sullivan, 1988). Multiple α- and β-tubulins are typically expressed in a cell, giving rise to isotypically diverse microtubules (Miller et al., 2010). Moreover, the complexity of microtubule arrays is further modulated by post-translational modifications. Tubulin is subject to several chemically diverse and evolutionarily conserved post-translational modifications: (1) cyclical removal and addition of the α-tubulin C-terminal tyrosine (resulting in "Glu-tubulin") (Barra, Rodriguez, Arce, & Caputto, 1973), (2) irreversible removal of the penultimate glutamate of α-tubulin (resulting in "Δ2-tubulin") (Paturle-Lafanechere et al., 1991), (3) acetylation of α-tubulin (L'Hernault & Rosenbaum, 1983, 1985), (4) polyglutamylation, and (5) polyglycylation of α- and β-tubulin (Alexander et al., 1991; Edde et al., 1990; Redeker et al., 1994; Redeker, Melki, Prome, Le Caer, & Rossier, 1992; Rudiger, Plessman, Kloppel, Wehland, & Weber, 1992).

Most post-translational modifications occur on the unstructured negatively charged tubulin C-terminal tails (Fig. 9.1) (Nogales, Wolf, & Downing, 1998; Sullivan, 1988). Tubulin tails decorate the microtubule exterior and can interact with motors and microtubule-associated proteins (MAPs) and modulate their activities (Garnham & Roll-Mecak, 2012; Janke & Bulinski, 2011; Wloga & Gaertig, 2010). Cytoplasmic linker protein-170, a plus end microtubule-tracking protein, preferentially binds tyrosinated tubulin (Bieling et al., 2008) and the microtubule-severing enzyme spastin preferentially severs polyglutamylated microtubules (Lacroix et al., 2010; Roll-Mecak & McNally, 2010; Roll-Mecak & Vale, 2008). Glutamylation also increases synaptic vesicle transport by kinesin-2 and targets MAP2 to dendritic microtubules (Ikegami et al., 2007). Acetylation of lysine 40 on α-tubulin is unique among tubulin modifications as it occurs inside the microtubule lumen (Nogales, Whittaker, Milligan, & Downing, 1999; Soppina, Herbstman, Skiniotis, & Verhey, 2012), close to the interprotofilament interface where it can affect microtubule stability (Cueva, Hsin, Huang, & Goodman, 2012; Topalidou et al., 2012).

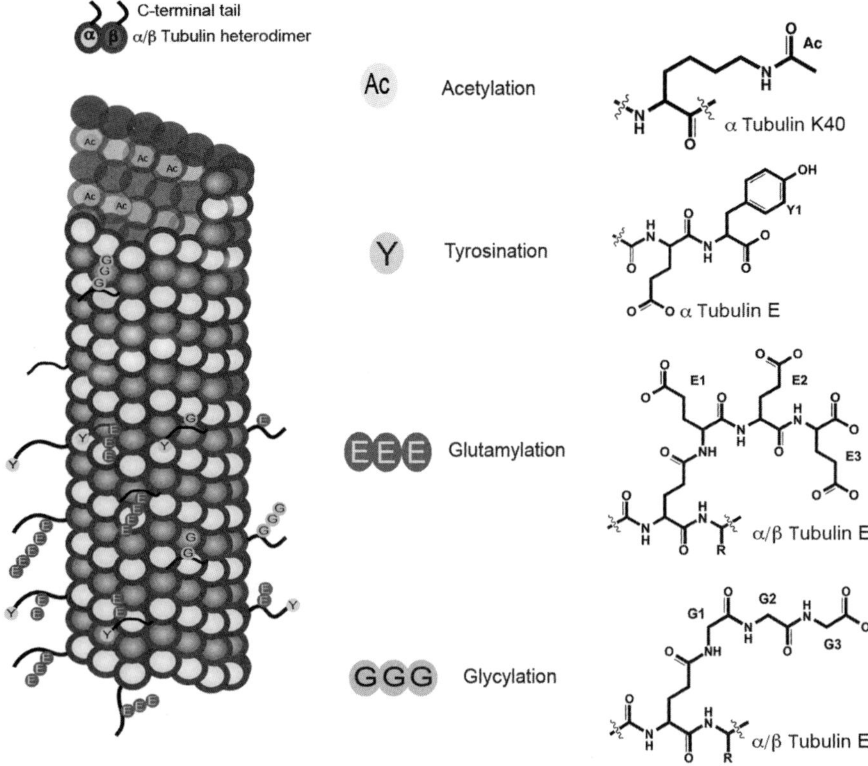

Figure 9.1 (Left) Schematic representation of a microtubule (α-tubulin, green; β-tubulin, blue). The unstructured C-terminal tails are shown in red. Tyrosination (cyan), glutamylation (red), and glycylation (brown) occur on the unstructured C-terminal tails. Acetylation on α-tubulin Lys 40 (orange) occurs in the lumen. (Right) Chemical structures of various tubulin modifications. The elongated Glu chain is thought to be linear and not branched (Redeker, Le Caer, Rossier, & Prome, 1991). *Note:* Glycylation is not discussed in this chapter. (For interpretation of the references to color in this figure legend, the reader is referred to the online version of this chapter.)

Tubulin post-translational modifications have been known for several decades and studies using modification specific antibodies revealed the differential localization of post-translationally modified microtubules in the cell as well as their markedly different stabilities. However, a mechanistic understanding of their effect on microtubule biophysical properties as well as the behavior of microtubule regulators has been lacking. This is partly due to the difficulty in obtaining unmodified and modified tubulin that carries only one specific modification. Tubulin has traditionally been purified from brain tissue through repeated cycles of polymerization and depolymerization

(Weisenberg, 1972). This approach cannot easily be applied to other sources with lower tubulin concentrations than brain tissue, since it is hard to reach critical tubulin concentrations for robust polymerization. As a consequence, brain has been the *de facto* source for tubulin purification for several decades. However, brain tubulin is a heterogeneous mixture of isoforms and contains abundant post-translational modifications such as polyglutamylation, detyrosination, and acetylation (Sullivan, 1988). Moreover, modification levels vary from preparation to preparation depending on how the brain tissue was harvested and stored prior to tubulin isolation. In order to investigate the effects of individual modifications on microtubule behavior as well as their effect on the recruitment and activity of cellular effectors, it is necessary to prepare unmodified homogeneous tubulin that can be modified "at will" with a single type of modification. This necessitates: (1) preparation of biochemical quantities of unmodified or "naïve" tubulin, (2) preparations of active tubulin modification enzymes, and (3) development of protocols for the controlled modification of naïve tubulin using these enzyme preparations.

The challenge in purifying milligram amounts of unmodified tubulin from various sources was recently overcome by the Hyman and Howard laboratories through the development of an affinity-based purification that uses the tubulin-binding TOG domains from MAP215 cross-linked to solid support (Widlund et al., 2012). Moreover, we now have an almost complete catalog of tubulin post-translational modification enzymes, making it possible for the first time to undertake a systematic dissection of the roles of post-translational modifications in modulating microtubule functions (Garnham & Roll-Mecak, 2012).

Here we describe protocols for obtaining differentially modified tubulin and microtubules using recombinantly expressed tubulin modification enzymes. The selectively modified microtubules obtained using the protocols described here can be used in biochemical and biophysical assays to evaluate the effects of individual post-translational modifications on microtubule dynamics and the behavior of motors and MAPs. We focus here on three chemically distinct post-translational modifications: acetylation, polyglutamylation, and tyrosination. Tubulin acetyltransferase (α-TAT) acetylates α-tubulin on Lys 40 in the microtubule lumen (Akella et al., 2010; Shida, Cueva, Xu, Goodman, & Nachury, 2010). Tubulin tyrosine ligase-like 7 (TTLL7), the most abundant tubulin polyglutamylase in neurons, adds glutamate chains to the tubulin C-terminal tails (Ikegami et al., 2006; van Dijk et al., 2007). Tubulin tyrosine ligase (TTL) catalyzes the readdition of the genomically encoded α-tubulin C-terminal tyrosine (Raybin & Flavin, 1977; Schroder, Wehland, & Weber, 1985).

2. PURIFICATION AND CHARACTERIZATION OF UNMODIFIED MICROTUBULES

tsA201 cells are HEK293 derivatives with low levels of tubulin post-translational modifications and thus an excellent source of unmodified tubulin. The tubulin used in the protocols described here was purified from tsA201 cells using a His-TOG1 (Slep & Vale, 2007) column following the recently published protocol of Widlund et al. (2012).

The tubulin isolated using this affinity purification method is highly pure as evaluated by SDS-PAGE and reverse-phase liquid chromatography–mass spectrometry (LC–MS) (Fig. 9.2A). To analyze the purified tubulin via LC–MS, mix 2 μl of 5 μ*M* tubulin with 10 μl of 0.1% trifluoroacetic acid (TFA) and centrifuge at 16,000 × *g* for 10 min at 4 °C. Load 12 μl of the sample onto a Zorbax 300SB-C18 column (50 mm × 2.1 mm) (Agilent) attached in-line with an Agilent 6224 electrospray ionization time-of-flight

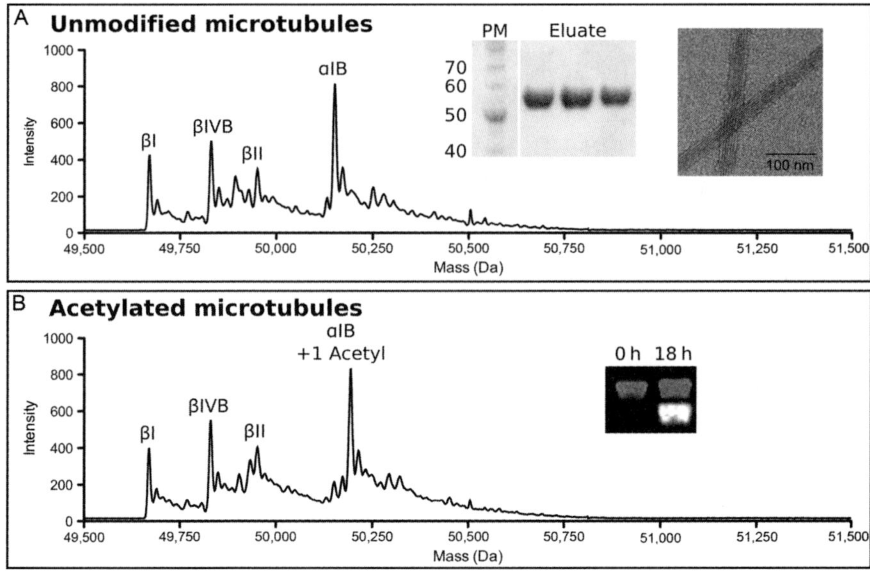

Figure 9.2 (A) Reverse-phase LC–MS, SDS-PAGE, and negative stain electron microscopy analysis of unmodified microtubules. Individual tubulin isoforms are labeled in the spectrum. PM, protein markers. Eluate indicates elution fractions from the TOG affinity column. (B) Reverse-phase LC–MS and Western blot analysis of unmodified microtubules acetylated by α-TAT. Tubulin isoforms labeled as in (A). Inset shows progression of acetylation monitored by Western blot using antibodies specific for acetylated tubulin. α-Tubulin, gray; acetylated tubulin, white. The two channels are offset for clarity.

LC–MS. Use a 0–70% acetonitrile gradient in 0.05% TFA at a 0.2 ml/min flow rate. The data can be analyzed using the Agilent MassHunter Workstation platform. The LC–MS analyses show the presence of one α- and three β-tubulin isoforms (α1B, βI, βII, and βIVB) (Fig. 9.2A) and the absence of post-translational modifications. The purified tubulin polymerizes robustly into microtubules using standard polymerization protocols (Section 4.1). Negative stain electron microscopy shows no aggregates or intermediate polymerization products (Fig. 9.2A).

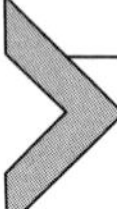

3. Purification of Tubulin Modification Enzymes: α-TAT, TTLL7, and TTL

Required reagents

Tobacco etch virus (TEV) protease. This can be expressed and purified according to published protocols (Kapust et al., 2001) or purchased from Sigma Aldrich.

Required equipment

Microfluidizer for cell disruption (we use the C3 Homogenizer from Avestin)

AKTA purifier (GE Healthcare)

3.1. Expression and purification of α-TAT

Solutions required for α-TAT purification

10 × phosphate buffered saline (PBS)

α-TAT resuspension buffer: 1 × PBS supplemented with 10 m*M* $MgCl_2$, 1 m*M* phenylmethylsulfonyl fluoride (PMSF), 5 m*M* dithiothreitol (DTT), protease inhibitor cocktail (Roche Applied Science)

α-TAT GST-A buffer: 50 m*M* Tris–HCl (pH 7.5), 500 m*M* NaCl, 10 m*M* $MgCl_2$, 5 m*M* DTT

α-TAT GST-B buffer: 50 m*M* Tris–HCl (pH 7.5), 100 m*M* NaCl, 10 m*M* $MgCl_2$, 5 m*M* DTT

α-TAT ion-exchange buffer A: 50 m*M* Tris–HCl (pH 7.5), 10 m*M* $MgCl_2$, 2 m*M* DTT

α-TAT ion-exchange buffer B: 50 m*M* Tris–HCl (pH 7.5), 10 m*M* $MgCl_2$, 2 m*M* DTT, 2 *M* NaCl

α-TAT gel filtration buffer: 50 m*M* Tris–HCl (pH 7.5), 200 m*M* NaCl, 5 m*M* $MgCl_2$, 2 m*M* tris(2-carboxyethyl)phosphine

Express *Mus musculus* α-TAT (residues 1–196) in *Escherichia coli* Rosetta2 (DE3)pLysS as an N-terminal cleavable GST fusion protein. Induce expression overnight (O/N) at 16 °C with 0.35 m*M* isopropyl β-D-1-thiogalactopyranoside (IPTG). Pellet cells via centrifugation and resuspend in α-TAT resuspension buffer. Lyse cells by microfluidization (three passes at 12,000 psi) while keeping the sample cold. Perform all the following steps at 4 °C. Supplement the lysate with 0.4 *M* NaCl and spin for 45 min at 31,000 × *g* to pellet cellular debris. Pass lysate supernatant over a gravity flow GST column equilibrated in α-TAT GST-A buffer. Wash resin with 10 column volumes (CVs) of α-TAT GST-A buffer. Resuspend GST resin with one CV of α-TAT GST-B buffer and rock slurry O/N following addition of TEV protease at a 1:50 molar ratio of protease:α-TAT. Load GST slurry flow-through onto a Q-sepharose column equilibrated in 5% α-TAT ion-exchange buffer B and collect flow-through. (*Note*: α-TAT does not bind Q-sepharose resin. This is a substractive purification step.) Load Q-sepharose flow-through onto an S75 gel filtration column (GE Healthcare) equilibrated in α-TAT gel filtration buffer. If the Q-sepharose flow-through is too dilute, first concentrate using an Amicon concentrator with a 5-kDa cutoff (Millipore). Pool fractions containing α-TAT and determine protein concentration using 280 nm absorbance. Concentrate to 4 mg/ml and flash freeze in α-TAT gel filtration buffer supplemented with 15% glycerol.

3.2. Expression and purification of the Glu-ligase TTLL7

Solutions required for TTLL7 purification

TTLL7 resuspension buffer: 50 m*M* Tris–HCl (pH 7.4), 200 m*M* NaCl, 10 m*M* $MgCl_2$, 1 m*M* PMSF

TTLL7 GST-A buffer: 50 m*M* Tris–HCl (pH 7.4), 500 m*M* NaCl, 10 m*M* $MgCl_2$, 2 m*M* DTT

TTLL7 GST-B buffer: 50 m*M* Tris–HCl (pH 7.4), 500 m*M* NaCl, 10 m*M* $MgCl_2$, 2 m*M* DTT, 20 m*M* reduced glutathione, 20 m*M* Tris–HCl (pH 8.8)

TTLL7 ion-exchange buffer A: 50 m*M* HEPES (pH 7.0), 10 m*M* $MgCl_2$, 2 m*M* DTT

TTLL7 ion-exchange buffer B: 50 m*M* HEPES (pH 7.0), 10 m*M* $MgCl_2$, 2 m*M* DTT, 2 *M* NaCl

TTLL7 gel filtration buffer: 20 m*M* HEPES (pH 7.0), 150 m*M* NaCl, 10 m*M* $MgCl_2$, 1 m*M* DTT

Express *Xenopus tropicalis* TTLL7 (residues 1–520) in *E. coli* Rosetta2(DE3) pLysS as a cleavable N-terminal GST fusion protein. Induce expression with 0.5 m*M* IPTG at 16 °C and harvest after 16 h. Pellet cells and resuspend in cold TTLL7 resuspension buffer. Lyse cells using a microfluidizer (three passes at 12,000 psi). Perform the following steps at 4 °C. Spin lysate for 45 min at 31,000 × *g* to pellet cellular debris. Run lysate supernatant over a GST column equilibrated in TTLL7 GST-A buffer. Wash column with 10 CVs of TTLL7 GST-A buffer and elute with two CVs of TTLL7 GST-B buffer. Pool TTLL7-containing fractions and load onto a heparin sepharose-6 column (GE Healthcare) equilibrated in 25% TTLL7 ion-exchange buffer B. Wash the column with two CVs of 25% TTLL7 ion-exchange buffer B and elute with a 25–50% gradient over 10 CVs. Pool fractions containing TTLL7, dilute 1:1 with TTLL7 ion-exchange buffer A, and digest O/N with TEV protease at a 1:50 protease:TTLL7 molar ratio. Following TEV digestion, load TTLL7 onto the heparin column equilibrated in 12.5% TTLL7 ion-exchange buffer B, wash with two CVs of 12.5% TTLL7 ion-exchange buffer B, and elute with a 12.5–50% gradient over eight CVs. Pool fractions containing TTLL7 and load onto a Superdex 75 gel filtration column equilibrated in TTLL7 gel filtration buffer. Pool TTLL7-containing fractions and determine concentration from absorbance at 280 nm. If needed, concentrate TTLL7 using a 30-kDa cutoff Amicon concentrator (Millipore). Protein can be flash frozen and stored at −80 °C following 15% glycerol addition.

3.3. Expression and purification of TTL

Solutions required for TTL purification

TTL resuspension buffer: 1× PBS, 10 m*M* $MgCl_2$, 1 m*M* PMSF, 5 m*M* DTT

TTL GST-A buffer: 50 m*M* Tris–HCl (pH 7.5), 500 m*M* NaCl, 10 m*M* $MgCl_2$, 5 m*M* DTT

TTL GST-B buffer: 50 m*M* Tris–HCl (pH 7.5), 100 m*M* NaCl, 10 m*M* $MgCl_2$, 5 m*M* DTT

TTL ion-exchange buffer A: 50 m*M* Tris–HCl (pH 7.5), 10 m*M* $MgCl_2$, 2 m*M* DTT

TTL ion-exchange buffer B: 50 m*M* Tris–HCl (pH 7.5), 10 m*M* $MgCl_2$, 2 m*M* DTT, 2 *M* NaCl

TTL hydrophobic column buffer A: 20 m*M* HEPES (pH 7.0), 10 m*M* $MgCl_2$, 5 m*M* DTT, 1 *M* ammonium sulfate

TTL hydrophobic column buffer B: 20 m*M* HEPES (pH 7.0), 10 m*M* $MgCl_2$, 5 m*M* DTT

Express full-length *M. musculus* TTL in *E. coli* Rosetta2(DE3)pLysS as a cleavable N-terminal GST fusion. Follow the α-TAT purification protocol up to and including the Q-sepharose step. Bring TTL Q-sepharose flow-through to 1 *M* ammonium sulfate and load onto a phenyl-sepharose column (GE healthcare) equilibrated in TTL hydrophobic column buffer A. Elute the protein using a 0–100% TTL hydrophobic column buffer B gradient over 10 CVs. Pool TTL-containing fractions and determine protein concentration using 280 nm absorbance. If required, concentrate TTL using a 10-kDa cutoff Amicon concentrator. TTL can be flash frozen and stored at −80 °C following addition of 15% glycerol.

4. GENERATION OF DIFFERENTIALLY MODIFIED TUBULIN AND MICROTUBULES

The following section describes the *in vitro* enzymatic modification of microtubules and tubulin using the purified enzymes obtained using protocols described in section 3. α-TAT and TTLL7 act preferentially on microtubules, while TTL only tyrosinates monomeric tubulin effectively. Thus, we use microtubules as substrates for the acetylation and glutamylation reaction and monomeric tubulin for the tyrosination reaction. Monomeric tubulin can also be acetylated and glutamylated by α-TAT and TTLL7, respectively, but with lower efficiency.

Required equipment

Optima MAX-XP table top ultracentrifuge (Beckman Coulter)

Electrospray ionization time-of-flight LC–MS (we use an Agilent 6224).

4.1. Preparation of taxol-stabilized microtubules

Required reagents and solutions

Taxol stock: 10 m*M* in DMSO, stored at −20 °C

5 × BRB80: 400 m*M* PIPES (pH 6.8), 5 m*M* $MgCl_2$, 5 m*M* EGTA. Store in the dark at 4 °C

2 × Polymerization buffer: 20% (v:v) DMSO, 2 m*M* GTP, 2 m*M* $MgCl_2$, 2 m*M* EGTA

Glycerol cushion: 60% glycerol in 1 × BRB80 supplemented with 10 μ*M* taxol

BRB80-DT: 1 × BRB80 supplemented with 1 m*M* DTT and 10 μ*M* taxol

Thaw frozen tubulin in 37 °C water bath and immediately place on ice. Preclear the tubulin to remove aggregates via ultracentrifugation at 436,000 × *g* for 10 min at 4 °C. (*Note*: It is important to prechill the rotor as well as the centrifuge tubes.) Remove the supernatant and place on ice. The pellet contains small amounts of tubulin aggregates. Remeasure the concentration of tubulin in the supernatant by Bradford assay. Mix tubulin 1:1 (v/v) with 2 × polymerization buffer. Mix thoroughly and incubate in a 37 °C water bath for 30–60 min. Supplement polymerization reaction with 5 μ*M* taxol and incubate in a 37 °C water bath for 15 min. Prewarm glycerol cushion and BRB80-DT at 37 °C. Overlay polymerization reaction on glycerol cushion (80 μl glycerol cushion: 100 μl polymerization reaction). Pellet microtubules by ultracentrifugation using prewarmed rotor at 109,000 × *g* for 10 min at 30 °C. Discard supernatant. Wash pellet and walls of the ultracentrifuge tube with BRB80-DT. This step removes any unpolymerized tubulin on the tube walls. Resuspend microtubule pellet in 1 × BRB80-DT. (*Note*: Cut the pipette tip when mixing microtubules to minimize shearing.) Measure tubulin concentration in 6 *M* guanidine hydrochloride using an extinction coefficient of 115,000 M^{-1}.

4.2. Generation of acetylated microtubules using α-TAT

This protocol generates 100% acetylated tubulin or microtubules using recombinant α-TAT.

Required solutions

Acetyl-coA stock: 100 m*M* in water. Store at −20 °C.

Acetylation buffer: 1 × BRB80 supplemented with 250 m*M* KCl, 1 m*M* DTT, 5 μ*M* taxol, 100 μ*M* acetyl-coA.

Incubate microtubules with α-TAT at a 1:1 molar ratio in acetylation buffer at room temperature (RT) O/N. The extent of post-translational modification can be monitored by reverse-phase LC–MS as well as Western blot using a tubulin acetylation specific antibody (below). Reverse-phase LC–MS analysis (described in Section 1) shows complete acetylation of α1B tubulin by α-TAT, indicated by the +42 Da mass shift observed in the mass spectra [50,193 Da (acetylated) vs. 50,151 Da (nonmodified)] (Fig. 9.2B). No additional species with a mass shift are visible underscoring the specificity of the modification. The acetylation reaction can also be performed with nontaxol stabilized microtubules at 37 °C in the presence of 20% glycerol and 1 m*M* GTP (Kormendi, Szyk, Piszczek, & Roll-Mecak, 2012).

Detection of tubulin acetylation by Western blot

We use infrared dye (IRDye®) conjugated secondary antibodies (Li-Cor) to detect tubulin post-translational modifications on Western blots. Two different secondary IRDye antibodies are used simultaneously: one secondary antibody detects the tubulin antibody while the other detects the antibody specific for the tubulin post-translational modification.

Required reagents and solutions

1 × PBS

1 × PBS-T; 0.1% (v:v) Tween-20 in 1 × PBS

Blocking agent: 4% milk powder in 1 × PBS

Dilution buffer: 4% milk powder in 1 × PBS-T

Primary antibody 1 solution: mouse 6-11B (recognizes acetylated tubulin) (Sigma Aldrich) diluted 1:10,000 in dilution buffer

Primary antibody 2 solution: rabbit E-19R (recognizes α-tubulin N-terminus) (Santa Cruz Biotech) diluted 1:1000 in dilution buffer

Secondary antibody 1: IRDye 680LT goat (polyclonal) antimouse IgG (Li-Cor)

Secondary antibody 2: IRDye 800LT goat (polyclonal) antirabbit IgG (Li-Cor)

Required equipment

Odyssey CLx (Li-Cor)

iBlot 7-min Blotting System (Life Technologies) (*Note*: Traditional transferring methods such as wet and semi-wet transfers will work as well)

Separate 125 ng each of naïve and acetylated tubulin on SDS-PAGE and transfer onto nitrocellulose membrane. All subsequent steps require rocking of the membrane. First, block O/N in blocking agent at 4 °C or for 2 h at RT. Perform all subsequent steps at RT. Incubate membrane in primary antibody 1 solution for 1 h. Wash twice with dilution buffer then incubate membrane in primary antibody 2 solution for 1 h. Wash five times with 1 × PBS-T for 5 min. Incubate with secondary antibody 1 and secondary antibody 2 simultaneously (both diluted 1:18,000 in dilution buffer) for 1 h. (*Note*: protect the blot from light as the IR antibodies are light sensitive.) Wash blot five times with 1 × PBS-T for 5 min, then twice with 1 × PBS. Image using Li-Cor Odyssey CLx (Li-Cor). The acetylated tubulin produces a strong signal in the 680 nm channel (white) that is not observed at the 0 h time point (Fig. 9.2B). The strong 800 nm signal (gray) represents total tubulin loaded. The anti-acetylated tubulin antibody can be calibrated using a tubulin sample that is 100% acetylated.

If desired, α-TAT can be removed after the acetylation reaction by a high salt wash. Pellet the acetylated microtubules by ultracentrifugation for 10 min at 109,000 × *g*. Resuspend pellet in 100 μl of BRB80-DT supplemented with 350 m*M* NaCl and incubate at 37 °C for 10 min. (*Note*: It is important to keep buffers warm in order not to depolymerize microtubules.) Overlay the resuspended pellet on 100 μl glycerol cushion

(1 × BRB80, 350 m*M* NaCl, 60% glycerol, and 10 μ*M* taxol). Pellet microtubules by ultracentrifugation at 109,000 × *g* for 15 min, at 30 °C. Remove supernatant and wash pellet with 200 μl of BRB80-DT supplemented with 350 m*M* NaCl followed by 200 μl of BRB80-DT buffer. Resuspend pellet in BRB80-DT to desired volume and concentration. Removal of the enzyme can be verified by SDS-PAGE or reverse-phase LC–MS.

4.3. Generation of polyglutamylated microtubules using TTLL7

The following protocol describes the polyglutamylation of unmodified microtubules with recombinantly expressed TTLL7. Varying the incubation time with TTLL7 controls the length of polyglutamate chains added to α- and β-tubulin.

Required solutions

Glutamate stock: 100 m*M* in water, pH adjusted to 7.0, stored at −20 °C.

Adenosine triphosphate (ATP) stock: 100 m*M* in water, pH adjusted to 7.0, stored at −20 °C.

Glutamylation buffer: 20 m*M* HEPES (pH 7.0), 50 m*M* NaCl, 5 m*M* $MgCl_2$, 1 m*M* ATP, 1 m*M* glutamate, 1 m*M* DTT.

Use taxol-stabilized microtubules as the substrate for TTLL7. If addition of glutamate chains primarily on β-tubulin is desired, incubate TTLL7 with unmodified microtubules at 1:10 molar ratio of enzyme:substrate for 1 h at RT in glutamylation buffer. Reverse-phase LC–MS analysis of the reaction reveals multiple mass increments of 129 Da, indicating the addition of glutamate residues. Closer inspection reveals chains consisting of 12, 3, and 2 glutamates attached to βIVB, βI, and βII tubulin, respectively, while α1B tubulin is monoglutamylated only sparingly (Fig. 9.3, top panel). Polyglutamate chains of increasing length can be added to both α- and β-tubulin by increasing the incubation time. Incubation for 6 h produces chains of up to 4, 17, and 22 glutamates on α1B, βI, and βIVB, respectively (Fig. 9.3B), while an 18 h incubation generates chains up to 10, 28, and 35 glutamates on α1B, βI, and βIVB, respectively (Fig. 9.3C). No unmodified βII tubulin remains after 6 h; however, the polyglutamylated moieties are not visible in the spectrum because of low signal.

The extent of glutamylation can also be monitored by Western blot using the monoclonal GT335 antibody (Adipogen, 1:2000 dilution) that recognizes the branch point created during the addition of the first glutamate residue of a growing glutamate chain (Fig. 9.1) (Wolff, 1992). Follow the protocol detailed in Section 4.2. Figure 9.3 shows the signal in the 680 nm channel (white) intensifies over the course of the reaction,

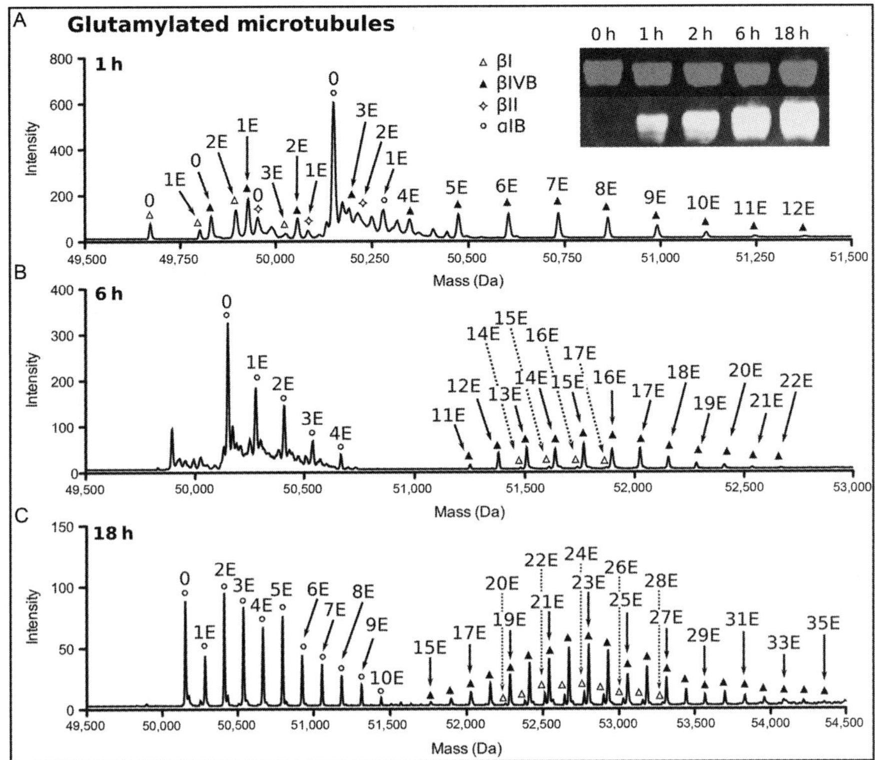

Figure 9.3 Reverse-phase LC–MS and Western blot analysis of unmodified microtubules glutamylated by TTLL7. (A), 1 h, (B), 6 h, (C), 18 h incubation. Tubulin isoforms are indicated by symbols. The number of glutamates added to each isoform by TTLL7 is indicated. Only every other glutamate added to βI and βIVB at 18 h is labeled for clarity. Inset shows the progression of tubulin glutamylation monitored by Western blot at the indicated time points. α-Tubulin, gray, glutamylated tubulin, white. The two channels are offset for clarity.

indicating increasing levels of glutamylation. The 800 nm channel (gray) monitors total tubulin loaded.

If desired, TTLL7 can be removed after the glutamylation reaction by a high salt wash (see protocol described in Section 4.2 for the removal of α-TAT from modified microtubules).

4.4. Generation of tyrosinated tubulin using TTL

This protocol generates 100% tyrosinated tubulin using recombinant TTL.

TTL acts preferentially on the tubulin monomer and is inefficient at modifying tubulin already incorporated into microtubules (Raybin &

Flavin, 1975; Szyk, Deaconescu, Piszczek, & Roll-Mecak, 2011). Thus, in order to generate tyrosinated microtubules, monomeric tubulin is first tyrosinated and then polymerized into microtubules. Mammalian brain tubulin contains high levels of several detyrosinated α-tubulin isoforms (Gundersen, Kalnoski, & Bulinski, 1984) (Fig. 9.4A). LC–MS analysis of porcine brain tubulin (Cytoskeleton) identified detyrosinated α1A and α1B tubulin, both present in multiple glutamylation states, with up to three glutamates attached to each (Fig. 9.4A). The unmodified tubulin isolated from tsA201 cells is 100% tyrosinated (Fig. 9.1A), thus this protocol is applicable to brain tubulin or other tubulin preparations that have high levels of detyrosinated tubulin.

Required solutions

Tyrosine stock: 30 m*M* in water, stored at −20 °C.

ATP stock: 100 m*M* in water, pH adjusted to 7.0, stored at −20 °C.

Tyrosination buffer: 1 × BRB80, 1 m*M* DTT, 0.3 m*M* Tyr, 2 m*M* ATP.

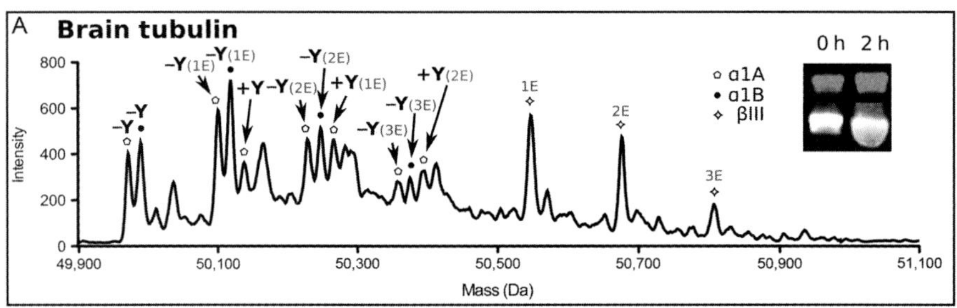

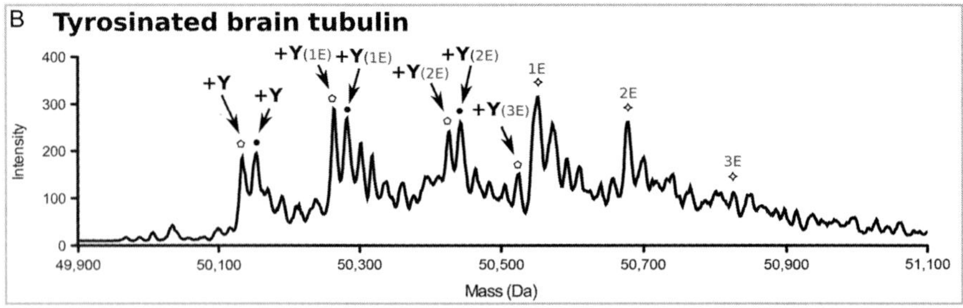

Figure 9.4 (A) Reverse-phase LC–MS analysis of porcine brain tubulin. Tubulin isoforms are indicated by their respective symbols. Additional α- and β-tubulin isoforms are present; however, they elute at different time points and are not shown. The tyrosination status of each isoform is indicated in bold black font, while glutamylation is indicated in gray. The inset shows the progression of tyrosination monitored by Western blot using antibodies specific for tyrosinated tubulin. α-Tubulin, gray, tyrosinated tubulin, white. The two channels are offset for clarity. (B) Reverse-phase LC–MS analysis of porcine brain tubulin tyrosinated by TTL. Isoforms labeled as in (A).

Incubate TTL with tubulin at a 1:50 ratio for 2 h at RT in tyrosination buffer. Reverse-phase LC–MS analysis reveals all detyrosinated α-tubulin isoforms are converted to the tyrosinated form after 2 h (Fig. 9.4B). The extent of tubulin tyrosination can also be monitored by Western blot with a monoclonal antibody specific for tyrosinated tubulin (TUB-1A2, Sigma; 1:1000 dilution) and following the protocol in Section 3.1. Figure 9.4 shows the signal in the 680 nm channel (white) intensifies over the course of the reaction, indicating increasing levels of tyrosination. The signal in the 800 nm channel (gray) represents total tubulin. The anti-tyrosinated tubulin antibody can be calibrated using a tubulin sample that is 100% tyrosinated.

If removal of TTL from the reaction is desired, cycle the tubulin once to remove TTL as well as any nonpolymerization competent tubulin and aggregates (see below).

Tubulin cycling

Supplement the tyrosinated tubulin with 1 m*M* GTP and 33% (v:v) glycerol. Allow tubulin to polymerize for 40 min at 37 °C. Layer polymerized tubulin on a 60% glycerol cushion in 1 × BRB80. Pellet the microtubules by ultracentrifugation using a prewarmed rotor at 109,000 × *g* for 10 min at 37 °C. Aspirate supernatant and cushion. Rinse microtubule pellet with 1 × BRB80. Incubate pellet on ice for 5 min. Resuspend pellet in 1 × BRB80 supplemented with 1 m*M* DTT. The volume of buffer is chosen based on the desired final tubulin concentration. Incubate on ice for 30 min, intermittently pipetting up and down. Centrifuge sample at 109,000 × *g* for 10 min at 4 °C to remove aggregates and microtubules that did not depolymerize. Collect supernatant and freeze in small aliquots for future use in microtubule dynamic assays or to generate taxol-stabilized microtubules (as described in Section 4.1). For example, one could use differentially tyrosinated tubulin to investigate the role of tyrosination on the kinetics of association to the microtubule of plus end tracking proteins (Bieling et al., 2008).

5. CONCLUSIONS

Here we describe the purification of three recombinant tubulin-modifying enzymes—α-TAT, TTLL7, and TTL—as well as protocols for enzymatic modification of naïve and brain tubulin using these enzymes to produce differentially modified microtubules. These different "flavors" of microtubules can be used in a wide range of biochemical and biophysical assays to systematically dissect the specific effects of acetylation, tyrosination, and glutamylation on microtubule dynamics as well as cellular effectors. For example, does one modification change the time a motor remains associated with the microtubule? Are motors specialized for different tubulin modifications? Does a post-translational modification bias a motor toward one microtubule over another at a junction? Lastly, it has been known for a long time that acetylated and polyglutamylated microtubules have increased stabilities in cells; however, it is not yet clear whether this increased stability is due to these modifications or is an indirect effect of regulators recruited to these microtubules. The ability to make unmodified (Widlund et al., 2012) and homogenously modified (this chapter) microtubules finally enables the investigation of the direct effects of post-translational modifications on microtubule dynamics as well as the identification through proteomic approaches of microtubule regulators that are differentially recruited to post-translationally modified microtubules.

REFERENCES

Akella, J. S., Wloga, D., Kim, J., Starostina, N. G., Lyons-Abbott, S., Morrissette, N. S., et al. (2010). MEC-17 is an alpha-tubulin acetyltransferase. *Nature*, *467*(7312), 218–222.

Alexander, J. E., Hunt, D. F., Lee, M. K., Shabanowitz, J., Michel, H., Berlin, S. C., et al. (1991). Characterization of posttranslational modifications in neuron-specific class III beta-tubulin by mass spectrometry. *Proceedings of the National Academy of Sciences of the United States of America*, *88*(11), 4685–4689.

Barra, H. S., Rodriguez, J. A., Arce, C. A., & Caputto, R. (1973). A soluble preparation from rat brain that incorporates into its own proteins (14C)arginine by a ribonuclease-sensitive system and (14C)tyrosine by a ribonuclease-insensitive system. *Journal of Neurochemistry*, *20*(1), 97–108.

Bieling, P., Kandels-Lewis, S., Telley, I. A., van Dijk, J., Janke, C., & Surrey, T. (2008). CLIP-170 tracks growing microtubule ends by dynamically recognizing composite EB1/tubulin-binding sites. *Journal of Cell Biology*, *183*(7), 1223–1233.

Cueva, J. G., Hsin, J., Huang, K. C., & Goodman, M. B. (2012). Posttranslational acetylation of alpha-tubulin constrains protofilament number in native microtubules. *Current Biology*, *22*(12), 1066–1074.

Edde, B., Rossier, J., Le Caer, J. P., Desbruyeres, E., Gros, F., & Denoulet, P. (1990). Posttranslational glutamylation of alpha-tubulin. *Science*, *247*(4938), 83–85.

Garnham, C. P., & Roll-Mecak, A. (2012). The chemical complexity of cellular microtubules: Tubulin post-translational modification enzymes and their roles in tuning microtubule functions. *Cytoskeleton (Hoboken)*, *69*(7), 442–463.

Gundersen, G. G., Kalnoski, M. H., & Bulinski, J. C. (1984). Distinct populations of microtubules: Tyrosinated and nontyrosinated alpha tubulin are distributed differently in vivo. *Cell*, *38*(3), 779–789.

Howard, J., & Hyman, A. A. (2003). Dynamics and mechanics of the microtubule plus end. *Nature*, *422*, 753–758.

Ikegami, K., Heier, R. L., Taruishi, M., Takagi, H., Mukai, M., Shimma, S., et al. (2007). Loss of alpha-tubulin polyglutamylation in ROSA22 mice is associated with abnormal targeting of KIF1A and modulated synaptic function. *Proceedings of the National Academy of Sciences of the United States of America*, *104*(9), 3213–3218.

Ikegami, K., Mukai, M., Tsuchida, J., Heier, R. L., Macgregor, G. R., & Setou, M. (2006). TTLL7 is a mammalian beta-tubulin polyglutamylase required for growth of MAP2-positive neurites. *Journal of Biological Chemistry*, *281*(41), 30707–30716.

Janke, C., & Bulinski, J. C. (2011). Post-translational regulations of the microtubule cytoskeleton: Mechanisms and functions. *Nature Reviews Molecular Cell Biology*, *12*, 773–786.

Kapust, R. B., Tözsér, J., Fox, J. D., Anderson, D. E., Cherry, S., Copeland, T. D., et al. (2001). Tobacco etch virus protease: Mechanism of autolysis and rational design of stable mutants with wild-type catalytic proficiency. *Protein Engineering*, *14*(12), 993–1000.

Kormendi, V., Szyk, A., Piszczek, G., & Roll-Mecak, A. (2012). Crystal structures of tubulin acetyltransferase reveal a conserved catalytic core and the plasticity of the essential N terminus. *Journal of Biological Chemistry*, *287*(50), 41569–41575.

Lacroix, B., van Dijk, J., Gold, N. D., Guizetti, J., Aldrian-Herrada, G., Rogowski, K., et al. (2010). Tubulin polyglutamylation stimulates spastin-mediated microtubule severing. *Journal of Cell Biology*, *189*(6), 945–954.

L'Hernault, S. W., & Rosenbaum, J. L. (1983). Chlamydomonas alpha-tubulin is post-translationally modified in the flagella during flagellar assembly. *Journal of Cell Biology*, *97*(1), 258–263.

L'Hernault, S. W., & Rosenbaum, J. L. (1985). Chlamydomonas alpha-tubulin is post-translationally modified by acetylation on the epsilon-amino group of a lysine. *Biochemistry*, *24*(2), 473–478.

Miller, L. M., Xiao, H., Burd, B., Horwitz, S. B., Angeletti, R. H., & Verdier-Pinard, P. (2010). Methods in tubulin proteomics. *Methods in Cell Biology*, *95*, 105–126.

Nogales, E. (2000). Structural insights into microtubule function. *Annual Review of Biochemistry*, *69*, 277.

Nogales, E., Whittaker, M., Milligan, R. A., & Downing, K. H. (1999). High-resolution model of microtubule. *Cell*, *96*, 79–88.

Nogales, E., Wolf, S. G., & Downing, K. H. (1998). Structure of the alpha beta tubulin dimer by electron crystallography. *Nature*, *391*(6663), 199–203.

Paturle-Lafanechere, L., Edde, B., Denoulet, P., Van Dorsselaer, A., Mazarguil, H., Le Caer, J. P., et al. (1991). Characterization of a major brain tubulin variant which cannot be tyrosinated. *Biochemistry*, *30*(43), 10523–10528.

Raybin, D., & Flavin, M. (1975). An enzyme tyrosylating alpha-tubulin and its role in microtubule assembly. *Biochemical and Biophysical Research Communications*, *65*(3), 1088–1095.

Raybin, D., & Flavin, M. (1977). Enzyme which specifically adds tyrosine to the alpha chain of tubulin. *Biochemistry*, *16*(10), 2189–2194.

Redeker, V., Le Caer, J. P., Rossier, J., & Prome, J. C. (1991). Structure of the polyglutamyl side chain posttranslationally added to alpha-tubulin. *Journal of Biological Chemistry*, *266*(34), 23461–23466.

Redeker, V., Levilliers, N., Schmitter, J. M., Le Caer, J. P., Rossier, J., Adoutte, A., et al. (1994). Polyglycylation of tubulin: A posttranslational modification in axonemal microtubules. *Science*, *266*(5191), 1688–1691.

Redeker, V., Melki, R., Prome, D., Le Caer, J. P., & Rossier, J. (1992). Structure of tubulin C-terminal domain obtained by subtilisin treatment. The major alpha and beta tubulin isotypes from pig brain are glutamylated. *FEBS Letters*, *313*(2), 185–192.

Roll-Mecak, A., & McNally, F. J. (2010). Microtubule-severing enzymes. *Current Opinion in Cell Biology*, *22*(1), 96–103.

Roll-Mecak, A., & Vale, R. D. (2008). Structural basis of microtubule severing by the hereditary spastic paraplegia protein spastin. *Nature*, *451*(7176), 363–367.

Rudiger, M., Plessman, U., Kloppel, K. D., Wehland, J., & Weber, K. (1992). Class II tubulin, the major brain beta tubulin isotype is polyglutamylated on glutamic acid residue 435. *FEBS Letters*, *308*(1), 101–105.

Schroder, H. C., Wehland, J., & Weber, K. (1985). Purification of brain tubulin-tyrosine ligase by biochemical and immunological methods. *Journal of Cell Biology*, *100*(1), 276–281.

Shida, T., Cueva, J. G., Xu, Z., Goodman, M. B., & Nachury, M. V. (2010). The major alpha-tubulin K40 acetyltransferase alphaTAT1 promotes rapid ciliogenesis and efficient mechanosensation. *Proceedings of the National Academy of Sciences of the United States of America*, *107*(50), 21517–21522.

Slep, K. C., & Vale, R. D. (2007). Structural basis of microtubule plus end tracking by XMAP215, CLIP-170, and EB1. *Molecular Cell*, *27*(6), 976–991.

Soppina, V., Herbstman, J., Skiniotis, G., & Verhey, K. J. (2012). Luminal localization of a-tubulin K40 acetylation by Cryo-EM analysis of Fab-labeled microtubules. *PLoS One*, 7(10).

Sullivan, K. F. (1988). Structure and utilization of tubulin isotypes. *Annual Review of Cell Biology*, *4*, 687–716.

Szyk, A., Deaconescu, A. M., Piszczek, G., & Roll-Mecak, A. (2011). Tubulin tyrosine ligase structure reveals adaptation of an ancient fold to bind and modify tubulin. *Nature Structural and Molecular Biology*, *18*(11), 1250–1258.

Topalidou, I., Keller, C., Kalebic, N., Nguyen, K. C., Somhegyi, H., Politi, K. A., et al. (2012). Genetically separable functions of the MEC-17 tubulin acetyltransferase affect microtubule organization. *Current Biology*, *22*(17), 1057–1065.

van Dijk, J., Rogowski, K., Miro, J., Lacroix, B., Edde, B., & Janke, C. (2007). A targeted multienzyme mechanism for selective microtubule polyglutamylation. *Molecular Cell*, *26*(3), 437–448.

Weisenberg, R. C. (1972). Microtubule formation in vitro in solutions containing low calcium concentrations. *Science*, *177*, 1104–1105.

Widlund, P. O., Podolski, M., Reber, S., Alper, J., Storch, M., Hyman, A. A., et al. (2012). One-step purification of assembly-competent tubulin from diverse eukaryotic sources. *Molecular Biology of the Cell*, *23*, 4393–4401.

Wloga, D., & Gaertig, J. (2010). Post-translational modifications of microtubules. *Journal of Cell Science*, *123*(Pt. 20), 3447–3455.

Wolff, A., de Néchaud, B., Chillet, D., Mazarguil, H., Desbruyères, E., Audebert, S., Eddé, B., Gros, F., & Denoulet, P. (1992). Distribution of glutamylated alpha and beta-tubulin in mouse tissues using specific monoclonal antibody, GT335. *Eur J Cell Biol*, *59*(2), 425–432.

SECTION III

Molecular Motor Ensembles on Natural and Engineered Cargoes

CHAPTER TEN

Engineering Defined Motor Ensembles with DNA Origami

Brian S. Goodman, Samara L. Reck-Peterson[1]
Department of Cell Biology, Harvard Medical School, Boston, Massachusetts, USA
[1]Corresponding author: e-mail address: reck-peterson@hms.harvard.edu

Contents

Abstract

Many cytoskeletal motors function in groups to coordinate the spatial and temporal positioning of cellular cargo. While methods to study the biophysical properties of single motors are well established, methods to understand how multiple motors work synergistically or antagonistically are less well developed. Here, we describe a three-dimensional synthetic cargo structure made using DNA origami, which can be used to template defined numbers and types of cytoskeletal motors with programmable geometries and spacing. We describe methods for building the DNA origami structure, covalently attaching motors to DNA, forming the motor–DNA origami structure complex, and single-molecule assays to examine the motile properties of motor ensembles.

Methods in Enzymology, Volume 540
ISSN 0076-6879
http://dx.doi.org/10.1016/B978-0-12-397924-7.00010-8

1. INTRODUCTION

Cytoskeletal motors and the dynamic filaments they move on generate the organization critical for a cell to maintain homeostasis, grow, divide, and communicate with neighboring cells. Motors carry out these functions either by directly transporting cargos or by generating pulling forces (Hirokawa, Niwa, & Tanaka, 2010; Vale, 2003). In this review, we focus on microtubule-based motors, but the same methods could be applied to the actin-based myosin motors. Cytoplasmic dynein and kinesin are the microtubule-based motors that move cargos toward the minus and plus ends of microtubules, respectively. The ultimate spatial and temporal distribution of cargos depends on the arrangement, number, and type of motors on the cargos (Bryantseva & Zhapparova, 2012; Welte, 2004). While great strides have been made in understanding the mechanisms of single cytoskeletal motor proteins (Block, 2007; Vale, 2003), much less is known about the mechanisms governing motor ensembles, which are important in physiological contexts. Bridging the gap between observations of single motors and motor ensembles is critical to understand the biophysical mechanisms that govern how multiple motors coordinate or interfere to produce the complex motility observed in cells. While some methods to study motor ensembles *in vitro* have been described (Bieling, Telley, Piehler, & Surrey, 2008; Mallik, Petrov, Lex, King, & Gross, 2005; Vale, Malik, & Brown, 1992; Vershinin, Carter, Razafsky, King, & Gross, 2007), it has not been possible to control the absolute numbers of motors using these methods. The use of DNA nanotechnology to build synthetic cargos offers a platform for exquisite control over motor number, type, and spacing (Derr et al., 2012; Rogers, Driver, Constantinou, Kenneth Jamison, & Diehl, 2009). DNA origami (Shih & Lin, 2010), building three-dimensional (3D) shapes out of DNA, is modular and allows for sophisticated structures to be designed, built, and modified rapidly and easily. Alternatively, a duplex DNA scaffold can be used to template one or two motors as described in Chapter 11 by Rogers and colleagues.

In this chapter, we describe techniques to template motor proteins on a DNA origami structure. These methods can be used to understand the biophysical mechanisms of motor proteins *in vitro* with single-molecule motility assays (Fig. 10.1). We provide methods for folding and purifying the 3D DNA origami structure, attaching oligonucleotides (oligos) to motor proteins, forming the motor–DNA origami structure complex, and single-molecule

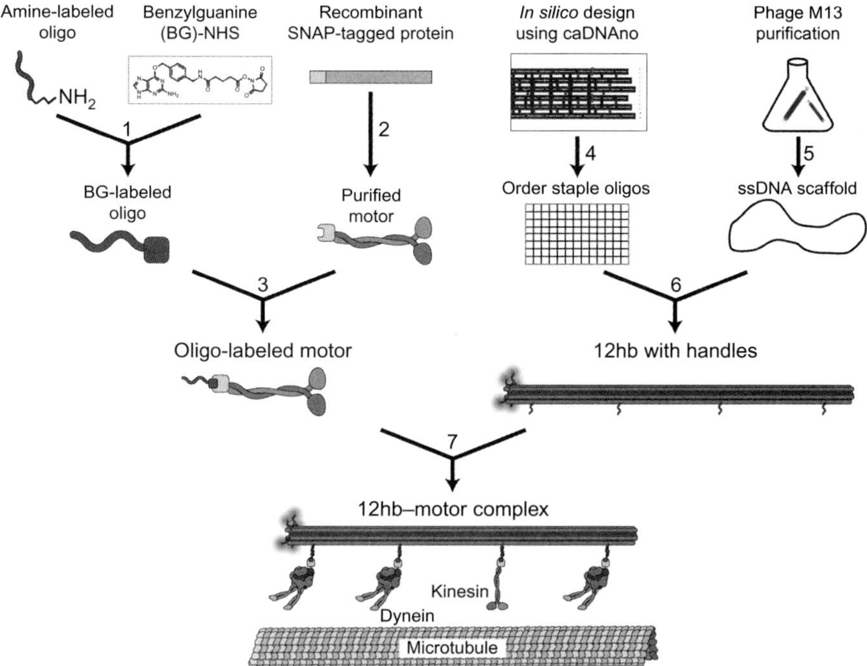

Figure 10.1 Production and assembly of motor–DNA origami structure complexes. (1) Oligo-labeled motors are generated by first coupling amine-labeled oligos to benzylguanine (BG, the SNAP ligand) through NHS ester chemistry. (2) Recombinant SNAP-labeled motor proteins are purified using affinity purification. (3) BG-oligos react with SNAP-tagged motors to generate oligo-labeled motors. (4) The 12hb DNA origami structure is designed using caDNAno software, which generates a list of oligo staples appropriate for a given scaffold sequence. (5) An M13 phage DNA purification generates the ssDNA-scaffolding strand. (6) Staple oligos are mixed with the scaffold strand, and a folding procedure generates the 12hb DNA origami structure. (7) Oligo-labeled motors are mixed with the 12hb DNA origami structure (bearing handle sites complimentary to the motor oligos) to generate a 12hb–motor complex. (See the color plate.)

assays to determine the motile properties of motor ensembles. Throughout we discuss practical and theoretical considerations and methods for validation and quality control. Updates to these techniques can be found online (https://reck-peterson.med.harvard.edu/protocols).

2. BUILDING THE DNA ORIGAMI STRUCTURE

There is an expanding list of nanoscale devices that can be built using 2D and 3D DNA origami (Douglas, Bachelet, & Church, 2012;

Langecker et al., 2012; Rothemund, 2006; Shih & Lin, 2010). Software, tutorials, and other resources for sharing and building DNA origami structures exist, creating a lower barrier for adapting this technology (Douglas et al., 2009). While DNA origami can be used to build structures of varying geometries (Dietz, Douglas, & Shih, 2009; Douglas et al., 2009; Ke, Voigt, Gothelf, & Shih, 2012; Shih & Lin, 2010), the 12-helix bundle (12hb) described here is a simple, straightforward design (Fig. 10.2A) (Derr et al., 2012). It is composed of six inner helixes surrounded by six outer helixes. All origami structures are constructed by folding a long, single-stranded (ss) piece of scaffold DNA by the annealing of short (~20–60 nucleotides) ss oligos, called staples, which bind to noncontiguous sequences on the scaffolding strand. In our design, ss oligos protruding from the structure, called handles, offer sites for motor, fluorophore, or biotin attachment. Handles can be designed every 14 nm on any of the six outer faces by including nonhomologous sequence to the 3′ end of the staple strands. Staples that are positioned in the structure at handle locations are referred to as either positive handles (those that contain a handle sequence) or negative handles (those lacking a handle sequence). Thus, there are 15 sites on each of the six faces of the outer helixes that allow for up to 90 uniquely addressable motor attachment sites. We avoid placing handles on the ends of the origami structure (positions 0 and 14; Fig. 10.2A), as these staples are the least stably associated with the structure (B. Goodman, N. Derr, W. Shih, & S. Reck-Peterson, unpublished data).

The ssDNA scaffold used to fold the 3D origami structure is purified from M13 phage as previously described (Bellot, McClintock, Chou, & Shih, 2013). 273 short oligo staples are required to fold the DNA origami structure. The sequence of the M13 genome we use (p8064; Bellot et al., 2013) and the complementary oligo staples can be found online (https://reck-peterson.med.harvard.edu/protocols). Any chemical moiety conjugated to antihandle oligos (an oligo that is complementary to the handle sequence) can be added to the structure, provided it can withstand the high temperatures associated with the folding protocol. In our hands, fluorophores, biotin, and photocleavable linkers remain functional after the folding process. New protocols for enhanced folding have been developed and could be considered for optimizing yield and fidelity (Sobczak, Martin, Gerling, & Dietz, 2012).

The methods described here have been optimized for our 12hb structure, although similar methods can be used to fold and purify other DNA

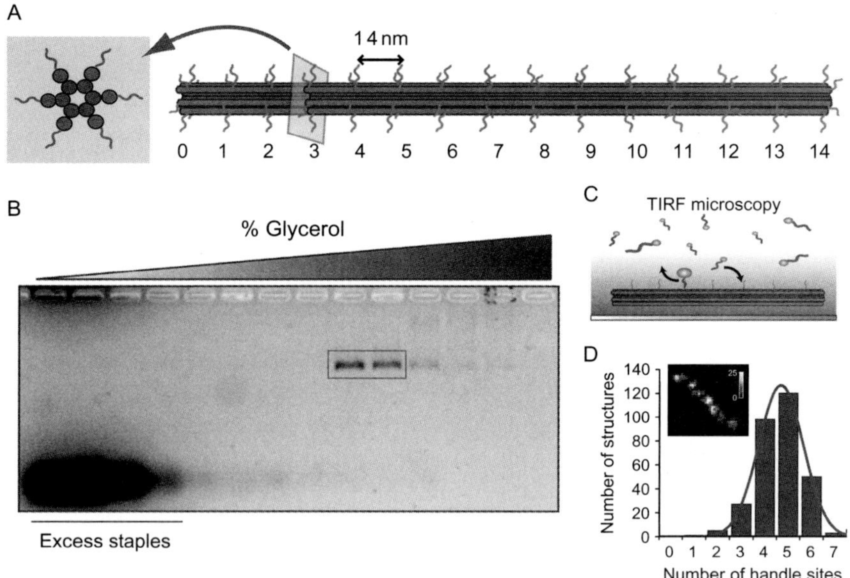

Figure 10.2 Design and analysis of the 12hb DNA origami structure. (A) Design of the 12hb DNA origami structure. The 12hb is composed of six inner helixes and six outer helixes. The structure is 14 nm in diameter and 225 nm in length with optional handle sites every 14 nm. (B) 12hb folding is analyzed by glycerol gradients. After folding, the 12hb is sedimented through a 10–45% glycerol gradient as described (Lin, Perrault, Kwak, Graf, & Shih, 2013) and fractions from the gradient are run on a 2% agarose TBE gel. The excess staples remain in the lower percentage glycerol fractions, while the denser folded 12hb is found in the higher percentage glycerol fractions (box). (C) Handle strand incorporation into the 12hb is assessed by a DNA PAINT experiment (Derr et al., 2012; Jungmann et al., 2010). In this technique, short (10 bp), fluorescent antihandles base pair with the handles of surface-immobilized 12hb structures (Jungmann et al., 2010). Due to the short stretch of homology, stochastic binding of the labeled antihandles generates the ability to perform high-precision localization over many frames of a movie. (D) A DNA PAINT experiment of a 12hb bearing seven handles. A histogram of the number of handle sites observed reveals that the majority of structures have five or more assessable handles. The inset shows a pseudo-image of the handle locations generated from a movie of a single 12hb structure. (See the color plate.)

origami structures. Magnesium concentrations for folding different structures must be optimized empirically as described earlier (Douglas, Dietz, et al., 2009). New origami structures can be designed using caDNAno software (cadnano.org) (Douglas, Marblestone, et al., 2009), and additional design principles are discussed in the literature (Castro et al., 2011; Dietz et al., 2009; Ke et al., 2012; Shih & Lin, 2010).

2.1. Folding the DNA origami structure

1. Mix the following reagents at room temperature:
   - 2.5 μl of 20× folding buffer.
   - 5 μl of 1 μ*M* p8064 scaffold (final scaffold concentration is 100 n*M*). For methods for purifying the p8064 scaffold, see Bellot et al. (2013).
   - Core staples to a final concentration of 600 n*M* (Table S1 in Derr et al., 2012).
   - Negative handle staples (Table S2 in Derr et al., 2012) to a final concentration of 600 n*M*.
   - Positive handle staples (Table S3 in Derr et al., 2012) to a final concentration of 10 μ*M*. All staple sequences can be found online (https://reck-peterson.med.harvard.edu/protocols).
   - Fluorophore-conjugated oligos complementary to the handle strands, referred to as antihandles, can be included to label the chassis. If included, they should be added to a final concentration of 10 μ*M*.
   - Add ddH_2O to bring the final volume to 50 μl.
2. Using a thermal cycler, run the following protocol:
   - 80 °C for 5 min.
   - 14 additional cycles of 5 min with 1 °C temperature decrease per cycle.
   - 65 °C for 30 min.
   - 35 additional cycles of 30 min with 1 °C temperature decrease per cycle.
   - Hold at 4 °C.
3. Proceed directly to the purification of the folded origami structure to remove excess staples (recommended). Alternatively, the folded structure can be stored at 4 °C (do not freeze the folded structures) for several days.
4. Notes:
   - Either a positive or a negative handle staple (but not both) should be used at each handle site position. This ensures that the 12hb structure folds properly as the positive and negative handle staples also have sequence required for binding the scaffolding strand.
   - Core staples should have a final concentration of 6× the p8064 scaffold.
   - All positive handle staples should have a final concentration of 100× the p8064 scaffold.

- Folding and purification can be monitored using 2% agarose gel electrophoresis. Gels and running buffers should be supplemented with 11 m*M* $MgCl_2$. Gels may need to be run in water baths to prevent melting of the DNA origami structures. Melted structures appear as a smear rather than a tight band.

Solutions

1. 20 × Folding buffer
   - 100 m*M* Tris, pH 8.0
   - 20 m*M* EDTA
   - 320 m*M* $MgCl_2$

2.2. Purification of the folded DNA origami structure

1. The folded 12hb origami structure is purified away from excess staple strands using velocity gradient centrifugation as described earlier (Lin et al., 2013).
2. Use Amicon 100K MWCO filters (Millipore) for buffer exchange, glycerol removal, and to concentrate the DNA origami structures.
3. The folded structure should be stored in buffer with at least 11 m*M* $MgCl_2$ at 4 °C. Lower $MgCl_2$ concentrations or temperatures can lead to unfolding.

2.3. Determining the integrity of the folded structure

To determine the integrity of the folded DNA origami structure, several methods can be used. Here we briefly describe agarose gel electrophoresis, transmission electron microscopy (TEM), and DNA PAINT.

1. Agarose gel electrophoresis can be used to visualize the folded chassis compared to the initial input in the folding reaction. The folded chassis will run faster in the gel. Run the folded chassis on a 2% agarose, 1 × TBE gel containing 11 m*M* $MgCl_2$, and 0.7 μg/ml ethidium bromide. The $MgCl_2$ should be added after boiling the agarose in TBE buffer. Run the gel at a constant voltage of 70 V for 90 min (Fig. 10.2B). Running the gel at a voltage >70 V will cause the gel to overheat, causing the DNA origami structure to denature and run as a smeared band. We routinely place our gel apparatus in an icebox to dissipate heat and exchange the running buffer if the gel is run for longer than 2 h.
2. TEM techniques can be used to assess the structure using previously described methods (Douglas, Marblestone, et al., 2009).

3. DNA PAINT techniques can be used to observe handle incorporation (Fig. 10.2C) as described previously (Derr et al., 2012; Jungmann et al., 2010). To immobilize the DNA origami on a cover slip, biotin-labeled handles are included during the folding reaction. The biotin-labeled origami structures can then be adhered to avidin-coated cover slips. Previously, using the DNA PAINT method we found that handles are incorporated with 80% efficiency. Modifying staple length and other design principles may improve the efficiency of handle incorporation (Sobczak et al., 2012).

Solutions

1. 10× TBE buffer
   - 450 m*M* Tris, pH 8.1
   - 450 m*M* Boric acid
   - 10 m*M* EDTA
2. Agarose gel
   - 2% Agarose
   - 1× TBE buffer
   - Boil to dissolve agarose
   - Add $MgCl_2$ to 11 m*M*
3. Gradient buffers
   - 1× TBE buffer
   - 11 m*M* $MgCl_2$
   - 15–45% glycerol
4. Agarose gel running buffer
   - 1× TBE buffer
   - 11 m*M* $MgCl_2$

3. COVALENT ATTACHMENT OF OLIGONUCLEOTIDES TO MOTOR PROTEINS

We use the SNAP-tag technology (NEB) to attach oligos to motor proteins (Keppler et al., 2003). The motor proteins we express and purify are fusion proteins with the SNAP-tag (Derr et al., 2012; Qiu et al., 2012). The SNAP-tag is appended to the cargo-binding region (or "tail") of the motor. The first step in linking a motor to DNA is to link antihandle oligos (oligos that will base pair with positive handle staples on the origami structure) to the SNAP ligand, benzylguanine (BG). The BG-oligo conjugate is then linked to the motor-SNAP-tag fusion protein to produce an oligo-linked motor. The oligo-linked motors are then attached to the 12hb origami structure via base pairing.

3.1. Conjugating BG-NHS to an amine-oligonucleotide

1. Using a needle, add nonaqueous DMSO to a vial containing dry BG-NHS (NEB) to a final concentration of 20 m*M*. All reagents should be at room temperature to reduce condensation. Unused BG-NHS can be stored in a tube sealed with parafilm for up to 4 months at −20 °C.
2. Resuspend the amine-functionalized oligos in water to a final concentration of 2 m*M*. Take a small sample (~1 μl) and dilute 1:3000 for a gel sample (pre).
3. Mix and incubate the following at room temperature for 30 min:
   - 4 μl of 2 m*M* amine-oligo in aqueous solution
   - 8 μl of 200 m*M* HEPES, pH 8.5
   - 12 μl of 20 m*M* BG-NHS in DMSO
4. Run the reacted mixture through a 0.1-μm filter (Ultrafree-MC with Durapore membrane; Millipore) to remove any precipitate from the solution. Take a small sample (~1 μl) and dilute 1:500 for a gel sample (post).
5. Run 5 μl of "pre" and "post" oligos on a 20% polyacrylamide TBE gel (Life Sciences) at 200 V for 65 min as specified (http://www.lifetechnologies.com/order/catalog/product/EC6315BOX).
6. Stain the gel with Sybr Gold (Invitrogen) following the standard protocol (Fig. 10.3A; http://probes.invitrogen.com/media/pis/mp11494.pdf).
7. Use appropriate software (e.g., ImageJ) to quantify the percentage of oligo that was successfully linked to the BG ligand.

3.2. Purification of BG-oligonucleotides

Purification of BG-oligos away from unreacted BG-NHS is critical because unreacted BG-NHS can competitively inhibit downstream labeling of the protein through reaction with the SNAP-tag.

1. Prepare four gel filtration columns (Micro Bio-Spin 6 Columns; Bio-Rad). For Bio-Rad spin column information, see http://www.bio-rad.com/prd/en/US/LSR/SKU/732-6221/Micro-Bio-Spin-6-Columns and http://www.bio-rad.com/webroot/web/pdf/lsr/literature/LIT-507G.pdf. Use four columns for each 24 μl of oligo. For larger-scale BG-oligo preparations, we have used size-exclusion chromatography columns prepared with Bio-Gel P-6DG Gel (Bio-Rad) in conjunction with an FPLC system.

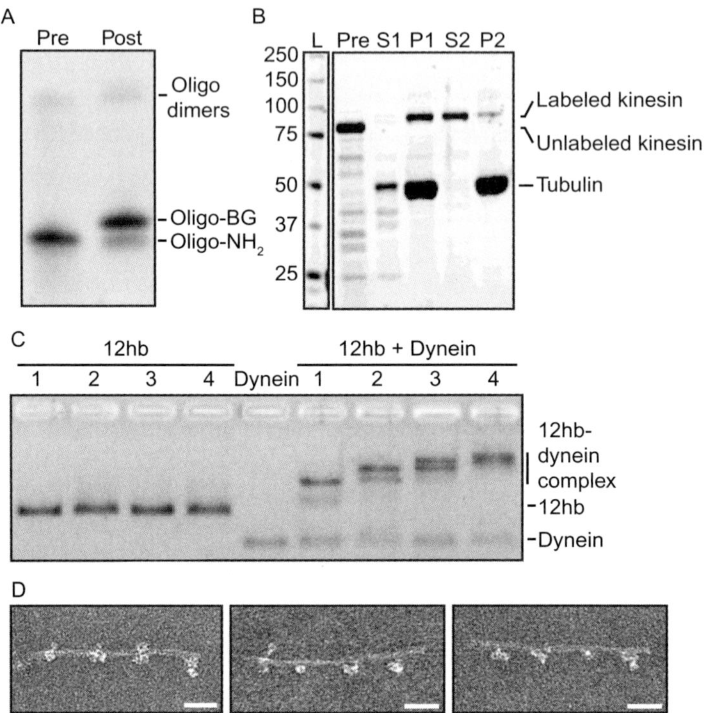

Figure 10.3 Oligo labeling of motors and formation of the 12hb–motor complex. (A) Labeling of oligos with BG is assessed by 10% agarose TBE gel electrophoresis. A mobility shift of the oligo after addition of BG is observed. Oligo dimers from cross-linking during oligo synthesis are also visible at the top of the gel. "Pre" and "post" refer to samples that were taken before or after the reaction, respectively. (B) Labeling of purified SNAP-tagged kinesin motors with BG-oligos is assessed by 4–12% PAGE. SNAP-tagged kinesin motors (pre) are mixed with BG-oligos and microtubules in the presence of AMPPNP and centrifuged. Most of the BG-oligo-SNAP-motor complex is found in the pellet (P1) fraction. ATP is added to the pellet fraction to release the motor from microtubules and the mixture is centrifuged a second time. Most of the motor is now found in the supernatant (S2) fraction. A small gel shift of the BG-oligo-labeled SNAP-tagged kinesin is observed. (C) Attachment of oligo-labeled motors to the 12hb structure is assessed by 2% agarose TBE gel electrophoresis. 12hb origami structures with 1, 2, 3, or 4 oligo-dynein attachment sites reveal an upward mobility shift that corresponds to the programmed handle number when mixed with antihandle oligo-labeled dynein. TAMRA-labeled 12hb structures are visualized. (D) Attachment of oligo-labeled motors to the 12hb origami structure can also be analyzed by transmission electron microscopy (Derr et al., 2012). Negatively stained dynein motors can be seen bound to a 12hb with four dynein attachment sites. Scale bars: 40 nm.

2. Remove excess packing buffer with the following protocol:
   - Prepare and assemble the column as directed by the product manual.
   - Centrifuge for 2 min at 1000 × *g* at 4 °C. Discard the packaging buffer.
   - Exchange the column buffer with the desired buffer for protein labeling by adding 500 μl of buffer to the gel and centrifuging for 1 min at 1000 × *g* at 4 °C. Discard the buffer and repeat the column washing procedure three additional times.
3. To purify the BG-oligos from excess free BG-NHS, apply 24 μl of the sample to the first column. Centrifuge for 1 min at 1000 × *g* at 4 °C and collect the eluate containing the BG-oligos. Repeat this procedure three additional times.
4. Use a UV spectrophotometer to measure the final oligo concentration.
5. Store the BG-oligos at −20 °C.

3.3. Labeling SNAP-tagged motor proteins with BG-oligonucleotides

We use two different strategies for labeling motors with oligos. For SNAP-ZZ-tagged dynein purified from yeast, we label the protein while it is attached to IgG sepharose beads during the purification (Qiu et al., 2012). For SNAP-6×HiS-tagged kinesin (Case, Pierce, Hom-Booher, Hart, & Vale, 1997), we use a microtubule affinity step after the nickel column purification because the positive charge on the nickel column binds negatively charged oligos (Derr et al., 2012). It is critical to remove the unreacted BG-oligo from the motor-oligo product to ensure that the unreacted oligo does not compete with the motor-oligo for binding to the 12hb origami structure.

1. Protocols for polymerizing tubulin to form microtubules can be found online (http://mitchison.med.harvard.edu/protocols.html).
2. Mix the following components. The final concentrations are listed in parentheses.
   - 50 μl of 3.5 μ*M* kinesin (2 μm)
   - 22 μl of ~240 μ*M* BG-oligo (60 μ*M*)
3. Incubate the mixture at room temperature for 15 min. Add the following *in order*. The final molar concentration of each component is listed in parentheses.
   - 0.85 μl of 0.1 *M* AMPPNP (1 m*M*)
   - 0.17 μl of 10 m*M* taxol prepared in DMSO (20 μ*M*)
   - 0.85 μl of 660 units/ml Apyrase (6.6 units/ml)

- 11 μl of 50 μ*M* microtubules (6.5 μ*M*). It is important that taxol is in the buffer before the microtubules are added.
- ddH_20 to a final volume of 85 μl

4. Mix by gently flicking the tube (to avoid microtubule shearing) and incubate at room temperature for 15 min.
5. Centrifuge the mixture through 110 μl of a 60% glycerol cushion in a TLA100 (Beckman) or equivalent rotor at 108,000 × *g* at 25 °C for 15 min.
6. Remove the top layer and a small amount (~10 μl) of the top of the cushion. Save 4 μl for SDS-PAGE analysis (S1).
7. Gently wash the sides of the tube and glycerol cushion by adding 110 μl of BRB12 + taxol above the glycerol cushion and removing the wash and some of the cushion. Repeat three times until the entire cushion is removed without disturbing the pellet.
8. Wash the pellet three times with BRB12 + taxol without disturbing the pellet.
9. Resuspend the pellet containing kinesins bound to microtubules in 50 μl of release buffer. Save 3 μl for SDS-PAGE analysis (P1).
10. Incubate the mixture for 5 min at room temperature.
11. Pellet the microtubules by spinning at 278,000 × *g* for 6 min at 25 °C in a TLA 100 rotor (Beckman) or equivalent.
12. Remove the supernatant and add 10 μl of BRB12 + taxol containing 60% sucrose to a final concentration of 10% sucrose. Save 3.6 μl (post-sucrose addition) of the supernatant for SDS-PAGE analysis (S2).
13. The final kinesin-oligo-containing supernatant should be aliquoted, flash-frozen in liquid nitrogen, and stored at −80 °C. There will be a mixture of labeled and unlabeled kinesin if the reaction has not gone to completion, but only the labeled kinesin will bind to the DNA origami structure.
14. Resuspend the pellet in 50 μl of BRB12. Save 3 μl for a gel sample (P2).
15. Analyze the fractions by SDS-PAGE and SYPRO Red (Life Sciences) staining (Fig. 10.3B) alongside a protein concentration standard to measure the labeling efficiency (the percent of kinesin that shifts to a higher molecular weight due to covalent linkage to the ~8 kDa oligo) and yield (the final concentration of S2 sample by densitometry compared to a protein standard).

Solutions

1. BRB12
   - 12 m*M* PIPES, pH 6.8

- 1 m*M* EGTA
- 2 m*M* $MgCl_2$

2. BRB80
   - 80 m*M* PIPES, pH 6.8
   - 1 m*M* EGTA
   - 2 m*M* $MgCl_2$
3. BRB12 + taxol
   - 1 × BRB12
   - 1 m*M* DTT
   - 20 μ*M* Taxol
4. 60% Glycerol cushion
   - 60% Glycerol
   - 1 × BRB12
   - 1 m*M* DTT
   - 20 μ*M* Taxol
5. Release buffer
   - 30 m*M* KCl
   - 1 × BRB80
   - 1 m*M* DTT
   - 5 m*M* ATP-$MgCl_2$
   - 20 n*M* Taxol
6. 60% Sucrose BRB12
   - 60% Sucrose
   - 1 × BRB12
   - 1 m*M* DTT

4. FORMING THE MOTOR–DNA ORIGAMI COMPLEX

The motor–DNA complex is formed by mixing the origami structure with the oligo-linked motors. Base pairing of the complementary positive handle and antihandle sequences links the motors to the 12hb.

1. Mix the following on ice for 30 min:
   - 5 μl of 10 n*M* 12hb origami structure
   - 5 μl of >800 n*M* oligo-labeled motor
   - (Optional) 5 μl of >800 n*M* of a second oligo-labeled motor (e.g., opposite polarity motor or mutant motor).
2. To assess complex formation, run the 12hb alone, the motor alone, and the motor–12hb complex on a 2% agarose gel in 1 × TBE supplemented with 11 m*M* $MgCl_2$ and 0.5% lithium dodecyl sulfate (LDS). Gel

samples are mixed with 4 × NuPAGE LDS sample buffer (Life Sciences) and incubated on ice for 5 min.

3. Run the gel at 70 V for 90 min at room temperature (Fig. 10.3C).
4. Complexes can also be observed on an EM grid with standard negative stain techniques (Fig. 10.3D) as previously described (Derr et al., 2012).
5. If excess motors will interfere with downstream procedures, a gel filtration step can be performed using Sephacryl S-500 HR resin (GE Healthcare) in a microchromatography spin column (Bio-Rad):
   - Dilute the 12hb–motor complex in Dynein assay buffer (DAB) to a final volume of 50 μl and apply to a 450 μl Sephacryl S-500 HR column.
   - Centrifuge for 10 s at 1000 × *g* and collect the supernatant.

Solutions

1. DAB
   - 30 m*M* HEPES, pH 7.4
   - 50 m*M* K-acetate
   - 2 m*M* Mg-acetate
   - 1 m*M* EGTA, pH 7.5
   - 10% Glycerol

5. SINGLE-MOLECULE MOTILITY ANALYSIS OF MOTOR–DNA ORIGAMI COMPLEXES

To determine the motile properties of motor–DNA origami complexes, we use single-molecule motility assays, which can provide the velocity, run length, and stepping behavior of individual motor-driven cargos. For analysis of microtubule-based motors, taxol-stabilized microtubules are immobilized to a cover slip via biotin–streptavidin (described in detail in Gennerich & Reck-Peterson, 2011). Fluorescently labeled motor–cargo complexes are then observed moving along microtubules in the presence of ATP using time-lapse, total internal reflection fluorescence (TIRF) microscopy (described in detail in Derr et al., 2012).

One important consideration is the positioning of the fluorescent dyes on the 12hb structure and interpretation of the data based on these positions. In our experiments, we have positioned 5 oligo-linked dyes at the end of the 225-nm-long DNA origami structure. Because the structure can rotate around a single microtubule-motor attachment, if a motor at the end opposite the fluorophore attachment site is the only motor making microtubule contact, jumps of 450 nm in the fluorescent signal can occur. This becomes particularly important when interpreting bidirectional movements or high-precision stepping experiments.

A variation of the standard motility assay involves attaching photocleavable positive handle oligos to the 12hb structure such that the role of individual motor types can be assessed dynamically. We used this experimental design to determine whether a tug-of-war was occurring between dynein and kinesin motors attached to the same DNA cargo structure (Derr et al., 2012). In this experiment, either dyneins or kinesins are attached to the 12hb structure by photocleavable handles. These handles have a nitrobenzyl spacer group (Integrated DNA Technologies) inserted just prior to the portion of the handle that will hybridize to the motor antihandle, providing a means to sever any designated motor from the origami structure with 405-nm laser light. In our experiments, we used photocleavage and release of one motor type to determine the role of the remaining motor type (Fig. 10.4A) (Derr et al., 2012). In this work, cargos containing photocleavable kinesins were labeled with TAMRA, and cargos with photocleavable dyneins were labeled with Cy5. The following protocol outlines a photocleavage experiment.

1. Prepare a flow cell with biotinylated and fluorophore-labeled microtubules immobilized on the surface via a biotin–streptavidin sandwich as previously described (Gennerich & Reck-Peterson, 2011).
2. Dilute the 12hb–motor complex to single-molecule conditions (~10–100 p*M*) in motility buffer containing 2.5 mg/ml casein and 1 × oxygen scavenger mix in a final volume of 20 μl.
3. Flow the above mixture into the motility chamber and seal the chamber with vacuum grease.
4. Image the chamber immediately as the oxygen scavenger system will acidify the buffer after ~1 h. Buffer with low buffer capacity such as BRB12 can only be used for ~20 min.
5. We use an Olympus IX-81 inverted objective TIRF microscope (Qiu et al., 2012) equipped with 405-, 561-, and 640-nm lasers (Coherent). A dual-band sputtered emission filter (z561/635rpc and etCy3/Cy5m, Chroma) in the main optical path is used to image rapidly between the 561- and 640-nm channels. The 561-nm laser is controlled by an acousto-optical tunable filter (NEOS; 10 ms response time) and the 640-nm laser is controlled by a fast mechanical shutter (SmartShutter, Sutter; 25 ms response time).
6. For experiments with yeast dynein and human kinesin (Derr et al., 2012), we have imaged the 561-nm (cargo bearing photocleavable kinesins) and 640-nm channels (cargo bearing photocleavable dyneins)

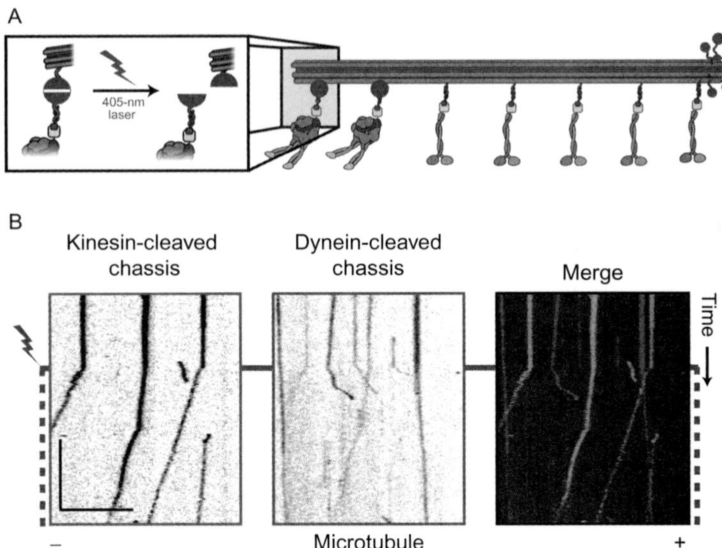

Figure 10.4 Photocleavable oligos allow release of motors from the 12hb. (A) To program the removal of specific motors from the 12hb structure, photocleavable linkers are added to select handle strands (circles). Illumination with 405-nm light induces photocleavage of the motors bearing the photocleavable linker (inset). (B) Two types of 12hb structures with different fluorophores were designed for simultaneous imaging. Kinesin cleavable, mixed-motor 12hb structures (left) were labeled with Cy5 and had photocleavable linkers on the kinesin handle strands. Dynein cleavable, mixed-motor 12hb structures (center) were labeled with TAMRA and contained dynein handle strands with photocleavable linkers. After 1 min of imaging both the TAMRA and Cy5 channels, a 405-nm laser was pulsed in between frames to induce cleavage (dotted line) showing that 12hb structures move in the expected direction when the opposing motor is cleaved. Different colored fluorophores allow the simultaneous observation of the two different motor ensembles during the same experiment (right). Vertical scale bar: 1 min; horizontal scale bar: 10 μm. (See the color plate.)

at 1 image/s, with 100 ms exposure times in each channel with 1.5 mW of power measured at the objective. To induce photocleavage, after 1 min of imaging the 405-nm laser (0.5 mW at the objective) is pulsed at 1 image/s, with 400 ms exposure times. Imaging is continued for a total of 4 min. As a control, we image motor–cargo complexes lacking the photocleavage handles.

7. To measure the kinetics of photocleavage, experiments with immobile motor–cargo complexes are performed in motility chambers lacking ATP (Derr et al., 2012).

Solutions

1. DAB + casein
   - Supplement DAB with casein to a final concentration of 2.5 mg/ml.
2. Oxygen scavenger mix (Yildiz et al., 2003)
   - 50 μl of 1 × DAB
   - 15 μl of 20 mg/ml glucose catalase (Roche)
   - 7.5 mg glucose oxidase (*Aspergillus niger*; Sigma)
   - Vortex the mixture and then centrifuge it at 16,100 × *g* for 5 min at 4 °C.
   - Remove the supernatant and aliquot, freeze in liquid nitrogen, and store at −80 °C.

6. DATA ANALYSIS CONSIDERATIONS

Single-molecule behavior of the motor–cargo complexes is analyzed by generating kymographs (position vs. time plots, Fig. 10.4B) using ImageJ. Automated particle tracking software can also be used (Jaqaman et al., 2008; Ruhnow, Zwicker, & Diez, 2011). Analysis of motor ensembles, as opposed to individual motors, raises some challenges, including very long run lengths and bidirectional behavior. Motor–cargo complexes containing multiple dynein or kinesin motors that we have studied typically generate run lengths >10 μm (Derr et al., 2012). Thus, to achieve accurate run lengths, we analyze long microtubules (typically >65 μm) for sufficient amounts of time (Derr et al., 2012). To generate long microtubules, we avoid pipetting microtubules through small-gauge pipette tips and mix microtubule solutions by gently flicking the tubes. Raising the ionic strength of the buffer can also be used to shorten the run lengths of dynein and kinesin ensembles as both motors' microtubule binding is based on ionic interactions (Redwine et al., 2012). To analyze potential bidirectional motility of motor–cargo complexes containing both dyneins and kinesins, it is important to consider the position of the fluorescent labels on the DNA origami cargo as described in Section 5.

7. SUMMARY AND FUTURE DIRECTIONS

It is likely that motor ensembles drive most cargo transport *in vivo* (Gross, 2004). Thus, the ability to control the number, type, and spatial arrangement of motors on a synthetic cargo structure is an important tool for determining the biophysical properties governing movements by motor ensembles. For example, by using DNA origami to program motor ensembles of one to seven identical polarity motors (either dyneins or kinesins), we determined that motor number had very little affect on directional velocity (Derr et al., 2012). We also programmed DNA origami structures to bind varying ratios of dynein and kinesin. Many of these mixed-motor ensembles were immotile; however, by using a photocleavage experiment, we found that this apparent tug-of-war could be resolved by removing one motor species (Derr et al., 2012).

DNA origami is an ideal method for these types of experiments because it offers precise engineering control of shape at scales that are biologically relevant (Douglas et al., 2012; Douglas, Dietz, et al., 2009; Douglas, Marblestone, et al., 2009; Langecker et al., 2012; Shih & Lin, 2010). Spherical objects can now be designed, which may more accurately resemble organellar cargos (Erickson, Jia, Gross, & Yu, 2011). DNA origami also allows the incorporation of a variety of chemistries to gain dynamic control. In addition to photocleavable groups, *cis–trans* isomerization groups, disulfide bridges, DNA aptamers, and other controllable chemistries could be applied in the future to create dynamic systems. The rapidly decreasing costs of oligos and freely available software to design DNA origami structures should make this method feasible and approachable for many laboratories.

DNA origami methods are a valuable addition to the biophysicist's toolbox for the study of groups of actin- or microtubule-based molecular motors. We envision future experiments that will determine, in greater detail, the biophysical properties of motor ensembles. For example, linking origami structures to objects that can be trapped by optical tweezers will allow force production to be probed in the context of varying numbers or types of motors. Understanding how regulatory proteins affect motor ensembles will also be an important area where synthetic DNA technology can aid in modeling the complexity found in physiological contexts. Finally, while methods to build DNA or RNA structures *in vivo* or introduce folded nucleic acid structures into cells are in their infancy (Guo, 2010; Pinheiro, Han, Shih, & Yan, 2011), breakthroughs in these areas will open doors for using DNA origami to probe motor function in cells.

ACKNOWLEDGMENTS

We thank Bret Redwine for comments on the manuscript and Nathan Derr who along with B. S. G. developed the protocols described. S. L. R.-P. is funded by NIH Grant RGM100947A.

REFERENCES

Bellot, G., McClintock, M. A., Chou, J. J., & Shih, W. M. (2013). DNA nanotubes for NMR structure determination of membrane proteins. *Nature Protocols*, *8*(4), 755–770.

Bieling, P., Telley, I. A., Piehler, J., & Surrey, T. (2008). Processive kinesins require loose mechanical coupling for efficient collective motility. *EMBO Reports*, *9*(11), 1121–1127.

Block, S. M. (2007). Kinesin motor mechanics: Binding, stepping, tracking, gating, and limping. *Biophysical Journal*, *92*(9), 2986–2995.

Bryantseva, S. A., & Zhapparova, O. N. (2012). Bidirectional transport of organelles: Unity and struggle of opposing motors. *Cell Biology International*, *36*(1), 1–6.

Case, R. B., Pierce, D. W., Hom-Booher, N., Hart, C. L., & Vale, R. D. (1997). The directional preference of kinesin motors is specified by an element outside of the motor catalytic domain. *Cell*, *90*(5), 959–966.

Castro, C. E., Kilchherr, F., Kim, D.-N., Shiao, E. L., Wauer, T., Wortmann, P., et al. (2011). A primer to scaffolded DNA origami. *Nature Methods*, *8*(3), 221–229.

Derr, N. D., Goodman, B. S., Jungmann, R., Leschziner, A. E., Shih, W. M., & Reck-Peterson, S. L. (2012). Tug-of-war in motor protein ensembles revealed with a programmable DNA origami scaffold. *Science (New York, NY)*, *338*(6107), 662–665.

Dietz, H., Douglas, S. M., & Shih, W. M. (2009). Folding DNA into twisted and curved nanoscale shapes. *Science (New York, NY)*, *325*(5941), 725–730.

Douglas, S. M., Bachelet, I., & Church, G. M. (2012). A logic-gated nanorobot for targeted transport of molecular payloads. *Science (New York, NY)*, *335*(6070), 831–834.

Douglas, S. M., Dietz, H., Liedl, T., Högberg, B., Graf, F., & Shih, W. M. (2009). Self-assembly of DNA into nanoscale three-dimensional shapes. *Nature*, *459*(7245), 414–418.

Douglas, S. M., Marblestone, A. H., Teerapittayanon, S., Vazquez, A., Church, G. M., & Shih, W. M. (2009). Rapid prototyping of 3D DNA-origami shapes with caDNAno. *Nucleic Acids Research*, *37*(15), 5001–5006.

Erickson, R. P., Jia, Z., Gross, S. P., & Yu, C. C. (2011). How molecular motors are arranged on a cargo is important for vesicular transport. *PLoS Computational Biology*, 7(5), e1002032.

Gennerich, A., & Reck-Peterson, S. L. (2011). Probing the force generation and stepping behavior of cytoplasmic Dynein. *Methods in Molecular Biology (Clifton, NJ)*, *783*, 63–80.

Gross, S. P. (2004). Hither and yon: A review of bi-directional microtubule-based transport. *Physical Biology*, *1*(1–2), R1–R11.

Guo, P. (2010). The emerging field of RNA nanotechnology. *Nature Nanotechnology*, *5*(12), 833–842.

Hirokawa, N., Niwa, S., & Tanaka, Y. (2010). Molecular motors in neurons: Transport mechanisms and roles in brain function, development, and disease. *Neuron*, *68*(4), 610–638.

Jaqaman, K., Loerke, D., Mettlen, M., Kuwata, H., Grinstein, S., Schmid, S. L., et al. (2008). Robust single-particle tracking in live-cell time-lapse sequences. *Nature Methods*, *5*(8), 695–702.

Jungmann, R., Steinhauer, C., Scheible, M., Kuzyk, A., Tinnefeld, P., & Simmel, F. C. (2010). Single-molecule kinetics and super-resolution microscopy by fluorescence imaging of transient binding on DNA origami. *Nano Letters*, *10*(11), 4756–4761.

Ke, Y., Voigt, N. V., Gothelf, K. V., & Shih, W. M. (2012). Multilayer DNA origami packed on hexagonal and hybrid lattices. *Journal of the American Chemical Society*, *134*(3), 1770–1774.

Keppler, A., Gendreizig, S., Gronemeyer, T., Pick, H., Vogel, H., & Johnsson, K. (2003). A general method for the covalent labeling of fusion proteins with small molecules in vivo. *Nature Biotechnology*, *21*(1), 86–89.

Langecker, M., Arnaut, V., Martin, T. G., List, J., Renner, S., Mayer, M., et al. (2012). Synthetic lipid membrane channels formed by designed DNA nanostructures. *Science (New York, NY)*, *338*(6109), 932–936.

Lin, C., Perrault, S. D., Kwak, M., Graf, F., & Shih, W. M. (2013). Purification of DNA-origami nanostructures by rate-zonal centrifugation. *Nucleic Acids Research*, *41*(2), e40.

Mallik, R., Petrov, D., Lex, S. A., King, S. J., & Gross, S. P. (2005). Building complexity: An in vitro study of cytoplasmic dynein with in vivo implications. *Current Biology*, *15*(23), 2075–2085.

Pinheiro, A. V., Han, D., Shih, W. M., & Yan, H. (2011). Challenges and opportunities for structural DNA nanotechnology. *Nature Nanotechnology*, *6*(12), 763–772.

Qiu, W., Derr, N. D., Goodman, B. S., Villa, E., Wu, D., Shih, W., et al. (2012). Dynein achieves processive motion using both stochastic and coordinated stepping. *Nature Structural & Molecular Biology*, *19*(2), 193–200.

Redwine, W. B., Hernández-López, R., Zou, S., Huang, J., Reck-Peterson, S. L., & Leschziner, A. E. (2012). Structural basis for microtubule binding and release by dynein. *Science (New York, NY)*, *337*(6101), 1532–1536.

Rogers, A. R., Driver, J. W., Constantinou, P. E., Kenneth Jamison, D., & Diehl, M. R. (2009). Negative interference dominates collective transport of kinesin motors in the absence of load. *Physical Chemistry Chemical Physics*, *11*(24), 4882–4889.

Rothemund, P. W. K. (2006). Folding DNA to create nanoscale shapes and patterns. *Nature*, *440*(7082), 297–302.

Ruhnow, F., Zwicker, D., & Diez, S. (2011). Tracking single particles and elongated filaments with nanometer precision. *Biophysical Journal*, *100*(11), 2820–2828.

Shih, W. M., & Lin, C. (2010). Knitting complex weaves with DNA origami. *Current Opinion in Structural Biology*, *20*(3), 276–282.

Sobczak, J. P. J., Martin, T. G., Gerling, T., & Dietz, H. (2012). Rapid folding of DNA into nanoscale shapes at constant temperature. *Science (New York, NY)*, *338*(6113), 1458–1461.

Vale, R. D. (2003). The molecular motor toolbox for intracellular transport. *Cell*, *112*(4), 467–480.

Vale, R. D., Malik, F., & Brown, D. (1992). Directional instability of microtubule transport in the presence of kinesin and dynein, two opposite polarity motor proteins. *The Journal of Cell Biology*, *119*(6), 1589–1596.

Vershinin, M., Carter, B. C., Razafsky, D. S., King, S. J., & Gross, S. P. (2007). Multiple-motor based transport and its regulation by Tau. *Proceedings of the National Academy of Sciences of the United States of America*, *104*(1), 87–92.

Welte, M. A. (2004). Bidirectional transport along microtubules. *Current Biology*, *14*(13), R525–R537.

Yildiz, A., Forkey, J. N., McKinney, S. A., Ha, T., Goldman, Y. E., & Selvin, P. R. (2003). Myosin V walks hand-over-hand: Single fluorophore imaging with 1.5-nm localization. *Science (New York, NY)*, *300*(5628), 2061–2065.

CHAPTER ELEVEN

Construction and Analyses of Elastically Coupled Multiple-Motor Systems

Arthur Rogers, Pamela E. Constantinou, D. Kenneth Jamison, Jonathan W. Driver, Michael R. Diehl[1]

Departments of Chemistry and Bioengineering, Rice University, Houston, Texas, USA
[1]Corresponding author: e-mail address: diehl@rice.edu

Contents

Abstract

Precision analyses of the collective motor behaviors have become important to dissecting mechanisms underlying the trafficking of subcellular commodities in eukaryotic cells. Here, we describe a synthetic approach to create structurally defined multiple protein complexes containing two elastically coupled motor molecules. Motors are connected using a simple DNA-scaffolding molecule and DNA-conjugated, artificial protein polymers that function as tunable elastic linkers. The procedure to self-assemble these components produces complexes in high synthetic yield and allows individual multiple-motor systems to be interrogated at the single-complex level. Methods to evaluate cooperative motor responses in a static optical trap are also discussed. While enabling the average transport properties of single/noninteracting and coupled motors to be compared, these procedures can provide insight into the extent to which motors cooperate productively via load sharing as well as the roles loading-rate-dependent phenomena play in collective motor functions.

Methods in Enzymology, Volume 540
ISSN 0076-6879
http://dx.doi.org/10.1016/B978-0-12-397924-7.00011-X

1. INTRODUCTION

The development of single-molecule techniques has opened important avenues to resolve foundational principles underlying how molecular motors function while transporting subcellular objects directionally in the viscoelastic environment of the cytoplasm (Carter & Cross, 2005; Howard, Hudspeth, & Vale, 1989; Schnitzer, Visscher, & Block, 2000). Yet, the motions of a large variety of vesicles, organelles, and signaling complexes are driven by teams of interacting motors (Rogers, Tint, Fanapour, & Gelfand, 1997; Ross, Wallace, Shuman, Goldman, & Holzbaur, 2006). Collective motor functions may be important during specific transport challenges requiring the production of large forces. Regulated competitions among oppositely polarized motors and even collections of motors that move along different types of filament tracks can support bidirectional modes of cargo motion and are essential components of intracellular trafficking processes (Gennerich & Schild, 2006). The importance of these collective behaviors has motivated the development of new classes of biophysical assays aimed at elucidating how interactions among and between motors influence cargo motility and are regulated.

In vitro analyses of collective motor behaviors have been examined primarily by controlling the average density of motors bound to solid supports (e.g., glass slides in filament gliding assays and polystyrene beads in cargo tracking/optical trapping experiments) (Beeg et al., 2008; Bieling, Telley, Piehler, & Surrey, 2008; Leduc, Ruhnow, Howard, & Diez, 2007). This strategy has been employed to characterize collective motor-stepping behaviors (Leduc et al., 2007) and to evaluate how motor-support linkage elasticity affects processive motor motion (Bieling et al., 2008). Similar approaches have been applied to examine collective motor force production (Beeg et al., 2008; Schroeder, Mitchell, Shuman, Holzbaur, & Goldman, 2010). For example, a study by Schroeder et al. (2010) demonstrated the maximum force reached in an optical trap by a cargo functionalized with multiple dynein, and myosin V motors can be predictive of whether cargos will continue move along the same filament or switch transport directions at actin–microtubule filament junctions. These studies highlight the importance of collective motor functions in intracellular trafficking. Nevertheless, additional control over the composition and organization of each motor system is required to determine how trafficking responses depend on motor copy number, ratio and organization, as well as the mechanical (elastic) properties of cargos. Direct quantification of these dependencies is important to defining how sensitively transport processes can be regulated by modulating motor numbers and activities. It is also important for determining how motor mutants associated with disease alter the composite trafficking behaviors of cargos whose motions are driven simultaneously by collections of wild-type motors.

Several new synthetic approaches have been developed that provide control over the local coupling of motors on a cargo. These methods typically employ a multivalent macromolecular scaffold that can function as a template to define the nanometer-scale organization of motor complexes. Scaffolds have been designed using engineered protein polymers (Diehl, Zhang, Lee, & Tirrell, 2006), antibodies (Xu, Shu, King, & Gross, 2012), and self-assembled DNA nanostructures (Derr et al., 2012; Furuta et al., 2013). Here, we describe a method to prepare motor complexes containing two elastically coupled motor molecules using a simple DNA scaffold (Jamison, Driver, Rogers, Constantinou, & Diehl, 2010; Lu et al., 2012; Rogers, Driver, Constantinou, Jamison, & Diehl, 2009). We also review procedures to characterize the force-dependent dynamics of these motor systems at the single-complex level using optical trapping techniques (Jamison, Driver, & Diehl, 2012; Jamison et al., 2010). Given the richness and complexities of collective motor behaviors, even if only two-motor molecules are involved in cargo transport, we feel these assays are best performed hand-in-hand with detailed mechanical and chemo-kinetic modeling procedures (Driver et al., 2010; Klumpp & Lipowsky, 2005). Details surrounding the computational framework of these procedures are described elsewhere (Driver et al., 2011; Uppulury et al., 2012). Here, we highlight measurements needed to implement these theories and to gain mechanistic insight into the multiple-motor force generation.

2. SYNTHETIC STRATEGY: A MODULAR PLUG-AND-PLAY APPROACH

2.1. Basic strategy

DNA-templated protein complexes/assemblies containing two elastically coupled motors are synthesized using modular building blocks that associate via noncovalent linkages (Fig. 11.1). They are composed of three basic components: a partially hybridized DNA polymer that functions as a molecular scaffold and provides control over intermotor spacing, two DNA-conjugated artificial protein polymers (Z_R-ELS$_6$-ssDNA) (Fig. 11.1A) that

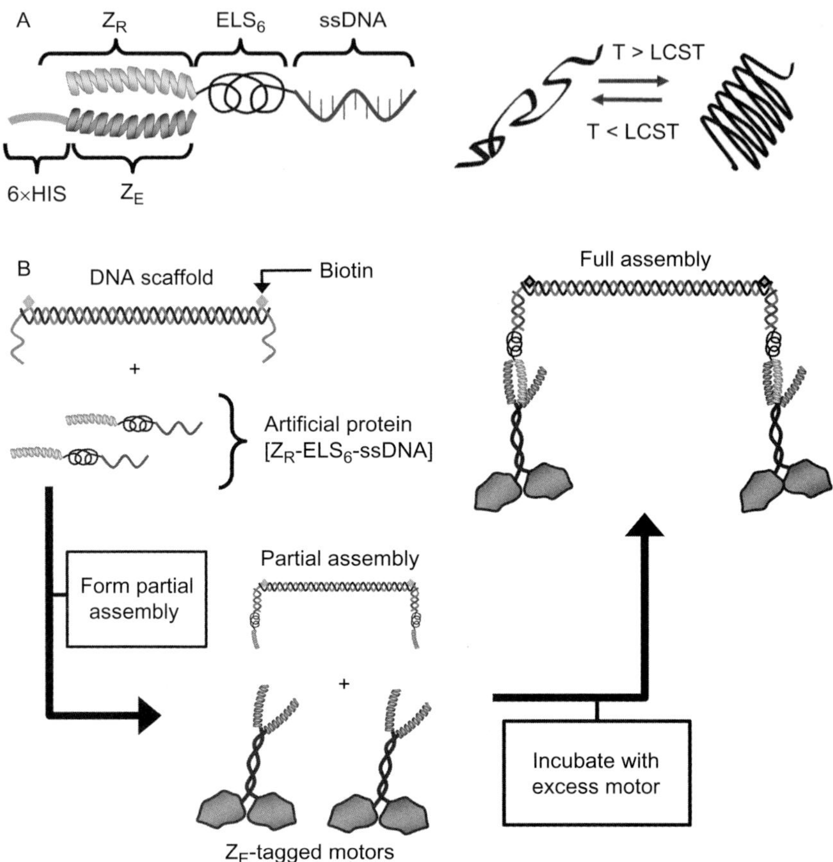

Figure 11.1 Self-assembly of DNA-templated multiple-motor complexes. (A) The domain structure of the polymer linkers that connect motors to the DNA scaffold. The thermally driven condensation of the elastin-like ELS_6 polymer domain is depicted at the right. (B) The protocol to assemble two-motor complexes. Once fully formed, these assemblies can be labeled with quantum dots and anchored to bead surfaces. (See the color plate.)

attach to the unhybridized ends of the DNA scaffold and function as genetically tunable elastic linkages, and recombinant motors (e.g., k560-Z_E) possessing terminal acidic leucine zipper fusions (Z_E) that associate strongly with a basic zipper complement (Z_R) within the artificial protein linker. Intermotor distances and the elasticity of motor connections are tuned by controlling the lengths of the DNA and Z_R-ELS_6-ssDNA, respectively. Each of these components is synthesized and purified to homogeneity. This strategy was adopted since the overall synthetic yield of complex formation will depend multiplicatively on the purity of each individual component.

2.2. DNA scaffold design and synthesis

Most of the DNA scaffolds employed in Jamison et al. (2010), Lu et al. (2012), and Rogers et al. (2009) were formed from two 170-bp oligonucleotides that hybridize partially to create a central dsDNA duplex flanked by two, 22-bp ssDNA "overhangs." The length of this duplex determines the spacing between motor attachment points. Biotin and other small molecules can be incorporated into one or both terminal ends of the duplex to facilitate labeling with quantum dots or the anchoring of complexes to solid supports (e.g., streptavidin-functionalized beads). Our scaffolds typically contain 148 matched bp (14 helical turns) to set an intermotor distance of ~50 nm, the persistence length of DNA. The overhangs extending from each end of the scaffold hybridize with the ssDNA tags conjugated to the Z_R-ELS_6-ssDNA polymers. Two thiamine bases are placed at the junction of the central duplex and the overhangs; these bases remain unhybridized after the full assembly is formed so that junctions between the polymer/motors linkage and scaffold remain rotationally flexible.

Strand sequences are designed at the domain level using software programs such as SEQUIN to minimize potential secondary structures and undesired interactions between domains (e.g., between the ssDNA domains used for motor attachment). The 170-base DNA strands were each created from four separate ssDNA oligomers (~40 bp) that are PAGE-purified and ligated together via splint ligation (Liu, Barrick, Szostak, & Roberts, 2000). This approach minimizes problems associated with strand sequence and length errors, which occur more prevalently during the synthesis of long oligonucleotides. The resultant full-length ssDNA strands are purified and thermally annealed to form the complete scaffold. Duplex formation and the absence of excess ssDNA are confirmed using nondenaturing PAGE-gel analysis. Strand stoichiometries can be adjusted to account for potential ssDNA concentration determination errors. Annealed scaffolds can also be gel-purified, if necessary.

In principle, it is possible to prepare complexes possessing longer or shorter central. However, bundled duplexes may be required to generate scaffolds longer than 50 nm. In Chapter 14, Goodman and Reck-Peterson describe an approach called DNA origami that allows large and rigid scaffolds to be created. Flexible DNA scaffolds can also be prepared by increasing the number of unmatched bases between the DNA overhangs. However, these elements will exhibit significant strain-induced stiffening. As shown in Lu et al. (2012), the net mechanical properties of a complex may therefore be effectively equivalent to fully duplexed scaffolds.

2.3. Artificial protein polymer linkers

While motors can be connected to the DNA scaffolds by simply conjugating them directly with ssDNA tags and this strategy has been used successfully to create multiple-motor complexes (Derr et al., 2012; Furuta et al., 2013), our approach employs Z_R-ELS_6-ssDNA polymers as intermediate, noncovalent linkages between motors and the DNA scaffold (Fig. 11.1A). This strategy was chosen since it allows motors to be coupled with ssDNA tags near quantitatively during a single incubation step, and alleviates the need for post-expression processing procedures (e.g., DNA conjugation and conjugate purification steps) that can potentially affect the motor activity levels. The Z_R-ELS_6-ssDNA polymers also function as tunable elastic linkages and allow much more compliant multiple-motor complexes to be prepared compared to those integrating direct motor–scaffold linkages. Such control is required to examine the impact of intermotor mechanical coupling strength on collective motor responses.

The design of the polymer linkers used previously to create multiple kinesin-1 complexes is as follows. The ssDNA tag used for motor anchoring to the scaffold overhangs is conjugated to a C-terminal cysteine using maleimide chemistry. Motors bind to the polymer's N-terminus via a Z_E/Z_R zipper complex that associates in a parallel orientation. The Z_R-ELS_6-ssDNA polymers are therefore best suited for constructing assemblies of motors possessing N-terminal tail domains, as is the case for kinesin-1. Building multiple-motor complexes containing motors with C-terminal catalytic domains may require the cloning of polymers possessing an N-terminal cysteine and a C-terminal Z_R peptide.

The ELS component of the polymers possesses a poly(VPGV$_\alpha$G) sequence and is a polypeptide mimetic of the protein elastin. Single-molecule pulling experiments have demonstrated several mimetics of elastin behave as near-ideal elastomers with reversible force-extension curves (Urry et al., 2002). Additionally, solutions of elastin-like polypeptides undergo thermally responsive phase transitions where the polymer solutions transition from soluble to condensed phases as the temperature is raised above the polymer's lower critical solution temperature (LCST) (Urry, 1988). This process is accompanied by a reduction in polymer extension length (Fig. 11.2A) (Urry et al., 2002). The LCST depends on the net hydrophobicity of the elastin mimetic and can be controlled by replacing the V_α amino acids in the polymer backbone with hydrophobic or charged residues. Our early polymer scaffolds incorporated phenylalanine (F) substitutions to set the LCST slightly below room temperature; thus, motor coupling could

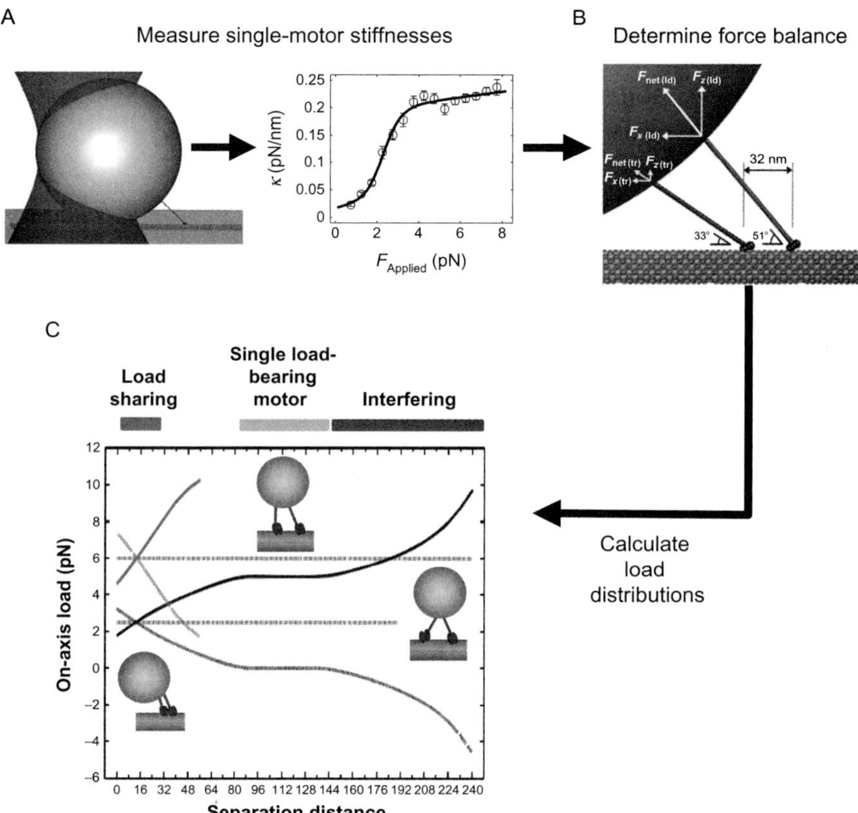

Figure 11.2 Mechanical analyses to assess configuration-dependent load distributions within two-motor complexes. (A) Nonlinear stiffness of single motor–bead linkages can be measured in an optical trap by performing power spectral analyses of bead trajectories. (B) These measurements are used as inputs in a mechanical model to calculate bead positions, and motor extension lengths for different bound configurations assuming force balance in each configurations. (C) Repeating this process for a range of applied loads and binding configurations allows binding-state-dependent load distributions to be calculated. Details needed to implement these calculations are described in Driver et al. (2011). (For color version of this figure, the reader is referred to the online version of this chapter.)

be tuned in temperature-dependent motility assays (Diehl et al., 2006). The Z_R-ELS_6-ssDNA used for two-kinesin tracking (Rogers et al., 2009) and trapping experiments (Jamison et al., 2012, 2010) was designed to remain uncondensed up to and above 37 °C. These polymers incorporated two serine (S) substitutions, yielding the ELS fragment sequence: (VPGVG VPGSG VPGVG VPGSG VPGVG).

The Z_R-ELS_6-C proteins can be expressed in *E. coli* from engineered genes. These constructs are generated by first building a gene encoding for a Z_R-ELS_1-C polymer using PCR and synthetic oligonucleotides whose sequences are designed using publicly available software (DNAWorks). The ELS fragment can then be polymerized into larger ELS_N polypeptide units in sequential cloning reactions that employ recursive and directional ligation procedures (McDaniel, MacKay, Quiroz, & Chilkoti, 2010).

Proteolytic degradation of the Z_R-ELS_6-C polymer occurs when it is expressed in *E. coli* due to the high charge and low homodimeric affinity ($K_D \sim 10^{-3}$ *M*) of the Z_R peptide. This issue can be circumvented by coexpressing the Z_R-ELS_6-C polymer with a 6×His-Z_E peptide. With this adaptation, full-length linker polymers can be expressed at 37 °C using standard media. The exceptionally high heterodimeric affinity of the Z_E/Z_R complex ($K_D \sim 10^{-15}$ *M*) also allows the 6×His-Z_E to be used as a purification tag under denaturing conditions. For example, the polymers can be bound to Ni^{2+}-NTA resin in urea-based cell lysis buffers (8 *M* urea, 10 m*M* Tris–HCl, 100 m*M* NaH_2PO_4, pH 8.0). The columns can then be washed with the same buffer, and the Z_R-ELS_6-C protein can be eluted using 6 *M* guanidine–HCl, 100 m*M* NaH_2PO_4, 10 m*M* Tris–HCl, pH 8.0. This buffer disassociates the Z_E/Z_R complex, allowing the Z_R-$(ELS)_6$-C to be eluted while leaving the majority of the Z_E-6×His proteins on the column. Excess Z_E-6×His protein can be removed by passing the solution through a second smaller Ni^{2+}-NTA column (~1/10 the first column volume). The resultant polymer is dialyzed into distilled water and lyophilized for storage. The protein purity can be verified using SDS-PAGE and matrix-assisted laser desorption/ionization time-of-flight mass spectroscopy.

DNA conjugation can be performed using amine-modified oligonucleotides and the heterobifunctional cross-linking reagent succinimidyl 4-[*N*-maleimidomethyl] cyclohexane-1-carboxylate. Even though this chemistry is inefficient, it is inexpensive to implement, and the reaction can be scaled up to facilitate purification of the conjugate to near homogeneity using fast protein liquid chromatography and a HiTrap QXL ion exchange column. Afterward, the conjugates should be lyophilized, dispersed in Tris–acetate buffer, and aliquoted for storage at 20 °C. Repeated freeze thaw cycles should be avoided, as they can degrade the ssDNA conjugate.

2.4. Multiple-motor complex assembly

Two-motor assemblies are constructed in sequential incubation steps. In the first step, the Z_R-ELS_6-ssDNA polymers are connected to the DNA scaffold by thermally annealing both components at a 1:2 stoichiometry (Fig. 11.1). Synthetic yields can be quantified via PAGE-gel analyses, and the incubation procedure can be adjusted in the case either component is found to be in excess due to protein/DNA concentration determination errors. Stock solutions of partial assemblies can be aliquoted and stored at 20 °C. Complete two-kinesin assemblies should be prepared immediately before assays are performed. This step simply requires a slight excess of Z_E-tagged motors to be incubated with the scaffold (motor/Z_R ratio ~ 4–10).

3. TWO-MOTOR BIOPHYSICAL ASSAYS

3.1. Methods overview

The objectives of many multiple-motor assays are to evaluate the extent to which grouping motors enhances cargo run lengths, velocities, and detachment forces relative to those of single motors. Again, an advantage of the procedures described above is that they allow individual two-motor assemblies examined. Incorporating small-molecule fluorophores into the scaffolds or outfitting them with quantum dots allows enhancements in multiple-motor run lengths to be measured (Rogers et al., 2009). Motors can also be labeled near their catalytic domains to examine collective motor-stepping behaviors and distributions of the spacing between their motor domains when both motors are engaged in transport (Lu et al., 2012). These assays are performed similarly to those developed for tracking single-motor molecules. Alternatively, the DNA scaffolds can be anchored to beads for optical trapping assays that monitor multiple-motor force generation (Jamison et al., 2010).

3.2. Multiple-motor optical trapping analyses

Many multiple-motor assays have shown that the gains in run lengths, load-dependent velocities, and detachment forces resulting from collective motor functions can be relatively small, on average. For example, reported single and multiple kinesin-1 detachment force distributions are remarkably similar (Furuta et al., 2013; Jamison et al., 2010). This behavior occurs since the number of filament-associated motors is often lower than the total number of motors incorporated into a complex. Moreover, the number of motors engaged in force production is not necessarily the same as the number of filament-bound motors since a multiple-motor complex can bind its filament via a spectrum of configurations where the motors are positioned at different filament lattice sites. Given the vectorial nature of the forces experienced by each motor, the front or leading motor in a complex will bear the majority of the applied load on a cargo in many of these configurations. In these circumstances, cargo velocities and detachment behaviors may be indistinguishable from those of single-motor molecules. Structurally organized synthetic complexes provide important opportunities to examine these aspects of collective motor dynamics since they allow cooperative gains to be quantified and potential mechanistic sources leading to weak cooperative (negative cooperative) responses to be dissected more reliably than approaches where motor number is unknown.

Since the force production gains produced by grouping motors together can be quite small, specific analyses of features within optical trapping traces are often needed to evaluate whether the conditions supporting the trapping of individual motor complexes have been established, and to confirm complete complex formation. As with single-motor experiments, the percent of beads that are found to be capable of moving along microtubule filaments provides a useful metric to evaluate the probability transport can be attributed to the action of an individual, multiple-motor complex. Analyses of events indicating partial assembly detachment and transitions between low and high cargo velocity states can be examined to assess complex formation. The identification of these signatures provides a foundation to compare the detachment forces and velocities of two-motor complexes with those of single-motor molecules.

Measurements of the distribution of filament-bound configurations adopted by a complex under the applied load of the trap are also needed to gain mechanistic insight into collective motor responses. This key aspect of multiple-motor dynamics can be examined by characterizing the elasticities of individual motor–bead linkages and using this information to evaluate configuration-dependent distributions of forces between a complex's motors (Fig. 11.2). Importantly, these computational analyses have shown that multiple-motor complexes will exhibit significant deviations from steady state (Driver et al., 2011). These differences arise when load distributions between motors change slowly compared to the rate the total load of the trap changes due to cargo advancement. It is therefore important to consider the roles of this unique loading-rate-dependent phenomenon. Effects stemming from the spatially dependent properties of loads should also be considered. Our approach to characterizing these effects is outlined below.

(a) *Partial complex detachment.* The partial detachment of a bead due to the unbinding of only one motor in a complex can result in large-amplitude rearward displacements whose magnitudes depend on the positions of each motor bound to the filament prior to detachment. These events constitute useful signatures of multiple-motor engagement. Nevertheless, the size of these displacements will not necessarily equate to the scaffold lengths since the on-filament motor spacing, the force distribution between motors, motor–bead linkage elasticity, and the bead size also influence rearward displacement sizes. Additionally, as discussed by Driver et al. (2011), partial detachment events can affect force-dependent distributions of configurations adopted by a complex, and the role of these effects will depend on the trap stiffness since this parameter defines how rapidly loads change spatially.

(b) *Motor complex stiffness and load distributions.* Direct characterization of single- and multiple-motor spring constants (stiffness) is necessary for evaluating distributions of applied loads within a complex during transport (Fig. 11.2). While the Z_R-ELS_6-ssDNA linker is generally very compliant, it exhibits nonlinear force-extension behaviors due to strain-induced stiffening. Force-dependent, single-motor and motor complex stiffness can be estimated by performing power spectrum analyses on trajectory components that fall within specified bins, as described by Jamison et al. (2010). These measurements can then be used as inputs into mechanical models that employ energy minimization algorithms to calculate load distributions for different filament-bound configurations, the extent to which motor linkages stretch in each configuration, as well as the position of and torque on the bead in the trap (Driver et al., 2011). Dependencies of bead displacement sizes on the applied load and the bound configuration of a complex can also be predicted using this computational approach (Uppulury et al., 2012).

(c) *Detachment force distributions.* Evaluating the average force that a multiple-motor complex can produce is essential to determining how effectively motors can cooperate. This property is, in our view, best characterized by analyzing complete distributions of cargo unbinding events. Detachment force distributions are calculated by determining the peak force a cargo reached in a trap regardless of the time spent at that force or whether the cargo detached completely or partially from the filament. Of note, cargo detachment force and cargo stalling force distributions may be significantly different since stall force distributions only reflect moments in traces where the motors shared their applied load, and not the full spectrum of states that a complex is capable of adopting under load. Whether or not stalling forces are additive (e.g., two motors would stall at twice the force of one) can therefore have little bearing on average detachment forces.

As a key test, distributions of cargo detachment forces measured for all beads in all traces can be compared to those of a single bead undergoing multiple transport events to ensure ensemble average behaviors in a complex reflect the long-time average behaviors of the complexes. Such comparisons constitute useful tests to further confirm the successful preparation of multiple-motor complexes and establishment of conditions supporting interrogations of individual complexes (e.g., motor complex/bead concentration ratios).

(d) *Multiple-motor velocities and state transitions.* Load-dependent cargo velocities can change rapidly within an optical trapping trace since motor complexes will advance faster when both motors share their load (Fig. 11.3). Average cargo velocities will naturally depend on the

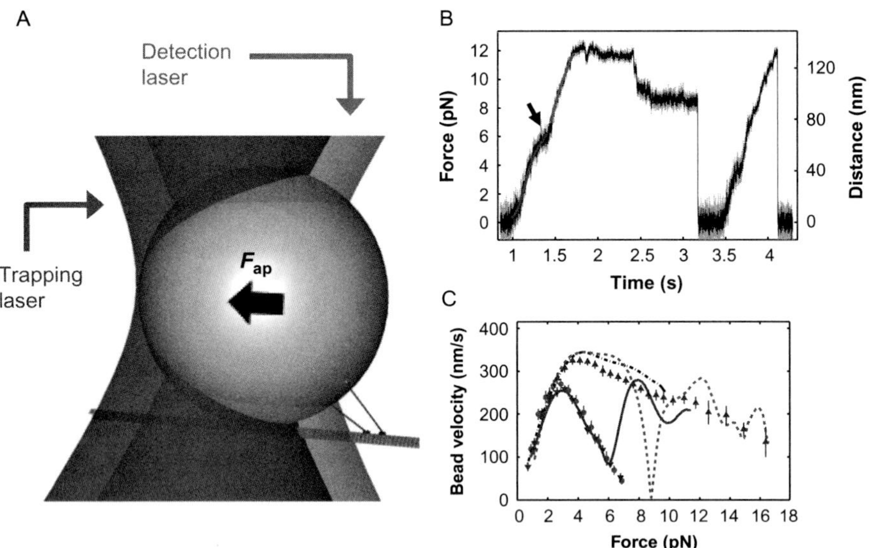

Figure 11.3 Analyses of multiple-motor state transitions. (A) Illustration of the optical trapping assay. (B) Bead velocities can change rapidly in an optical trap when a complex transitions between single (red) and two (blue) load-bearing motor states. (C) Configuration-dependent state transitions are also visible in individual force–velocity traces (solid and dashed lines). Thresholds can be defined to identify "slow" and "fast" components of traces to be calculated, allowing the average velocities of each component to be calculated (downward and upward pointing triangles, respectively). See Jamison et al. (2010) for details surrounding the procedures. (See the color plate.)

amount of time complexes spent in load-sharing states. In a static trap, bead displacement sizes and velocities will also depend on motor linkage elasticity. Since these linkages are compliant at low applied loads and stiffen at higher loads, force–velocity (F–V) relationships possess downward concave curvature (Fig. 11.3C). Measurements or assumptions of motor elasticity are therefore often used to generate F–V curves, characterizing the motion of the motor catalytic domain (Kojima, Muto, Higuchi, & Yanagida, 1997). While this treatment can be applied to multiple-motor data for certain comparisons (e.g., responses in static trapping and force-clamping experiments), it is important to recognize that the "compliance-corrected" velocities will not reflect the actual velocities of the motors since, as discussed above and in Driver et al. (2011), bead displacement sizes are dependent on how load distributions change when individual motors within a complex step, the resultant stretching of motor linkages, and off-center rotations of beads/cargos. Thus, it is often simpler to compare differences in single- and two-motor velocity data using the unadjusted (*elasticity-dependent*) F–V curves.

Transitions between single- and two-load-bearing motor states can be identified by examining the first and second derivative of F–V curves for individual optical trapping traces (Fig. 11.3B and C). Beads accelerate and decelerate during these transitions due to the changing load experienced by each motor. Thresholds can be set for $\mathrm{d}V/\mathrm{d}F$ to identify places in trajectories where these transitions occurred. These thresholds should be set sufficiently high so that no events are identified in single-motor data. The sign of the derivative indicates whether the number of engaged motors increased or decreased. Such analyses are useful to evaluate the percentage of time a complex spends in single-motor-like or cooperatively productive load-sharing states, and thus, provide insight into the force-dependent cooperative abilities of a complex. Of note, hidden Markov modeling approaches similar to those used for step size analyses (Syed, Müllner, Selvin, & Sigworth, 2010) could likely provide an alternative objective approach.

(e) *Transition rate analyses.* The classification of trajectory segments as being associated with single- or multiple load-bearing motor states also allows the rate complex transition between these general classes of states to be calculated (Fig. 11.3B and C). Here, transition rates can be calculated by simply dividing the number of transitions identified within a force bin by the total amount of time a complex remained in a state at that force.

Plots of force-dependent transition rates (e.g., *the partial complex attachment and detachment rates* $k_{ON[1M\to 2M]}(F)$ or $k_{OFF[2M\to 1M]}(F)$) can then be compared to predictions generated by baseline models that assume motors exhibit noncooperative behaviors (*equal load sharing and additive free energies of motor-filament binding: thus,* $k_{OFF[2M\to 1M]}(F) = 2k^{o}_{OFF}e^{F/2F_d}$, where k^{o}_{OFF} is a motor's unloaded detachment rate and F_d is a parameter called the critical detachment force). As discussed in Driver et al. (2011), deviations from these models provide mechanistic insight into how intermotor interactions and loading-rate-dependent effects influence collective motor force production. Of note, seemingly unusual nonmonotonic detachment rates can emerge due to these effects where the application of a load appears to enhance the free energy gain associated with motor-filament binding (see Figs 6 and 3 in Beeg et al., 2008; Driver et al., 2011). Yet, these responses do not necessarily imply phenomena such as force-induced slip-catch bond behavior is occurring since they may simply reflect load distributions between motors change in time as a complex transports cargo against the spatially and temporally varying loads of an optical trap.

4. CONCLUDING REMARKS AND FUTURE PERSPECTIVES

We have described procedures to create and analyze the dynamics of individual multiple-motor complexes using optical trapping techniques. These procedures offer advantages for dissecting the intricate dynamics of multiple-motor systems and evaluating factors influencing their ability to cooperate productively under applied loads, particularly when deviations from idealized behaviors and loading-rate-dependent effects play are influential. Key aspects of these methods can be applied to examine the behaviors of larger complexes containing more than two motors as well as different types of motors and their accessory factors. Such extensions have already yielded significant results (Derr et al., 2012). In this way, these types of synthetic approaches can now provide important avenues to build increasingly complex experimental systems that recapitulate the organization of motors on endogenous cargos.

ACKNOWLEDGMENTS

These procedures were developed with support from the NSF (MCB-0643832), NIH (GM094489-01), and the Welch Foundation (C-1625).

REFERENCES

Beeg, J., Klumpp, S., Dimova, R., Gracia, R. S., Unger, E., & Lipowsky, R. (2008). Transport of beads by several kinesin motors. *Biophysical Journal*, *94*, 532–541.

Bieling, P., Telley, I. A., Piehler, J., & Surrey, T. (2008). Processive kinesins require loose mechanical coupling for efficient collective motility. *EMBO Reports*, *9*, 1121–1127.

Carter, N. J., & Cross, R. A. (2005). Mechanics of the kinesin step. *Nature*, *435*, 308–312.

Derr, N. D., Goodman, B. S., Jungmann, R., Leschziner, A. E., Shih, W. M., & Reck-Peterson, S. L. (2012). Tug-of-war in motor protein ensembles revealed with a programmable DNA origami scaffold. *Science*, *338*, 662–666.

Diehl, M. R., Zhang, K., Lee, H. J., & Tirrell, D. A. (2006). Engineering cooperativity in biomotor-protein assemblies. *Science*, *311*, 1468–1471.

Driver, J. W., Jamison, K. D., Uppulury, K., Rogers, A. R., Kolomeisky, A. B., & Diehl, M. R. (2011). Productive cooperation among processive motors depends inversely on their mechanochemical efficiency. *Biophysical Journal*, *101*, 386–395.

Driver, J. W., Rogers, A. R., Jamison, K. D., Das, R. K., Kolomeisky, A. B., & Diehl, M. R. (2010). Coupling between motor proteins determines dynamic behaviors of motor protein assemblies. *Physical Chemistry Chemical Physics*, *12*, 10398–10405.

Furuta, K., Furuta, A., Toyoshima, Y. Y., Amino, M., Oiwa, K., & Kojima, H. (2013). Measuring collective transport by defined numbers of processive and nonprocessive kinesin motors. *Proceedings of the National Academy of Sciences of the United States of America*, *110*, 501–506.

Gennerich, A., & Schild, D. (2006). Finite-particle tracking reveals submicroscopic-size changes of mitochondria during transport in mitral cell dendrites. *Physical Biology*, *3*, 45–53.

Howard, J., Hudspeth, A. J., & Vale, R. D. (1989). Movement of microtubules by single kinesin molecules. *Nature*, *342*, 154–158.

Jamison, K. D., Driver, J. W., & Diehl, M. R. (2012). Cooperative responses of multiple kinesins to variable and constant loads. *Journal of Biological Chemistry*, *287*, 3357–3365.

Jamison, K. D., Driver, J. W., Rogers, A. R., Constantinou, P. E., & Diehl, M. R. (2010). Two kinesins transport cargo primarily via the action of one motor: Implications for intracellular transport. *Biophysical Journal*, *99*, 2967–2977.

Klumpp, S., & Lipowsky, R. (2005). Cooperative cargo transport by several molecular motors. *Proceedings of the National Academy of Sciences of the United States of America*, *102*, 17284–17289.

Kojima, K., Muto, E., Higuchi, H., & Yanagida, T. (1997). Mechanics of single kinesin molecules measured by optical trapping nanometry. *Biophysical Journal*, *73*, 2012–2022.

Leduc, C., Ruhnow, F., Howard, J., & Diez, S. (2007). Detection of fractional steps in cargo movement by the collective operation of kinesin-1 motors. *Proceedings of the National Academy of Sciences of the United States of America*, *104*, 10847–10852.

Liu, R., Barrick, J. E., Szostak, J. W., & Roberts, R. W. (2000). Optimized synthesis of RNA-protein fusions for in vitro protein selection. *Methods in Enzymology*, *318*, 268–293.

Lu, H., Efremov, A. K., Bookwalter, C. S., Krementsova, E. B., Driver, J. W., Trybus, K. M., et al. (2012). Collective dynamics of elastically-coupled myosin V motors. *Journal of Biological Chemistry*, *287*, 27753–27761.

McDaniel, J. R., MacKay, J. A., Quiroz, F. G., & Chilkoti, A. (2010). Recursive directional ligation by plasmid reconstruction allows rapid and seamless cloning of oligomeric genes. *Biomacromolecules*, *11*, 944–952.

Rogers, A. R., Driver, J. W., Constantinou, P. E., Jamison, K. D., & Diehl, M. R. (2009). Negative interference dominates collective transport of kinesin motors in the absence of load. *Physical Chemistry Chemical Physics*, *11*, 4882–4889.

Rogers, S. L., Tint, I. S., Fanapour, P. C., & Gelfand, V. I. (1997). Regulated bidirectional motility of melanophore pigment granules along microtubules *in vitro*. *Proceedings of the National Academy of Sciences of the United States of America, 94*, 3720–3725.

Ross, J. L., Wallace, K., Shuman, H., Goldman, Y. E., & Holzbaur, E. L. F. (2006). Processive bidirectional motion of dynein-dynactin complexes in vitro. *Nature Cell Biology, 8*, 562–570.

Schnitzer, M. J., Visscher, K., & Block, S. M. (2000). Force production by single kinesin motors. *Nature Cell Biology, 2*, 718–723.

Schroeder, H. W., Mitchell, C., Shuman, H., Holzbaur, E. L. F., & Goldman, Y. E. (2010). Motor number controls cargo switching at actin-microtubule intersections *in vitro*. *Current Biology, 20*, 687–696.

Syed, S., Müllner, F. E., Selvin, P. R., & Sigworth, F. J. (2010). Improved hidden Markov models for molecular motors, part 2: Extensions and application to experimental data. *Biophysical Journal, 99*, 3696–3703.

Uppulury, K., Efremov, A. K., Driver, J. W., Jamison, D. K., Diehl, M. R., & Kolomeisky, A. (2012). How the interplay between mechanical and non-mechanical interactions affect multiple kinesin dynamics. *Journal of Physical Chemistry B, 116*, 8846–8855.

Urry, D. W. (1988). Entropic elastic processes in protein mechanisms. I. Elastic structure due to an inverse temperature transition and elasticity due to internal chain dynamics. *Journal of Protein Chemistry, 7*, 1–34.

Urry, D. W., Hugel, T., Seitz, M., Gaub, H. E., Sheiba, L., Dea, J., et al. (2002). Elastin: A representative ideal protein elastomer. *Philosophical Transactions of the Royal Society of London. Series B: Biological Sciences, 357*, 169–184.

Xu, J., Shu, Z., King, S. J., & Gross, S. P. (2012). Tuning multiple-motor travel via single motor velocity. *Traffic, 13*, 1198–1205.

CHAPTER TWELVE

Reconstitution of Cortical Dynein Function

Sophie Roth*, Liedewij Laan†, Marileen Dogterom*,[1]
*FOM Institute AMOLF, Science Park, Amsterdam, The Netherlands
†Fas Center for Systems Biology, Harvard University, Cambridge, Massachusetts, USA
[1]Corresponding author: e-mail address: dogterom@amolf.nl

Contents

Abstract

Cytoplasmic dynein is a major microtubule (MT)-associated motor in nearly all eukaryotic cells. A subpopulation of dyneins associates with the cell cortex and the interaction of this cortical dynein with MTs helps to drive processes such as nuclear migration, mitotic spindle orientation, and cytoskeletal reorientation during wound healing. In this chapter, we describe three types of assays in which interactions between cortical dynein and MTs are reconstituted *in vitro* at increasing levels of complexity. In the first 1D assay, MTs, nucleated from a centrosome attached to a surface, grow against dynein-coated gold barriers. In this assay configuration, the interactions between MTs and dynein attached to a barrier can be studied in great detail. In the second and third assays, a freely moving dynamic aster is placed in either a 2D microfabricated chamber or a 3D water-in-oil emulsion droplet, with dynein-coated boundaries. These assays

Methods in Enzymology, Volume 540
ISSN 0076-6879
http://dx.doi.org/10.1016/B978-0-12-397924-7.00012-1

can be used to study how cortical dynein positions centrosomes. Finally, we discuss future possibilities for increasing the complexity of these reconstituted systems.

1. INTRODUCTION

Cytoplasmic dynein (referred to as dynein in this chapter) is a large multisubunit molecular motor that generates force and performs minus-end-directed microtubule (MT)-based transport in eukaryotic cells. Dynein plays fundamental roles in cell division and in controlling the intracellular distribution and transport of organelles. Of particular relevance for the assays discussed in this chapter, dynein can be recruited to the cortex (reviewed in Dujardin & Vallee, 2002), where it interacts with MTs in two distinct ways: sidewise, that is, where it can move along the entire MT, or in a so-called end-on interaction with the distal tip of a MT (Adames & Cooper, 2000; Carminati & Stearns, 1997). These cortical-localized dyneins are thought to not only drive the positioning of MT-organizing centers but also potentially influence MT dynamics. To unravel dynein's activity at the cortex, mutations and downregulation of dynein or its cofactors have been used, confirming dynein's role in nuclear and centrosome positioning or spindle orientation in different *in vivo* systems (O'Connell & Wang, 2000; Robinson, Wojcik, Sanders, McGrail, & Hays, 1999; Sharp, Rogers, & Scholey, 2000; Swan, Nguyen, & Suter, 1999). However, dynein interacts with many proteins that regulate and affect its activity (reviewed in Kardon & Vale, 2009) and is also active in the cytoplasm. Thus, from whole-cell knockdown experiments, it is difficult to directly link dynein force generation at the cortex to movement or dynamics of MTs.

A bottom-up approach where known components are added in a minimal *in vitro* system is thus a useful complement to *in vivo* experiments. Using such reconstituted systems, one can address questions such as: (1) how do cortical dyneins interact with MTs in an end-on fashion? (2) how do they exert force and affect MT dynamics? and (3) how does dynein localization at the cortex influence positioning processes in contained environments of different sizes and shapes, which mimic different cell types? Our recent work in reconstituted systems has shed light on some aspects of cortical dynein's influence on MT dynamics and the positioning of centrosomes (Laan, Pavin, Husson, Romet-Lemonne, van Duijn, Lopez et al., 2012).

To build this type of *in vitro* reconstituted systems, three main conditions need to be fulfilled:

1. A cell-like compartment must be created.
2. Dyneins must be attached "cortically" (i.e., to the cell-like compartment boundary).
3. A "cytoplasm" containing dynamic MTs, and proteins of interest must be encapsulated.

In this chapter, we describe three different assays where we reconstitute functional interactions between cortical dynein and dynamic MT ends in 1D, 2D, and 3D confinements. For each assay, we explain in detail how to build the cell-like compartment, how to attach dynein to the "cortex," and how to grow MTs from centrosomes against boundaries covered with dynein. Section 2 describes the 1D assay, where dyneins are attached to a microfabricated wall and interact with MTs growing from a centrosome bound to the surface. In Section 3, centrosomes are freely moving in a 2D microfabricated chamber whose boundaries are covered with dyneins. These two sections describe methods used in our recent paper (Laan, Pavin, et al., 2012), which has also been published in a book chapter (Laan & Dogterom, 2010). Here, these methods are presented in more detail as a protocol. Section 4 describes a recent protocol developed in our lab, which involves the use of 3D water-in-oil (w/o) emulsion droplets, in which dyneins are linked to a lipid monolayer at the border of the droplet. This assay is described here for the first time, although some preliminary results were published in a recent review (Laan, Roth, & Dogterom, 2012). Finally, Section 5 explains the advantages and limitations of each assay and discusses interesting additional features that could be added in the future.

2. RECONSTITUTION OF CORTICAL DYNEIN FUNCTION IN 1D GEOMETRIES

Here, we describe an assay to mimic the interaction between growing MTs and a cell boundary covered with dyneins. Dynamic MTs are growing from a centrosome attached to the sample surface toward microfabricated gold barriers functionalized with dyneins (Fig. 12.2A).

2.1. Dynein modification and purification

To reconstitute dynein's cortical function, motors need to be selectively attached to the boundary of a cell-like compartment, which can be achieved by the addition of a tag to the protein. The tag needs to be located such that the motor domain is free to interact with MTs via its MT-binding domain. So far, two model systems, *Dictyostelium* (Koonce & Samso, 1996; Nishiura et al., 2004) and *Saccharomyces cerevisiae* (Reck-Peterson et al., 2006) have been employed as robust sources of recombinant dyneins for purification. Dyneins are large protein dimers composed of multiple subunits: heavy chains (HCs), intermediate chains, light intermediate chains, and light chains (Fig. 12.1A-1/2) (Pfister et al., 2006). Purification of the full-length holoenzyme with associated chains was achieved with *S. cerevisiae*, although the amount of protein obtained is small. As an alternative, Reck-Peterson and

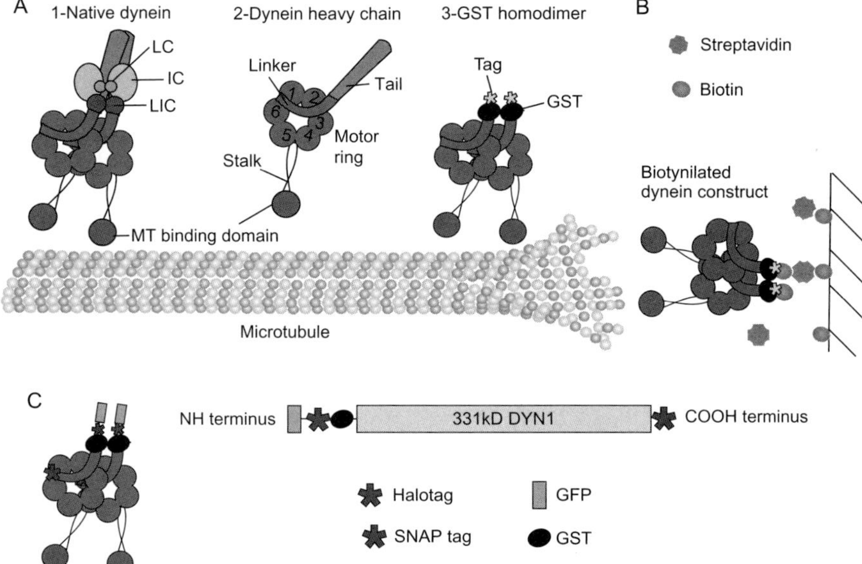

Figure 12.1 "Cortical" dyneins: attaching dyneins to the border of a cell-like confinement. (A) Dynein structure. 1: Native dynein heavy chain with its associated subunits, intermediate chain (IC), light intermediate chain (LIC), and light chain (LC). 2: Two-dimensional domain structure of the dynein heavy chain, with its motor ring and MT-binding domain. 3: Monomeric dynein artificially dimerized with GST. A tag at the NH terminus of the protein enables biotinylation. (B) "Cortical" dynein: attachment of a biotinylated dynein construct via biotin–streptavidin chemistry. The MT-binding domain is free to interact with MT ends. (C) Detail of one of our cortical dynein constructs: two-dimensional domain structure (left) and schematic showing of dynein heavy chain truncations and tags (right). A Halotag at the COOH terminus allows attachment of a fluorophore. GST is used to dimerize monomeric constructs at the NH terminus. A SNAP tag located between the GFP and the GST allows for biotinylation of the protein. (For color version of this figure, the reader is referred to the online version of this chapter.)

coworkers have developed artificially dimerized monomers of the motor domain of the *S. cerevisiae* dynein, which is 331 kDa in size (dynein HC minus its tail domain, Fig. 12.1A-3). This artificial dimer behaves similar to a full-length yeast dynein dimer, exhibiting processive minus-end-directed motility and stalling at a force (4.8 ± 1.0 pN) similar to full-length dynein (Reck-Peterson et al., 2006).

We purify two versions of these truncated dyneins, dimerized with glutathione *S*-transferase with the help of an affinity tag composed of two copies of the protein A IgG-binding domain, and a TEV cleavage site, as described (Reck-Peterson et al., 2006). Both constructs have a tag at their NH terminus that can be biotinylated (Fig. 12.1A-3). Biotinylation of the protein enables a specific and nearly permanent link via streptavidin to a cell-like boundary functionalized with biotin, as illustrated in Fig. 12.1B (*note*: the K_d of the biotin/streptavidin interaction is on the order of 10^{-14} *M*; Green, 1990).

The latest version of the construct we purify is detailed in Fig. 12.1C. In this version, two tags, SNAP and Halotag, enable us to both biotinylate the protein (SNAP biotin®, New England Biolabs) and label it with a fluorophore (Halotag® TMR, Promega). A GFP is also present at the NH terminus of the protein. The second construct has one Halotag at the NH terminus and no GFP. These biotinylated constructs will be referred to as "dynein" in the assays described below.

2.2. Microfabrication of gold barriers

The barriers, approximately 1 μm high, contain a thin layer of chromium (5 nm) that ensures good adhesion of the gold layer (750 nm). An overhang, made of a thick layer of chromium (250 nm), provides an extra feature to prevent MTs from growing over the barrier, enforcing end-on contact between the MT and the barrier (Fig. 12.2A and C). The fabrication process is detailed in Fig. 12.2B and explained below. All the microfabrication steps except evaporation are performed in a clean room (class ISO 6).

2.2.1 Materials

2.2.1.1 Clean room equipment

Coverslips No. 1 24 × 24 mm 170 μm (Menzel Glässer, Germany).

Delta 80 GYSET® Spin coater (Süss MicroTec, Germany).

Homemade turbo-pumped vacuum system with base pressure of 10^{-7} mbar is equipped with a resistance heating evaporation system

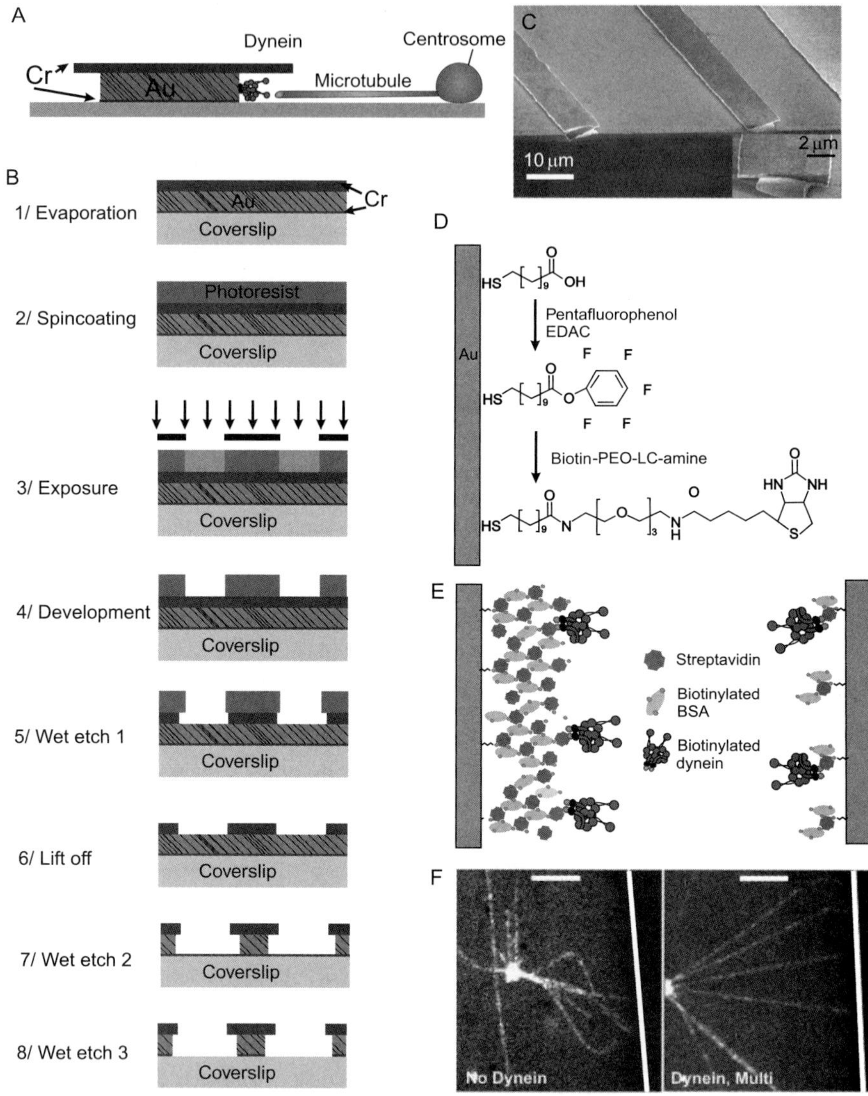

Figure 12.2 Reconstitution of cortical dynein function at gold barriers. (A) Scheme of the experiment. (B) Fabrication process of gold barriers. (C) SEM image of microfabricated gold barriers (image courtesy of Taberner, N.). (D) Functionalization of the gold barriers by thiol chemistry. (E) (Left) Dynein attachment to the gold barriers with multiple layers of BSA–streptavidin. (Right) Limiting the amount of dyneins on the wall by competing biotinylated BSA. (F) Spinning disk confocal image of microtubules grown from a centrosome attached to the surface and interacting with the gold barriers (left) or dynein attached to the gold barriers (right) (Laan, Pavin, et al., 2012). Scale bar: 10 μm. The gold wall is represented by a line on the right of the picture.

for tungsten boats loaded with chromium or gold. 2510 Ultrasonic Cleaner (Branson, USA).
MJB3 mask aligner for UV exposure (Süss MicroTec).
Binary chromium/soda lime mask (Delta Mask, The Netherlands).
FEI XL30 SFEG electron microscope.

2.2.1.2 Reagents

Hexamethyldisilazane (HMDS) primer (Microresist, Germany).
Shipley Microposit® S1813 positive UV-resist (Microresist).
Microposit® MF®-319 developer (Microresist).
Standard gold etchant (Sigma-Aldrich, USA).
Standard chromium etchant (Sigma-Aldrich).

2.2.2 Protocol

2.2.2.1 Evaporation of metals (Fig. 12.2B-1)

Start by cleaning the glass coverslips with base piranha (NH_4OH:H_2O_2 in 3:1 at 75 °C) for 15 min. Rinse, first in double-distilled water (ddH_2O), then in isopropanol. Blow-dry with N_2 flow. Evaporate a 5-nm layer of chromium, followed by 750 nm of gold and finally 250 nm of chromium in an evaporation chamber under pressure below 10^{-6} mbar with a deposition rate of 0.06 nm/s.

2.2.2.2 Photolithography (Fig. 12.2B-2–4)

Coat the samples with HMDS using evaporation under vacuum for 1 h. Use a standard protocol of photolithography with the following steps: spin S1813 photoresist to obtain a 1.2-μm thick layer, soft bake, UV expose through the chromium mask, hard bake, and develop with MF319 developer. The different temperatures and times can be adjusted following the recommendations of Microposit for the photoresist S1813.

Comment: HMDS helps for a good contact between chromium and the photoresist.

2.2.2.3 Wet etching (Fig. 12.2B-5–8)

Immerse the samples in chromium etchant until the first layer of chromium is completely dissolved (Fig. 12.2B-5). The time of immersion can be determined empirically by checking the chromium layer under a microscope. Usually, it takes between 30 s and 1 min. Remove the photoresist by sonicating the samples in acetone (Fig. 12.2B-6). Place the samples in gold etchant and shake carefully until all gold that is not protected by the chromium layer is removed (Fig. 12.2B-7). Finally, immerse the slide in chromium etchant to remove the thin layer of chromium on the bottom, but keeping the thick layer of chromium on top (Fig. 12.2B-8). Clean the sample again with base piranha to make it ready for functionalization steps. The sample will be very hydrophilic.

2.3. Functionalization of gold walls

Gold barriers are specifically labeled with biotin using thiol chemistry (Fig. 12.2D, Dogterom, Felix, Guet, & Leibler, 1996; Romet-Lemonne, VanDuijn, & Dogterom, 2005).

2.3.1 Reagents

11-Mercapto-1-undecanoic acid (Sigma-Aldrich) (MDA); dilute before use in ethanol.

1-Ethyl-3-(3-(dimethylamino)propyl)carbodiimide (Molecular Probes, E-2247) (EDAC); dilute before use in ethanol.

Pentafluorophenol (Sigma-Aldrich) (PFP); dilute before use in ethanol.

Biotin-PEO-LC-amine (biotin-PLA) (Pierce Inc.); dilute before use in ethanol.

Comment: Thiols can form disulfide bonds in the presence of oxygen. They should thus not be stored in the presence of oxygen.

2.3.2 Protocol thiol chemistry (Fig. 12.2D)

Clean the samples containing gold barriers with base piranha right before the functionalization steps. Clean with ethanol. Immerse in 200 m*M* MDA for 3 h. Rinse with ethanol. Prepare a solution of 100 m*M* EDAC and 200 m*M* PFP in ethanol, and immerse the samples for 20 min. Rinse with ethanol. Immerse in 0.1 m*M* biotin-PLA for 20 min and finally rinse with ethanol.

To remove non-specific interactions of the thiol groups with glass surfaces, immerse each slide sequentially in 2 *M* NaCl solution for 7 min, 0.1% Tween solution for 15 min, and 0.1% Triton X-100 solution for 7 min, and thoroughly rinse with ddH_2O. The slides can be stored in ethanol for several weeks.

Comment: The attachment of biotin to gold can be tested by evaluating the specific binding of fluorescent streptavidin to gold structures. Note that an alternative one-step technique for the thiol chemistry is described elsewhere (Taberner, Weber, You, Dries, Piehler, & Dogterom, 2014).

2.4. Dynamic MTs growing against functionalized gold barriers

2.4.1 Materials

All reagents for surface treatment and MT assay are dissolved in MRB80 buffer (80 m*M* K-pipes, 4 m*M* $MgCl_2$, 1 m*M* EGTA, pH 6.8) at the stated stock concentration, filtered with 0.2 μm membranes, flash frozen, and stored at −80 °C.

2.4.1.1 Surface treatment

Albumin, biotin-labeled bovine (Sigma-Aldrich) (biotinylated BSA); 20 mg/ml solution.
Bovine serum albumin (BSA) (Sigma-Aldrich); 25 mg/ml solution.
Streptavidin (Sigma-Aldrich); 1 mg/ml.
Alexa Fluor® 488 streptavidin (Invitrogen); 1 mg/ml.
κ-Casein from bovine milk (Sigma-Aldrich); 5 mg/ml.
Poly-L-lysine polyethylene glycol (SurfaceSolutions, Switzerland) (PLL-PEG); 0.2 mg/ml.

2.4.1.2 MT assay

Tubulin from bovine brain (Cytoskeleton, Inc., USA); 100 μ*M* solution.
Rhodamine-labeled tubulin from porcine brain (Cytoskeleton, Inc.); 50 μ*M* solution.
Fluorescent HiLyte 488 tubulin from porcine brain (Cytoskeleton, Inc.); 50 μ*M* solution.
Biotinylated tubulin from porcine brain (Cytoskeleton, Inc.); 50 μ*M* solution.
Guanosine 5′-triphosphate sodium salt hydrate (GTP) (Sigma-Aldrich); 50 m*M* solution.
Glucose oxidase from *Aspergillus niger* (Sigma-Aldrich); 20 mg/ml dissolved in 200 m*M* DL-dithioltheitol (Sigma-Aldrich) with 10 mg/ml catalase from bovine liver (Sigma-Aldrich) (glucose oxidase 50 ×).
D-(+)-Glucose (Sigma-Aldrich).
Methylcellulose 4000 cP (Sigma-Aldrich); 1% solution.
Adenosine 5′-triphosphate, disodium salt hydrate (ATP) (Sigma-Aldrich); 50 m*M* solution.
Purified centrosomes from human lymphoblastic KE37 cell lines as described in Moudjou and Bornens (1998).
Purified dyneins as described in Section 2.1.

2.4.1.3 Imaging

Leica microscope with a 100 × 1.3 NA oil-immersion objective.
Spinning disk confocal head from Yokogawa.
Cooled EM-CCD camera (C9100, Hamamatsu Photonics).

2.4.2 Protocol

2.4.2.1 Flow cell and centrosome attachment

Prepare a flow cell with the microfabricated gold barriers coverslip and a cleaned glass slide using double-sided TESA® tape. Subsequent solutions can be flown in by putting absorbent paper on the other side of the cell. Importantly, avoid drying out any part of the flow cell.

Flow a solution of centrosomes in MRB80 and incubate for 5 min. The centrosomes will nonspecifically adhere to the glass surfaces. Remove the unbound centrosomes by rinsing with two flow cell volumes of MRB80.

2.4.2.2 Dynein attachment (Fig. 12.2E)

Biotinylated dynein motor proteins are specifically attached via biotin–streptavidin linkage and through blocking of the other surfaces.

Incubate for 5 min with 0.1 m*M* PLL-PEG to passivate the coverslip surface. Wash with MRB80 and incubate with 1 mg/ml κ-casein for 5 min. Rinse with MRB80. Perform alternating 5 min incubations of a streptavidin mix (0.5 mg/ml streptavidin, 1 mg/ml κ-casein, 5 mg/ml BSA in MRB80) (three times) and a biotinylated BSA mix (1.5 mg/ml biotinylated BSA, 1 mg/ml κ-casein, 5 mg/ml BSA in MRB80) (two times). Wash thoroughly with MRB80 in between incubations. Further, passivate the surfaces with 1.2 mg/ml κ-casein for 10 min. Incubate with a mix containing 20 n*M* dynein, 1 mg/ml κ-casein, 5 mg/ml BSA in MRB80 for 5 min. Rinse with MRB80.

Comment: To achieve efficient binding of dynein, we experienced that multiple layers of streptavidin and biotin are needed, probably because of steric effects (Fig. 12.2E; Romet-Lemonne et al., 2005). The protocol given describes how to create those layers (Fig. 12.2E, left panel). Alternatively, to vary dynein concentration at the wall, only one layer of streptavidin could be used, and biotinylated dynein could be further diluted by adding competitive biotinylated BSA (Fig. 12.2E, right panel).

2.4.2.3 Tubulin mix and imaging

Prepare the tubulin mix on ice in MRB80: 15 μ*M* tubulin, 1 μ*M* fluorescent tubulin, 1 m*M* GTP, 1 m*M* ATP, 0.8 mg/ml κ-casein, 0.1% methyl cellulose, glucose oxidase 1 ×, glucose 50 m*M*. Introduce in the flow cell. Seal and examine at 25 °C using spinning disk confocal microscopy.

Comment: It is possible to vary the tubulin concentration (keeping the ratio tubulin/fluorescent tubulin constant) and the temperature, both of will affect MT nucleation and dynamics (Pedigo & Williams, 2002). Good conditions for nucleating exclusively from the centrosome can vary from one preparation of purified centrosomes to another. They have to be empirically determined for each preparation. Too high temperatures typically lead to free nucleation, resulting in numerous small free MTs around the centrosome. At the proper temperature, 25 °C for our centrosome preparation, MTs grow long and exclusively from the centrosome.

2.4.2.4 Typical results (Fig. 12.2F)

With this assay, the details of the interaction between dynamic MTs and dyneins in an end-on configuration can be studied. For example, we showed that when dyneins are present at the barrier, MTs are captured, and pulling forces are generated (Laan, Pavin, et al., 2012). Time-lapse movies are typically taken with a 3 s time interval to assess the dynamics of MTs encountering the barrier.

3. RECONSTITUTION OF CORTICAL DYNEIN FUNCTION IN 2D GEOMETRIES

Here, we describe the protocol we developed to study MT-based positioning processes in confining 2D geometries (Laan, Pavin, et al., 2012; Romet-Lemonne et al., 2005). We designed a 2D experiment where dynamic MTs growing from a centrosome are confined in microfabricated chambers whose walls are coated with dyneins (Fig. 12.3A). The motors are selectively attached to the walls via gold-thiol chemistry. The number of dyneins at the walls can be varied by varying the thickness of the gold layer at the wall of the chamber, thus changing the likelihood that MT-wall interactions generate pulling forces.

3.1. Dynein modification and purification

Details on the dynein constructs we purify are described in Section 2.1.

3.2. Microfabrication of quasi 2D chambers with gold walls

The chambers, of about ~2.5–2.7 μm high, are composed of a gold layer sandwiched in between two chromium layers (5 nm) that provide a proper

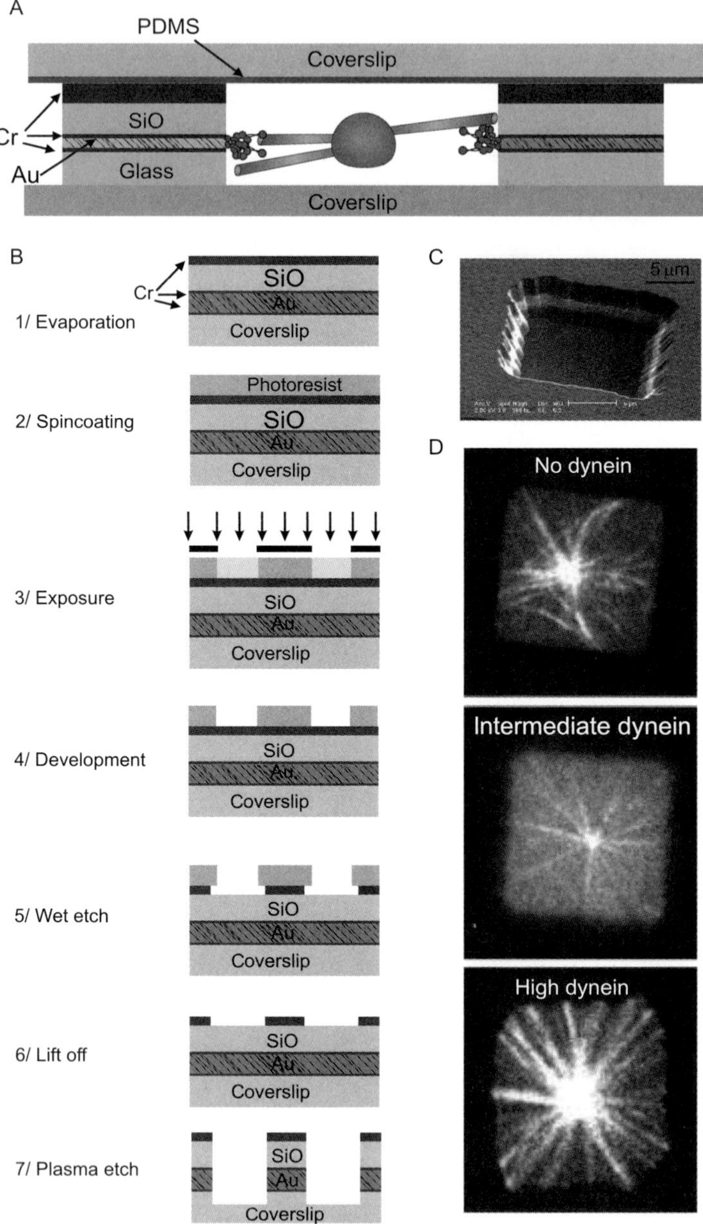

Figure 12.3 Reconstitution of cortical dynein function in microfabricated chambers. (A) Scheme of the experiment. (B) Fabrication process of 2D microfabricated chambers. (C) SEM picture of a microfabricated chamber. (D) Typical positioning experiment in a 15-μm square chamber, without, with intermediate or with high dynein amounts (Laan, Pavin, et al., 2012).

adhesion between the coverslip below and a layer of silicon mono oxide above. A final chromium layer of 400 nm is deposited on top and a polydimethylsiloxane (PDMS) lid provides a good sealing of the chambers. The chambers we used are squares of 10 or 15 μm wide. The fabrication process is detailed in Fig. 12.3B and explained below.

3.2.1 Materials

3.2.1.1 Special equipment

- Clean room equipment, as listed in Section 2.2.1.1 .
- Plasmalab plasma-etcher.80+ ICP.

3.2.1.2 Reagents

- As described in Section 2.2.1.2.
- Sylgard 184 elastomer kit (Dow corning, USA) base and curing agent. Prepare right before use by mixing curing agent and base at 1:10 ratios (PDMS).

3.2.2 Protocols

3.2.2.1 Evaporation of metals (Fig. 12.3B-1)

Clean the glass coverslips with base piranha for 15 min. Rinse first in Millipore deionized water and then in isopropanol, and blow-dry with N_2 flow. Evaporate a 5-nm layer of chromium, followed by 100 or 700 nm of gold, 5 nm of chromium, 1200 or 900 nm of silicon monoxide, and finally 400 nm of chromium in an evaporation chamber under pressure below 10^{-6} mbar with a deposition rate of 0.06 nm/s.

3.2.2.2 Photolithography (Fig. 12.3B-2–4)

Coat the samples with HMDS using evaporation under vacuum for 1 h. Use a standard protocol of photolithography with the following steps: spin S1813 photoresist to obtain a 1.2-μm layer, soft bake, UV expose through the chromium mask, hard bake, and develop with MF319 developer. The different temperatures and times can be adjusted following the recommendations of Microposit for the photoresist S1813.

Comment: HMDS ensures a good contact between chromium and the photoresist.

3.2.2.3 Wet etching (Fig. 12.3B-5–6)

Immerse the samples in chromium etchant till the first layer of chromium is completely removed (Fig. 12.3B-5). The time of immersion can be determined empirically by checking the chromium layer under a microscope. Usually, it takes between 30 s and 1 min. Remove the photoresist by sonication of the samples in acetone (Fig. 12.2B-6).

3.2.2.4 Plasma etching (Fig. 12.3B-7)

Etch the samples with a CHF_3^-/SF_6^- Argon plasma till all remaining layers are removed.

The results are microfabricated chambers that consist of a sandwich of 1200/900 nm glass, 5 nm chromium, 100/700 nm gold, 5 nm chromium, 1200/900 silicon monoxide, and ~100 nm chromium. To remove possible contamination from the gold layer, end the process with 15 min exposure to oxygen plasma. Check the samples for defects in an electron microscope (Fig. 12.3C).

3.2.2.5 Fabrication of PDMS lids

The microfabricated chambers are closed with a PDMS lid to achieve good sealing. Firmly squeeze a droplet of PDMS between a piece of transparency slide and a 24 × 60 mm coverslip. Cure at 100 °C for 1 h. Remove the transparency slide from the coverslip. This will leave an approximately 80 μm flat layer of PDMS on the coverslip. This is thin enough to allow for microscopic observation with a high magnification oil-immersion objective through the microfabricated chambers, looking from either side. The PDMS coverslips can be stored before usage in a closed box for maximum 1 week.

Comment: The slides can be used for maximum 1 week because exposure to water in the air creates swelling of the PDMS.

3.3. Functionalization of gold walls

Gold walls functionalization is described in Section 2.3.

3.4. Dynamic MTs growing from centrosomes in microfabricated chambers

3.4.1 Materials

- Surface treatment and MT assay material as described in Section 2.4.1.1 and 2.4.1.2.
- PDMS lid as described in Section 3.2.2.5.
- Microfabricated chambers with gold walls as described in Section 3.2.
- Imaging with spinning disk confocal as described in Section 2.4.1.3.

3.4.2 *Protocol*

3.4.2.1 Flow cell and dynein attachment

Rinse the functionalized microfabricated chambers and the PDMS lid with ethanol, then MRB80. Blow-dry using a N_2-flow and immerse in a mix of κ-casein (2 mg/ml) and BSA (5 mg/ml) in MRB80 for 15 min. Blow-dry again. Construct a temporary flow cell, placing double-sided Tesa® tape in between the PDMS coverslip and the microfabricated chamber coverslip. Place a metal block as a weight on top of the flow cell to keep the two coverslips tightly together. Block the surfaces by introducing a solution of κ-casein (2 mg/ml) and BSA (5 mg/ml) in MRB80 for 10 min. Wash with MRB80 and introduce the biotinylated dynein. Before introducing the tubulin solution, place the sample on a metal block at 4 °C to prevent MT growth.

3.4.2.2 Tubulin mix, sealing, and imaging

Make the following tubulin mix on ice: centrosomes, 22 μ*M* tubulin, 1.6 μ*M* Rhodamine tubulin, 1 m*M* GTP, 1 m*M* ATP, 1 × glucose oxidase, 0.5 mg/ml κ-casein, and 16% sucrose in MRB80. Introduce into the flowcell and wait for 4 min. Remove carefully the Teflon tape and press firmly the PDMS coverslip on the microfabricated chamber coverslip for 2 min to create a good seal. Seal the edges of the microfabricated chamber coverslip with hot candle wax. Typically, 20% of the microfabricated chambers would be properly sealed. Examine the sample at 25 °C using spinning disk confocal microscopy by imaging through the PDMS layer. (This way, microfabricated chambers that are not well sealed can be easily visualized. MTs will grow outside of the well.)

3.4.2.3 Typical results (Fig. 12.3D)

Using this assay, we were able to show that MT asters are faithfully positioned in the center of square chambers when high amounts of dyneins are attached to the chamber walls. Without dynein at the walls, growing MTs create pushing forces that cause MTs to bend and buckle, which eventually leads to destabilization of the central position (Laan, Pavin, et al., 2012; Romet-Lemonne et al., 2005).

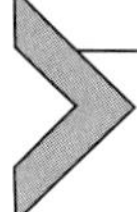

4. RECONSTITUTION OF CORTICAL DYNEIN FUNCTION IN 3D GEOMETRIES

Here, we describe the protocol we developed to study MT-based positioning processes in confining 3D w/o emulsion droplets (Fig. 12.4A). MTs growing from a centrosome are confined in an emulsion droplet whose surface is covered by biotin lipids. Biotinylated dyneins are selectively attached to biotin lipids via biotin–streptavidin linkages.

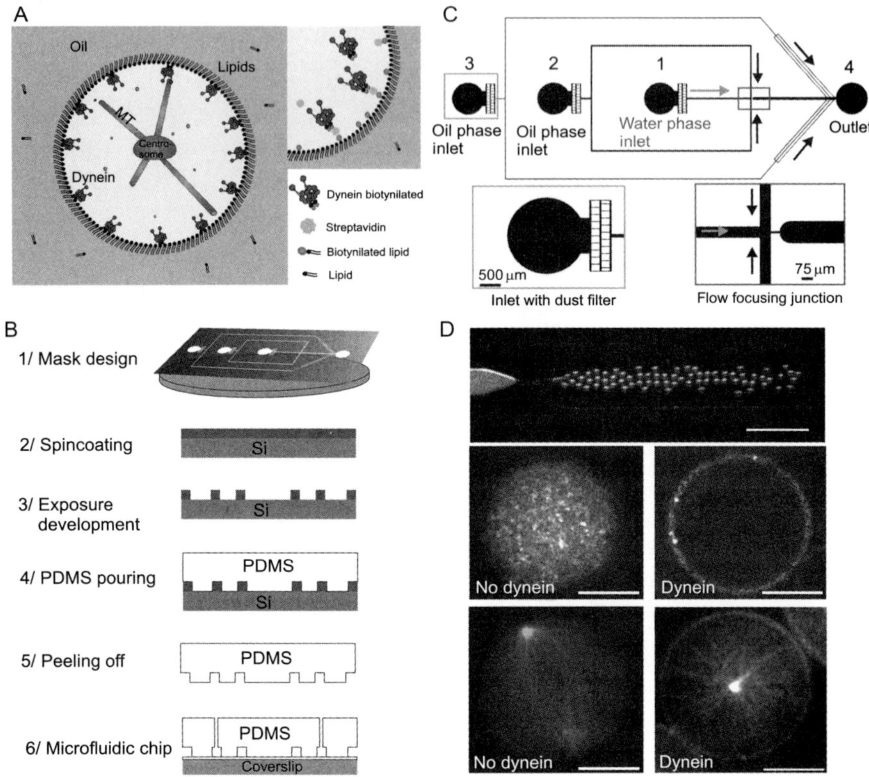

Figure 12.4 Reconstitution of cortical dynein function in w/o emulsion droplets. (A) Scheme of the experiment. (B) Fabrication of microfluidic chips. (C) Flow focusing design. Inlet 1 is dedicated to the water phase, inlet 2 and 3 to the oil phase. Small channels of 10 μm width at the end of the inlets are used as dust filters (zoom in the down left side). Channels 1 and 2 meet in a standard flow focusing junction, where the droplets are created (zoom in the down right side). Channel 3 is used to dilute the droplets before observation. (D) (Top) Droplet formation: fluorescent image of dextran FITC in the inner buffer. Scale bar: 100 μm. (Middle) MT seeds localize in the middle of the droplet in the absence of dynein (left) and at the border in the presence of cortical dynein (right). Scale bar: 8 μm. (Bottom) Spinning disk confocal image of single centrosomes confined in a w/o emulsion droplet in the absence (left) or the presence of dyneins at the border (both dynein and the microtubules are fluorescently labeled) (Laan, Roth, & Dogterom, 2012). Scale bar: 8 μm.

4.1. Dynein modification and purification

Details on the dynein construct we purify are described in Section 2.1.

4.2. W/o Emulsion droplets

W/o emulsions are aqueous microdroplets in an oil phase containing an excess of solubilized phospholipids and/or surfactants. The phospholipids and/or surfactants spontaneously form a monolayer surrounding the aqueous microdroplets, stabilizing the w/o emulsions. To functionalize the interior of the droplet with biotinylated dyneins, biotin lipids are inserted in the oil (Fig. 12.4A).

Two general methods can generate such w/o emulsions. In the first one, a drop of buffer is broken in the oil/lipid mixture by shear or impact stresses generated by manual or mechanical agitation. This method, fast and easy, generates droplets with a broad distribution of droplet sizes. We chose not to use this method because of the impact of such mechanical agitation on tubulin, which we have found leads to protein denaturation.

Alternatively, droplets can be generated from two immiscible liquids in a microfluidic device (digital microfluidics) (Atencia & Beebe, 2005; Stone, Strook, & Ajdari, 2004; Teh, Lin, Hung, & Lee, 2008), via flow focusing (Anna, Bontoux, & Stone, 2003; Nguyen et al., 2007; Zhou, Yue, & Feng, 2006), using T-junctions (Garstecki, Fuerstman, Stone, & Whitesides, 2006; Xu, Li, Tan, Wang, & Luo, 2006) or coflowing (Xu, Li, Lan, & Luo, 2008). Using such techniques, droplets can be generated that are highly monodisperse in size (Nisisako, Torii, Takahashi, & Takizawa, 2006), with a rate reaching many thousands per second (Yobas, Martens, Ong, & Ranganathan, 2006).

Three main ingredients are necessary for mastering w/o emulsions:

- The microfluidic chip needs to be hydrophobic.
- The composition of oil/internal buffer must be tuned to give stable w/o emulsions.
- The geometry of the channels and the flow rate of the liquids need to be adjusted to obtain the wanted droplet size.

Here, we describe the setup that we have built to create droplets from a small starting volume of an inside buffer (10 μl; necessary to save purified proteins), and that retains the polymerization activity of tubulin.

4.2.1 Microfluidic device fabrication

Generally, the design of a microfluidic chip is first printed on a photomask with high resolution, the areas of the channels being transparent while the rest is opaque (Fig. 12.4B-1). The design is then transferred by photolithography to a silicon wafer with SU-8 photoresist (Fig. 12.4B-2–3). The silicon wafer with SU-8 pattern is then used as a mold to make microfluidic PDMS chips (Fig. 12.4B-4–6). The fabrication process is explained in Fig. 12.4B and is detailed below.

4.2.1.1 Photomask

Photomasks from film substrate are ordered from Selba S.A. (Versoix, Switzerland). We adapted our design from standard flow focusing designs, as detailed in Fig. 12.1C. With such a design and our setup, we can produce droplets that have our wanted droplet size: ~15 μm diameter.

4.2.1.2 SU-8 mold

SU-8 molds are realized in a clean room, with a standard protocol of photolithography: spin coat, soft bake, UV exposure, postexposure bake, development, hard bake.

We used SU-8 3025 (MicroCHEM, USA) photoresist and its associated developer and followed the recommendations of MicroCHEM to create channels with a thickness of 40 μm.

Comment: A last hard bake step of at least 1 h was shown to be very important to prevent SU-8 channels from peeling off from wafers during microfluidic chip fabrication, as well as preventing PDMS from sticking to the surface.

4.2.1.3 Microfluidic chips

To obtain w/o emulsions, we need our microfluidic devices to be hydrophobic: droplets of buffer must not wet the surface of the microchannels.

- Mix 10 wt. PDMS prepolymer RVT615 from MOMENTIVE (Lubribond, The Netherlands) with 1 wt. corresponding curing agent in a boat-like vessel.
- Transfer to 50-ml falcon tubes (BD Falcon) and centrifuge 300 rcf for 5 min to remove large air bubbles from the PDMS mixture.
- Place a nitrogen-cleaned wafer with SU-8 mold on a circular piece of aluminum foil (diameter slightly larger than the wafer). Push up the borders to form a cup around the mold.
- Pour the PDMS mixture.
- Place it in a vacuum chamber to get rid of all air bubbles in the PDMS.
- Cure for 1 h at 100 °C.
- Peel off the PDMS from the wafer mold.
- Punch with Harris Unicore cutting tips to make inlets (0.5 mm) and outlets (0.75 mm).
- Treat both the PDMS slab and PDMS-coated coverslip (as explained in Section 4.2.3) with a corona discharger for a few seconds (Corona treater, model BD-20AC Electro-Technic Products Inc.).
- Bond the PDMS slab to the PDMS-coated coverslip by placing one on top of the other (Channels facing inside). Do not press, to avoid channel collapse.
- Postbake at 100 °C overnight.
- The microfluidic chips can be stored for months.

4.2.2 Microfluidic setup

Our microfluidic setup is pressure regulated. Liquids are placed in Micrew® tubes (0.5 ml for the inside buffer, 1 ml for the oil phase) and mounted on a FLUIWELL accessory (Fluigent, France). PEEK tubes with outside diameter of 510 μm and inside diameter of 125 μm (Cluzeau Info. Labo, France) directly connect the Micrew® tubes in the FLUIWELL to the inlets of the microfluidic chip. Liquids are driven with a pressure controller MFCS-FLEX-4C-1000 mbar (Fluigent, Paris). Droplet formation is observed with a Leica PMIRB-inverted microscope in bright field.

4.2.3 Chambers of observation

4.2.3.1 Material and equipment

- Parafilm
- Momentive RTV615 (Lubribond)
- Glass slides and coverslips (Menzel Glässer)
- Spin coater (Suss MicroTec)

4.2.3.2 Protocol

We make standard flow channels by melting parafilm in between two RTV615-coated glass slides/coverslips. The hydrophobicity of RTV615 assures the w/o emulsions to remain in a spherical shape. RTV615 coating is realized by mixing 1:10 weight RVT615 A with B component and spinning it on the glass slides/coverslips using a spin coater (200 rpm for 5 s, then 3000 rpm for coverslips, and 1600 rpm for slides for 30 s), followed by curing 1 h at 100 °C.

4.2.4 Droplet generation

- Prepare one tube of Micrew® 0.5 ml filled with buffer and two tubes of Micrew® 1 ml filled with oil phase.
- Connect the two oil phase tubes to channels labeled 2 and 3 in Fig. 12.4C.
- Flow both tubes of oil phase in the microfluidic chip till the whole chip is filled.
- Reduce oil phase flow.
- Connect the inside buffer to channel labeled 1 in Fig. 12.4C.
- Adjust the flows to generate droplets of the wanted size.
- The droplets will go out through the output channels automatically.
- When the droplets have the correct size, wait to have enough of them and pipet them gently into the chamber of observation.

4.3. Dynamic MTs growing from centrosomes in w/o emulsion droplets

Comment: In contrast to the two other assays, dyneins cannot be attached to the wall before the tubulin mix is introduced. All components (streptavidin, dynein, and tubulin) must be mixed together in the inside buffer. The amount of streptavidin/dynein must thus be properly adjusted. The proper localization of dynein at the border can be checked by introducing short pieces of stabilized MTs (MT seeds) inside the droplet. In the presence of dynein at the border, the MT seeds will stick to the border (Fig. 12.4D). Also, note that dyneins are able to move since the lipids they are attached to can diffuse in the monolayer.

4.3.1 Materials

4.3.1.1 Special equipment

Airfuge® Air-driven ultracentrifuge (Beckman Coulter, USA)
Homemade temperature controller
Valap (Vaseline, lanolin, paraffin wax melted at equal concentrations)
Microfluidic chip for droplet generation (made as described in Section 4.2)
Chamber of observation (made as described in Section 4.2)
Tubes Micrew® 0.5 ml and 1 ml (VWR, The Netherlands)
Microfluidic setup as described in Section 4.2

4.3.1.2 Imaging

Same imaging using spinning disk confocal microscopy as described in Section 2.4.1.3

4.3.1.3 Material for oil phase

1,2-Dioleoyl-*sn*-glycero-3-phospho-L-serine (sodium salt) (Avanti Polar Lipids Alabaster, AL) (DOPS)
1,2-Dipalmitoyl-*sn*-glycero-3-phosphoethanolamine-*N*-(biotinyl) (Avanti Polar Lipids Alabaster, AL) (biotin PE)
Span® 80 (Sigma-Aldrich)
Mineral oil (M5904, Sigma-Aldrich)

4.3.1.4 Proteins and reagents

Same reagents as MT assay material in Section 2.4.1.2 are supplemented with the following reagents, prepared the same way.

Dextran, Alexa Fluor® 647, 10,000 MW, Anionic, Fixable (Life technologies): 25 mg/ml.

Phospho(enol)pyruvic acid monosodium salt hydrate ≥97% (enzymatic) (P0564, Sigma-Aldrich) (PEP): 200 m*M*.

Pyruvate kinase/lactic dehydrogenase enzymes from rabbit muscle (P0294, Sigma-Aldrich) (PK), aliquoted and stored at −20 °C.

4.3.2 Protocol

- Prepare the oil phase fresh the day of the experiment:

  Mix the lipids in chloroform at a molar ratio DOPS/biotin PE of 99:1. Dry them under nitrogen flow and dissolve in mineral oil at a total concentration of 0.5 mg/ml. Add Span 80 at 2 wt%. Sonicate the mixture for 30 min.
- Fill two Micrew® tubes 1 ml with the oil phase and one tube 0.5 ml with MRB80.
- Flow the liquids in the microfluidic chip as described in Section 4.2.4 "Droplet generation" and find the good flow rates to create the wanted droplet size.
- When the flow rates are found, take out the MRB80 inside buffer channel from the microfluidic chip and empty the MRB80 PEEK tube. Keep a slow flow of oil phase in the microfluidic device.
- Prepare the spinning disk microscope and set the temperature controller to 25 °C.
- Place an aliquot of purified centrosomes at 37 °C. We experienced that this step later helps to nucleate MT growth from the centrosomes.
- During this time, prepare 10 μl of the following protein mix in MRB80 on ice:

  Tubulin (34 μ*M*), Fluorescent Hilyte 488 tubulin (3.4 μ*M*), GTP (5 m*M*), streptavidin (40 n*M*), ATP (1 m*M*), glucose (50 m*M*), glucose oxidase (50 times dilution), PEP 160 m*M*, PK/LDH, Alexa Fluor® 647dextran (20 μ*M*), and dynein (60 n*M*). Airfuge the mix (without centrosomes!) at 4 °C for 5 min at 30 psi (200,000 rcf). This spin is not necessary for the success of the experiment but will eliminate aggregates of proteins.
- Place the protein mix on top of the centrosome aliquot at 4 °C and mix by pipetting up and down.
- Place the mix in a Micrew® tube 0.5 ml.
- Connect the PEEK tube to the microfluidic chip after having filled it completely.

- Generate droplets.
- Pipet the droplets in the chamber of observation.
- Seal with valap.
- Image with spinning disk confocal at 25 °C.

4.3.3 Typical results

Typical results are shown in Fig. 12.4D. Centrosome position is drastically decentered when no dynein is present at the border.

Comment: In 3D, it is much more difficult to visualize MTs than in 2D. To assess MT growth at the beginning of the experiment, movies can be taken in one plane of the droplet with 2 s time intervals. Another indication of MT growth comes from observing centrosome movement, which is strongly reduced upon MT growth. After 30 min at 25 °C, MTs become longer and thus more clearly visible.

5. DISCUSSION AND PERSPECTIVES

We have described three assays to place dyneins in a cortical configuration either on a barrier (Section 2), on the walls of a 2D microfabricated chamber (Section 3), or at the border of a 3D w/o emulsion droplet (Section 4). These different geometries provide different levels of information. The use of 1D geometries (barriers) in combination with centrosomes bound to the surface enables the study of single MTs interacting with dyneins at the barrier. The dynamics of MTs at the wall can be tracked in the presence/absence of dyneins. Though not detailed in this chapter, optical tweezers can be used to directly assess force generation (Laan & Dogterom, 2010; Laan, Pavin, et al., 2012). 2D microfabricated chambers enable another level of study, where one can assess the influence of confinement on centrosome positioning. The 3D w/o emulsion droplets better represent the three-dimensional nature of living cells and at the same time allow for potentially relevant mobility of dynein at the "cortex." However, 3D imaging and data treatment are more challenging than for 2D microfabricated chambers.

We use these assays to unravel the nature of MT end interactions with cortical dyneins and to understand centrosome positioning influenced by MT dynamics and dynein pulling at the cortex (Laan, Pavin, et al., 2012; Laan, Roth, & Dogterom, 2012). The versatility of these reconstituted systems allows for variations in the components in the reaction (molecular motors, MTs-associated proteins, etc.), the nature of the dynein construct, or the cell shape confinement, thereby mimicking specific *in vivo* configurations. One can also build up the system in step-by-step manner to explore additional regulation and complexity of cortical dynein function. Dynein's activity is *in vivo*, for example, tightly regulated by its cofactor dynactin, and Lys1/NudE (Kardon, Reck-Peterson, & Vale, 2009). These cofactors increase dynein's processivity and ability to remain bound under load (Kardon et al., 2009; King & Schroer, 2000; McKenney, Vershinin, Kunwar, Vallee, & Gross, 2010; Waterman-Storer, Karki, & Holzbaur, 1995). It could thus be interesting to use a recombinant full-length dynein and study the influence of these cofactors on (1) cortical dynein's interaction with single MTs and (2) positioning processes, where dynein is expected to work against higher loads. Also, dyneins could be placed at specific locations at the border, mimicking, for example, cortical dynein accumulation in a circumferential belt at the position of the spindle during mitosis in MDCK cells. In this case, dyneins are thought to influence the orientation of the spindle (Busson, Dujardin, Moreau, Dompierre, & De Mey, 1998). In a more elaborate version, a dynein construct could be modified to have, instead of a biotin tag, an FRKB protein, which upon addition of a dimerizer rapamycin can form the ternary complex FRB–rapamycin–FRKB (DeRose, Miyamoto, & Inoue, 2013). This would enable, in combination with caged rapamycin (Karginov et al., 2011) and FRB-biotin linked to the lipids, to trigger by light and in time the localization of dyneins at the border.

The geometry of the confinement could also be varied. Indeed, cell shape influences MT distribution, and thus the overall balance of forces driven by dyneins at the cortex during positioning processes (Pavin, Laan, Ma, Dogterom, & Julicher, 2012). In 2D, microfabricated chambers can be designed in any desired shape, simply by modifying the design of the mask. 3D w/o emulsion droplets could also be easily deformed by constraining them in channels, making them, for example, cylindrical to match the shape of fission yeast. Although suitable to mimic cells with a hard wall, our oil phase could also be adapted to lower the surface tension between oil and water, by varying the surfactants (Hashimoto, Garstecki, Stone, & Whitesides, 2008). This would enable deformations of the droplets by pushing MTs, and thus the study of the feedback that exists between cell shape and MT distribution, and its influence on centrosome positioning. Thus, in summary, reconstituted systems provide rich tools for mechanistically understanding biological processes that involve cortical dynein activity.

ACKNOWLEDGMENTS

We thank Sam Reck-Peterson's group, and, in particular, Sirui Zou, for help with dynein purification. We thank UMR144-CNRS at Institut Curie and, in particular, Claude Celati for help with the centrosome purification. We thank P. Tabeling's group, ESPCI, Paris, and, in particular, Bingqing Shen, for help with microfluidics. This work is part of the research program of the Foundation for Fundamental Research on Matter (FOM), which is part of the Netherlands Organization for Scientific Research (NWO). L. L. thanks the Human Frontiers Science Program for funding. M. D. acknowledges support from a NW0-ALW VICI Grant.

REFERENCES

Adames, N. R., & Cooper, J. A. (2000). Microtubule interactions with the cell cortex causing nuclear movements in Saccharomyces cerevisiae. *Journal of Cell Biology, 149*, 863–874.

Anna, S., Bontoux, N., & Stone, H. (2003). Formation of dispersions using "flow focusing" in microchannels. *Applied Physics Letters, 82*, 364–366.

Atencia, J., & Beebe, D. J. (2005). Controlled microfluidic interfaces. *Nature, 437*, 648–655.

Busson, S., Dujardin, D., Moreau, A., Dompierre, J., & De Mey, J. (1998). Dynein and dynactin are localized to astral microtubules and at cortical sites in mitotic epithelial cells. *Current Biology, 8*, 541–544.

Carminati, J. L., & Stearns, T. (1997). Microtubules orient the mitotic spindle in yeast through dynein-dependent interactions with the cell cortex. *Journal of Cell Biology, 138*, 629–641.

DeRose, R., Miyamoto, T., & Inoue, T. (2013). Manipulating signaling at will: Chemically-inducible dimerization (CID) techniques resolve problems in cell biology. *Pflugers Archiv: European Journal of Physiology, 465*, 409–417.

Dogterom, M., Felix, M. A., Guet, C. C., & Leibler, S. (1996). Influence of M-phase chromatin on the anisotropy of microtubule asters. *Methods in Cell Biology, 133*, 125–140.

Dujardin, D., & Vallee, R. (2002). Dynein at the cortex. *Current Opinion in Cell Biology, 14*, 44–49.

Garstecki, P., Fuerstman, M., Stone, H., & Whitesides, G. (2006). Formation of droplets and bubbles in a microfluidic T-junction—Scaling and mechanism of break-up. *Lab on a Chip, 6*, 437–446.

Green, N. (1990). Avidin and streptavidin. *Methods in Enzymology, 185*, 51–67.

Hashimoto, M., Garstecki, P., Stone, H., & Whitesides, G. (2008). Interfacial instabilities in a microfluidic Hele-Shaw cell. *Soft Matter, 4*, 1403–1413.

Kardon, J., Reck-Peterson, S., & Vale, R. (2009). Regulation of the processivity and intracellular localization of Saccharomyces cerevisiae dynein by dynactin. *Proceedings of the National Academy of Sciences of the United States of America, 106*, 5669–5674.

Kardon, J., & Vale, R. (2009). Regulators of the cytoplasmic dynein motor. *Nature Reviews Molecular Cell Biology, 10*, 854–865.

Karginov, A., Zou, Y., Shirvanyants, D., Kota, P., Dokholyan, N., Young, D., et al. (2011). Light regulation of protein dimerization and kinase activity in living cells using photocaged Rapamycin and engineered FKBP. *Journal of the American Chemical Society, 133*, 420–423.

King, S., & Schroer, T. (2000). Dynactin increases the processivity of the cytoplasmic dynein motor. *Nature Cell Biology, 2*, 20–24.

Koonce, M. P., & Samso, M. (1996). Overexpression of cytoplasmic dynein's globular head causes a collapse of the interphase microtubule network in Dictyostelium. *Molecular Biology of the Cell, 7*, 935–948.

Laan, L., & Dogterom, M. (2010). Chapter 31—In vitro assays to study force generation at dynamic microtubule ends. In W. Leslie & J. C. John (Eds.), *Methods in cell biology: Vol. 95. Microtubules, in vitro* (pp. 617–639). Amsterdam: Elsevier.

Laan, L., Pavin, N., Husson, J., Romet-Lemonne, G., van Duijn, M., Lopez, M. P., et al. (2012). Cortical dynein controls microtubule dynamics to generate pulling forces that position microtubule asters. *Cell, 148*, 502–514.

Laan, L., Roth, S., & Dogterom, M. (2012). End-on microtubule-dynein interactions and pulling-based positioning of microtubule organizing centers. *Cell Cycle, 11*, 3750–3757.

McKenney, R., Vershinin, M., Kunwar, A., Vallee, R., & Gross, S. (2010). LIS1 and NudE induce a persistent dynein force-producing state. *Cell, 141*, 304–314.

Moudjou, M., & Bornens, M. (1998). Method of centrosome isolation from cultured animal cells. In J. E. Celis (Ed.), *Cell biology: A laboratory handbook, Vol. 2*. London: Academic Press, pp. 111–119.

Nguyen, N., Ting, T., Yap, Y., Wong, T., Chai, J., Ong, W., et al. (2007). Thermally mediated droplet formation in microchannels. *Applied Physics Letters, 91*, 084102.

Nishiura, M., Kon, T., Shiroguchi, K., Ohkura, R., Shima, T., Toyoshima, Y. Y., et al. (2004). A single-headed recombinant fragment of Dictyostelium cytoplasmic dynein can drive the robust sliding of microtubules. *Journal of Biological Chemistry, 279*, 22799–22802.

Nisisako, T., Torii, T., Takahashi, T., & Takizawa, Y. (2006). Synthesis of monodisperse bicolored janus particles with electrical anisotropy using a microfluidic co-flow system. *Advanced Materials, 18*, 1152–1156.

O'Connell, C., & Wang, Y. (2000). Mammalian spindle orientation and position respond to changes in cell shape in a dynein-dependent fashion. *Molecular Biology of the Cell, 11*, 1765–1774.

Pavin, N., Laan, L., Ma, R., Dogterom, M., & Julicher, F. (2012). Positioning of microtubule organizing centers by cortical pushing and pulling forces. *New Journal of Physics, 14*. 105025.

Pedigo, S., & Williams, J. R. C. (2002). Concentration dependence of variability in growth rates of microtubules. *Biophysical Journal, 83*, 1809–1819.

Pfister, K., Shah, P., Hummerich, H., Russ, A., Cotton, J., Annuar, A., et al. (2006). Genetic analysis of the cytoplasmic dynein subunit families. *PLoS Genetics, 2*, 11–26.

Reck-Peterson, S. L., Yildiz, A., Carter, A. P., Gennerich, A., Zhang, N., & Vale, R. D. (2006). Single-molecule analysis of dynein processivity and stepping behavior. *Cell, 126*, 335–348.

Robinson, J., Wojcik, E., Sanders, M., McGrail, M., & Hays, T. (1999). Cytoplasmic dynein is required for the nuclear attachment and migration of centrosomes during mitosis in Drosophila. *Journal of Cell Biology, 146*, 597–608.

Romet-Lemonne, G., VanDuijn, M., & Dogterom, M. (2005). Three-dimensional control of protein patterning in microfabricated devices. *Nano Letters, 5*, 2350–2354.

Sharp, D. J., Rogers, G. C., & Scholey, J. M. (2000). Cytoplasmic dynein is required for poleward chromosome movement during mitosis in Drosophila embryos. *Nature Cell Biology, 2*, 922–930.

Stone, H., Strook, A., & Ajdari, A. (2004). Engineering flows in small devices: Microfluidics toward a lab-on-a-chip. *Annual Review of Fluid Mechanics, 36*, 31.

Swan, A., Nguyen, T., & Suter, B. (1999). Drosophila Lissencephaly-1 functions with Bic-D and dynein in oocyte determination and nuclear positioning. *Nature Cell Biology, 1*, 444–449.

Taberner, N., Weber, G., You, C., Dries, R., Piehler, J., & Dogterom, M. (2014). Reconstituting functional microtubule-barrier interactions. In M. Piel & M. Thery (Eds.), *Methods in cell biology: Vol. Micropatterning in cell biology part A*. Amsterdam: Elsevier.

Teh, S., Lin, R., Hung, L., & Lee, A. (2008). Droplet microfluidics. *Lab on a Chip*, *8*, 198–220.

Waterman-Storer, C., Karki, S., & Holzbaur, E. (1995). The p150(glued) component of the dynactin complex binds to both microtubules and the actin related protein centractin (arp-1). *Proceedings of the National Academy of Sciences of the United States of America*, *92*, 1634–1638.

Xu, J., Li, S., Lan, W., & Luo, G. (2008). Microfluidic approach for rapid interfacial tension measurement. *Langmuir*, *24*, 11287–11292.

Xu, J., Li, S., Tan, J., Wang, Y., & Luo, G. (2006). Preparation of highly monodisperse droplet in a T-junction microfluidic device. *AIChE Journal*, *52*, 3005–3010.

Yobas, L., Martens, S., Ong, W., & Ranganathan, N. (2006). High-performance flow-focusing geometry for spontaneous generation of monodispersed droplets. *Lab on a Chip*, *6*, 1073–1079.

Zhou, C., Yue, P., & Feng, J. (2006). Formation of simple and compound drops in microfluidic devices. *Physics of Fluids*, *18*. 092105.

CHAPTER THIRTEEN

Reconstitution of Microtubule-Dependent Organelle Transport

Pradeep Barak, Ashim Rai, Alok Kumar Dubey, Priyanka Rai, Roop Mallik[1]

Department of Biological Sciences, Tata Institute of Fundamental Research, Colaba, Mumbai, Maharashtra, India

[1]Corresponding author: e-mail address: roop@tifr.res.in

Contents

Abstract

Microtubule (MT)-based motor proteins transport many cellular factors to their functionally relevant locations within cells, and defects in transport are linked to human disease. Understanding the mechanism and regulation of this transport process in living cells is difficult because of the complex *in vivo* environment and limited means to manipulate the system. On the other hand, *in vitro* motility assays using purified motors attached to beads does not recapitulate the full complexity of cargo transport *in vivo*. Assaying motility of organelles in cell extracts is therefore attractive, as natural cargoes are being examined, but in an environment that is more amenable to manipulation. Here, we describe the purification and *in vitro* MT-based motility of phagosomes from *Dictyostelium* and lipid droplets from rat liver. These assays have the potential to address diverse questions related to endosome/phagosome maturation, fatty acid regulation, and could also serve as a starting point for reconstituting the motility of other types of organelles.

Methods in Enzymology, Volume 540
ISSN 0076-6879
http://dx.doi.org/10.1016/B978-0-12-397924-7.00013-3

1. INTRODUCTION

Motor proteins drive active ATP-dependent motion of organelles along cytoskeletal filaments, a process that is necessary for protein homeostasis, secretion, and for many other vital cellular functions (Barlan, Rossow, & Gelfand, 2013; Vale, 2003). The dynein and kinesin motors transport membranous organelles as well as other cargo over long distances along microtubules (MTs). Multiple motors, in complex with motor-associated regulatory proteins, usually comprise the machinery that transports a single organelle (Gross, Vershinin, & Shubeita, 2007; Mallik & Gross, 2004). The emergent behavior of these multimotor complexes are poorly understood and cannot be necessarily inferred from the properties of single-motor proteins (Mallik, Rai, Barak, Rai, & Kunwar, 2013; Rai, Rai, Ramaiya, Jha, & Mallik, 2013).

A number of proteins, such as dynactin, Lis1, kinesin accessory proteins, miro–milton, JIP, etc., interact with MT motors to regulate their motion (Verhey & Hammond, 2009; Vallee, McKenney, & Ori-McKenney, 2012). This regulation controls many motor properties such as ATPase rate, affinity for cargo or filament, force generation, geometrical arrangement of motors on the cargo, as well as other properties. The sum total of this regulation is manifested as the motion of organelles inside cells, which can be easily observed. However, the interpretation of this motion is often difficult because: (1) The microarchitecture of the dense cytoskeletal network is unknown and likely modulates the motion in unpredictable manner, (2) Additional protein factors besides motors and their binding partners may influence motion (e.g., MT binding proteins). Furthermore, interfering with the function of specific motors/motor-associated proteins *in vivo* by RNAi or dominant negative constructs may lead to unknown pleiotropic effects.

To address some of these problems, reductionist *in vitro* approaches have been adopted in which purified motors are coated on artificial plastic beads which then move as cargo along single cytoskeletal filaments adhered to a coverslip. Such assays have revealed basic parameters of single-motor function such as processivity, velocity, force, step-size, and force-velocity correlations (Mallik, Carter, Lex, King, & Gross, 2004; Rice, Purcell, & Spudich, 2003; Svoboda, Schmidt, Schnapp, & Block, 1993). However, the artificial motor-cargo attachment in these assays prevents a direct extrapolation of these results to *in vivo* transport. An alternative approach lies in using the motor-driven motility of organelles in cell extracts, which permits interrogation of endogenous motor complexes on native membranes. Organelle motility has been reproduced in cell extract for neuronal vesicles (Brady, Lasek, & Allen, 1982; Vale, Schnapp, Reese, & Sheetz, 1985), endosomes (Murray, Bananis, & Wolkoff, 2000), pigment granules (Rogers, Tint, Fanapour, & Gelfand, 1997), pollen tube organelles (Romagnoli, Cai, & Cresti, 2003), mRNP complexes (Sladewski, Bookwalter, Hong, & Trybus, 2013), virus particles (Lee, Murray, Wolkoff, & Wilson, 2006), and lipid droplets (LDs; Barak, Rai, Rai, & Mallik, 2013; Bartsch, Longoria, Florin, & Shubeita, 2013). The reduced complexity in these assays (compared to *in vivo* transport) provides an opportunity to ask questions that otherwise may not be possible inside of cells. For example, drugs, antibodies, or peptides that inhibit a specific motor can be added to identify the motor involved in a transport function and investigate the consequences of its inhibition (Barak et al., 2013). Once a motor is identified, pull-down experiments using purified organelles and antibodies against the motors can be used to identify motor-interacting regulatory proteins. To interrogate how these proteins regulate motors, a function-blocking peptide that mimics the motor-binding domain of the regulator can be added to the organelle transport assay. Furthermore, to probe connections between metabolism and motor activity, organelles can be isolated from animals subjected to specific metabolic conditions and their motility investigated and compared (Barak et al., 2013). Last but not the least, quantitative force measurements of motors on organelles by optical trapping is now possible in cell extract (Barak et al., 2013). This permits a controlled interrogation of the biophysical properties of motors at near-cellular complexity on the surface of organelles.

Cell extract assays are not a substitute for studying *in vivo* motility, but rather a powerful companion and a way to build up to the complexity of intracellular transport (Mallik et al., 2013). However, there are challenges. Cytosolic factors important for transport can get diluted/inactivated in cell extract, and the ionic concentrations may also be different. Therefore, efforts must be made to check that motility of the organelle in cell extract is similar to its motility *in vivo* (at least to first approximation). Here, we present methods to reconstitute motility of latex bead phagosomes (LBPs) in *Dictyostelium* cell extracts and LDs purified from rat liver. We have used these assays to address questions such as mechanical competition (tug-of-war) between opposing motors (Soppina, Rai, & Mallik, 2009; Soppina, Rai, Ramaiya, Barak, & Mallik, 2009), transition from single-molecule to collective motor function (Rai et al., 2013; Soppina, Rai, & Mallik, 2009; Soppina, Rai, Ramaiya, et al., 2009), and possible role of motors in maintaining lipid homeostasis in the liver (Barak et al., 2013).

2. ISOLATION AND MOTILITY OF LBPs FROM *DICTYOSTELIUM*

Most organelles move along MTs in a bidirectional (back-and-forth) manner, with frequent reversals between directed segments of motion (Welte, 2004). This motion can be regulated in time through cellular cues, resulting in bias of net transport towards plus or minus direction. Understanding the molecular basis of this regulation is important, because it governs the spatiotemporal distribution of each particular organelle. A large body of evidence now attributes bidirectional transport to the activities of oppositely directed teams of kinesin and dynein motors on single organelles (Welte, 2004). To understand bidirectional transport, it is important to measure the activities of these motor teams in real-time, for example by studying the forces generated by motor teams using an optical trap (Rice et al., 2003).

Phagosome motility can be conveniently measured by having cells engulf latex beads from the medium. These LBPs (Fig. 13.1) acquire a bilayer

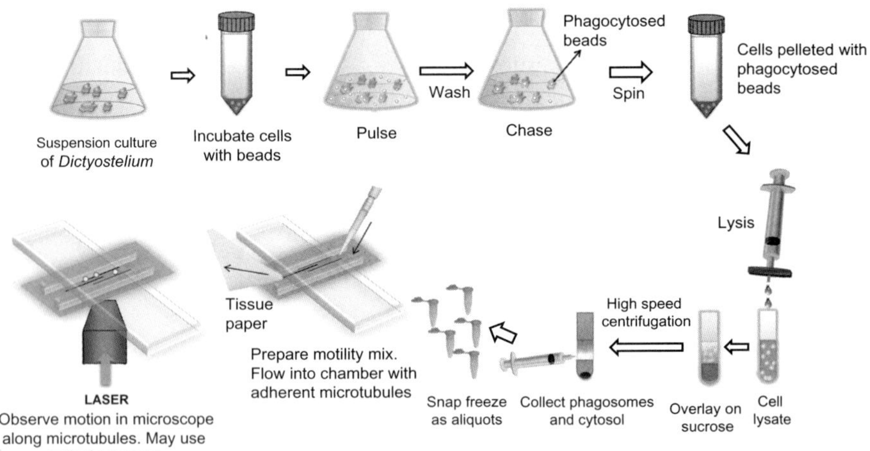

Figure 13.1 Preparation and motility of latex bead phagosomes (LBPs) from *Dictyostelium*. The steps of phagocytosis and phagosome isolation are shown. LBP aliquots are frozen and can be used later to assay motility in a flow cell. An optical trap can be used to assay force generation by motors that are attached endogenously on the LBPs. See text for details. (For color version of this figure, the reader is referred to the online version of this chapter.)

membrane through a maturation process inside cells (Desjardins & Griffiths, 2003) and bind endogenous motor proteins on their surface, allowing them to move along MTs. When extracted from cells, they also move in an *in vitro* motility assay (Blocker et al., 1997; Rai et al., 2013). The motility of LBPs in a *Dictyostelium* cell extract is similar to that inside of *Dictyostelium* cells, suggesting that the motor machinery on extracted LBPs is largely intact and native like. Because of their engulfed bead, LBPs are highly refractile, which makes them easily distinguishable from other cellular organelles and allows large forces to be exerted on them by an optical trap (Rai et al., 2013). Their shape is spherical and their size precisely known, making it easy to calibrate optical traps on LBPs for force measurements both inside and outside cells. The duration for which LBPs mature inside cells (pulse + chase; see Fig. 13.1) can be controlled precisely. Thus, changes in motility as a function of phagosome maturation can be examined. Several pathogenic organisms (e.g., salmolella, tuberculosis bacterium) arrest phagosome maturation, making it interesting to look for associated changes in MT-dependent motility. Because of their high buoyancy, LBPs can be separated easily from cytosol and other organelles in cell extract, which permits biochemical detection of motors/motor-associated proteins on the LBP (Blocker et al., 1997; Desjardins & Griffiths, 2003). These advantages make *Dictyostelium* cell extract an attractive model system for studying the MT-based motility of LBPs.

In this section, we describe protocols to purify motile LBPs in the early and late stages of phagocytic maturation from *Dictyostelium* cells. *Dictyostelium* is an established model system to study the phagocytic pathway and MT-dependent motion (Koonce, 2000; Pollock, de Hostos, Turck, & Vale, 1999; Soppina, Rai, & Mallik, 2009; Soppina, Rai, Ramaiya, et al., 2009). Being accessible to genetic manipulation and RNAi, *Dictyostelium* has been used extensively for functional and structural studies on motor proteins (Kon et al., 2012). Purified LBPs move vigorously on *in vitro* polymerized MTs, with the motion being different for early phagosomes (bidirectional motion) versus late phagosomes (largely minus-directed motion). If healthy cells are used, and the steps described below are followed, we find that 60–70% of LBPs move when placed on MTs using an optical trap. Biochemical quantities of LBPs can be purified easily, because *Dictyostelium* cells are highly phagocytic and can be grown in large quantity in suspension culture using inexpensive medium and equipment. The MT-dependent motility of LBPs in *Dictyostelium* is driven by an Unc104 kinesin (DdUnc104) and cytoplasmic dynein (Pollock et al., 1999;

Pollock, Koonce, de Hostos, & Vale, 1998; Soppina, Rai, & Mallik, 2009; Soppina, Rai, Ramaiya, et al., 2009). LBPs are also motile inside of mammalian cells, and it is possible to reconstitute their motion along MTs *in vitro* (Blocker et al., 1997). However, the incidence of motility of mammalian LBPs appears lower compared to the *Dictyostelium* extract.

2.1. Materials

2.1.1 Buffers

Sorensen's buffer: 15 mM KH_2PO_4, 2 mM Na_2HPO_4 (pH 6.0).

Lysis buffer (*LB-30*): 30 mM Tris–HCl (pH 8), 4 m MEGTA, 3 m MDTT, 5 mM benzamidine, 10 μg/mL soybean trypsin inhibitor, 5 μg/mL TPCK/TAME, 10 μg/mL leupeptin, 10 μg/mL pepstatin A, 10 μg/mL chymostatin, and 5 mM PMSF (phenylmethyl-sulfonylfluoride) containing 30% (w/v) sucrose.

Lysis buffer (*LB-25*): As above, except for 25% (w/v) sucrose.

2.1.2 Cells and culture medium

Dictyostelium AX-2 strain cells at a cell density of 4–6 × 10^6cells/mL. Total cell number required is 4–6 × 10^8 cells. Cells may be obtained from *Dictyostelium* stock center (Northwestern University, Chicago, IL, USA). For culturing cells, we use HL-5 medium with glucose (Formedium Cat# HLG0102) according to manufacturer specifications.

2.1.3 Other requirements

Carboxylated polystyrene beads of 500 nm diameter or larger are used, for example Polybead ® Carboxylate Microspheres (750 nm diameter, Cat# 07759-15, Polysciences, USA).

Polycarbonate filter for lysing cells (5 μm pore size, 25 mm diameter; Sartorius Cat# 16517E).

Other materials and equipment: syringe (2 mL), eppendorf table top centrifuge, Beckman centrifuge, Beckman table top ultracentrifuge, Beckman JA-10 rotor and tubes, Beckman MLS-50 rotor and tubes, and liquid nitrogen (for snap freezing)

2.2. Purification protocol

2.2.1 Preparation of beads for phagocytosis

Beads should be used within 12 months of manufacture. Prepare fresh from stock solution for each phagocytosis assay. 250 μL latex beads are spun (11000 × *g*, 5 min, 4 °C) to pellet the beads. The supernatant is removed, the beads are resuspended in 1 mL of HL-5 medium, and the wash step is repeated twice. Washing with HL-5 medium improves phagocytic efficiency. Beads are finally resuspended in 0.5 mL of ice-cold Sorensen's buffer. The bead solution is vortexed briefly, sonicated for 10 min in a bath sonicator, and stored on ice till further use.

2.2.2 Phagocytosis

100 mL of *Dictyostelium* AX-2 cells are harvested at a cell density of 4–6 × 10^6cells/mL and centrifuged to pellet the cells (900 × *g*; 22 °C; 5 min).The supernatant is removed and the cell pellet is immediately kept on ice. It is important that the cells be kept on ice after pelleting. Preferably, the whole procedure should be carried out in a cold room. The cell pellet is resuspended in 5 mL of ice-cold Sorensen's buffer. Beads prepared as above are added to the cell suspension, mixed gently, and kept on a rotator at 4 °C for 15 min. This preincubation of beads with cells is important for synchronization of phagocytosis in all the cells.

After this time, the bead-cell suspension is added to 100 mL of HL-5 and kept in a 500-mL conical flask at 22 °C in a shaking incubator (this is the "pulse" duration). Depending on whether early or late phagosomes are being prepared, the cells are pulsed for different durations of time.

Early phagosome prep: 5 min pulse.

For *Dictyostelium*, the early phagosome stage has been shown to last for less than 5 min. After this duration, the late phagosomal marker Rab7 appears on phagosomes (Rupper, Grove, & Cardelli, 2001). Therefore, maintaining the pulse duration for exactly 5 min is important when preparing early phagosomes.

Late phagosome prep: 15 min pulse.

The 15-min pulse maximizes the bead uptake and phagosome yield, while still maintaining a homogeneous yield of late phagosomes. The pulse is stopped by quickly adding cells to 330 mL of ice-cold Sorensen's buffer kept in a JA-10 rotor tube. Adding the cells to ice-cold buffer immediately after pulse and/or chase is important to arrest the maturation of phagosomes. The cell suspension is washed to remove beads that were not phagocytosed by spinning at 900 × *g* for 5 min at 4 °C. The cell pellet is resuspended in 5 mL of ice-cold HL-5 medium.

The phagocytosed beads are now "chased" inside cells to allow maturation. The cell suspension is added to 100 mL of HL-5 kept in a 500-mL conical flask at 22 °C in a shaking incubator. The chase duration can be changed to allow phagosomes to mature to different stages, as described below.

Early phagosomes: No chase, only 5 min pulse is sufficient.

Late phagosomes: 45 min chase.

Chase is stopped by adding the cell suspension to 330 mL of ice-cold Sorensen's buffer kept in JA-10 rotor tube. Cells are centrifuged (900 × *g* for 5 min at 4 °C) to obtain a cell pellet. The cell pellet is kept on ice and washed by resuspending in 50 mL of ice-cold Sorensen's buffer and spinning at 900 × *g* for 5 min at 4 °C to remove beads that are not phagocytosed. The washing of cells with ice-cold Sorensen's buffer is repeated twice.

2.2.3 *Cell lysis and phagosome isolation*

The cell pellet is weighed, kept on ice, and resuspended in lysis buffer (LB/30% sucrose). As a typical example, 1 mL lysis buffer is used with 1 gm cell pellet. Cells are lysed on ice by passing once through a 5-μm polycarbonate filter using a 2-mL syringe. Proper cell lysis is critical to obtain a motile fraction of phagosomes. The efficiency of lysis should be checked by observing the lysate under a phase contrast microscope at 20 × magnification. Cells that are not lysed would appear white, while lysed cells appear dark. Ensure that > 80% of cells are lysed.

The cell lysate is now layered over a 1-mL cushion of LB/25% sucrose in an MLS-50 rotor tube and centrifuged at 100,000 × *g* for 20 min at 4 °C. After the spin, the tube is kept on ice. The phagosome–cytosol mixture present above the LB/25% sucrose cushion is collected, immediately snap frozen in liquid nitrogen as 40 μL aliquots and then stored in liquid nitrogen for further use. Frozen aliquots show robust motility for ~7 days.

2.3. Motility assay

The *in vitro* motility assay for LBPs is very similar to LDs, and has therefore been described together with LDs later. Wherever necessary, important distinctions between the two assays have been pointed out.

3. ISOLATION AND MOTILITY OF LDs FROM RAT LIVER

Almost all cells contain cytosolic LDs, organelles that store esterified fat in the form of triacyl glycerol and cholesterol-esters bounded by a monolayer of phospholipids (Farese & Walther, 2009; Goodman, 2008). A limited set of proteins embed in the monolayer phospholipid membrane, and may in turn recruit other proteins. Interestingly, *in vitro* experiments show that the Arf1–COP1 complex assembles on LDs to pinch off small portions of the protective phospholipid monolayer, thereby increasing the reactivity of the LD and promoting its interaction with cytosolic proteins (Thiam et al., 2013). Indeed, a large number of proteins (Brasaemle, Dolios, Shapiro, & Wang, 2004) including motors (Welte, 2009) associate in transient manner with LDs. This transient association may be a reason why LDs are highly motile (Welte, 2009), moving to and interacting with many other cellular organelles (e.g., mitochondria, ER, peroxisomes, endosomes). This motion of LDs may regulate dynamics of lipid in the liver. Molecules such as Arf1 and Rab18, that associate with LDs regulate lipid homeostasis and also belong to a family of GTPases that are master-regulators of organelle transport. Impaired LD dynamics is associated with disorders in lipid metabolism and insulin resistance in insulin-responsive tissues like adipose tissue, cardiac and skeletal muscles, and liver (Farese & Walther, 2009; Goodman, 2008; Herker & Ott, 2012). Dramatic examples of LD dynamics can be seen in their reorganization in response to exercise and fasting in adipocytes and hepatocytes. The dynamics of hepatic LDs is largely unexplored, but has implications for fatty liver diseases and also for understanding the lifecycle of HIV and dengue virus (Herker & Ott, 2012).We have recently explored a possible regulatory connection between motor-dependent LD motion and metabolism. Specifically, we have reported differences in the MT-dependent motility of LDs extracted from the liver of normal versus fasted rats (Barak et al., 2013). The observed motility of LDs, when combined with biochemical assays and imaging of LD motion in hepatocytes, may reveal connections between motor protein activity and metabolic signals, particularly in the context of fatty acid regulation.

LDs exhibit MT-based motility in *Drosophila* embryos (Shubeita et al., 2008), embryo extract (Bartsch et al., 2013), cultured cells (Targett-Adams et al., 2003; Wang et al., 2010; Welte, 2009), and in rat liver extract (Barak et al., 2013). LDs are also suitable for quantitative examination of motor protein activity/force generation because they are spherical and highly refractile, allowing large calibrated optical forces to be exerted on them by an optical trap (Barak et al., 2013; Bartsch et al., 2013; Shubeita et al., 2008). Lastly, their buoyancy allows highly pure and large quantities of LDs to be isolated from tissues and cells for biochemical assays using a sucrose gradient (Barak et al., 2013; Brasaemle & Wolins, 2006). Here, we present protocols to reconstitute robust motility of LDs from the liver of normally fed rats. A large fraction (~75%) of these LDs move when brought in contact with MTs using an optical trap (Barak et al., 2013). We however caution that only occasional motility of LDs has been observed in hepatocytes in cell culture or in primary culture hepatocytes from rat liver. A possibility exists that the *ex vivo* LD motion is an artifact. However this appears unlikely, considering the reduction in LD motion upon fasting. Further, we observed no effect of a salt wash on the LD motion (Barak et al., 2013) suggesting that the motion is not an artifact from contaminating motors binding to LDs through weak electrostatic interactions during purification. It is possible that the *in vitro* LD motion reflects an unknown stimulus present in the liver under normal feeding conditions.

3.1. Materials

MEPS buffer

Consists of magnesium sulfate (5 mM), EGTA (5 mM), pipes (35 mM), and sucrose. Buffers with different molarity of sucrose (0.25, 0.5, 0.9, 1.2, 1.4, and 2.5 M) are prepared before the start of the experiment. Adjust all buffers to pH 7.2.

Lysis buffer

MEPS buffer with 0.9 M sucrose is used as lysis buffer. For purification of LDs, 10 gm of liver is lysed in 15 mL of lysis buffer in a Dounce homogenizer using three strokes. These details are kept constant for every LD purification.

Protease inhibitors

1× (or 2× during lysis step) Roche protease inhibitor cocktail (Cat# 11697498001, Roche Diagnostics) with DTT (4 mM), pepstatin A (8 μg/mL). PMSF is added to a final concentration of 4 mM final wherever mentioned. Roche protease inhibitor cocktail is prepared according to manufacturer's instructions.

Gel filtration

S-200 Sephacryl beads (GE healthcare, Cat# 17-0584-01) is washed thrice in 1× PBS, then twice in lysis buffer. Final suspension of beads is done in lysis buffer with all the protease inhibitors (1× without PMSF).

Centrifugation

Low speed centrifugation is done in an Eppendorf 5810 swinging bucket centrifuge in falcon tubes. Ultra high speed centrifugation is done in a Beckman SW41 swinging bucket rotor.

Animals

3- to 4-month-old male Sprague Dawley rats are used to isolate liver and prepare LDs. Rats are bred and maintained in an animal house facility according to protocols approved by Institutional Animal Ethics

Committee (IAEC), formulated for the purpose of control and supervision of experiments on animals in India.

3.2. Purification of LDs from rat liver

The protocol (Fig. 13.2) is an adaptation of published methods to purify endosomes from rat liver (Murray et al., 2000). Anesthetize a 3- to 4-month-old male Sprague Dawley rat with intraperitoneal injection of 40 mg/kg sodium thiopentone. Dissect rat to expose abdomen and liver. Perfuse liver through hepatic portal vein with 20–30 mL of ice-cold PBS to clear blood from the liver and also to chill the liver rapidly. Remove liver from abdominal cavity and wash with ice-cold PBS in a petri dish kept on ice. Retain only 10 gm of liver in the petri dish and discard the rest. Add 15 mL of ice-cold 0.9 M MEPS buffer with protease inhibitors (PI). PI here consists of 2 × Roche protease inhibitor cocktail with 8 μg/mL of pepstatin A, 4 mM DTT and 4 mM PMSF. Mince liver with clean razor/scissors into

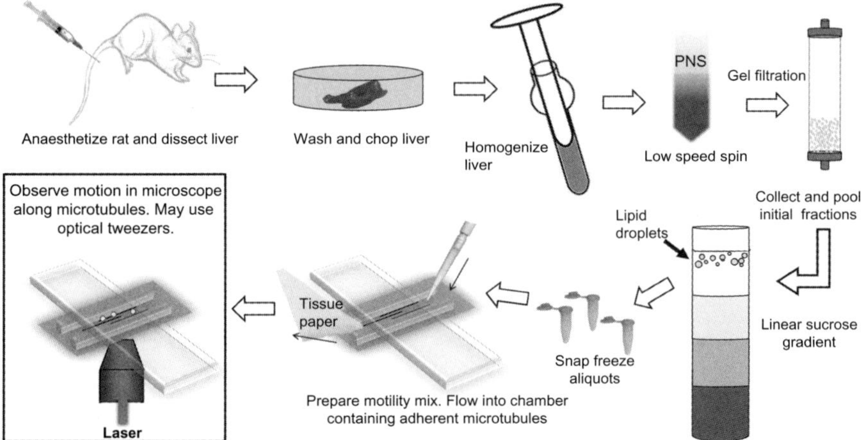

Figure 13.2 Preparation and motility of lipid droplets from rat liver. The liver is dissected out from a rat, perfused and washed, and then homogenized in a Dounce homogenizer. The homogenate is centrifuged in a falcon tube to obtain post nuclear supernatant (PNS). PNS is loaded on a S-200 gel filtration column and 4 mL of the whitish flow-through containing organelles is collected. This is loaded at the bottom of a sucrose step gradient to isolate buoyant LDs in the top fraction. LDs are snap-frozen in 40 μL aliquots. For motility, an aliquot is thawed, mixed gently with ATR (=ATP regenerating system; see text) and pipetted with constant wicking from other side using a tissue paper into a flow cell containing MTs stuck to a coverslip. Motility of LDs can be observed at 37 °C in a DIC microscope. An optical trap may be used to position single LDs on an MT. (For color version of this figure, the reader is referred to the online version of this chapter.)

small pieces. Transfer to 15 mL Dounce homogenizer and lyse with three loose strokes to obtain liver homogenate. Spin the liver homogenate at 1811 × *g*, 4 °C, 10 min in Eppendorf swinging table top rotor to obtain 15 mL post nuclear supernatant (PNS). PNS contains cytosol and all other organelles including LDs. Add protease inhibitors (1 × Roche PI cocktail with 8 μg/mL of pepstatin A, 4 mM DTT) to PNS. Do not add PMSF at this step.

To separate organelles from cytosol, PNS is loaded onto an S-200 gel filtration column (bed volume 40 mL) in cold room at 4 °C. This separates large organelles including LDs from cytosolic proteins within few minutes. This step is crucial for motility as liver cytosol has high protease and ATPase activity. The separation must be rapid, and any block in the column at this stage kills the motility of LDs. We have observed that motile LDs lose motility upon adding separately prepared liver cytosol to them.

Collect first 4 mL of flow-through (whitish front in column) and supplement with PI (1 × Roche protease inhibitor cocktail with 8 μg/mL of pepstatin A and 4 mM DTT). Do not add PMSF. Adjust the 4-mL flow-through to >1.4 M sucrose by adding equal volume of 2.5 M MEPS (with 1 × PI, DTT, and pepstatin A). Load this at the bottom of two SW41 tubes (4 mL/tube). Overlay with 2 mL solutions (each) of 1.4 M MEPS, 1.2 M MEPS, 0.5 M MEPS in both the tubes. All the solutions should have PI with DTT and pepstatin A. Spin the gradient at 120,000 × *g* for 60 min at 4 °C in a SW41 rotor. The fraction on top will have whitish layer of LDs. Snap freeze 40 μL aliquots and store in liquid nitrogen. Frozen LD aliquots are motile upto 2 months.

3.3. Motility of LBPs and LDs

Motility assays require preparation of MTs and a flow cell chamber in which motility is observed. These steps are similar for LBPs and LDs.

3.3.1 Preparation of MTs and flow cell

Tubulin (~10 mg/mL) is purified from goat brain using two cycles of polymerization and depolymerization followed by removal of MAPs in high molarity PIPES buffer (Castoldi & Popov, 2003). Purified tubulin is polymerized at 37 °C in a water bath to obtain MTs in the presence of GTP (1 mM) and taxol (20 μM). For motility assays, this MT solution is diluted in BRB80 buffer containing GTP (1 mM) and taxol (20 μM). In motility assays using a purified motor, the beads move in a predetermined direction depending on the motor of choice. However, organelles can move

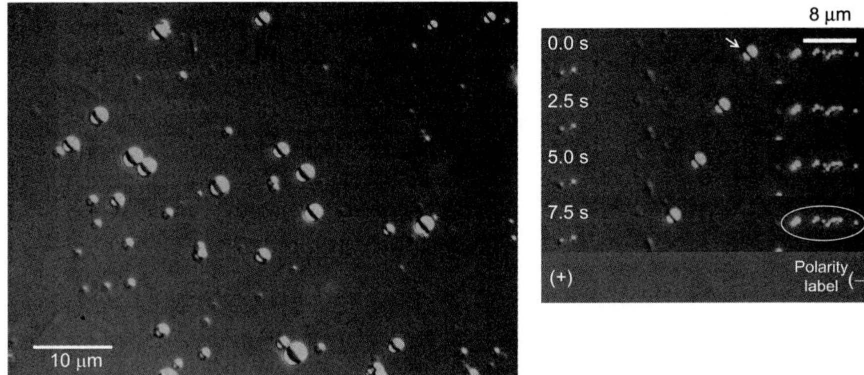

Figure 13.3 Lipid droplets isolated from rat liver and their motility. *Left panel*: A field of view showing LDs (refractile spheres) isolated from the liver of a normally fed rat under differential interference contrast imaging (100 × objective). *Right panel*: Motion of a LD (arrow in image at 0 s) along a polarity labeled MT (label on the right; indicated by a white circle in the image at 7.5 s). The LD moves toward the plus end of the MT which is horizontal in the field of view and is stuck to the coverslip. The MT is not visible because it is in low contrast and is slightly below the horizontal plane in focus in this view.

bidirectionally due to presence of motors of opposing polarity. To identify the active motor at any instant, motility must be assayed on MTs having a polarity label (Fig. 13.3). We have developed methods to polarity label the MT minus ends with non fluorescent avidin coated magnetic beads (Soppina, Rai, & Mallik, 2009; Soppina, Rai, Ramaiya, et al., 2009). If polarity labeling is being attempted according to Soppina, Rai, and Mallik (2009) and Soppina, Rai, Ramaiya, et al. (2009), then the casein used to block coverslips should be first dialysed to remove trace amounts of biotin. This reduces nonspecific binding of magnetic beads to the coverslip.

Motility is observed in a flow cell (volume ~20 µL) prepared by sticking a microscope coverslip to a slide using two strips of double-stick tape (3 M; Cat# 137DM-2) placed 3–5 mm apart (Barak et al., 2013; Mallik et al., 2004; Rice et al., 2003; Soppina, Rai, & Mallik, 2009; Soppina, Rai, Ramaiya, et al., 2009). Coverslips are immersed overnight in acetone, and then cleaned in a plasma cleaner (Harrick Plasma, USA, PDC-32G-2) at medium RF value and 1300 mT oxygen pressure for 30 min. Cleaned coverslips are coated with poly-L-lysine (Sigma, P8920) to stick MTs to the coverslip. Polylysine solution (300 µL poly-L-lysine + 4 mL methanol + 36 mL ethanol) is prepared in a 100-mL beaker. Plasma cleaned coverslips are incubated in polylysine solution for 30 min and then dried in an oven for 10 min at 100 °C. These coverslips are used to prepare flow cells using double-stick tape.

To image in an inverted microscope, the coverslip is kept at the bottom with MTs sticking to its inner surface. The above protocol should show 5–10 MTs stuck to the coverslip when observed through a camera (100 × objective; 50 μm × 50 μm field of view). The coverslip surface is blocked for 30 min using casein (5.5 mg/mL in BRB80 buffer) to prevent nonspecific binding and inactivation of motors. The casein solution is supplemented with GTP (1 mM) and taxol (20 μM) before use.

3.3.2 Preparation of motility mix and observation of motility

For LDs, motility is not affected till ~2 months. We do not use LDs after 2 months, as they start sticking to the surface of the coverslip. We do not use phagosomes for more than a week for motility. After 1 week, we do observe a reduction in the motile fraction of LBPs.

Motility mix for LBPs

The frozen LBPs are taken out from liquid nitrogen, quickly thawed, and kept immediately on ice. LBPs are diluted fourfold in LB-15% sucrose. ATP regenerating system (ATR = 20 mM ATP, 20 mM $MgCl_2$, 40 mM creatine phosphate and 40 U/mL creatine kinase) is added to a final ATP concentration of ~1 mM.

Motility mix for LDs

Rapidly thaw 40 μL of frozen LD fraction and gently mix with 2.2 μL ATR. Final ATP concentration is ~1 mM.

Observing Motility

Motility mix (for LBPs or LDs) is introduced into the flow cell containing MTs by pipetting from one end with constant suction from the other end using a small piece of tissue paper (Figs. 13.1 and 13.2). The flow cell is then sealed from both ends with vaseline before observation to prevent evaporation. Each slide is observed for 20 min or less at 37 °C in a microscope (100 × objective, DIC imaging) equipped with a temperature controller. Typically 4–5 LBPs or LDs should be seen in a field of view. LBPs/LDs diffusing in solution are captured using an optical trap and placed gently on MTs. We find that ~60% of LBPs move when placed on MTs using an optical trap. Around 75% of LDs purified from liver of normally fed rats move (see Fig. 13.3) over long distance along MTs (Barak et al., 2013).

Optical trapping is normally done on an inverted microscope. Being buoyant, LDs gradually float to the top of the flow cell (away from the objective focus). Thus, motility should be observed immediately after flowing LDs into the flow cell. The objective should be focussed slightly deep (4–5 μm) into the solution to trap a LD, and then bring it down onto the MTs. If an optical trap is not being used, it may be advantageous to work in an inverted geometry (upright microscope, coverslip on top, MTs sticking to the inner surface) so that LDs float up to the MTs, attach, and move.

3.4. Analysis of organelle transport

The analysis of motility has been described earlier (Barak et al., 2013; Rai et al., 2013; Soppina, Rai, & Mallik, 2009; Soppina, Rai, Ramaiya, et al., 2009), and depends on the experiments. Single organelles can be tracked with subpixel resolution (translating to ~5 nm precision in our setup) by video tracking at 30 frames/s. This permits parsing of motion into segments of constant velocity to obtain biophysical characteristics of motion (velocities and fractions of plus/minus motion, pause durations, frequency of reversals, etc.,). Video tracking can also detect steps taken by motors (Rai et al., 2013). Change in shape of deformable organelles such as endosomes can be quantified and interpreted in terms of force from opposing motor teams during motion (Soppina, Rai, & Mallik, 2009; Soppina, Rai, Ramaiya, et al., 2009). However, no discernible changes are observed for rigid cargoes like LBPs and LDs.

Force generated by motor teams against an optical trap is obtained by measuring with nanometre precision the displacement (X) of cargo from trap center (Rice et al., 2003). The optical trap works as a linear spring of stiffness K_{TRAP} upto ~150 nm from the trap center, yielding the instantaneous force $F_{MOTORS} = K_{TRAP} \times X$. We have described such measurements on LDs (Barak et al., 2013) and LBPs (Rai et al., 2013). K_{TRAP} can be obtained from thermal fluctuations of a trapped LBP/LD measured at 40 KHz (or higher temporal resolution) using a quadrant photodetector arrangement (Rice et al., 2003).

4. CONCLUSIONS

We have presented assays to reconstitute the motion of LBPs and LDs, organelles relevant to phagosome and lipid biology. In the overwhelming majority of *in vitro* motility experiments, purified motors coated on artificial plastic beads are observed. Many limitations of these assays can be overcome by replacing beads with motile organelles. Thus, native motor complexes could be interrogated in controlled assays to infer biological function and regulation of motors. In this direction, we are currently studying connections between kinesin activity on LDs and the high lipolytic flux that may exist in liver under normal dietary conditions.

ACKNOWLEDGMENTS

R. M. acknowledges funding through an International Senior Research Fellowship from the Wellcome Trust UK (grant WT079214MA), and also a Wellcome Trust—Department of Biotechnology Alliance Senior Fellowship (grant IA/S/11/2500255).

REFERENCES

Barak, P., Rai, A., Rai, P., & Mallik, R. (2013). Quantitative optical trapping on single organelles in cell extract. *Nature Methods*, *10*(1), 68–70.

Barlan, K., Rossow, M. J., & Gelfand, V. I. (2013). The journey of the organelle: Teamwork and regulation in intracellular transport. *Current Opinion in Cell Biology*, *25*(4), 483–488.

Bartsch, T. F., Longoria, R. A., Florin, E. L., & Shubeita, G. T. (2013). Lipid droplets purified from Drosophila embryos as an endogenous handle for precise motor transport measurements. *Biophysical Journal*, *105*(5), 1182–1191.

Blocker, A., Severin, F. F., Burkhardt, J. K., Bingham, J. B., Yu, H., Olivo, J. C., et al. (1997). Molecular requirements for bi-directional movement of phagosomes along microtubules. *The Journal of Cell Biology*, *137*(1), 113–129.

Brady, S. T., Lasek, R. J., & Allen, R. D. (1982). Fast axonal transport in extruded axoplasm from squid giant axon. *Science*, *218*(4577), 1129–1131.

Brasaemle, D. L., Dolios, G., Shapiro, L., & Wang, R. (2004). Proteomic analysis of proteins associated with lipid droplets of basal and lipolytically stimulated 3 T3-L1 adipocytes. *The Journal of Biological Chemistry*, *279*(45), 46835–46842.

Brasaemle, D. L., & Wolins, N. E. (2006). Isolation of lipid droplets from cells by density gradient centrifugation. *Current Protocols in Cell Biology*, Chapter 3: Unit 3 15.

Castoldi, M., & Popov, A. V. (2003). Purification of brain tubulin through two cycles of polymerization-depolymerization in a high-molarity buffer. *Protein Expression and Purification*, *32*(1), 83–88.

Desjardins, M., & Griffiths, G. (2003). Phagocytosis: Latex leads the way. *Current Opinion in Cell Biology*, *15*(4), 498–503.

Farese, R. V., Jr., & Walther, T. C. (2009). Lipid droplets finally get a little R-E-S-P-E-C-T. *Cell*, *139*(5), 855–860.

Goodman, J. M. (2008). The gregarious lipid droplet. *The Journal of Biological Chemistry*, *283*(42), 28005–28009.

Gross, S. P., Vershinin, M., & Shubeita, G. T. (2007). Cargo transport: Two motors are sometimes better than one. *Current Biology*, *17*(12), R478–R486.

Herker, E., & Ott, M. (2012). Emerging role of lipid droplets in host/pathogen interactions. *The Journal of Biological Chemistry*, *287*(4), 2280–2287.

Kon, T., Oyama, T., Shimo-Kon, R., Imamula, K., Shima, T., Sutoh, K., et al. (2012). The 2.8 a crystal structure of the dynein motor domain. *Nature*, *484*(7394), 345–350.

Koonce, M. P. (2000). Dictyostelium, a model organism for microtubule-based transport. *Protist*, *151*(1), 17–25.

Lee, G. E., Murray, J. W., Wolkoff, A. W., & Wilson, D. W. (2006). Reconstitution of herpes simplex virus microtubule-dependent trafficking in vitro. *Journal of Virology*, *80*(9), 4264–4275.

Mallik, R., Carter, B. C., Lex, S. A., King, S. J., & Gross, S. P. (2004). Cytoplasmic dynein functions as a gear in response to load. *Nature*, *427*, 649–652.

Mallik, R., & Gross, S. P. (2004). Molecular motors: Strategies to get along. *Current Biology, 14*, R971–R982.

Mallik, R., Rai, A. K., Barak, P., Rai, A., & Kunwar, A. (2013). Teamwork in microtubule motors. *Trends in Cell Biology, 23*, 575–582.

Murray, J. W., Bananis, E., & Wolkoff, A. W. (2000). Reconstitution of ATP-dependent movement of endocytic vesicles along microtubules in vitro: An oscillatory bidirectional process. *Molecular Biology of the Cell, 11*(2), 419–433.

Pollock, N., de Hostos, E. L., Turck, C. W., & Vale, R. D. (1999). Reconstitution of membrane transport powered by a novel dimeric kinesin motor of the Unc104/KIF1A family purified from Dictyostelium. *The Journal of Cell Biology, 147*, 493–506.

Pollock, N., Koonce, M. P., de Hostos, E. L., & Vale, R. D. (1998). In vitro microtubule-based organelle transport in wild-type Dictyostelium and cells overexpressing a truncated dynein heavy chain. *Cell Motility and the Cytoskeleton, 40*, 304–314.

Rai, A. K., Rai, A., Ramaiya, A. J., Jha, R., & Mallik, R. (2013). Molecular adaptations allow dynein to generate large collective forces inside cells. *Cell, 152*(1–2), 172–182.

Rice, S. E., Purcell, T. J., & Spudich, J. A. (2003). Building and using optical traps to study properties of molecular motors. *Methods in Enzymology, 361*, 112–133.

Rogers, S. L., Tint, I. S., Fanapour, P. C., & Gelfand, V. I. (1997). Regulated bidirectional motility of melanophore pigment granules along microtubules in vitro. *Proceedings of the National Academy of Sciences of the United States of America, 94*(8), 3720–3725.

Romagnoli, S., Cai, G., & Cresti, M. (2003). In vitro assays demonstrate that pollen tube organelles use kinesin-related motor proteins to move along microtubules. *Plant Cell, 15*(1), 251–269.

Rupper, A., Grove, B., & Cardelli, J. (2001). Rab7 regulates phagosome maturation in Dictyostelium. *Journal of Cell Science, 114*(Pt 13), 2449–2460.

Shubeita, G. T., Tran, S. L., Xu, J., Vershinin, M., Cermelli, S., Cotton, S. L., et al. (2008). Consequences of motor copy number on the intracellular transport of kinesin-1-driven lipid droplets. *Cell, 135*(6), 1098–1107.

Sladewski, T. E., Bookwalter, C. S., Hong, M. S., & Trybus, K. M. (2013). Single-molecule reconstitution of mRNA transport by a class V myosin. *Nature Structural & Molecular Biology, 20*(8), 952–957.

Soppina, V., Rai, A., & Mallik, R. (2009). Simple non-fluorescent polarity labeling of Microtubules for Molecular Motor assays. *Biotechniques, 46*, 297–303.

Soppina, V., Rai, A. K., Ramaiya, A. J., Barak, P., & Mallik, R. (2009). Tug-of-war between dissimilar teams of microtubule motors regulates transport and fission of endosomes. *Proceedings of the National Academy of Sciences of the United States of America, 106*(46), 19381–19386.

Svoboda, K., Schmidt, C. F., Schnapp, B. J., & Block, S. M. (1993). Direct observation of kinesin stepping by optical trapping interferometry. *Nature, 365*(6448), 721–727.

Targett-Adams, P., Chambers, D., Gledhill, S., Hope, R. G., Coy, J. F., Girod, A., et al. (2003). Live cell analysis and targeting of the lipid droplet-binding adipocyte differentiation-related protein. *The Journal of Biological Chemistry, 278*(18), 15998–16007.

Thiam, A. R., Antonny, B., Wang, J., Delacotte, J., Wilfling, F., Walther, T. C., et al. (2013). COPI buds 60-nm lipid droplets from reconstituted water-phospholipid-triacylglyceride interfaces, suggesting a tension clamp function. *Proceedings of the National Academy of Sciences of the United States of America, 110*(33), 13244–13249.

Vale, R. D. (2003). The molecular motor toolbox for intracellular transport. *Cell, 112*, 467–480.

Vale, R. D., Schnapp, B. J., Reese, T. S., & Sheetz, M. P. (1985). Movement of organelles along filaments dissociated from the axoplasm of the squid giant axon. *Cell, 40*(2), 449–454.

Vallee, R. B., McKenney, R. J., & Ori-McKenney, K. M. (2012). Multiple modes of cytoplasmic dynein regulation. *Nature Cell Biology*, *14*(3), 224–230.

Verhey, K. J., & Hammond, J. W. (2009). Traffic control: Regulation of kinesin motors. *Nature Reviews Molecular Cell Biology*, *10*(11), 765–777.

Wang, H., Wei, E., Quiroga, A. D., Sun, X., Touret, N., & Lehner, R. (2010). Altered lipid droplet dynamics in hepatocytes lacking triacylglycerol hydrolase expression. *Molecular Biology of the Cell*, *21*(12), 1991–2000.

Welte, M. A. (2004). Bidirectional transport along microtubules. *Current Biology*, *14*, R525–R537.

Welte, M. A. (2009). Fat on the move: Intracellular motion of lipid droplets. *Biochemical Society Transactions*, *37*(Pt 5), 991–996.

CHAPTER FOURTEEN

Reconstituting the Motility of Isolated Intracellular Cargoes

Adam G. Hendricks[*,†], **Yale E. Goldman**[*,†], **Erika L.F. Holzbaur**[*,†,1]

[*]Pennsylvania Muscle Institute, Perelman School of Medicine, University of Pennsylvania, Philadelphia, Pennsylvania, USA

[†]Department of Physiology, Perelman School of Medicine, University of Pennsylvania, Philadelphia, Pennsylvania, USA

[1]Corresponding author: e-mail address: holzbaur@mail.med.upenn.edu

Contents

Abstract

Kinesin, dynein, and myosin transport intracellular cargoes including organelles, membrane-bound vesicles, and mRNA along the cytoskeleton. These motor proteins work collectively in teams to transport cargoes over long distances and navigate around obstacles in the cell. In addition, several types of motors often interact on the same cargo to allow bidirectional transport and switching between the actin and microtubule networks. To examine transport of native cargoes in a simplified *in vitro* system, techniques have been developed to isolate endogenous cargoes and reconstitute their motility. Isolated cargoes can be tracked and manipulated with high precision using total internal reflection fluorescence microscopy and optical trapping. Through use of native cargoes, we can examine vesicular transport in a minimal system while retaining endogenous motor stoichiometry and the biochemical and mechanical characteristics of both motor and cargo.

Methods in Enzymology, Volume 540
ISSN 0076-6879
http://dx.doi.org/10.1016/B978-0-12-397924-7.00014-5

1. INTRODUCTION

Much of the progress to date in understanding motor protein function has resulted from efforts to reconstitute motility *in vitro* using purified motors (Kron & Spudich, 1986; Vale, Reese, & Sheetz, 1985). Single-molecule studies of individual motors have allowed the mechanochemistry of dynein, kinesin, and myosin motors to be studied in detail (Capitanio & Pavone, 2013; Park, Toprak, & Selvin, 2007; Selvin & Ha, 2008). However, many cellular processes are driven by the collective action of motors working in teams, for example, during muscle contraction, cell division, and intracellular transport. To study the collective dynamics of motors, researchers have bound multiple motors to beads (Schroeder et al., 2010, 2012; Vershinin, Carter, Razafsky, King, & Gross, 2007) or engineered DNA-based scaffolds to link multiple motors (Derr et al., 2012; Furuta et al., 2013; Jamison, Driver, Rogers, Constantinou, & Diehl, 2010). These assemblies enable tight control over the number of motors on a cargo and modulation of the physical attributes of the scaffold. However, the approaches used to link motors to artificial cargos do not model endogenous linkages, which also include cargo-bound components such as scaffolding molecules (Fu & Holzbaur, 2013; Spronsen et al., 2013; Wang et al., 2011). Further, the mechanical properties of the cargoes themselves are likely critical to motor function and not well understood. To address these questions, approaches have been developed to isolate endogenous vesicular cargoes and organelles and reconstitute their motility *in vitro* (Bananis et al., 2004; Fort et al., 2011; Hendricks, Holzbaur, & Goldman, 2012; Hendricks et al., 2010; Rogers, Tint, Fanapour, & Gelfand, 1997; Schnapp, Reese, & Bechtold, 1992; Schroer, Schnapp, Reese, & Sheetz, 1988).

Here, we describe methods to isolate two types of intracellular cargoes: axonal transport vesicles from mouse brain and latex-bead-containing phagosomes (LBCs) from macrophages. *In vitro* functional assays allow high-resolution measurements of the motility and forces of isolated vesicular cargoes along microtubules using total internal reflection fluorescence (TIRF) microscopy and optical trapping.

2. ISOLATION OF NEURONAL TRANSPORT VESICLES

The protocol below describes the isolation of axonal transport vesicles from mouse brain. To allow imaging using TIRF microscopy, a mouse line was used that expresses low levels of GFP-dynamitin, a subunit of the dynein–dynactin complex (Ross, Wallace, Shuman, Goldman, & Holzbaur, 2006). Isolated vesicles move bidirectionally along microtubules *in vitro*, driven by kinesin-1, kinesin-2, and cytoplasmic dynein. Biochemical analysis indicates that the vesicles are primarily late endosomes/lysosomes, with an average diameter of ~90 nm (Hendricks et al., 2010). Using similar methods, researchers have reconstituted the motility of endogenous organelles from a range of sources including squid axoplasm (Schroer et al., 1988; Schnapp et al., 1992), melanosomes from Xenopus melanophores (Rogers et al., 1997), and endosomes from rat liver (Bananis et al., 2004; Fort et al., 2011). Importantly, the axonal transport vesicles purified using the following protocol exhibit bidirectional motility that closely models motility in live cells in the absence of additional cytosolic factors (Hendricks et al., 2010); likely due to the gentle homogenization that can be used with brain tissue and the limited number of steps in the isolation procedure (Fig. 14.1).

- Buffer solutions:
  1. Motility assay buffer (MAB). 10 m*M* PIPES, 50 m*M* K-Acetate, 4 m*M* $MgCl_2$, 1 m*M* EGTA, pH 7.0.
  2. Protease inhibitors and DTT. All buffers are supplemented with protease inhibitors and DTT: 1 m*M* PMSF, 105 μ*M* leupeptin, 0.75 μ*M* pepstatin-A, 26.4 μ*M* *N*-*p*-Tosyl-L-arginine methyl ester, 10 m*M* DTT.
  3. Homogenization buffer. Protease inhibitors and DTT are added to MAB at twice the concentrations above to account for the tissue volume.
  4. Sucrose solutions. Prepare a 2.5-*M* (85.5%) sucrose solution by dissolving 42.8 g of sucrose into MAB to a total volume of 50 ml. Place on a rocker overnight at room temperature to ensure sucrose is fully dissolved. For two sucrose gradients, prepare 4 ml of 2.5 *M* sucrose, 12 ml of 1.5 *M* sucrose, and 8 ml of 0.6 *M* sucrose. Supplement sucrose solutions with protease inhibitors and DTT. Mix well.
- Brain homogenization:
  1. Chill a 10- to 15-ml glass tube homogenizer with a tight-fitting teflon pestle, centrifuges, rotors, and buffer solutions.
  2. Harvest three adult mouse brains and place them directly in 2 ml homogenization buffer on ice. Chop brains with a spatula.
  3. Homogenize in 3–4 passes with the homogenizer set at 60–80 rpm, taking care not to introduce air bubbles. Keep the homogenizer on ice at all times.

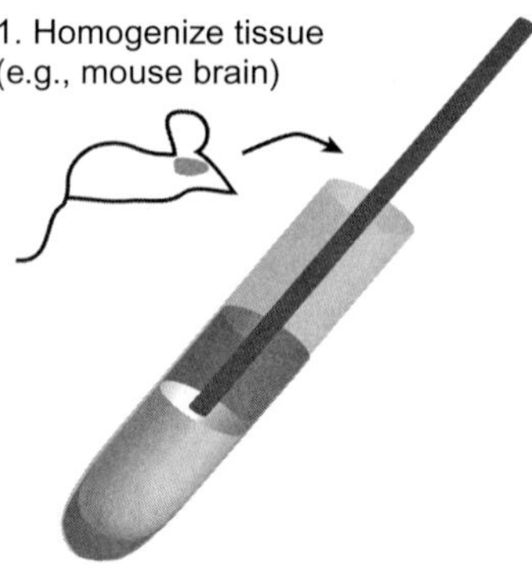

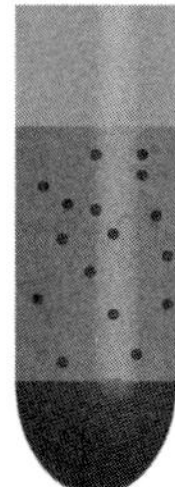

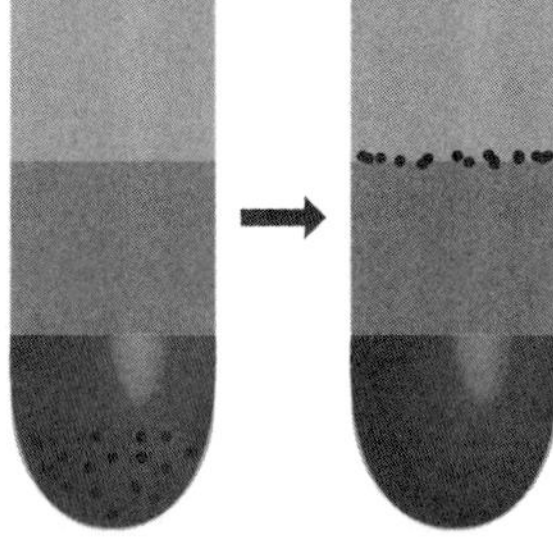

Figure 14.1 Isolation of neuronal transport vesicles. (1) Tissue is homogenized to disrupt cell membranes. (2) A low-speed spin pellets cell membranes, nuclei, and large organelles. (3) The high-speed spin separates membranous vesicles from soluble proteins. (4) Floatation on a sucrose step gradient selects for a subpopulation of vesicles similar in size and density. (For the color version of this figure, the reader is referred to the online version of this chapter.)

- Sucrose step gradient:
  1. Spin the homogenate at 17,200 × *g* for 30 min at 4 °C. Recover the supernatant.
  2. Spin the supernatant at 95,000 × *g* for 20 min at 4 °C. Discard the supernatant and recover the pellet in 200 μl homogenization buffer. Partially resuspend the pellet using a pipet and transfer to a ground glass homogenizer and gently resuspend the pellet.
  3. Add 400 μl of 2.5 *M* sucrose and mix well. Place into the bottom of a 13-ml, 14 by 89-mm tube (Ultra-Clear, Beckman SW41).
  4. Add 1.6 ml of 2.5 *M* sucrose. A quick spin can be used to remove bubbles.
  5. Using a pipette-aid, gently layer 6 ml of 1.5 *M* sucrose on top. Take care to add the sucrose in a slow, steady stream.
  6. Layer 4 ml of 0.6 *M* sucrose on top.
  7. Centrifuge in a swinging bucket rotor at 200,000 × *g* for 2 h at 4 °C.
  8. The vesicle fraction appears as a white fluffy layer at the 0.6 *M*/1.5 *M* sucrose interface. Remove the vesicle fraction by inserting an 18 G needle through the side of the tube. Take care to remove the vesicle fraction in the smallest volume possible (200–500 μl). Store on ice and protected from light. Vesicles exhibit robust motility for 2–3 days.

3. ISOLATION OF LATEX BEAD COMPARTMENTS

To examine the forces exerted by motors in intracellular transport, we use LBCs. Macrophage cells readily phagocytose latex beads. Once internalized, the beads are enclosed in a native organelle, which is transported by an endogenous complement of tightly bound motor proteins. LBCs are highly refractile and uniform in size, making them ideal for optical trapping studies (Blehm, Schroer, Trybus, Chemla, & Selvin, 2013; Hendricks et al., 2012; Rai, Rai, Ramaiya, Jha, & Mallik, 2013). In addition, because of their uniform density, LBCs can be separated from other cellular components with high purity on a sucrose gradient. Previous work took advantage of facile isolation to examine phagosome biochemistry (Desjardins et al., 1994) and motility (Al-Haddad et al., 2001; Blocker et al., 1996, 1997). The protocol below is based on the procedure in Vinet and Descoteaux (2009) with modifications aimed at maintaining the association and activity of the bound motor proteins (Fig. 14.2).

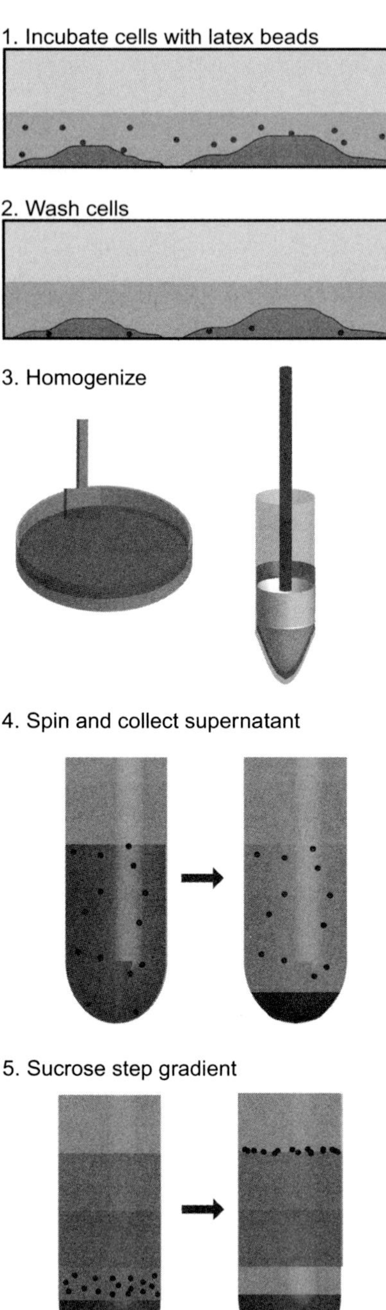

Figure 14.2 Isolation of latex-bead-containing phagosomes. (1) BSA-coated latex beads are incubated with cells and internalized via phagocytosis. (2) Excess beads are washed away. (3) A teflon homogenizer is used to gently disrupt cell membranes. (4) A low-speed spin pellets cell membranes and nuclei. (5) Phagocytosed latex beads are uniform in size and density, enabling separation from other intracellular cargoes on a sucrose step gradient. (For the color version of this figure, the reader is referred to the online version of this chapter.)

- Buffer solutions:
  1. MAB, protease inhibitors, DTT, and MgATP. Supplement MAB as for isolation of neuronal transport vesicles (see above) with the addition of 1 mM MgATP.
  2. Homogenization buffer. MAB + 8.5% sucrose, protease inhibitors, MgATP, and DTT.
  3. Sucrose solutions. Prepare 4 ml of 0.3 M (10%), 0.7 M (25%), 1.0 M (35%) sucrose solutions, and 10 ml of 1.8 M (62%) sucrose. Supplement with protease inhibitors, MgATP, and DTT.
- Cell culture: Mouse macrophage cells (J774A.1) are maintained in 10 cm dishes in complete medium (DMEM supplemented with 10% heat-denatured newborn calf serum, 1% glutamine) at 37 °C, 5% CO_2. Cells are grown to 70–80% confluence and passaged by scraping with a plastic policeman.
- Internalization of latex beads:
  1. Polystyrene beads (0.5–1 μm, carboxylated, 2.68% solids) are pelleted and resuspended in 1 mg/ml BSA in MAB. Sonicate beads for 1 min to disrupt aggregates. Dilute beads 1:50 in complete media.
  2. Replace cell media with bead-containing media. Incubate cells at 37 °C and 5% CO_2 for 90 min to allow the beads to be internalized and for the phagosomes to mature. Bead-containing phagosomes appear as refractile dots under bright field microscopy.
- Sucrose step gradient:
  1. Wash cells 3 × 5 min in PBS at room temperature on a shaker.
  2. Replace medium with 500 μl of homogenization buffer. Scrape cells with a plastic policeman and transfer the cells from two 10-cm dishes to a 2-ml conical glass homogenizer with a tight-fitting teflon pestle.
  3. Homogenize by hand until about 90% of cells are broken with little breakage of the nuclei, as monitored by light microscopy.
  4. Transfer homogenate to a 15-ml Falcon tube and spin at 2000 rpm for 5 min at 4 °C.
  5. Remove the supernatant (~1 ml) containing the LBCs. Add an equal volume of 62% sucrose and gently mix well.
  6. Load 3 ml of 62% sucrose into the bottom of a 13-ml, 14 by 89 mm tube (Ultra-Clear, Beckman SW41). Layer the LBC suspension on top, followed by 2 ml of 35% sucrose, 2 ml of 25% sucrose, and 2 ml of 10% sucrose.
  7. Centrifuge in a swinging bucket rotor at 100,000 × g (24 krpm) for 1 h at 4 °C.

8. The LBC fraction appears as a thin band at the interface between the 10% and 25% sucrose layers. The LBC concentration can be estimated by measuring the optical density of the preparations at 600 nm. The extinction coefficient for latex beads is approximately 100 $(\text{mg/ml})^{-1}$ cm^{-1}. Store on ice and protected from light for up to 2–3 days.

4. *IN VITRO* MOTILITY ASSAYS

Isolated vesicles and LBCs maintain activity for 2–3 days on ice. Here, we describe the procedure for vesicle motility assays using Pluronic F-127 to block nonspecific interactions with the coverslip, which we find to be superior to protein-based blocking agents such as BSA or casein (Dixit & Ross, 2010). The same protocol is used for TIRF microscopy and optical trapping assays.

- Flow chamber: Construct a flow chamber using a cleaned and silanized coverslip (see Dixit & Ross, 2010) and a glass slide, spaced by double-sided tape. Form two to three lanes (volume ~15 μl each) with vacuum grease to isolate the flow chambers from the tape, dispensed using a syringe with a shortened 18 G needle.
- Protocol:
  1. Flow 1 chamber volume of beta-tubulin antibody (Sigma, clone TUB 2.1, aliquoted and stored at –20 °C), diluted 1:100 in MAB. Incubate for 5 min.
  2. Wash with 2 chamber volumes of MAB using capillary action by placing a small drop at one end of the chamber and blotting with filter paper at the other end.
  3. Flow 1 chamber volume of F-127 (50 mg/ml in MAB), incubate for 5 min.
  4. Wash with 2 chamber volumes of MAB.
  5. Flow 1 chamber volume of polarity-marked microtubules (see Katsuki, Muto, & Cross, 2011), diluted to 0.02 mg/ml in MAB supplemented with 20 μ*M* taxol (TMAB). For TIRF assays, use dim microtubules (labeling ratio of ~1:50 labeled:total tubulin) to avoid microtubule fluorescence bleeding into the channel used to image the vesicles. Flow the microtubules quickly to align them. Incubate for 30 s. Repeat.

6. Wash with 2 chamber volumes of TMAB. Image the microtubules to check density.
7. Dilute vesicles or LBCs to desired concentration in TMAB (1:2–1:4) and add 10 m*M* DTT, 1 mg/ml BSA, 0.5 μg/ml glucose oxidase, 470 U/ml catalase, 15 mg/ml glucose, and the desired concentration of MgATP. Flow into chamber.
8. Image vesicle motility using TIRF microscopy or measure the forces produced by LBCs using an optical trap.

5. IMAGING AND ANALYSIS

- Subpixel resolution tracking. Fluorescently labeled vesicles are imaged using TIRF microscopy (Fig. 14.3A). To obtain fluorescently labeled vesicles, our lab isolates vesicles from transgenic GFP-dynamitin mice. Vesicles can also be labeled using a lipophylic dye (Bananis et al., 2004). Vesicles move bidirectionally along microtubules *in vitro* and maintain association with the microtubule for several seconds on average. Kymograph analysis, a 1D projection of the intensities along the microtubule in time, clearly displays the bidirectional motility (Fig. 14.3B). The binding and unbinding rates of the vesicles, run lengths, and average velocities can be estimated from the kymographs. To obtain high-resolution data, automated tracking algorithms are used (Ruhnow, Zwicker, & Diez, 2011), which allow subpixel spatial resolution and unbiased analysis of vesicle motility. Briefly, the algorithm localizes the vesicles in each frame of the movie by fitting the intensity to a 2D Gaussian (Yildiz et al., 2003). Then, trajectories are constructed from the tracked positions in each frame. When analyzing bidirectional cargoes, tracking parameters must be adjusted to avoid fragmented trajectories. In particular, avoid using the directionality of a cargo to link successive positions.
- Optical trapping. LBCs are large, refractile, and uniform in size, making them well suited for optical trapping studies (Fig. 14.3C). One advantage of using isolated LBCs is that assays are performed in aqueous medium rather than the viscoelastic environment present in living cells. As such, the optical trap can be calibrated using standard power spectral methods (Tolic-Norrelykke et al., 2006). Multiple kinesin and dynein motors

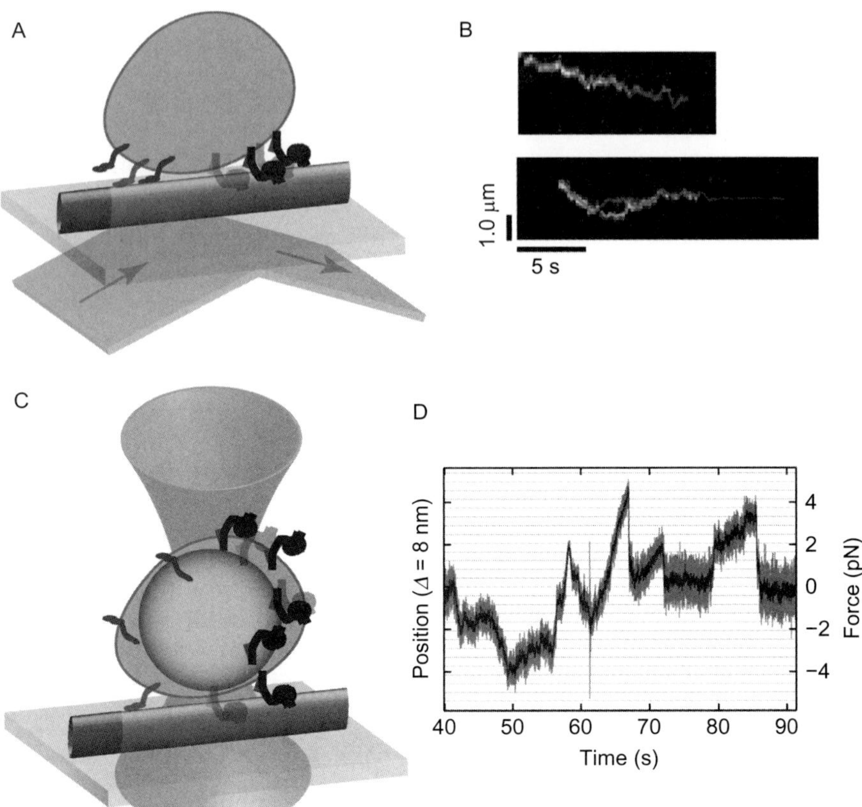

Figure 14.3 TIRF microscopy and optical trapping show that isolated cargoes move bidirectionally along microtubules *in vitro*, driven by an endogenous complement of stably bound motor proteins. (A) Isolated neuronal transport vesicles are imaged using TIRF microscopy (blue: kinesin-1 and kinesin-2, green: dynein, red/orange: microtubule). Blue arrows indicate incident and totally reflected laser illumination. (B) Vesicles move bidirectionally along microtubules. Red and orange lines indicate subpixel resolution trajectories from automated tracking analysis. (C) The forces exerted by motor proteins on LBCs are measured with an optical trap. (D) Kinesin-1, kinesin-2, and cytoplasmic dynein exert forces on LBCs along microtubules. Plus-end-directed forces are driven by kinesin-1 and kinesin-2 (plotted as positive forces), while dynein drives motility toward the minus-end (negative forces). Thin gray lines indicate 8 nm intervals. (See the color plate.)

drive the motility of LBCs (Fig. 14.3D), which can exert forces up to ~12 pN. Optical trap stiffnesses of ~0.05 pN/nm are typically used to ensure that measurements are recorded within the linear range of the optical trap.

6. TROUBLESHOOTING

Several steps in these procedures are critical to maintaining the activity of vesicle-bound motors:

- Homogenization. To minimize damage to the vesicles and LBCs, tissue or cells must be homogenized in a gentle manner. We have found that teflon homogenizers work well, operated at low speed (60–80 rpm) to prevent heating of the sample. The homogenizer is chilled prior to use, and the sample is kept on ice throughout homogenization procedure.
- Sucrose gradients. To pour the sucrose gradients, use a pipette-aid to layer the sucrose solutions in a slow, steady stream. Handle the gradients gently once poured. If the sucrose gradient results in inadequate separation, too much sample may have been loaded onto the gradient.
- Maturation. Many intracellular cargoes, including LBCs, undergo time-dependent biochemical changes that influence the complement of motors bound to the cargo and their motility (Blocker et al., 1997). For this reason, it is important to tightly control the time between bead internalization and homogenization.
- Nonspecific binding. In our experience, problems with *in vitro* motility assays often stem from nonspecific binding to the coverslip, which results in low motility and increased background. To effectively block nonspecific interactions, use fresh silanized coverslips and blocking agents (F-127, BSA, casein) and allow adequate time (at least 5 min) for blocking agents to adhere to the coverslip. Use the smallest amount of tubulin antibody adequate to adhere microtubules to the coverslip.

7. COMPARISON TO MEASUREMENTS IN LIVING CELLS

The motility of isolated vesicular cargoes can be compared to transport in living cells, in order to: (1) assess the degree to which isolated cargoes reconstitute the dynamics of intracellular cargoes and (2) to examine how the cellular environment, including complex cytoskeletal networks and motor-binding partners and effectors, influences transport. For example, we compared the motility of isolated neuronal transport vesicles to Lysotracker-positive cargoes in cortical neurons, and found the motility was similar with respect to velocities, directionality, and relative fractions of stationary, diffusive, and processive motility (Hendricks et al., 2010). However, rare long, directed runs were observed in living cells but not *in vitro*, suggesting soluble regulatory factors present in cells cause a subset of vesicles to be more processive. Using LBCs, we measured the forces on cargoes *in vitro* and in living cells and found that individual motors exerted similar forces *in vitro*, but that larger numbers of motors were able to engage the cytoskeleton in the cell (Hendricks et al., 2012). These results indicate that isolated cargoes replicate many aspects of intracellular transport and point to ways in which the cell regulates motor proteins to achieve proper spatiotemporal localization of vesicular cargoes.

8. OUTLOOK

The methods described here allow reconstitution of the motility of isolated intracellular cargoes and have enabled identification of the motors that drive specific vesicle populations (Al-Haddad et al., 2001; Bananis et al., 2004; Fort et al., 2011), estimation of the number of motors driving vesicular motility, and have provided insights into the mechanisms of bidirectional motility along microtubules (Blocker et al., 1997; Hendricks et al., 2010, 2012). Future work will focus on how bidirectional transport is regulated to achieve targeted trafficking in the cell, where scaffolding molecules (Fu & Holzbaur, 2013; Spronsen et al., 2013; Wang et al., 2011), binding partners, and effectors (Kardon & Vale, 2009) have been implicated in modulating the activity of cargo-bound motors. Further, increasing evidence suggests a role for the cytoskeleton in regulating transport, through posttranslational modifications, cytoskeletal-associated proteins, and organization of the cytoskeletal network (Brawley & Rock, 2009; Cai, McEwen, Martens, Meyhofer, & Verhey, 2009; Hodges et al., 2012). By reconstituting the motility of native cargoes *in vitro*, researchers can now systematically discriminate among candidate regulatory mechanisms.

ACKNOWLEDGMENTS

The authors thank Jennifer Ross, Karen Wallace, and Mariko Tokito for helpful advice and technical assistance. This work was supported by National Institutes of Health grants GM087253 (to E. L. F. H. and Y. E. G.) and GM089077 (to A. G. H.).

REFERENCES

Al-Haddad, A., Shonn, M. A., Redlich, B., Blocker, A., Burkhardt, J. K., Yu, H., et al. (2001). Myosin va bound to phagosomes binds to f-actin and delays microtubule-dependent motility. *Molecular Biology of the Cell*, *12*(9), 2742–2755.

Bananis, E., Nath, S., Gordon, K., Satir, P., Stockert, R. J., Murray, J. W., et al. (2004). Microtubule-dependent movement of late endocytic vesicles in vitro: Requirements for dynein and kinesin. *Molecular Biology of the Cell, 15*(8), 3688–3697.

Blehm, B. H., Schroer, T. A., Trybus, K. M., Chemla, Y. R., & Selvin, P. R. (2013). In vivo optical trapping indicates kinesin's stall force is reduced by dynein during intracellular transport. *Proceedings of the National Academy of Sciences of the United States of America, 110*(9), 3381–3386.

Blocker, A., Severin, F. F., Burkhardt, J. K., Bingham, J. B., Yu, H., Olivo, J. C., et al. (1997). Molecular requirements for bi-directional movement of phagosomes along microtubules. *Journal of Cell Biology, 137*(1), 113–129.

Blocker, A., Severin, F. F., Habermann, A., Hyman, A. A., Griffiths, G., & Burkhardt, J. K. (1996). Microtubule-associated protein-dependent binding of phagosomes to microtubules. *Journal of Biological Chemistry, 271*(7), 3803–3811.

Brawley, C. M., & Rock, R. S. (2009). Unconventional myosin traffic in cells reveals a selective actin cytoskeleton. *Proceedings of the National Academy of Sciences of the United States of America, 106*(24), 9685–9690.

Cai, D., McEwen, D. P., Martens, J. R., Meyhofer, E., & Verhey, K. J. (2009). Single molecule imaging reveals differences in microtubule track selection between kinesin motors. *PLoS Biology*, 7(10), e1000216.

Capitanio, M., & Pavone, F. S. (2013). Interrogating biology with force: Single molecule high-resolution measurements with optical tweezers. *Biophysical Journal, 105*(6), 1293–1303.

Derr, N., Goodman, B., Jungmann, R., Leschziner, A., Shih, W., & Reck-Peterson, S. (2012). Tug-of-war in motor protein ensembles revealed with a programmable DNA origami scaffold. *Science, 338*(6107), 662–665.

Desjardins, M., Celis, J. E., van Meer, G., Dieplinger, H., Jahraus, A., Griffiths, G., et al. (1994). Molecular characterization of phagosomes. *Journal of Biological Chemistry, 269*(51), 32194–32200.

Dixit, R., & Ross, J. L. (2010). Studying plus-end tracking at single molecule resolution using tirf microscopy. *Methods in Cell Biology, 95*, 543–554.

Fort, A. G., Murray, J. W., Dandachi, N., Davidson, M. W., Dermietzel, R., Wolkoff, A. W., et al. (2011). In vitro motility of liver connexin vesicles along microtubules utilizes kinesin motors. *Journal of Biological Chemistry, 286*(26), 22875–22885.

Fu, M.-M., & Holzbaur, E. L. F. (2013). Jip1 regulates the directionality of app axonal transport by coordinating kinesin and dynein motors. *Journal of Cell Biology, 202*(3), 495–508.

Furuta, K., Furuta, A., Toyoshima, Y. Y., Amino, M., Oiwa, K., & Kojima, H. (2013). Measuring collective transport by defined numbers of processive and nonprocessive kinesin motors. *Proceedings of the National Academy of Sciences of the United States of America, 110*(2), 501–506.

Hendricks, A. G., Holzbaur, E. L. F., & Goldman, Y. E. (2012). Force measurements on cargoes in living cells reveal collective dynamics of microtubule motors. *Proceedings of the National Academy of Sciences of the United States of America, 109*(45), 18447–18452.

Hendricks, A. G., Perlson, E., Ross, J. L., Schroeder, H. W., 3rd., Tokito, M., & Holzbaur, E. L. F. (2010). Motor coordination via a tug-of-war mechanism drives bidirectional vesicle transport. *Current Biology, 20*(8), 697–702.

Hodges, A. R., Krementsova, E. B., Bookwalter, C. S., Fagnant, P. M., Sladewski, T. E., & Trybus, K. M. (2012). Tropomyosin is essential for processive movement of a class v myosin from budding yeast. *Current Biology, 22*(15), 1410–1416.

Jamison, D. K., Driver, J. W., Rogers, A. R., Constantinou, P. E., & Diehl, M. R. (2010). Two kinesins transport cargo primarily via the action of one motor: Implications for intracellular transport. *Biophysical Journal, 99*(9), 2967–2977.

Kardon, J. R., & Vale, R. D. (2009). Regulators of the cytoplasmic dynein motor. *Nature Reviews Molecular Cell Biology*, *10*(12), 854–865.

Katsuki, M., Muto, E., & Cross, R. A. (2011). Preparation of dual-color polarity-marked fluorescent microtubule seeds. *Methods in Molecular Biology*, *777*, 117–126.

Kron, S. J., & Spudich, J. A. (1986). Fluorescent actin filaments move on myosin fixed to a glass surface. *Proceedings of the National Academy of Sciences of the United States of America*, *83*(17), 6272–6276.

Park, H., Toprak, E., & Selvin, P. R. (2007). Single-molecule fluorescence to study molecular motors. *Quarterly Reviews of Biophysics*, *40*(1), 87–111.

Rai, A. K., Rai, A., Ramaiya, A. J., Jha, R., & Mallik, R. (2013). Molecular adaptations allow dynein to generate large collective forces inside cells. *Cell*, *152*(1), 172–182.

Rogers, S. L., Tint, I. S., Fanapour, P. C., & Gelfand, V. I. (1997). Regulated bidirectional motility of melanophore pigment granules along microtubules in vitro. *Proceedings of the National Academy of Sciences of the United States of America*, *94*(8), 3720–3725.

Ross, J. L., Wallace, K., Shuman, H., Goldman, Y. E., & Holzbaur, E. L. F. (2006). Processive bidirectional motion of dynein-dynactin complexes in vitro. *Nature Cell Biology*, *8*, 562–570.

Ruhnow, F., Zwicker, D., & Diez, S. (2011). Tracking single particles and elongated filaments with nanometer precision. *Biophysical Journal*, *100*(11), 2820–2828.

Schnapp, B. J., Reese, T. S., & Bechtold, R. (1992). Kinesin is bound with high affinity to squid axon organelles that move to the plus-end of microtubules. *Journal of Cell Biology*, *119*(2), 389–399.

Schroeder, H. W., III, Mitchell, C., Shuman, H., Holzbaur, E. L., & Goldman, Y. E. (2010). Motor number controls cargo switching at actin-microtubule intersections in vitro. *Current Biology*, *20*(8), 687–696.

Schroeder, H. W., III, Hendricks, A. G., Ikeda, K., Shuman, H., Rodionov, V., Ikebe, M., et al. (2012). Force-dependent detachment of kinesin-2 biases track switching at cytoskeletal filament intersections. *Biophysical Journal*, *103*(1), 48–58.

Schroer, T. A., Schnapp, B. J., Reese, T. S., & Sheetz, M. P. (1988). The role of kinesin and other soluble factors in organelle movement along microtubules. *Journal of Cell Biology*, *107*(5), 1785–1792.

Selvin, P. R., & Ha, T. (Eds.), (2008). *Single-molecule techniques: A laboratory manual* (pp. 3–36). Cold Spring Harbor, NY: Cold Spring Harbor Laboratory Press.

Spronsen, M., Mikhaylova, M., Lipka, J., Schlager, M. A., van den Heuvel, D. J., Kuijpers, M., et al. (2013). TRAK/Milton motor-adaptor proteins steer mitochondrial trafficking to axons and dendrites. *Neuron*, 77(3), 485–502.

Tolic-Norrelykke, S. F., Schaffer, E., Howard, J., Pavone, F. S., Julicher, F., & Flyvbjerg, H. (2006). Calibration of optical tweezers with positional detection in the back focal plane. *Review of Scientific Instruments*, 77, 103101.

Vale, R. D., Reese, T. S., & Sheetz, M. P. (1985). Identification of a novel force-generating protein, kinesin, involved in microtubule-based motility. *Cell*, *42*, 39–50.

Vershinin, M., Carter, B. C., Razafsky, D. S., King, S. J., & Gross, S. P. (2007). Multiple-motor based transport and its regulation by tau. *Proceedings of the National Academy of Sciences of the United States of America*, *104*(1), 87–92.

Vinet, A. F., & Descoteaux, A. (2009). Large scale phagosome preparation. *Methods in Molecular Biology*, *531*, 329–346.

Wang, X., Winter, D., Ashrafi, G., Schlehe, J., Wong, Y. L., Selkoe, D., et al. (2011). Pink1 and parkin target miro for phosphorylation and degradation to arrest mitochondrial motility. *Cell*, *147*(4), 893–906.

Yildiz, A., Forkey, J. N., McKinney, S. A., Ha, T., Goldman, Y. E., & Selvin, P. R. (2003). Myosin V walks hand-over-hand: Single fluorophore imaging with 1.5-nm localization. *Science*, *300*, 2061–2065.

SECTION IV

Building Higher Order Networks and Interactions

CHAPTER FIFTEEN

Reconstitution of Contractile Actomyosin Arrays

Michael Murrell*, Todd Thoresen†,‡, Margaret Gardel†,‡,1

*Departments of Biomedical Engineering and Materials Science and Engineering, University of Wisconsin, Madison, Wisconsin, USA

†Department of Physics, Institute for Biophysical Dynamics, University of Chicago, Chicago, Illinois, USA

‡James Franck Institute, University of Chicago, Chicago, Illinois, USA

[1]Corresponding author: e-mail address: gardel@uchicago.edu

Contents

Abstract

Networks and bundles comprised of F-actin and myosin II generate contractile forces used to drive morphogenic processes in both muscle and nonmuscle cells. To elucidate the minimal requirements for contractility and the mechanisms underlying their contractility, model systems reconstituted from a known set of purified proteins *in vitro* are needed. Here, we describe two experimental protocols our lab has developed to reconstitute 1D bundles and quasi-2D networks of actomyosin that are amenable to quantitative biophysical measurement. These assays have enabled our discovery of the mechanisms of contractility in disordered actomyosin assemblies and of a mechanical feedback between contraction and F-actin severing.

Methods in Enzymology, Volume 540
ISSN 0076-6879
http://dx.doi.org/10.1016/B978-0-12-397924-7.00015-7

1. INTRODUCTION

The action of myosin II motors on the actin cytoskeleton generates contractile forces that are used in a myriad of morphogenic processes including cell division and migration, as well as the formation and maintenance of multicellular tissue. While the necessity of actomyosin for contractility in these processes is nearly universally conserved, its organization varies widely. In striated muscle, actomyosin is found in sarcomeres with highly regulated F-actin polarity and length. Contraction of sarcomeric actomyosin is well described by the sliding filament models of muscle contraction (Huxley, 2004). By contrast the actomyosin in smooth muscle and nonmuscle cells is typically arranged in bundles or network that lack sarcomeric organization and are "disorganized" with respect to F-actin polarity, lengths, and orientation. In such disordered arrangements, new physical models are needed to describe contractility.

As part of an effort to understand the physico-chemical origins of striated and smooth muscle contractility, reconstitutions of actin and myosin II have been studied for over 60 years. Solutions of actin and myosin were isolated from muscle tissue and supplemented with ATP which resulted in changes in its optical and mechanical properties (Dainty et al., 1944; Szent-Györgyi, 1945, 1947, 1950). These actomyosin solutions exhibited "superprecipitation", an increase in actomyosin density and viscosity, which implied F-actin network "contractility" (Spicer, 1951). Further studies outlined the requisite components of the contractile machinery by establishing stoichiometric relationships between actin, myosin, and actin-binding proteins (Ebashi & Ebashi, 1964; Janson, Kolega, & Taylor, 1991; Stossel, Hartwig, Yin, Zaner, & Stendahl, 1982; Watanabe & Yasui, 1965; Weber & Winicur, 1961). Thus, these early studies identified a minimal "parts list" for reconstructing contractile actomyosin assemblies.

Missing from these early studies was recapitulation of actomyosin organizations found in cells (e.g., bundles or 2D cortex rather than dilute 3D gels) and a mechanistic understanding of how contractility is regulated in diverse actomyosin organizations. Here, we describe methods for constructing actomyosin bundles and networks that are more faithful mimics to those found in living cells. These assays have enabled our studies of contractility, self-organization, and feedback in disordered actomyosin bundles and networks (Lenz, Thoresen, Gardel, & Dinner, 2012; Murrell & Gardel, 2012; Stachowiak et al., 2012; Thoresen, Lenz, & Gardel, 2011, 2013).

2. REAGENTS

2.1. Protein purification and filament formation

Myosin purification: All purification takes place at 4 °C. Native smooth muscle myosin is purified from fresh chicken gizzards essentially as described previously (Bárány, 1996), except myosin is actively phosphorylated using myosin light chain kinase prior to storage in order to minimize heterogeneity due to phosphorylation-dependent configuration. Nonmuscle myosin was purified similarly using expired human platelets obtained at a local blood bank. Skeletal muscle myosin was purified as described previously (Margossian & Lowey, 1982; Pollard, 1982) or purchased from Cytoskeleton, Inc. Fluorescent labeling of myosin is performed using a maleimide dye (Molecular Probes, Invitrogen) that readily reacts with available cysteine residues as described previously (Thoresen et al., 2011); the labeling methods are identical between skeletal, smooth, and nonmuscle isoforms with a typical labeling ratio of 3.6 dye per myosin dimer. Myosin is concentrated using Amicon Ultra-15 centrifugal filters (Millipore, 100 kDa cutoff) to a high concentration (~18 mg/mL) in Myosin Storage Buffer (5 m*M* Pipes, pH 7.0, and 0.45 *M* KCl), then flash frozen in liquid nitrogen for long-term storage at −80 °C.

Myosin thick filament formation: Flash-frozen aliquots of fluorescently labeled and phosphorylated myosin are rapidly thawed. To separate the fraction of myosin dimers that binds with high affinity to F-actin in saturating ATP (and presumed to be enzymatically dead) from the fraction that binds with weak affinity to F-actin in saturating ATP (and presumed to be enzymatically active), myosin dimers are mixed with phalloidin-stabilized F-actin at a 1:5 myosin:actin molar ratio in Spin-down Buffer (20 m*M* MOPS, pH 7.4, 500 m*M* KCl, 4 m*M* $MgCl_2$, 0.1 m*M* EGTA, 500 μ*M* ATP) and centrifuged for 30 min at 100,000 × *g*. The supernatant contains myosin with low affinity to F-actin, whereas the high-affinity binding fraction cosediments with the F-actin pellet. Myosin protein concentrations are determined spectroscopically using an extinction coefficient at 280 nm of 0.56 mL/mg/cm compared to a myosin-free sample that includes nucleotide.

For bundles described in Section 3, myosin thick filaments are formed by diluting myosin in Assay Buffer with varied KCl concentration to control thick filament size. By varying the KCl concentration from 100 to 200 m*M* KCl, the average lengths of skeletal muscle myosin thick filaments changed from 1.5 to 0.5 μm and smooth muscle myosin filaments changed from 1.2 to 0.7 μm (Thoresen et al., 2013). Further modulation of myosin filament lengths is possible by altering the final salt conditions and/or the rate of salt dilution (Thoresen et al., 2013). For the cortex assay described in Section 4, myosin dimers are added directly to the actin cortex contained in F-buffer and myosin thick filaments polymerize *in situ*.

Actin purification: Actin is purified from rabbit skeletal muscle acetone powder and stored in G-buffer (pH 8.0, 2 m*M* Tris–HCl, 0.2 m*M* ATP, 0.2 m*M* $CaCl_2$ 0.2 m*M* DTT, 0.005% NaN_3) at −80 °C. Biotinylated actin is prepared using EZ-Link NHS-PEO_4 biotinylation kit (Thermo Scientific).

Actin filament preparation: To form biotinylated actin filaments, biotinylated G-actin is mixed with unlabeled G-actin in a 1:10 stoichiometry prior to polymerization. G-actin is polymerized by addition of F-buffer (10 m*M* imidazole, pH 7.0, 1 m*M* $MgCl_2$, 50 m*M* KCl, 2 m*M* EGTA, 0.5 m*M* ATP). Fluorescently labeled phalloidin is added in a 2:1 mole ratio (phalloidin:actin) to both stabilize F-actin and for visualization in fluorescence imaging.

2.2. Microscopy

Fluorescence imaging is performed using a Ti-E microscope body (Nikon) fitted with a CSU-X spinning disc confocal head (Yokogawa), a HQ2 CoolSnap CCD camera (Roper Scientific), and a 60 × 1.2NA water immersion objective lens (Nikon). The instrument is controlled with Metamorph software (MDS Analytical Technologies).

3. RECONSTITUTED ACTOMYOSIN BUNDLES

Actomyosin bundles are assembled and contracted within a flow cell through a particular sequential addition of components. The bundles formed by the following method contain ~5 F-actin per cross-section and are 5–50 μm in contour length. Because bundles form without the enforcement of F-actin polarity, they are considered polarity disordered. The F-actin and myosin are both fluorescently labeled to facilitate direct observation by confocal microscopy during myosin II activity. The ends of bundles are tethered to avidin beads coupled to an elastic substrate to permit force measurement. Measurement of bundle contraction after a breakage event permits measurement of unloaded contraction velocity. The unloaded contraction rate and tension generated are modulated by varying myosin II isoform, filament density, and quantity (Thoresen et al., 2011, 2013).

3.1. Reagents

3.1.1 Avidin beads

3 μm diameter polystyrene carboxylate beads (Polysciences) are biotinylated using EZ-Link NHS-PEO_4 biotinylation kit (Thermo Scientific) and subsequently incubated in 5 mg/mL neutravidin (Thermo Scientific) to coat the beads. The beads are repeatedly spun down (15,000 $\times g$, 5 min) and resuspended 10 times in PBS to remove unbound neutravidin. The beads are then briefly sonicated and stored at 4 °C undergoing constant rotation.

3.1.2 Acrylamide gels

A polyacrylamide (PAA) gel is polymerized on the coverslip surfaces (22 mm diameter #1½, EMS). Predetermined acrylamide and bisacrylamide concentrations are used to form gels with a known shear elastic moduli which can be confirmed through measurement with a stress-controlled rheometer (Aratyn-Schaus, Oakes, Stricker, Winter, & Gardel, 2010). 1 mg/mL biotinylated bovine serum albumin (BSA), formed by reacting BSA (Sigma) with NHS-Biotin (Thermo Scientific) is covalently attached to the gel surface using sulfo-SANPAH (Thermo Scientific). Cover slips are washed in PBS, stored at 4 °C and used within 2 weeks.

3.1.3 Sample chamber

A flow chamber customized for imaging with high numerical aperture objectives and small (~30 μL) exchange volumes was designed and obtained from Chamlide (www.chamlide.com). A picture of the flow chamber is in Fig. 15.1A.

3.2. Actomyosin bundle construction

The sequential addition and incubation of sample components is used to template the assembly of bundles existing predominately within a single confocal imaging plane and tethered to surface-bound beads at their ends to facilitate force measurement. First, a 10–20-μm thick PAA gel is formed on a coverslip and biotinylated BSA is covalently attached to the top surface as described above. This biotinylated BSA-PAA substrate provides a surface largely inert to nonspecific myosin or actin binding and facilitates traction force microscopy. The substrate is then loaded into the custom flow

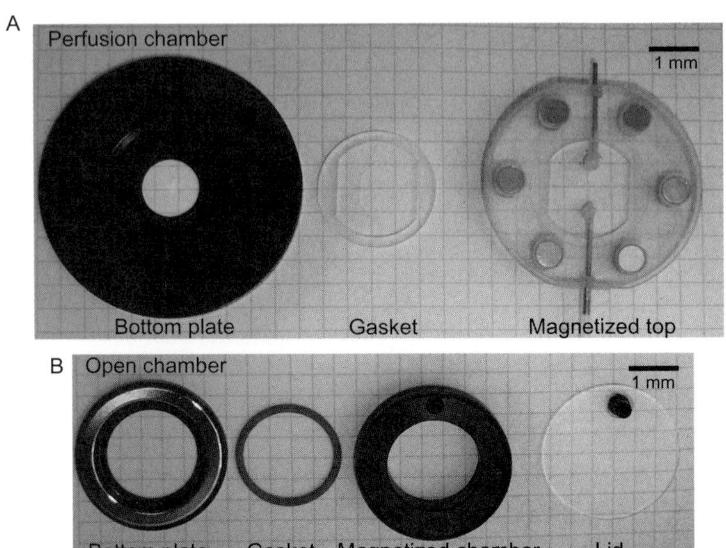

Figure 15.1 Photographs of custom sample chambers used in experiments. In (A), a perfusion chamber consists of a bottom anodized aluminum plate that holds a glass coverslip, a rubber gasket, and a magnetized top plate with molded inlet/outlet for fluid exchange. In (B), an open chamber comprised of an anodized aluminum bottom plate to hold coverslip with magnetized ring, a rubber gasket, magnetized chamber, and colorless top. (For color version of this figure, the reader is referred to the online version of this chapter.)

chamber. After assembling the perfusion chamber, water is perfused through the sample to maintain hydration of PAA gel prior to starting the experiment.

A dilute suspension of 3 μm diameter neutravidin beads in Wash Buffer (20 m*M* MOPS, pH 7.4, 50 m*M* KCl, 4 m*M* $MgCl_2$, 0.1 m*M* EGTA) is perfused into the perfusion chamber and incubated for ~10 min to allow for the beads to sediment and bind to the biotinylated-BSA surface (Fig. 15.2A, panel 1). Unbound beads are then removed by further perfusion of Wash Buffer. We aim for an average distance between beads >10 μm. Phalloidin-stabilized F-actin containing 10% biotinylated G-actin is gently sheared to a length of ~5–6 μm, diluted to 1 μ*M* in Assay Buffer (20 m*M* MOPS, pH 7.4, 100 m*M* KCl, 4 m*M* $MgCl_2$, 0.1 m*M* EGTA, 0.7% methylcellulose, 0.25 mg/mL glucose, 0.25% β-mercaptoethanol, 0.25 mg/mL glucose oxidase, 35 μg/mL catalase) and perfused into the chamber. Over the course of 30 min, F-actin binds to the neutravidin beads. A majority of free, unbound F-actin is removed by perfusion of two chamber volumes

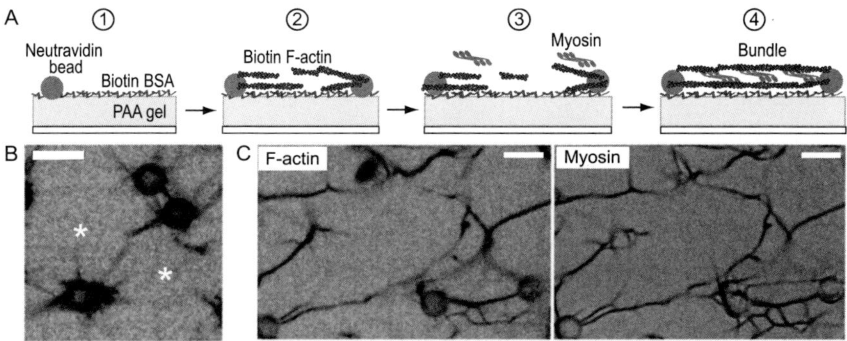

Figure 15.2 (A) Schematic illustrating the sequential process used for templated bundle assembly. (1) Biotinylated-bovine serum albumin is coupled to the surface of a PAA gel affixed to a glass coverslip. Neutravidin beads (gray circles) bind to the biotinylated-bovine serum albumin. (2) Biotinylated F-actin (red) is introduced and binds to beads. A dilute suspension of F-actin remains. (3) Myosin thick filaments (green) suspended in nucleotide-free Assay Buffer (black) are introduced. (4) F-actin cross-linking by myosin filaments mediates bundle formation. (B) Inverted contrast image of F-actin asters visualized with Alexa 568-phalloidin before myosin perfusion. Dark circles are F-actin-coated beads. Asterisks indicate free F-actin ends. Scale bar is 5 μm. (C) Inverted contrast images of F-actin visualized with Alexa 568-phalloidin (left) and OG-labeled myosin (right) illustrating network of bundles formed after 30 min incubation of F-actin asters with myosin thick filaments. Scale bar is 5 μm. (For interpretation of the references to color in this figure legend, the reader is referred to the online version of this chapter.)

of Assay Buffer. The remaining bead-bound F-actin provides sites to template the assembly of actomyosin bundles (Fig. 15.2A, panel 2, B). Since biotinylated G-actin is randomly incorporated into F-actin during polymerization, free F-actin ends emanating from beads are likely of random polarity. The formation of F-actin asters is not sensitive to small changes in wash steps, but is extremely sensitive to air bubbles within the flow chamber.

Preformed myosin II thick filaments are then perfused in with Assay Buffer lacking ATP. A dialysis against Storage buffer supplemented with 0.2 m*M* EGTA is performed to ensure complete removal of nucleotide using "drop dialysis" technique with 2.5 nm VSWP membrane (Millipore) and gentle stirring for 2 h at 4 °C. It is crucial that all free nucleotide is removed to prevent motor catalysis during bundle formation. Myosin II filaments cross-link F-actin bound to beads to F-actin remaining in solution to form a quasi-2D network of bundles bound to the beads (Fig. 15.2A, panels 3 and 4, C, and D). Myosin motor activity is then initiated by the introduction of Assay Buffer containing 0.1–1 m*M* ATP.

3.3. Force measurement

Because the stiffness of the underlying PAA gel substrate is tunable, and its elastic properties are well known, forces from an individual contracting bundle can be directly measured by measuring the Hookian displacements of the streptavidin beads linking the bundle to the gel, concurrent with the observation of bundle dynamics. Using traction force reconstruction with point forces to calculate force from a displacement field on the top surface of a PAA gel, the force is related to the local gel displacement by an effective spring constant, k_{eff}. As expected, k_{eff} varies linearly with the PAA gel stiffness (Thoresen et al., 2011) (Fig. 15.3). Assuming deformation of the polystyrene bead ($G' \sim 10^9$ Pa) is negligible compared to that of the soft PAA gel ($G' = 54$–600 Pa), the bead displacement is then multiplied by the k_{eff} to measure the force produced during contraction. A significant caveat to this approach is that this does not consider effects of poor bead attachment to the substrate and/or its rotation within the soft gel. Improvements to the forces measurements are ongoing.

4. BIOMIMETIC CORTEX

The reconstituted actomyosin cell cortex is created adjacent to a standard glass coverslip, in a layer-by-layer assembly of lipids and proteins (Fig. 15.4). The order of assembly is as follows: (1) formation of a bilayer membrane on the coverslip, (2) the addition of membrane-F-actin attachment factors that bind the membrane, (3) the addition of F-actin that couples to the surface of the membrane, (4) the addition of F-actin cross-linking proteins, and finally, (5) the addition of myosin II dimers that assemble into filaments *in situ*. The resultant network is highly disordered and quasi-2D. The membrane, F-actin, and myosin II are fluorescently labeled, and thus can be observed during contraction. The extent of network contraction is modulated by changing the extent of F-actin membrane coupling, cross-linking, and concentration of myosin II motors (Murrell & Gardel, 2012).

4.1. Reagents

4.1.1 Sample chamber

An open chamber with ~500 μL sample volume amenable to high numerical optics was designed and obtained from Chamlide (www.chamlide.com); a picture of the chamber is shown in Fig. 15.1B. After cortex assembly, the chamber is covered with a coverslip to prevent evaporation of contents.

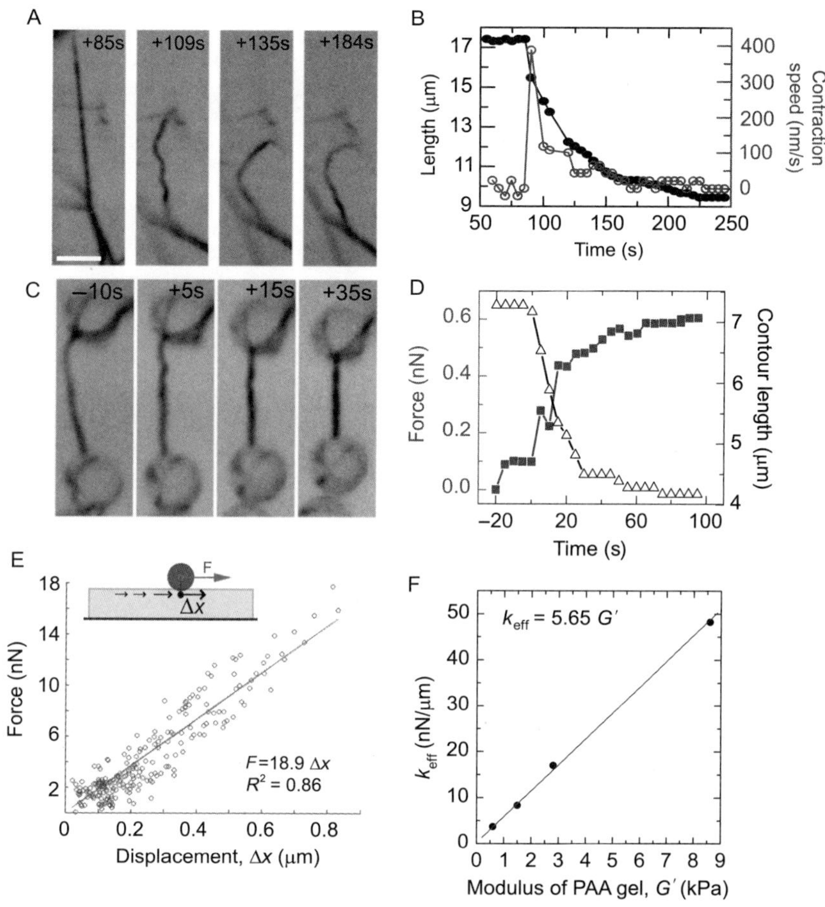

Figure 15.3 Contraction of tethered and untethered bundles. (A) Time-lapse series of inverted contrast, OG-myosin images in a contracting bundle with $R_{M:A} = 1.4$. Times are in seconds before (negative times) or after (positive times) addition of 0.1 m*M* ATP. A connection to a neighboring bundle breaks between 60 and 65 s (arrow), following which contraction of both the untethered bundle (asterisk) and tethered bundle (dashed line) resume. Scale bar is 5 μm. (B) Contour length (left axis, solid circles) and contraction speed (right axis, open circles) of the bundle indicated by the dashed line in (A). (C) Time-lapse series of inverted contrast OG-myosin images illustrating the contraction of a bundle tethered to beads at both ends. Bundle shown contains $R_{M:A} = 1.4$. Scale bar, 5 μm. (D) Bundle contour length (open triangles, right axis) and force (left axis, closed squares) versus time for bundle contraction shown in (C). (E) The calibration of force (in nN) as a function of bead displacement (in mm) to obtain the effective spring constant k_{eff} for a PAA gel with $G' = 2.8$ kPa obtained from Traction Force Reconstruction from Point Forces. (F) The effective spring constant as a function of G'. (For color version of this figure, the reader is referred to the online version of this chapter.)

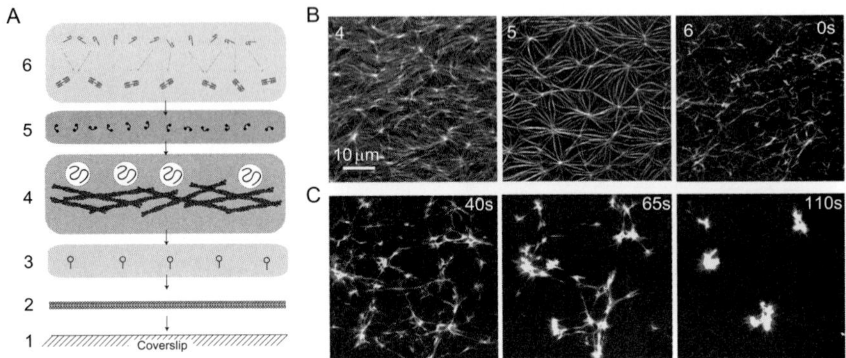

Figure 15.4 Assembly of a contractile model cortex. (A) Schematic of stepwise cortex assembly: (1) Piranha-treated coverslip, (2) phospholipid bilayer formed on coverslip, (3) incubation with F-actin-membrane attachment factors, (4) crowding of F-actin (F-actin in red, methylcellulose in circles), (5) F-actin cross-linking, and (6) myosin II addition. (B) Fluorescence images of (left) F-actin without cross-linker (Stage 4), (middle) with 30 n*M* α-actinin (Stage 5), and (right) 40 n*M* skeletal muscle myosin II (Stage 6). (C) Time course of contraction, 40, 65, and 110 s after the addition of myosin thick filaments. Red is actin, green is myosin. (See the color plate.)

4.1.2 Coverslips

Coverslips are hydroxylated with a 1:3 mixture of 30% hydrogen peroxide (Sigma) and sulfuric acid (Piranha Etching) to make them sufficiently hydrophilic for membrane attachment. Slowly, the peroxide is added to the acid within a Pyrex container containing 25-mm coverslips, which is stirred slowly for 15 min. The coverslips are then washed repeatedly in water, and then stored in methanol. Due to the quick decomposition of the hydrogen peroxide, Piranha solution must be freshly prepared and never stored. Proper safety precautions should be taken as Piranha solutions are volatile and generate large amounts of heat and gas (e.g., http://web.mit.edu/cortiz/www/PiranhaSafety.doc). Consult your local lab safety officials.

4.1.3 Construction of lipid bilayer

The bilayer is formed on the hydroxylated coverslips in the following series of steps:

1. Add 2.5 mg lipids (in combinations listed in Table 15.1) dissolved in chloroform to a glass vial, and then dry them under N_2 gas for 5 min.
2. Add 5 mL Vesicle Buffer (100 m*M* NaCl, 0.1 m*M* EDTA, pH 7.3) to the glass vial and cover with parafilm or a nonscrew plastic top. Vortex

Table 15.1 Compositions of the lipid bilayer membrane

Lipid	For attachment	No attachment
Egg phosphatidyl choline (EPC)	91%	99.8%
1,2-Di-(9Zoctadecenoyl)-*sn*-glycero-3-([*N*-(5-amino-1-carboxypentyl)iminodiacetic acid]succinyl) (NTA-Ni)	8.8%	0
Oregon Green 1,2-dihexadecanoyl-*sn*-glycero-3-phosphoethanolamine (OG-DHPE)	0.2%	0.2%

the lipid solution for 10 s to resuspend the lipids. After vortexing, the vesicle solution is cloudy, reflecting the formation of multilamellar vesicles which scatter light.

3. Sonicate the vesicle solution in a bath sonicator until the solution becomes clear. Use either a ring stand and test tube clamp, or other methods for suspending the glass vial a few millimeters into the water bath. When sonication is complete, the vesicle solution is clear because multilamellar vesicles are broken into small, unilamellar vesicles (SUVs) which are approximately 50–100 nm in diameter. The vesicle solution can be kept at 4 °C for up to 1 week, although is sonicated prior to reuse.
4. The clear SUV solution is then added to the Chamlide chamber sandwiching the Piranha-treated coverslip, which is immersed in 0.5–1 mL of Vesicle Buffer. The SUV solution is allowed to incubate in the chamber for 15 min in the dark to prevent bleaching of the fluorescent lipid. The SUVs will bind to the coverslip surface, rupture, and then fuse with each other to form a flat bilayer.
5. After the incubation, the solution is then washed repeatedly with Vesicle Buffer to remove the excess vesicles which have not adhered to the surface. Then, the bilayer is immersed in approximately 0.5 mL of 1 × F-buffer that does not contain ATP.
6. At this point, the chamber is mounted on the microscope and FRAP is performed to assess the quality of the bilayer. A circular spot of approximately 10 μm in diameter is bleached for 5 s using a 491 nm Mosaic laser. The bilayer is then imaged (Oregon Green DHPE) every 2 s for approximately 30 s. The fluorescence recovers completely in approximately 10–15 s.
7. The quality of the bilayer is further assessed by the uniformity of the fluorescence and the lack of any large aggregates of lipid which may have

been leftover from the SUV deposition. Often, there are scratches on the surface of the coverslip to which no lipid will bind. During the experiment, we avoid imaging the contraction of the network near these regions, as actin and myosin will stick to the coverslip.

4.2. Construction and contraction of F-actin cortex

4.2.1 Introduction of F-actin

The F-actin is not polymerized within the sample chamber, but separately in a 0.5 mL microcentrifuge tube and then added to the bilayer. This is done as to not adhere incompletely polymerized F-actin to the bilayer, and thereby influence the kinetics of polymerization and equilibrium length of the filaments. The steps are as follows:

1. In a 500 μL microcentrifuge tube, 2.0 μ*M* dark actin and 0.6 μ*M* fluorescent actin (Alexa 568, Molecular Probes) are combined with 4 μ*M* dark phalloidin (Cytoskeleton) in 1 × F-buffer. The solution is supplemented with 0.5% methylcellulose (14,000 MW, Sigma), 9% glucose oxidase/catalase (Calbiochem), and glucose. This solution is incubated on ice for 1.5–2 h.
2. Afterwards, this 500 μL polymerization mix is then added to the 500 μL solution that immerses the bilayer, dividing the concentration of methylcellulose, F-actin, and ATP by 2, leaving 0.25% methylcellulose, 1.3 μ*M* actin, and 0.250 m*M* ATP. The F-actin is allowed to accumulate on the surface of the bilayer for 15 min (Fig 15.4A). Although minimal bundling is observed in this 2D assay, a 0.2% methylcellulose solution has been shown to initiate bundle formation in 3D assays (Kohler, Lieleg, & Bausch, 2008).

4.2.2 Attachment of F-actin to membrane

For the attachment of F-actin to the bilayer membrane, we utilize FimA2 (pET-21a-MBP-FimA1A2p-His, gift of Dave Kovar, University of Chicago), a mutant construct of the actin cross-linker Fimbrin containing a single F-actin-binding domain (Skau et al., 2011). FimA2 has a His-tag; therefore, it can simultaneously bind the nickel lipid in the membrane as well as a single F-actin. It is added following bilayer membrane formation and prior to F-actin addition. It is added at varying concentrations that correspond to different degrees of immobilization of F-actin. We choose three concentrations: 10, 100, and 1000 n*M* FimA2 for low, medium, and high levels of adhesion. Below 10 n*M* FimA2, F-actin is completely mobile on the membrane surface as can be seen by the fluctuation of the F-actin. By the same metric, at 1 μ*M* FimA2, F-actin is completely immobile. Regardless of concentration, FimA2 is incubated on the membrane for 15 min.

1. Add volume of FimA2 to 0.5 mL of ATP-free F-buffer. Incubate on membrane for 15 min.
2. Wash repeatedly with ATP-free F-buffer. The stability of the nickel–His bond is strong, such that very little FimA2 leaves the surface of the membrane.

4.2.3 F-actin cross-linking

The presence of F-actin cross-linking proteins links individual F-actin, thereby increasing the length scale of contraction by myosin activity (Janson et al., 1991). In addition, cross-linking proteins can change the architecture of the network itself, from primarily filamentous to highly bundled. The type of cross-linker may also join filaments based on their polarity.

To cross-link our F-actin network, we include the cortical F-actin cross-linker α-actinin. α-Actinin is known to bind F-actin without bias on the orientation of the filaments. Furthermore, at low concentrations (1:300 [α-actinin]:[actin]) links F-actin isotropically, but can bundle F-actin at high concentrations (1:30 [α-actinin]:[actin]). Thus, the architecture as well as the connectivity of the network is modified by α-actinin.

1. After the sedimentation of the F-actin, add the desired volume of α-actinin to 100 μL F-buffer, and add this volume to the center of the chamber. The protein will diffuse throughout the chamber and bind the F-actin network.
2. Regardless of the concentration, incubate the α-actinin in the chamber for 15 min (Fig 15.4B). If 1:30 [α-actinin]:[actin] concentration, bundling can be observed to assess the completion and spatial uniformity of cross-linking.

4.2.4 Myosin II addition

After cross-linking, varying concentrations (10–100 n*M*) of skeletal muscle myosin II dimers are added to the sample chamber. The myosin dimers are small and diffuse quickly through the chamber. They polymerize into thick filament assemblies within minutes, bind the F-actin network, and induce contraction (Fig 15.4C). For highly cross-linked samples (1:300 [α-actinin]:[actin]), the F-actin network is highly connected. When myosin is added, the entire network contracts with a length scale larger than the microscope field of view and may be centered anywhere across the 25-mm coverslip (Fig. 15.5).

To spatially control the contraction, we have successfully used the property that the myosin II ATPase inhibitor, blebbistatin, is inactivated by short exposure (100 ms) to low power (>0.1 mW/μm^2) light with wavelength <500 nm (Sakamoto, Limouze, Combs, Straight, & Sellers, 2005). When using this approach, 40 μM blebbistatin is added after the formation of a lipid bilayer. Thus, when the F-actin is crowded to the surface of the bilayer, the solution is well mixed. After the myosin is added, it polymerizes, accumulates on the F-actin surface, and is weakly bound to F-actin, but its mechanochemical activity is inhibited. This method is therefore suitable for highly cross-linked networks, as the presence of blebbistatin-inhibited myosin itself introduces a small degree of cross-linking but is minor compared to the binding of passive cross-linkers at high concentrations such as α-actinin. Then, upon illumination of the network with the 491 nm light, the myosin inhibition is released, and the network contracts, centered within out field of view.

4.2.5 Sources of variability

Local organization of F-actin: F-actin that is crowded to the surface of the bilayer membrane in the absence of adhesion orders itself quasi-nematically (Fig. 15.4). Thus, there are local regions of F-actin alignment, which may vary across a 60× field of view. As we expect that network architecture may modulate contractility, the contraction of the network will vary across the field of view as well. Adhesion of the F-actin to the membrane abrogates this variability. As the nematic alignment of F-actin is thermal, very low adhesion would be required to eliminate this effect.

Size of myosin thick filaments: The myosin is added as dimers in solution after the sedimentation of F-actin. During incubation within the chamber, the myosin polymerizes into thick filaments, as it transitions from its 0.45 *M* KCl buffer into a 50 m*M* KCl buffer. However, the size of the thick filaments will vary with the dimer concentration added. Thus, in the future, myosin thick filaments should be preformed and introduced into the chamber fully polymerized.

5. FUTURE DIRECTIONS

Over the past several years, we have successfully used these assays to identify the requirements and regulation of actomyosin contractility in both bundles and networks. By systematically changing the myosin density, we

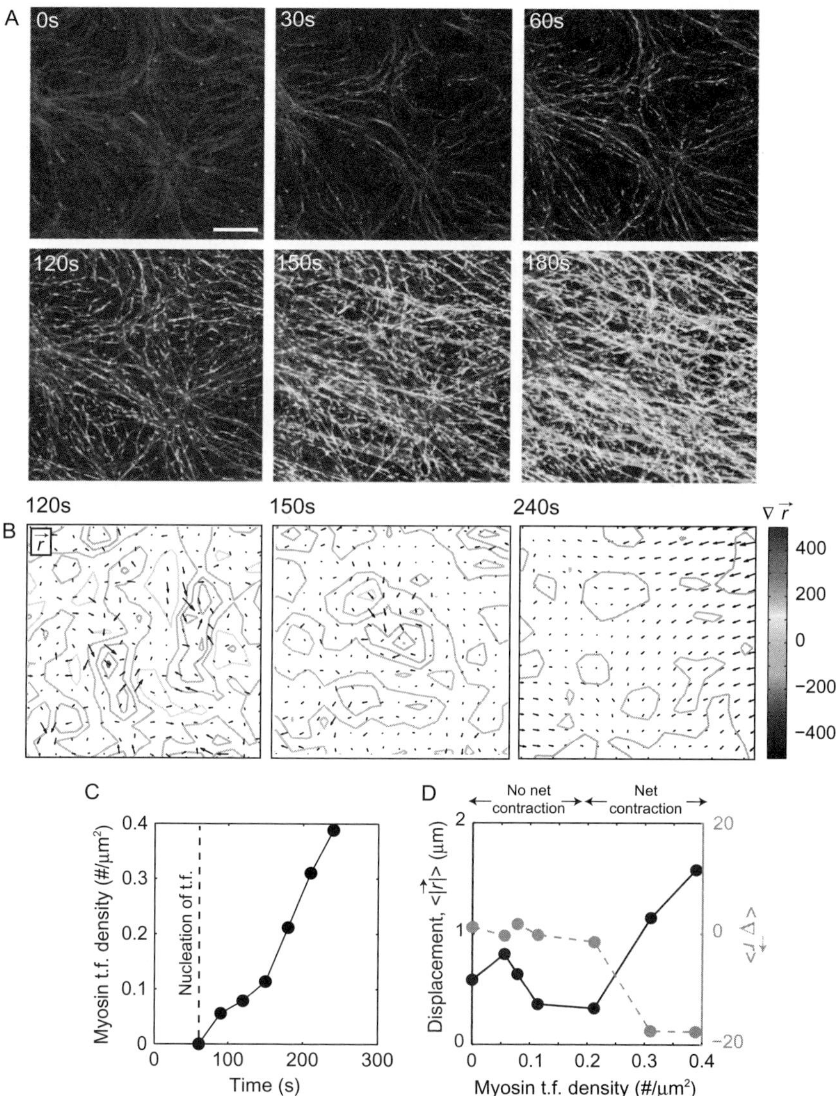

Figure 15.5 Quantification of F-actin network contractility. (A) F-actin (red) and smooth muscle myosin II (green) within a contracting model cortex. Myosin accumulates over time. Scale bar is 10 μm. (B) Overlay of F-actin displacement (*r*, black arrows) and divergence of F-actin displacement (colored contours) for the contracting network in (A). Hot colors indicate positive divergence (expansion) and cool colors indicate negative divergence (contraction). (C) Myosin thick filament density ρ, over time. (D) Mean divergence (green) and speed (blue) of the F-actin displacement (*r*) as a function of myosin thick filament density, *r*. (See the color plate.)

have identified a critical myosin density necessary for contraction (Thoresen et al., 2011) and determined how the myosin filament properties (size and isoform composition) regulate the contraction rate (Thoresen et al., 2013). We have also explored how myosin-driven stresses can result in self-organization of a sarcomere-like structure within bundles (Stachowiak et al., 2012). Finally, we have demonstrated the importance of F-actin bending and buckling in facilitating contraction in disordered actomyosin arrays (Lenz, Gardel, & Dinner, 2012; Lenz, Thoresen, Gardel, & Dinner, 2012; Murrell & Gardel, 2012). The semi-flexibility of F-actin is crucial to break the symmetry between tensile and compressive stresses generated in disordered actomyosin bundles (Lenz, Gardel, & Dinner, 2012) and networks to facilitate robust contractility over a wide range of network architectures (Lenz, Gardel, & Dinner, 2012; Lenz, Thoresen, Gardel, & Dinner, 2012; Murrell & Gardel, 2012). Moreover, we have found that myosin II-generated filament bending results in F-actin severing, thus providing a putative mechanism for coordinating contraction and actin filament polymerization dynamics in contractile systems.

We can now explore the phase space of these motor-filament bundles and networks to determine the regulation of contractility by changing parameters involving the filaments (bending rigidity, length, density), motors (myosin II isoform and thick filament size), and accessory proteins (F-actin assembly factors and cross-linkers, cross-linkers between the membrane and F-actin network). We anticipate such experiments will identify how contractility is spatially regulated in the cellular cortex to support diverse morphological processes. Through addition of F-actin assembly factors, we can attempt to reconstitute the steady-state dynamic contractile arrays observed to understand the coordination of F-actin assembly, contraction and disassembly observed in diverse systems such as the lamella and cytokinetic ring. Moreover, we anticipate these experiments will shed light on how strains and forces within contractile arrays are used to drive changes in the association of regulatory factors such as α-actinin (Aratyn-Schaus, Oakes, & Gardel, 2011) and zyxin (Smith et al., 2010). Finally, this assay can be utilized to explore the interplay between myosin-driven actin dynamics and membrane organization, a crucial interface that determines cellular response to chemical and physical stimuli from the external environment (Kapus & Janmey, 2013). Eventually, as these processes are revealed, these studies will facilitate the reconstitution of artificial cells by compartmentalization of crucial factors inside vesicles (Carvalho et al., 2013).

ACKNOWLEDGMENT

M. G. is funded by a Burroughs Wellcome Career Award at the Scientific Interface and the Packard Foundation. We thank Patrick McCall for a careful reading of the chapter.

REFERENCES

Aratyn-Schaus, Y., Oakes, P. W., & Gardel, M. L. (2011). Dynamic and structural signatures of lamellar actomyosin force generation. *Molecular Biology of the Cell, 22*, 1330–1339.

Aratyn-Schaus, Y., Oakes, P. W., Stricker, J., Winter, S. P., & Gardel, M. L. (2010). Preparation of compliant matrices for quantifying cellular contraction. *Journal of Visualized Experiments,* (46).

Bárány, M. (1996). *Biochemistry of smooth muscle contraction.* San Diego: Academic Press.

Carvalho, K., Tsai, F. C., Lees, E., Voituriez, R., Koenderink, G. H., & Sykes, C. (2013). Cell-sized liposomes reveal how actomyosin cortical tension drives shape change. *Proceedings of the National Academy of Sciences of the United States of America, 110*(41), 16456–16461.

Dainty, M., Kleinzeller, A., Lawrence, A. S., Miall, M., Needham, J., Needham, D. M., et al. (1944). Studies on the anomalous viscosity and flow-birefringence of protein solutions: III Changes in these properties of myosin solutions in relation to adenosine triphosphate and muscular contraction. *Journal of General Physiology, 27*(4), 355–399.

Ebashi, S., & Ebashi, F. (1964). A new protein factor promoting contraction of actomyosin. *Nature, 203*, 645–646.

Huxley, H. E. (2004). Fifty years of muscle and the sliding filament hypothesis. *European Journal of Biochemistry, 271*(8), 1403–1415.

Janson, L. W., Kolega, J., & Taylor, D. L. (1991). Modulation of contraction by gelation/solation in a reconstituted motile model. *Journal of Cell Biology, 114*(5), 1005–1015.

Kapus, A., & Janmey, P. (2013). Plasma membrane—Cortical cytoskeleton interactions: A cell biology approach with biophysical considerations. *Comprehensive Physiology, 3*, 1231–1281.

Kohler, S., Lieleg, O., & Bausch, A. R. (2008). Rheological characterization of the bundling transition in F-actin solutions induced by methylcellulose. *PLoS One, 3*(7), e2736.

Lenz, M., Gardel, M. L., & Dinner, A. R. (2012). Requirements for contractility in disordered cytoskeletal bundles. *New Journal of Physics, 14*(3), 033037.

Lenz, M., Thoresen, T., Gardel, M. L., & Dinner, A. R. (2012). Contractile units in disordered actomyosin bundles arise from F-actin buckling. *Physical Review Letters, 108*(23), 238107.

Margossian, S. S., & Lowey, S. (1982). Preparation of myosin and its subfragments from rabbit skeletal muscle. *Methods in Enzymology, 85*(Pt. B), 55–71.

Murrell, M. P., & Gardel, M. L. (2012). F-actin buckling coordinates contractility and severing in a biomimetic actomyosin cortex. *Proceedings of the National Academy of Sciences of the United States of America, 109*(51), 20820–20825.

Pollard, T. D. (1982). Myosin purification and characterization. *Methods in Cell Biology, 24*, 333–371.

Sakamoto, T., Limouze, J., Combs, C. A., Straight, A. F., & Sellers, J. R. (2005). Blebbistatin, a Myosin II inhibitor, is photoinactivated by blue light. *Biochemistry, 44*(2), 584–588.

Skau, C. T., Courson, D. S., Bestul, A. J., Winkelman, J. D., Rock, R. S., Sirotkin, V., et al. (2011). Actin filament bundling by fimbrin is important for endocytosis, cytokinesis, and polarization in fission yeast. *Journal of Biological Chemistry, 286*(30), 26964–26977.

Smith, M. A., Blankman, E., Gardel, M. L., Luettjohann, L., Waterman, C. M., & Beckerle, M. C. (2010). A zyxin-mediated mechanism for actin stress fiber maintenance and repair. *Developmental Cell, 19*(3), 365–376.

Spicer, S. S. (1951). Gel formation caused by adenosine triphosphate in actomyosin solutions. *Journal of Biological Chemistry, 190*(1), 257–267.

Stachowiak, Matthew R., McCall, Patrick M., Thoresen, T., Balcioglu, Hayri E., Kasiewicz, L., Gardel, Margaret L., et al. (2012). Self-organization of Myosin II in reconstituted actomyosin bundles. *Biophysical Journal, 103*(6), 1265–1274.

Stossel, T. P., Hartwig, J. H., Yin, H. L., Zaner, K. S., & Stendahl, O. I. (1982). Actin gelation and structure of cortical cytoplasm. *Cold Spring Harbor Symposia on Quantitative Biology, 46*(Pt. 2), 569–578.

Szent-Györgyi, A. (1945). Studies on muscle. *Acta Physiologica Scandinavica, 9*, 1–116.

Szent-Györgyi, A. (1947). *Chemistry of muscular contraction.* New York: Academic Press.

Szent-Gyorgyi, A. (1950). Actomyosin and muscular contraction. *Biochimica et Biophysica Acta, 4*(1–3), 38–41.

Thoresen, T., Lenz, M., & Gardel, M. L. (2011). Reconstitution of contractile actomyosin bundles. *Biophysical Journal, 100*(11), 2698–2705.

Thoresen, T., Lenz, M., & Gardel, M. L. (2013). Thick filament length and isoform composition determine self-organized contractile units in actomyosin bundles. *Biophysical Journal, 104*(3), 655–665.

Watanabe, S., & Yasui, T. (1965). Effects of magnesium and calcium on the super precipitation of Myosin B. *Journal of Biological Chemistry, 240*, 105–111.

Weber, A., & Winicur, S. (1961). The role of calcium in the superprecipitation of actomyosin. *Journal of Biological Chemistry, 236*, 3198–3202.

CHAPTER SIXTEEN

Directed Actin Assembly and Motility

Rajaa Boujemaa-Paterski[1], Rémi Galland, Cristian Suarez, Christophe Guérin, Manuel Théry, Laurent Blanchoin

Institut de Recherches en Technologies et Sciences pour le Vivant, iRTSV, Laboratoire de Physiologie Cellulaire et Végétale, CNRS/CEA/INRA/UJF, Grenoble, France

[1]Corresponding author: e-mail address: rajaa.paterski@cea.fr

Contents

Abstract

The actin cytoskeleton is a key component of the cellular architecture. However, understanding actin organization and dynamics *in vivo* is a complex challenge. Reconstitution of actin structures *in vitro*, in simplified media, allows one to pinpoint the cellular biochemical components and their molecular interactions underlying the architecture and dynamics of the actin network. Previously, little was known about the extent to which geometrical constraints influence the dynamic ultrastructure of these networks. Therefore, in order to study the balance between biochemical and geometrical control of complex actin organization, we used the innovative methodologies of UV and laser patterning to design a wide repertoire of nucleation geometries from which we assembled branched actin networks. Using these methods, we were able to reconstitute complex actin network organizations, closely related to cellular architecture, to precisely direct and control their 3D connections. This methodology mimics the actin networks encountered in cells and can serve in the fabrication of innovative bioinspired systems.

Methods in Enzymology, Volume 540
ISSN 0076-6879
http://dx.doi.org/10.1016/B978-0-12-397924-7.00016-9

1. INTRODUCTION

The actin cytoskeleton is critical for cellular integrity and is involved in many fundamental phenomena ranging from cell morphogenesis to cell division and motility. It is composed of extremely complex arrays of specialized and dynamic structures that overlap and interconnect continuously overtime. During the past decades, tremendous efforts have been focused on understanding the molecular mechanisms underlying actin cytoskeleton dynamics and elucidating physical properties of actin organization (Blanchoin, Boujemaa-Paterski, Sykes, & Plastino, 2014; Fletcher & Mullins, 2010; Mogilner & Zhu, 2012; Pollard, Blanchoin, & Mullins, 2000). In particular, reconstituted biomimetic systems, able to mimic motility of bacteria, viruses, or vesicles in cytoplasmic extracts or in a purified media, opened up a generation of assays, which significantly deepened our understanding of the essential biochemical components necessary for actin-based force generation (Bernheim-Groswasser, Wiesner, Golsteyn, Carlier, & Sykes, 2002; Cameron, Footer, Van Oudenaarden, & Theriot, 1999; Frishknecht et al., 1999; Giardini, Fletcher, & Theriot, 2003; Loisel, Boujemaa, Pantaloni, & Carlier, 1999; Theriot, Rosenblatt, Portnoy, Goldschmidt-Clermont, & Mitchison, 1994; Vignjevic et al., 2003). However, most of these assays were based on a steady-state balance between unpolymerized and polymerized actin and allowed a macroscopic description of actin assembly and dynamics around the motile particle (Loisel et al., 1999; Romero et al., 2004). Recently, a new generation of motility assays made of a large pool of unpolymerized actin, similar to physiological conditions, unveiled the link between molecular mechanisms of assembly and macroscopic properties of actin branched organization (Achard et al., 2010; Akin & Mullins, 2008; Dayel et al., 2009; Kawska et al., 2012; Reymann et al., 2011). However, how geometrical boundaries influence network dynamics and architecture remains poorly understood.

The innovative technology of micropatterning paves the way for investigations aimed at deciphering the balance existing between geometrical and biochemical constraints in controlling the architectural organization and dynamics of actin. Indeed, extremely simplified media allowing directed growth of actin networks from functionalized patterned geometries has demonstrated that actins self-organize into specific architectures with respect to both biochemical and geometrical constraints (Reymann et al., 2010, 2012). Interestingly, these methods revealed that the network's architecture governs its own selective interaction with regulatory proteins (Reymann et al., 2010, 2012). Specifically, these studies have demonstrated that, under specific geometrical constraints, actin filaments can self-assemble into parallel, filopodia-like structures, entangled networks, or an antiparallel organization that becomes contractile in the presence of molecular motors. With further improvement in micropatterning and by increasing the number of regulatory proteins used in the assays, it became possible to reconstitute the assembly of complex branched actin architectures similar to those assembled at the leading edge of motile cells or to control and direct 3D actin assembly (Galland et al., 2013). Here, we will first describe how to reconstitute actin-based bead motility in physiologically relevant medium made of a large pool of actin monomers. Second, we will describe how to design UV-micropatterned motifs that can drive the assembly of geometrically controlled actin networks. Finally, we will detail the laser-assisted micropatterning method that allows the reconstitution of directed 3D connections linking adjacent cross-linked actin networks.

2. RECONSTITUTION OF ACTIN-BASED MOTILITY IN A G-ACTIN-BUFFERED MEDIUM

In the following section, we present how to reconstitute motility in a G-actin-buffered medium. Capping protein (CP) is mandatory for symmetry breaking and motility (Iwasa & Mullins, 2007; Loisel et al., 1999). However, at the concentration used in motility assays (Achard et al., 2010; Akin & Mullins, 2008; Loisel et al., 1999; Reymann et al., 2011), which is usually 20–50 n*M*, CP activates spontaneous nucleation of actin monomers in solution. Indeed, CP nucleates actin by stabilizing the nuclei at their barbed end leaving them free to elongate from their pointed end (Schafer, Jennings, & Cooper, 1996). Therefore, in order to confine actin nucleation and elongation to the functionalized bead surface and avoid CP-mediated nucleation in the medium, we saturate actin monomers with profilin rather using free actin monomers; profilin–actin complexes are unable to spontaneously nucleate or undergo pointed-end elongation. Interestingly, profilin–actin complexes are capable of elongating from the barbed ends via nucleation by the Arp2/3 complex at the bead surface (Pantaloni and Carlier, 1993; Fig. 16.1).

2.1. Particles functionalization

2.1.1 Material

- Polystyrene carboxylate microspheres (Polysciences)

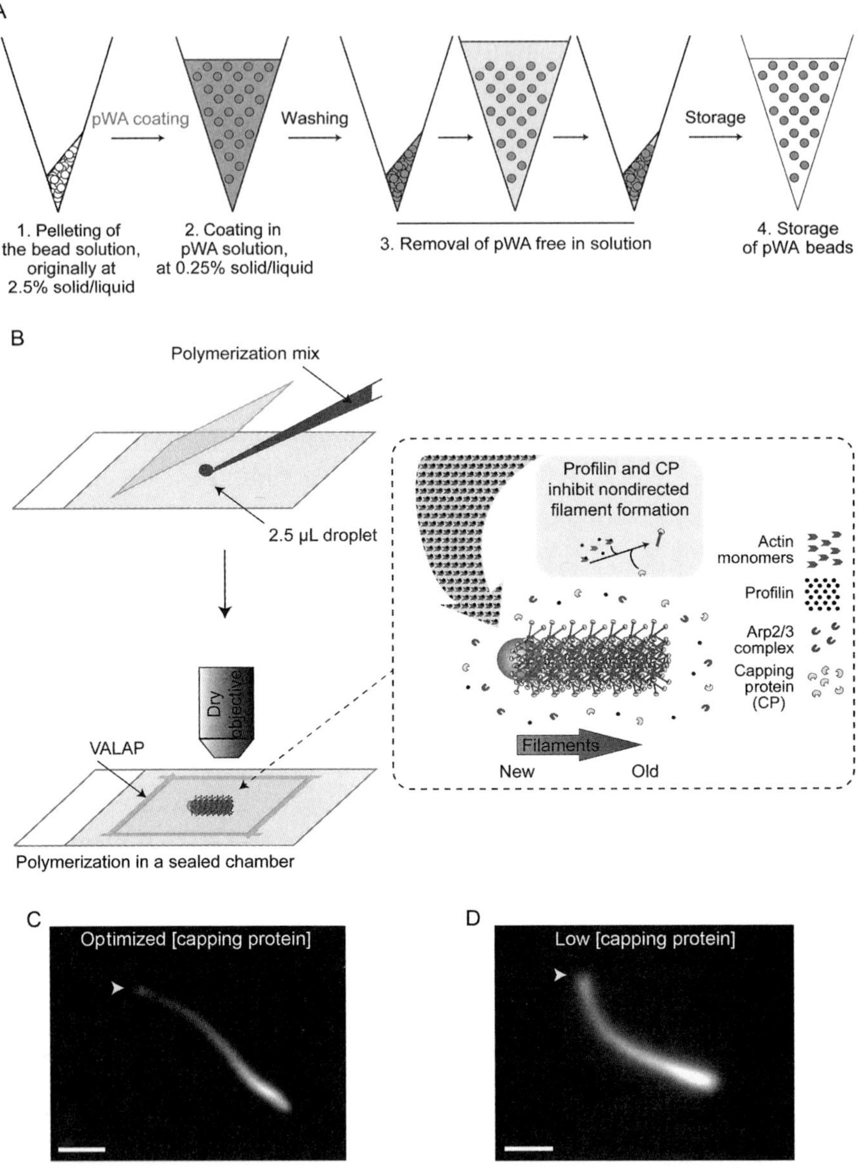

Figure 16.1 Actin-based motility in a G-actin buffered medium. (A) Schematic representation of the functionalization of polystyrene beads with pWA, an actin nucleation promoting factor. The crucial step of this procedure is the removal of pWA free in solution after the coating in order to prevent nondirected actin assembly in the medium. (B) The polymerization medium is placed between slide and coverslip in a sealed chamber, and bead motility is followed using epifluorescence microscopy. The polymerization medium contains a threefold excess of profilin (black) over actin (blue) in order

- BSA, Sigma A7030, resuspended in MilliQ water solution to make a 10% (w/v) solution and stored at −20 °C
- X buffer containing 10 mM HEPES (pH 7.5), 0.1 M KCl, 1 mM $MgCl_2$, 1 mM ATP, and 0.1 mM $CaCl_2$

2.1.2 Equipment

- Centrifuge
- Thermomixer (Eppendorf)

2.1.3 Methods

As illustrated in Fig. 16.1A, bead coating with NPFs is performed as follows:

1. Pellet 5 μL polystyrene carboxylated beads (16,000 × g, 2 min, room temperature).
2. Incubate beads pellet in usually 50 μL NPFs (900 rpm, 15 min, 20 °C in a thermomixer block, or even on a rotator at room temperature). Usually, we use 2 μM pWA fragment of WASP protein, purified as described (Achard et al., 2010).
3. Pellet the NPFs-beads (16,000 × g, 2 min, 4 °C).
4. Carefully discard the supernatant. This step is crucial as the presence of free NPFs in solution will activate the Arp2/3 complex-mediated actin assembly and competes with polymerization on beads.
5. Incubate the functionalized NPFs-beads in 100–300 μL of 1% BSA in X buffer for at least 15 min on ice in order to saturate the space left free on beads surface and more importantly to get rid of free NPFs remaining in the pellet volume. Therefore, adjust the volume of BSA solution according of the adsorption efficiency of the NPFs to set the good balance between free NPFs removal and NPFs density on beads.
6. Centrifuge (16,000 × g, 2 min, 4 °C) and resuspend the bead pellet in 0.1% BSA in X buffer.
7. Store at 4 °C for up to 3 days.

---

to efficiently inhibit spontaneous actin nucleation in the medium, and the amount of capping protein (yellow) required for reconstituted motility blocks spontaneous nucleation coming from the small amount of actin monomers not bound to profilin. Actin nucleation and elongation is directed to pWA-coated beads through the recruitment and activation of the Arp2/3 complex (red). (C and D) Reconstituted actin-based motility of 2 μm beads in a medium containing 6 μM actin monomers, 18 μM profilin, 150 nM Arp2/3 complex, and 20 nM capping protein (C) or 10 nM capping protein (D). Scale bars are 10 μm. (See the color plate.)

2.2. The motility assay

2.2.1 Material

- Glasses and coverslips (Agar Scientific).
- G buffer containing 2 m*M* Tris–HCl (pH 7.5), 0.2 m*M* ATP, 0.5 m*M* DTT, 0.1 m*M* $CaCl_2$, 1 m*M* NaN_3.
- 10% BSA solution and X buffer as mentioned in the former section.
- 3 × Exchanging buffer: EGTA, 600 μ*M*, and $MgCl_2$ at three times (10 μ*M*+ *C*) where *C* is the concentration of the actin solution to be Ca–Mg exchanged, and *C* is usually tens of micromolar.
- ATP solution containing 24 m*M* $MgCl_2$, 12 m*M* ATP, 20 m*M* DTT, 0.88 m*M*DABCO (1,4-diazabicyclo[2.2.2]octane solution from SIGMA).
- Methylcellulose, 1500 cPs (Sigma) resuspended in MilliQ water solution to make a 2% solution and stored at room temperature for a week.
- VALAP soft wax made of equal weights of Vaseline, Lanolin, and Paraffin wax (VWR). VALAP is maintained molten on heater block at 50–55 °C. The VALAP wax is a biologically inert material and therefore preferred to nail polish for coverslip sealing.

2.2.2 Equipment

- A dry block heater (VWR), in case of using VALAP to seal the motility chamber.
- Epifluorescence microscope. Images can be taken using dry 40 × objective, a motorized *XY* stage, and a CCD camera. The use of the motorized stage is useful as it allows acquiring actin dynamics on several beads under exactly the same biochemical conditions.

2.2.3 Methods

The following steps have to be performed in parallel:

(1) G-actin/profilin complex preparation

Extemporaneously prepare:

- Monomeric actin solution at a given *C* concentration (typically 3–6 μ*M*). According to the experiment's needs, the actin solution may contain nonlabeled or usually 5–10% Alexa labeled-actin.
- Mix the actin solution with a 1 × Exchanging buffer.
- Incubate for exactly 5 min on ice. At this step, the monomeric actin is Mg–ATP loaded.
- Add three times molar excess of profilin with respect to the final actin concentration.
- Incubate for at least 5 min on ice. At this step, actin monomers are mainly in the form of profilin–Mg–ATP–G-actin complex, ready to

be used. We recommend preparing the profilin/actin complex before every experiment.

(2) Motility medium

Prepare the following mix respecting the order and proportions:

- 2 μL of 10% BSA
- *x* μL Xb to make 15 μL in total
- 3 μL ATP mix
- 1 μL Arp2/3 complex to make 100–120 n*M* final
- 1 μL CP solution to make 20–30 n*M* final
- 0.5 μL NPFs-beads
- 2 μL of 2% methylcellulose
- *y* μL profilin–G-actin to make usually 1–4 μ*M* final
- Place 3 μL of the motility medium between glass and 20 × 20 coverslip (Fig. 16.1B)
- Check the onset of the reaction, seal with VALAP before evaporation
- Observe by phase contrast microscopy and/or by fluorescence microscopy (when using labeled proteins) (Fig. 16.1B). The presence of the singlet O_2 scavenger, DABCO, decreases the oxidative damage induced by photodynamic reactions

The bead motility assay is an extremely useful assay that allows the study of how biochemical parameters control actin network dynamics and architectural organization. We tested the role of CP in organizing the actin filaments into a network that is capable of propelling beads. By efficiently blocking barbed-end elongation, the CP protein restrains filament nucleation and elongation at the site of nucleation around the particle (Fig. 16.1B and C; Achard et al., 2010; Loisel et al., 1999). Therefore, decreasing the amount of CP in the medium leads to inefficient blocking of actin filament barbed ends initiated at the nucleating bead surface. Thus, actin filaments are able to elongate as the bead propulsion proceeds. This population of uncapped filaments gathers into bundles and consumes monomers but is only able to inefficiently produce bead propulsion (Fig. 16.1D; Pantaloni, Boujemaa, Didry, Gounon, & Carlier, 2000).

3. ASSEMBLY OF BRANCHED ACTIN NETWORKS ON A WIDE RANGE OF GEOMETRIES

Examination of actin network dynamics in living cells reveals how actin cytoskeleton structures and cell morphology are intimately related. During the past decades, a rich body of experimental work has revealed how the cell's morphogenetic processes are powered by actin cytoskeleton dynamics (Fletcher & Mullins, 2010; Pollard & Borisy, 2003). However, conversely, little is known about how cell boundaries regulate the geometrical organization and the dynamics of the actin cytoskeleton. To improve our understanding of such questions, we have developed a UV-patterning procedure (Reymann et al., 2010, 2012) to study how nucleation geometries may regulate network dynamics and architecture. In the following section, we discuss how the UV-patterning method can be applied to study effect of different nucleation geometries on the assembly of branched actin networks, similar to those present at the leading edge of motile cells.

3.1. Design and functionalization of UV-patterned surfaces

UV-micropatterning method had been extensively described (Reymann, Guérin, Théry, Blanchoin, & Boujemaa-Paterski, 2014). Hereafter is a brief description of the passivation of glass coverslips with a poly(L-lysine)-poly (ethylene glycol), PLL-PEG layer (Fig. 16.2).

3.1.1 Material

- 20×20-mm glass coverslips and cover glasses (Agar Scientific)
- Ethanol, MilliQ water, Hellmanex III, isopropanol
- Parafilm
- PLL-PEG powder stored at -20 °C (JenKem Technology)
- PLL-PEG stock solution: 1 mg/mL PLL-PEG in Hepes 10 m*M* pH 7.4 (store at 4 °C for a week). Optimized coating with a regular layer of PLL-PEG chains depends on the efficiency of the plasma-cleaning procedure and the freshness of the PLL-PEG stock solution
- $10\times$ KMEI buffer containing 500 m*M* KCl, 10 m*M* $MgCl_2$, 10 m*M* EGTA, and 100 m*M* imidazole, pH 7.0
- $1\times$ KMEI prepared in G buffer (as in Section 1.1.2)
- Petri dishes (for storage of cleaned and passivated glass surfaces)

3.1.2 Equipment

- Plasma cleaner Femto (Diener electronic)
- UV ozone oven, UVO cleaner (Jelight)
- A chrome mask transparent to UV at the sites of micropatterns (TOPPAN, Photomask) and a vacuum mask holder, as described in Azioune, Storch, Bornens, Thery, and Piel (2009)

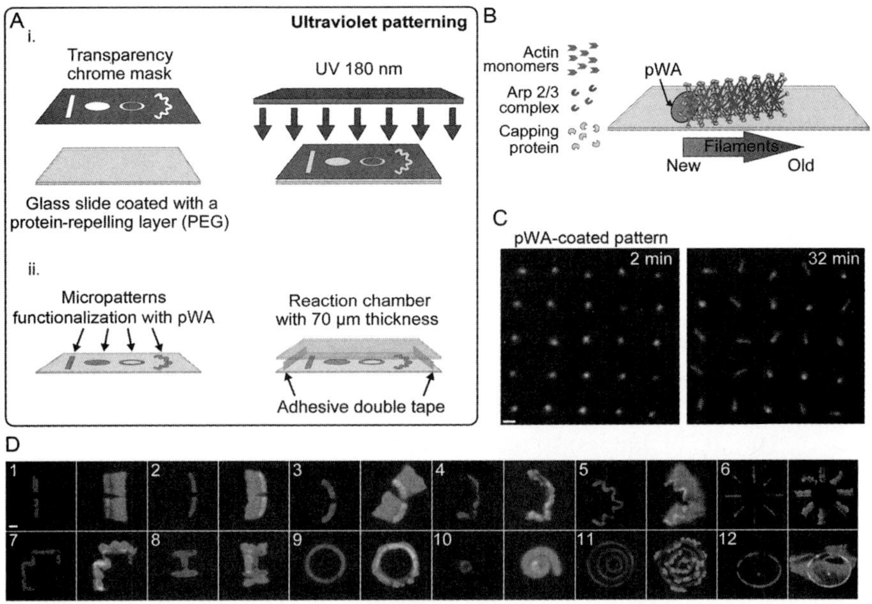

Figure 16.2 Reconstituted actin-based motility using UV patterning. (A) A chrome photomask containing transparent micropatterns is placed on a PLL-coated glass slide and exposed to ultraviolet light for 5 min (i). The micropatterned glass slide is incubated for 15 min with 1 μ*M* pWA prior to mounting in a reaction chamber (ii). (B) pWA-coated micropatterns are incubated with a motility medium containing actin monomers saturated with profilin (blue), Arp2/3 complex (red), and capping protein (yellow), resulting in the formation of actin filament networks. (C) Dot patterns (3 μm) are imaged by epifluorescence microscopy in a motility medium containing 6 μ*M* actin monomers (Alexa568 labeled), 18 μ*M* profilin, 150 n*M* Arp2/3 complex, and 25 n*M* capping protein. (D) Patterns with different geometries (red patterns, images from 1 to 12) are coated with pWA and fibrinogen 546 and then incubated with a mix containing 4 μ*M* actin monomers (Alexa488 labeled), 12 μ*M* profilin, 150 n*M* Arp2/3 complex, and 25 n*M* capping protein. The second image of the 12 pairs shows the actin network (green) developed after 1 h of polymerization. The width of each pattern line is 3 μm. Scale bars are 5 μm. (See the color plate.)

3.1.3 Methods

(1) Surface cleaning and PLL-PEG coating

In order to obtain a uniform layer of PLL-PEG, we first perform a drastic cleaning procedure that ensures a total removal of dusts and fats from the glass surfaces. Therefore, the washings in water are crucial as they allow removal of chemicals potentially harmful for proteins integrity and actin polymerization.

1. Slides are washed with 100% ethanol, wiped with absorbent paper, rinsed with MilliQ water, and then dried with a filtered airflow.
2. Slides are sonicated in acetone (30 min, room temperature), incubated in ethanol (10 min, room temperature), and then washed extensively with MilliQ water.
3. Slides are incubated in 2% Hellmanex (2 h, room temperature), washed extensively with MilliQ water, dried with a filtered airflow, and then stored at 4 °C in a clean box.
4. Cleaned slides are oxidized with O_2 plasma (3 min, 80–90 Watts).
5. Immediately after plasma oxidation, each coverslip is incubated with 100 μL of 0.1 mg/mL PLL-PEG solution on parafilm (30 min, room temperature). To gently lift up the coated coverslip, 100 μL of Hepes 10 m*M* (pH 7.4) is injected between glass and parafilm. If the coating was successful, the glass should come off dry, and the remaining tiny droplets at its edges can be wiped or blown off using pulsed air directed toward the noncoated side.
6. PLL-PEG-coated coverslips can be stored in Petri dishes at 4 °C for up to 2 weeks. The Petri dishes are sealed with parafilm to avoid desiccation.

(2) Surface patterning and functionalization

As previously described, we found that functionalization is the most efficient when performed immediately after UV insolation. As a control, patterns efficiently coated should initiate massive actin assembly within 5 min after the incubation with the polymerization medium detailed in the following section (Fig. 16.2A).

1. A chrome mask is washed successively with MilliQ water and isopropanol, dried with a filtered airflow.
2. The PLL-PEG coverslip is positioned on the vacuum mask holder facing upward, and the mask is placed on top of it. The vacuum as well as the absence of any dust ensures the tight contact between the coverslip and the mask, which is mandatory for obtaining reproducibly well-defined micropatterns.
3. After preheating the UV oven as recommended in the manufacturer's instructions, the PLL-PEG coated side of the coverslip is UV irradiated through the chrome mask for 5 min.
4. Immediately after UV patterning, each coverslip is incubated with 30 μL NPFs solution at a final concentration of 1–2 μ*M* (15 min on parafilm, room temperature). We usually use pWA fragment at 1 μ*M*. To remove the coverslip from the parafilm, 1 mL of

1 × KMEI buffer is injected between glass and parafilm. Each coverslip is then gently washed for 30 s in a small Petri dish containing cold 1 × KMEI. Functionalized coverslips must come off dry and the remaining droplet can be wiped or blown off using filtered-air flow directed toward the nontreated side.

5. Functionalized coverslips are stored for 48 h in a clean box at 4 °C.

3.2. Polymerization of branched actin filament network on UV-micropatterns

In order to study how geometry may control the actin dynamics and architectural organization of branched networks in a physiologically relevant medium, we polymerized dynamic actin network on micropatterned surfaces in the presence of the Arp2/3 complex and CP. CP is known to control filaments length and subsequently enhances branches density of Arp2/3-mediated networks (Blanchoin et al., 2000; Fig. 16.2B).

3.2.1 Material

- X buffer, 10 × KMEI buffer, G buffer, ATP mix, 10% BSA, 2% methylcellulose, VALAP, and freshly prepared profilin–Mg–ATP–actin (5–10% Alexa labeled), as in Section 1.1
- Adhesive double tape. We use a precut double tape 70 μm thick (LIMA)

3.2.2 Equipment

- 76 × 26 × 1.0-mm glass slides (Agar Scientific)
- A dry block heater (VWR) if using VALAP to seal the motility chamber
- Epifluorescence microscope, as described in Section 1.1

3.2.3 Methods

1. A flow cell chamber is freshly assembled as follows:
   - A glass slide is washed in water, wiped with ethanol and extensively rinsed with water just before use.
   - With a functionalized coverslip, a cleaned glass slide and precut adhesive double tapes assemble a stable flow chamber (Fig. 16.2A).
2. Immediately afterward, the motility medium, detailed in Section 1.1.2, is prepared in the same order and proportions excluding, however, the NPFs-coated beads.
3. Without delay, fill in the flow chamber with the motility medium.
4. Check the onset of the reaction and seal with VALAP before evaporation.

5. Follow assembly using fluorescence microscopy (Fig. 16.2 B, C, and D). Using this protocol, we were able to show that nucleation geometry can control the organization of branched networks. We polymerized these reticulated actin networks on a wide range of micropatterned nucleation geometries. The activation of the Arp2/3 complex and the recruitment of a preexisting filament at the pattern surface triggers branched actin filaments assembly. Since the CP blocks barbed-ends elongation, and the already assembled network serves as template for Arp2/3 nucleation activity at the pattern surface, the branched actin network grows off the pattern surface and adopts an organization consistent with the geometry of nucleation area (Fig. 16.2 B, C, and D).

4. RECONSTITUTION OF 3D CONNECTIONS OF STRUCTURED ACTIN NETWORKS

Combining the use of micropatterned geometries and the reconstituted motility assay performed in physiologically relevant medium, we were able to address two main questions. First, to what extent does geometry control dynamics and architecture of branched actin networks? Second, are we able to develop a new method based on actin filament self-organization to generate geometrically defined 3D connections of complex network organizations, which could exist *in vivo* or serve as a template for the engineering of bioinspired 3D systems? The following section explains how combining laser patterning and reconstituted biochemical media enables the control of such directed 3D connections of self-organized networks (Galland et al., 2013).

4.1. Design and functionalization of laser-patterned surfaces

Compared to UV-patterning, using lasers to pattern surfaces gives a significant advantage since it permits the high-precision positioning of motifs to be printed on two opposite walls of a preassembled chamber. This allows the precise direction of the interaction of networks growing off opposite patterns and the construction of 3D actin networks.

4.1.1 Material

- Coverslips (Agar Scientific)
- Adhesion promoter TI-Prime (MicroChemicals)
- 1% polystyrene (Sigma) in toluene (Sigma)
- 10 × KMEI buffer, G buffer as in Section 1.1
- PLL-PEG stock solution as in Section 1.2
- Adhesive double tape. We use a precut double tape 30 μm thick (LIMA)

4.1.2 Equipment

- Spin coater WS-400-6NPP (Laurell technologies)
- Heating plate
- Oxygen plasma oven (FEMTO, Diener Electronics)
- Nanoablation station: inverted microscope (TE2000-E, Nikon) equipped with a CFI S-Fluor oil objective (100 ×, NA 1.3, Nikon), a perfect focus system (Nikon), motorized stage (Marzhauser), and a dual-axis galvanometer that focalizes the laser beam on the sample on the field of the camera, including a telescope that adjust the laser focalization with the image focalization, and polarizer to control the laser power (iLasPulse device, Roper Scientific). The microscope uses a pulsed laser passively Q-switched laser (STV-E, TeamPhotonics) that delivers 300 ps pulses at 355 nm (energy per pulse 1.2 μJ, peak power 4 kW, variable repetition rate 0.01–2 kHz, average power ≤ 2:4 mW). The laser was scanned throughout the region of interest with a power set to 300 nJ. The laser displacement, exposure time, and repetition time were controlled using Metamorph software (Universal Imaging Corporation).

The microscope is moreover equipped with a fluorescence illumination system X-Cite 120PC Q (Lumen Dynamics) and QuantEM:512SC camera (Photometrics) to monitor the laser printing procedure (for further details refers to Vignaud et al., 2012; Fig. 16.1A).

4.1.3 Methods

1. Prepare polystyrene-PLL-PEG-coated coverslips as follows. For a detailed protocol, refer to Azioune et al. (2009).
   a. Cleaned coverslips were spin-coated (30 s, 1000 rpm) with TI-Prime, backed on a heating plate (5 min, 120 °C), and then spin-coated with 1% polystyrene (1000 rpm, 30 s).
   b. Polystyrene-coated coverslips were oxidized with oxygen plasma (30 s, 60 W), and then incubated with 0.1 mg/mL PLL-PEG (30 min, room temperature; as in Section 1.2.1).
2. Assemble flow chamber using adhesive double tape and two polystyrene-PLL-PEG-coated coverslips with their treated sides facing inwards. Stably mount it on the motorized stage.
3. Define the regions of interest, ROI, to be laser printed.
4. Printing a pair of superimposed motifs is performed manually:
   a. Focus on the image on the interface glass/PLL-PEG layer. Little tiny particles of dust remaining on the glass help focalization (they however should be rare if glasses preparation was good).

b. The laser beam should be then similarly focused. However, for an efficient oxidation of the PEG chains, we slightly unfocus the laser setting the focal plane above the glass surface in order to hit the PEG layer.

c. Select a predefined ROI and laser print it. Each laser spot is exposed for 5 ms at a repetition rate of 2000 Hz.

d. Once completed, *without changing the* (x,y) *coordinates*, move on the z direction and repeat steps a and b on the upper coverslip.

e. Given the geometry of the ROI already printed and thanks to the precise move of the laser beam to adequate (x,y) coordinates using the software, it is easy to precisely control the position of the next motif to be printed in the opposite glass. For instance, the Metamorph software offers the function "move to relative position," which enables precise movement of the motorized stage and laser repositioning on the coverslip. For the upper coverslip, each spot was exposed for 8 ms at a repetition rate of 2000 Hz.

f. Select the predefined ROI and laser print it.

5. Repeat step 3 as many times as needed.
6. Instantly after completion of laser printing procedure, fill the flow cell with NPFs solution (15 min, room temperature). We use to coat with 1 μM pWA fragment of WASP protein.
7. Wash the flow cell with 1 × KMEI buffer and store at 4 °C for 2–3 days.

4.2. Assembly of 3D connections of structured actin networks

4.2.1 *Material*

- Solutions and extemporarily prepared motility medium as described in Section 1.2.2.

4.2.2 *Equipment*

- A dry block heater (VWR) if using VALAP to seal the motility chamber.
- Confocal microscope: Eclipse TI-E Nikon inverted microscope equipped with a CSUX1-A1 Yokogawa confocal head, an Evolve EMCCD camera (Roper Scientific), a CFI Plan APO VC oil objective (60 ×, NA 1.4) or a CFI Plan Fluor oil objective (40 ×, NA 1.3) (Nikon), and a motorized stage MS 2000 (ASI imaging). The system was driven by Metamorph software (Universal Imaging Corporation).

4.2.3 *Methods*

- Fill in freshly functionalized laser-patterned flow cell with the motility medium, detailed in Section 1.1.2, respecting preparation order and proportions but excluding NPFs-coated beads.

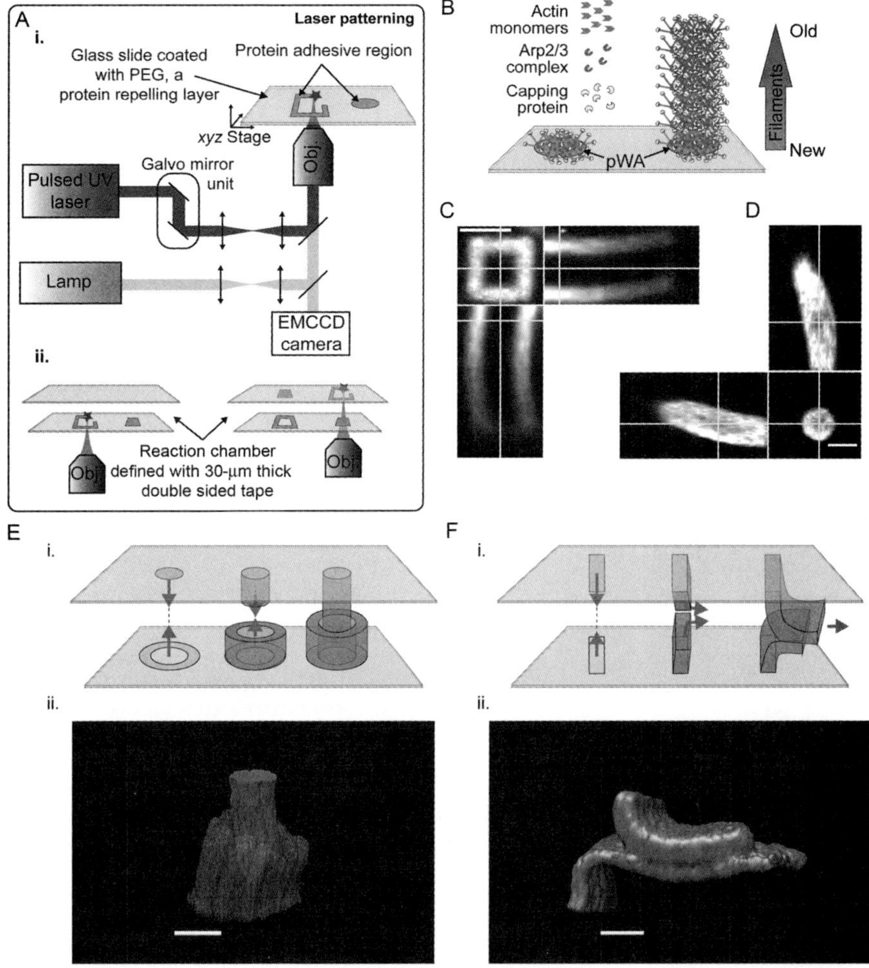

Figure 16.3 Laser patterning and oriented growth of actin structures in 3D. (A) Schematic representation of the laser-based patterning method. A pulsed UV laser is used to oxidize a protein-repellent layer of polyethylene glycol (PEG) to make it locally adhesive (i). Laser patterning of adhesive regions in a reaction chamber defined by two pegylated coverslips separated by 30–60 μm thick double-sided tape (ii). (B) Schematic representation of the growth of 3D actin structures from a nucleating pattern coated with the pWA in the presence of 3 μ*M* actin monomers (Alexa568 labeled), 9 μ*M* profilin, 50 n*M* Arp2/3 complex, and 60 n*M* capping protein. (C and D) 3D confocal images (planar, horizontal, and vertical cross sections) of actin structures polymerized from an empty square-shaped nucleating pattern (C) and from a disc-shaped nucleating pattern (D). (E) Schematic representation of the interaction of two actin structures growing face-to-face from two complementary-shaped nucleating patterns to form a 3D connection (i) and 3D reconstruction of confocal imaging of such a 3D connection (ii). (F) Schematic representation of the interaction of two actin structures growing face-to-face from two similar rectangular-shaped nucleating patterns that induce the formation of an actin structure in a plane parallel to the coverslip (i) and 3D reconstruction of confocal imaging of such a structure (ii). Scale bars are 10 μm. (See the color plate.)

- Check the onset of the reaction and seal the chamber with VALAP.
- Follow polymerization dynamics with a confocal microscope:
  - On each coverslip, in agreement with results with UV-patterned motifs, actin cross-linked structures grow normal to glass surface (Fig. 16.3B) and adopt the geometry of the pattern (Fig. 16.3C and D).
  - Judicious choice of pairs of patterns allows a fine control of the 3D connections between dynamic networks, which may lead to a "plug-and-socket-like connection" (Fig. 16.3E) or to "collision and release" connection (Fig. 16.3F).

5. CONCLUDING REMARKS

Investigations of the cell cytoskeleton reveal an extremely complex organization (Svitkina & Borisy, 1999; Xu, Babcock, & Zhuang, 2012). Specialized dynamic networks of a specific architecture and biochemical composition overlap and interconnect to build-up an entangled and reticulated web. The UV and laser-patterning procedures described here are valuable tools in deciphering how geometrical and biochemical constraints control the assembly of self-organized and dynamic actin structures. These reconstitution systems may pave the way for investigating the complex network interactions that occur *in vivo*. These methods are flexible and scalable, as one can biomimic more complex architectures using an unlimited variation of geometries, nucleation-promoting factors, or other regulatory proteins.

ACKNOWLEDGMENTS

This work was supported by grants from Human Frontier Science Program (HFSP RGP0004/2011 awarded to L. B.), Agence Nationale de la Recherche (ANR-08-BLANC-0012 awarded to L. B.), the "Chimtronique" programme of CEA to M. T. R. B. is a member of the Institut Universitaire de France. We thank Patrick Leduc and David Peyrade for insightful suggestions in building 3D connections and Chloé Zubieta for critical reading of the chapter.

REFERENCES

Achard, V., Martiel, J.-L., Michelot, A., Guérin, C., Reymann, A.-C., Blanchoin, L., et al. (2010). A "primer"-based mechanism underlies branched actin filament network formation and motility. *Current Biology*, *20*, 423–428.

Akin, O., & Mullins, R. D. (2008). Capping protein increases the rate of actin-based motility by promoting filament nucleation by the Arp2/3 complex. *Cell*, *133*, 841–851.

Azioune, A., Storch, M., Bornens, M., Thery, M., & Piel, M. (2009). Simple and rapid process for single cell micro-patterning. *Lab on a Chip*, *9*, 1640–1642.

Bernheim-Groswasser, A., Wiesner, S., Golsteyn, R. M., Carlier, M.-F., & Sykes, C. (2002). The dynamics of actin-based motility depend on surface parameters. *Nature*, *417*, 308–311.

Blanchoin, L., Amann, K. J., Higgs, H. N., Marchand, J. B., Kaiser, D. A., & Pollard, T. D. (2000). Direct observation of dendritic actin filament networks nucleated by Arp2/3 complex and WASP/Scar proteins. *Nature*, *404*, 1007–1011.

Blanchoin, L., Boujemaa-Paterski, R., Sykes, C., & Plastino, J. (2014). Actin dynamics, architecture and mechanics in cell motility. *Physiology Review*, *94*, 235–263.

Cameron, L. A., Footer, M. J., Van Oudenaarden, A., & Theriot, J. A. (1999). Motility of ActA protein-coated microspheres driven by actin polymerization. *Proceedings of the National Academy of Sciences of the United States of America*, *96*, 4908–4913.

Dayel, M. J., Akin, O., Landeryou, M., Risca, V., Mogilner, A., & Mullins, R. D. (2009). In silico reconstitution of actin-based symmetry breaking and motility. *PloS Biology*, *7*, 1000201.

Fletcher, D. A., & Mullins, R. D. (2010). Cell mechanics and the cytoskeleton. *Nature*, *463*(7280), 485–492.

Frishknecht, F., Moreau, V., Röttger, S., Gonfloni, S., Reckmann, I., Superti-Furga, G., et al. (1999). Actin-based motility of vaccinia virus mimics tyrosine kinase signalling. *Nature*, *401*, 926–929.

Galland, R., Leduc, P., Guerin, C., Peyrade, D., Blanchoin, L., & Thery, M. (2013). Fabrication of three-dimensional electrical connections by means of directed actin self organization. *Nature Materials*, *12*, 416–421.

Giardini, P. A., Fletcher, D. A., & Theriot, J. A. (2003). Compression forces generated by actin comet tails on lipid vesicles. *Proceedings of the National Academy of Sciences of the United States of America*, *100*, 6493–6498.

Iwasa, J. H., & Mullins, R. D. (2007). Spatial and temporal relationships between actin filament nucleation, capping, and disassembly. *Current Biology*, *17*, 395–406.

Kawska, A., Carvalho, K., Manzi, J., Boujemaa-Paterski, R., Blanchoin, L., Martiel, J. L., et al. (2012). How actin network dynamics control the onset of actin-based motility. *Proceedings of the National Academy of Sciences of the United States of America*, *109*, 14440–14445.

Loisel, T. P., Boujemaa, R., Pantaloni, D., & Carlier, M.-F. (1999). Reconstitution of actin-based motility of Listeria and Shigella using pure proteins. *Nature*, *67*(53), 613–616.

Mogilner, A., & Zhu, J. (2012). Cell polarity: Tension quenches the rear. *Current Biology*, *22*, R48–R51.

Pantaloni, D., Boujemaa, R., Didry, D., Gounon, P., & Carlier, M.-F. (2000). The Arp2/3 complex branches filament barbed ends: Functional antagonism with capping proteins. *Nature Cell Biology*, *2*, 385–391.

Pantaloni, D., & Carlier, M.-F. (1993). How profilin promotes actin filament assembly in the presence of thymosinß4. *Cell*, *75*, 1007–1014.

Pollard, T. D., Blanchoin, L., & Mullins, R. D. (2000). Molecular mechanisms controlling actin filament dynamics in nonmuscle cells. *Annual Review of Biophysics and Biomolecular Structure*, *29*, 545–576.

Pollard, T. D., & Borisy, G. G. (2003). Cellular motility driven by assembly and disassembly of actin filaments. *Cell*, *112*, 453–465.

Reymann, A.-C., Boujemaa-Paterski, R., Martiel, J.-L., Guérin, C., Cao, W., Chin, H. F., et al. (2012). Actin network architecture can determine myosin motor activity. *Science*, *336*, 1310–1314.

Reymann, A.-C., Guérin, C., Théry, M., Blanchoin, L., & Boujemaa-Paterski, R. (2014). Geometrical control of actin assembly and contractility. *Methods in Cell Biology*, *120*, 19–38.

Reymann, A.-C., Martiel, J.-L., Cambier, T., Blanchoin, L., Boujemaa-Paterski, R., & Théry, M. (2010). Nucleation geometry governs ordered actin networks structures. *Nature Materials, 9*, 827–833.

Reymann, A.-C., Suarez, C., Guerin, C., Martiel, J.-L., Staiger, C. J., Blanchoin, L., et al. (2011). Turnover of branched actin filament networks by stochastic fragmentation with ADF/cofilin. *Molecular Biology of the Cell, 22*, 2541–2550.

Romero, S., Le Clainche, C., Didry, D., Egile, C., Pantaloni, D., & Carlier, M.-F. (2004). Formin is a processive motor that requires profilin to accelerate actin assembly and associated ATP hydrolysis. *Cell, 119*, 419–429.

Schafer, D., Jennings, P. B., & Cooper, J. A. (1996). Dynamics of capping protein and actin assembly in vitro: Uncapping barbed ends by polyphosphoinositides. *Journal of Cell Biology, 135*, 169–179.

Svitkina, T. M., & Borisy, G. C. (1999). Arp2/3 complex and actin depolymerizing factor/cofilin in dendritic organization and treadmilling of actin filament array in lamellipodia. *Journal of Cell Biology, 145*, 1009–1026.

Theriot, J. A., Rosenblatt, J., Portnoy, D. A., Goldschmidt-Clermont, P. J., & Mitchison, T. J. (1994). Involvement of profilin in the actin-based motility of L. monocytogenes in cells and in cell-free extracts. *Cell, 76*, 505–517.

Vignaud, T., Galland, R., Tseng, Q., Blanchoin, L., Colombelli, J., & Théry, M. (2012). Reprogramming cell shape with laser nano-patterning. *Journal of Cell Science, 125*, 2134–2140.

Vignjevic, D., Yarar, D., Welch, M. D., Peloquin, J., Svitkina, T., & Borisy, G. G. (2003). Formation of filopodia-like bundles in vitro from a dendritic network. *Journal of Cell Biology, 160*, 951–962.

Xu, K., Babcock, H. P., & Zhuang, X. (2012). Dual-objective STORM reveals three-dimensional filament organization in the actin cytoskeleton. *Nature Methods, 9*, 185–188.

CHAPTER SEVENTEEN

In Vitro Reconstitution of Dynamic Microtubules Interacting with Actin Filament Networks

Magdalena Preciado López[*], Florian Huber[*], Ilya Grigoriev[†], Michel O. Steinmetz[‡], Anna Akhmanova[†], Marileen Dogterom[*,1], Gijsje H. Koenderink[*,1]

[*]FOM Institute AMOLF, Science Park, Amsterdam, The Netherlands
[†]Division of Cell Biology, Faculty of Science, Utrecht University, Padualaan, Utrecht, The Netherlands
[‡]Laboratory of Biomolecular Research, Paul Scherrer Institut, Villigen, Switzerland
[1]Corresponding authors: e-mail address: dogterom@amolf.nl; g.koenderink@amolf.nl

Contents

Abstract

Interactions between microtubules and actin filaments (F-actin) are essential for eukaryotic cell migration, polarization, growth, and division. Although the importance of these interactions has been long recognized, the inherent complexity of the cell interior hampers a detailed mechanistic study of how these two cytoskeletal systems influence each other. In this chapter, we show how *in vitro* reconstitution can be employed to study how actin filaments and dynamic microtubules affect each other's organization. While we focus here on the effect of steric interactions, these assays provide an ideal starting point to develop more complex studies through the addition of known F-actin–microtubule cross-linkers, or myosin II motors.

Methods in Enzymology, Volume 540
ISSN 0076-6879
http://dx.doi.org/10.1016/B978-0-12-397924-7.00017-0

1. INTRODUCTION

Interactions between microtubules and F-actin are essential for eukaryotic cells to execute processes such as migration, growth, division, and polarization (as reviewed in Rodriguez et al., 2003; Siegrist & Doe, 2007; Yarm, Sagot, & Pellman, 2001). Historically, however, actin and microtubules have been mostly studied independently, even though the importance of their interactions has been long recognized (Vasiliev et al., 1970). A chief obstacle in understanding how these cytoskeletal systems affect each other's organization lies in identifying when and where in the cell they interact. This is largely due to the inherent crowdedness of the cell, the high density and diversity of F-actin structures (stress fibers, filopodia, dendritic, and cortical networks), and the complicated trajectories of polymerizing microtubules (Fig. 17.1A). Furthermore, it is challenging to disentangle mechanical cross talk from biochemical interactions (Rodriguez et al., 2003), mediated through cross-linking proteins or cytoskeletal regulators such as the Rho family of GTPases (Etienne-Manneville & Hall, 2002; Waterman-Storer & Salmon, 1999).

Notwithstanding, several proteins and protein complexes capable of linking microtubules to F-actin have been identified (Fig. 17.1C). Some of these form passive cross-links (Geraldo, Khanzada, Parsons, Chilton, & Gordon-Weeks, 2008; Solinet et al., 2013); while others mediate active linkages, through molecular motors (Frey, Klotz, & Nick, 2009; Nadar, Lin, & Baas, 2012; Yin, Pruyne, Huffaker, & Brescher, 2000), actin filament nucleators (Mao, 2011; Shen et al., 2012), or proteins that localize to the growing plus ends of microtubules (+TIPs) (Akhmanova & Steinmetz, 2008; Applewhite et al., 2010; Jiang et al., 2012; Lee et al., 2000; Moseley et al., 2007; Wu, Kodama, & Fuchs, 2008).

It is likely that in addition to the molecular properties of cross-linking proteins (Fig. 17.1C), the local cytoskeletal architecture (Fig. 17.1B) plays a defining role in the outcome of a given actin–microtubule interaction. For instance, we hypothesize that microtubule growth will respond differently to a dendritic F-actin network interspersed with filopodia than to a comparatively rigid actin stress fiber, even if the same cross-linking proteins are at play. Reconstituted model systems using purified cellular components are ideally suited to test such hypotheses, as they can separately address biochemical and mechanical interactions and identify the minimum requirements to achieve a certain functional organization. Not surprisingly, *in vitro*

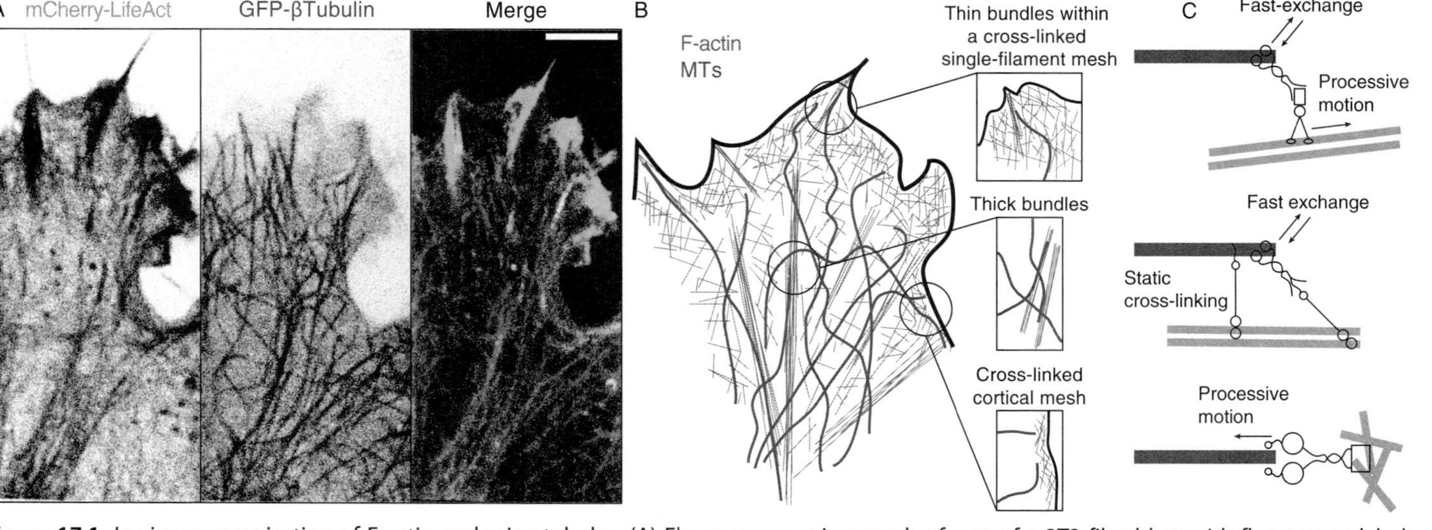

Figure 17.1 *In vivo* coorganization of F-actin and microtubules. (A) Fluorescence micrograph of part of a 3T3 fibroblast with fluorescent labels showing the distinct organization of F-actin (cyan) and microtubules (red). Actin forms rigid stress fibers and dense networks underneath the plasma membrane (top), whereas microtubules grow toward the cell periphery and form a sparse network. (B) Schematic showing how dynamic microtubules can encounter different F-actin architectures within a cell. (C) Conceptual drawings of the different types of F-actin–microtubule cross-linking proteins and protein complexes, classified based on their activities (e.g., processive motion) and exchange properties (e.g., slow vs. fast). Scale bar: 5 μm. (See the color plate.)

reconstitutions have played a fundamental role in our current understanding of the actin and microtubule cytoskeleton (as reviewed in Dogterom & Surrey, 2013; Huber et al., 2013; Mullins & Hansen, 2013).

To date, however, reconstitutions of composite microtubule/F-actin systems are still rare. Notable exceptions include work on the mechanical properties of passive composite networks (Brangwynne, Koenderink, MacKintosh, & Weitz, 2008; Lin, Koenderink, MacKintosch, & Weitz, 2010), molecular motor motility and cargo transport at F-actin–microtubule cross-roads (Ali et al., 2007; Schroeder, Mitchell, Shuman, Holzbaur, & Goldman, 2010), and F-actin–microtubule co-bundling (Huang et al., 2007; Selden & Pollard, 1986; Sider et al., 1999). Given the mounting evidence that microtubule growing ends, through end-binding (EB) proteins, act as hubs of biochemical activity within the cell (Akhmanova & Steinmetz, 2008; Lansbergen & Akhmanova, 2006; Tamura & Draviam, 2012), they are likely to play a defining role in the regulation of actin–microtubule coorganization. However, with the exception of one study in *Xenopus laevis* egg extracts (Waterman-Storer et al., 2000), and some promising work coupling actin filament nucleation to EBs (Breitsprecher et al., 2012), we are unaware of any *in vitro* reconstitution studies of the interaction between dynamic microtubules and F-actin.

In this chapter, we describe a set of assays that focus on steric interactions between dynamic microtubules and actin filaments arranged in predefined geometries. First, we address how microtubule growth and organization respond to quasi two-dimensional linear arrays of F-actin bundles (mimicking stress fibers), and second, how isotropic F-actin networks (mimicking cytosolic or cortical networks) respond to a radial array of microtubules growing within them. In both cases, we include EB proteins, which autonomously enhance microtubule dynamics through increased growth speeds and catastrophe rates (Komarova, De Groot, Grigoriev, Gouveia, Munteanu, Schober et al., 2009; Vitre et al., 2008), while potentially sensitizing the microtubule tip to obstacles (Janson, de Dood & Dogterom, 2003). We describe the procedures for reconstitution and fluorescence imaging and provide suggestions on how to analyze the resulting actin–microtubule coorganization by automated image analysis. As explained in Section 3, this work provides a starting point for more complex assays including actin-microtubule cross-linkers such as the spectraplakins MACF/Shot (Applewhite et al., 2010; Wu et al., 2008) or the Myo2/Kar9/Bim1 system of *Saccharomyces cerevisiae* (Lee et al., 2000; Yin et al., 2000).

2. EXPERIMENTAL METHODS

2.1. Flow-cell preparation

Both F-actin and microtubules are immobilized on thoroughly cleaned and functionalized glass coverslips.

2.1.1 Glass cleaning

We clean coverslips in "base piranha" solution, a safer alternative to "acid piranha" solution that only becomes reactive when heated to 60 °C. Storing the cleaned coverslips in a 0.1 *M* KOH aqueous solution activates OH groups on the glass surface (Gell et al., 2010), giving the surface a net negative charge which is needed for later functionalization steps.

1. Under the fume hood, mix Milli-Q water, 30% H_2O_2, and 30% NH_4OH at a 5:1:1 ratio in a glass beaker. The beaker should be large enough to fit a Teflon rack holding the coverslips and filled with enough solution to fully cover the glass.
2. Heat to 75 °C on a hot plate with continuous stirring using a magnetic bar.
3. Place coverslips in the Teflon rack and rinse twice, 5 min each, with Milli-Q water (Merck Millipore, Billerica, MA, USA) in a bath sonicator.
4. Once the base piranha solution has reached a temperature of 75 °C (bubbles can be seen), insert the Teflon rack with the coverslips.
5. Incubate at 75 °C for 15 min in the fume hood.
6. Rinse the coverslips as in step 3.
7. Store at room temperature (RT) in a 0.1 *M* KOH solution (for up to 1 month) until use.

   Tip: "base piranha" is a dangerous reaction, so wear gloves and work carefully.

   Tip: The "base piranha" solution can be left in the hood overnight. Dispose according to your own laboratory safety protocols.

The glass microscope slides can be cleaned in a less rigorous fashion:

1. Place the slides in a Teflon rack in a beaker with a 0.1% (v/v) aqueous solution of Hellmanex™ and bath sonicate for 20 min.
2. Rinse the slides twice in Milli-Q water, 5 min each, in a bath sonicator.
3. Sonicate the slides in 100% acetone for 20 min. Rinse.
4. Sonicate the slides in a 70% ethanol aqueous solution. Rinse.
5. Store at RT in a 0.1 *M* KOH solution until use.

2.1.2 Flow-cell construction

We assemble flow-cells with ~10-μl channels cut out from Parafilm® sandwiched between a cleaned glass slide and coverslip (Fig. 17.2A).

1. Rinse a clean glass slide and coverslip for 5 min in Milli-Q water, in a bath sonicator.
2. Using a glass slide as a ruler cut out three ~2-mm-wide and ~2.5-cm-long channels from Parafilm, spaced far enough that they will fit comfortably within a coverslip.
3. Blow-dry the glass with N_2 gas and place the Parafilm between coverslip and glass slide. Cut out excess film.
4. Melt the Parafilm on a hot plate set at 120 °C for 10 s, while applying even pressure by placing a glass slide and an even metal surface on top of the coverslip.
5. Begin the surface functionalization right away (Sections 2.4.1 or 2.5.1).

For assays in which we need to exchange the solutions during the experiment, we use glass slides perforated with ~1-mm-diameter holes spaced ~20 mm apart (to fit within a 22 mm coverslip; Fig. 17.2B). The Parafilm channels and coverslip are mounted on these glass slides such that the holes become the entry and exit points of the channel. In this way, all solutions are flowed in and out from the side of the glass slide, while the flow-cell is mounted on the microscope stage (Fig. 17.3A).

2.2. Stabilized actin filaments and microtubule seeds

Traditionally, *in vitro* assays with F-actin and microtubules have been performed in radically different buffers (Olmsted & Borisy, 1975; Pardee & Spudich, 1982). We found that microtubule polymerization does not proceed in the canonical F-buffer (20 m*M* Hepes, pH 7.4 with KOH, 1 m*M* EGTA, 2 m*M* $MgCl_2$, and 50 m*M* KCl), likely due to its low ionic strength and pH (Olmsted & Borisy, 1975). On the contrary, MRB80, a variant of the well-known BRB80 buffer (Deery & Brinkley, 1983) typically used for microtubules, has sufficiently high ionic strength to trigger actin polymerization. We thus use this buffer in all composite F-actin/microtubule assays.

2.2.1 Phalloidin-stabilized actin filaments

We purify G-actin from rabbit skeletal muscle (Gentry et al., 2012) and store it at 4 °C in G-buffer (2 m*M* Tris–HCl, pH 7.8, 0.2 m*M* Na_2ATP, 0.2 m*M* $CaCl_2$, 5 m*M* dithiothreitol (DTT)). After 4 weeks, we clarify the protein by centrifugation at 149,000 × *g* for 10 min and dialyze the supernatant overnight against G-buffer. In all the assays presented here, we use

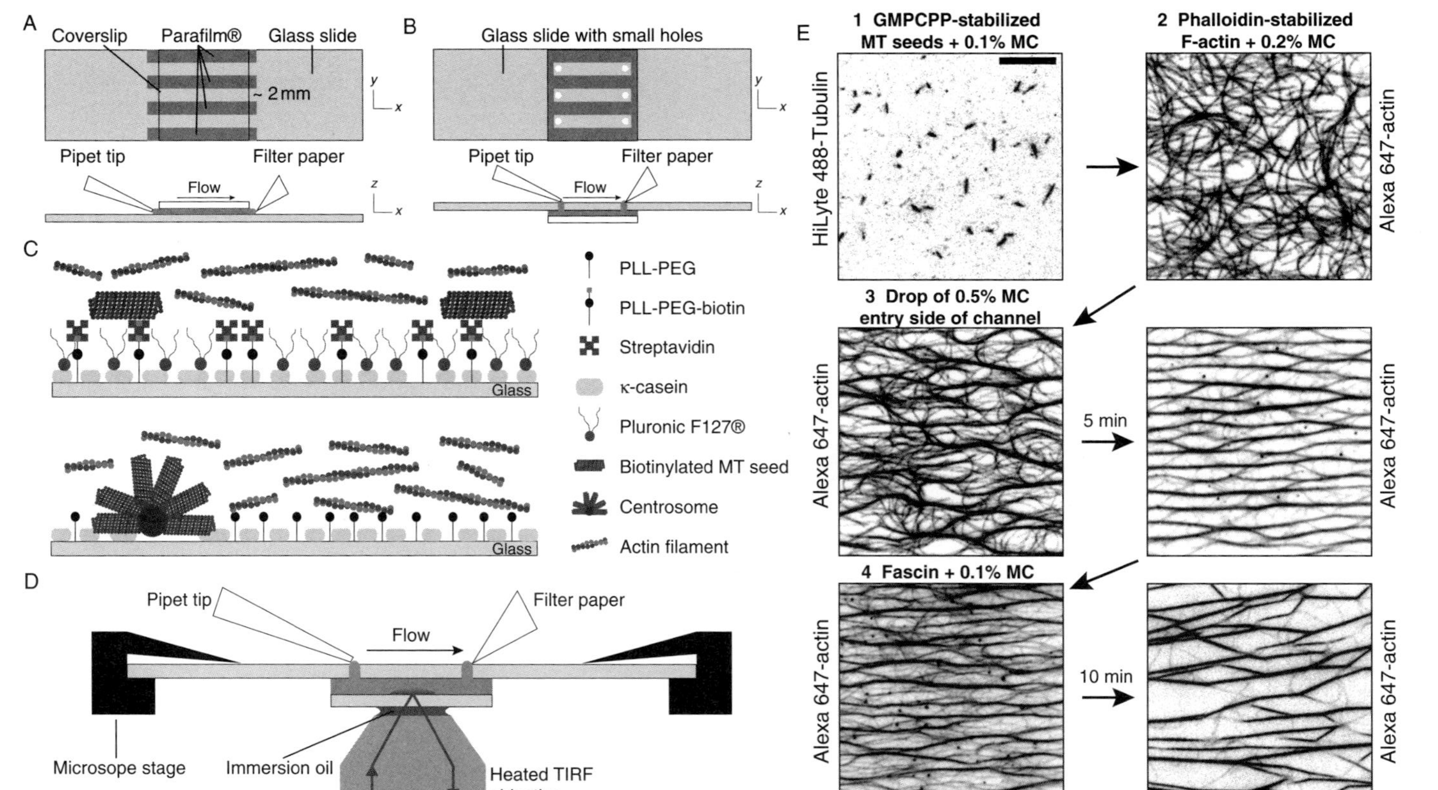

Figure 17.2 Flow-cell preparation and glass surface functionalization. Schematic of the two types of flow-cells we use for *in vitro* F-actin–microtubule interaction experiments with: (A) isotropic F-actin solutions and radial microtubule arrays and (B) random arrays of microtubules and aligned F-actin bundles. (C) Schematic of the surface functionalization steps for the F-actin bundle array experiments (top) and radial microtubule array experiments (bottom). (D) Schematic of the flow-cell described in (B) that can be used on a TIRF microscope stage. (E) TIRF images showing sequential deposition of a random array of microtubule seeds, then long actin filaments, then a methylcellulose (MC) solution to align and bundle the actin filaments, and finally, a fascin solution to exchange MC for fascin (Section 2.4). (For color version of this figure, the reader is referred to the online version of this chapter.)

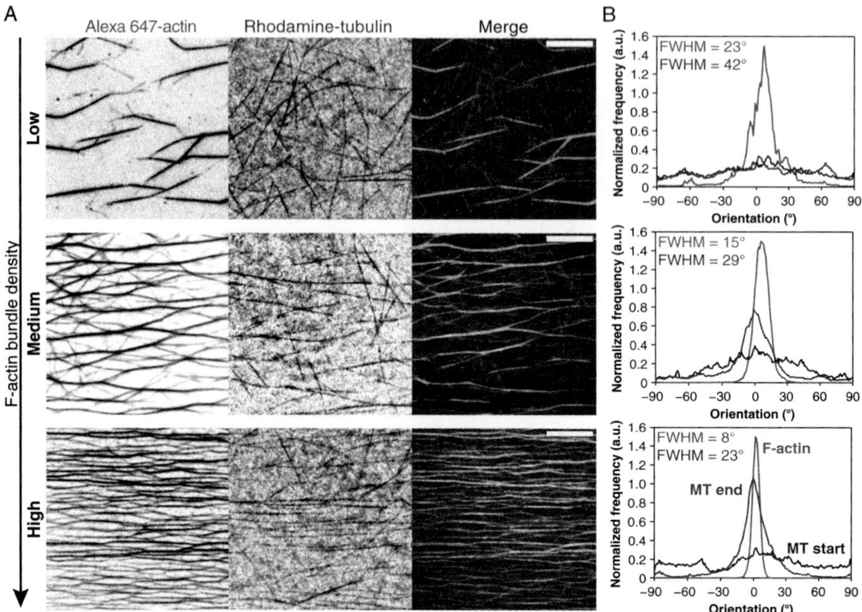

Figure 17.3 TIRF images of microtubule growth within parallel arrays of F-actin bundles. (A) (Left) Three F-actin bundle arrays of different bundle surface density. (Middle) Resulting microtubule arrays after ~1 h of growth within the corresponding F-actin bundle arrays. (Right) Merged panels. (B) Normalized histograms of F-actin bundle and microtubule orientation corresponding to the examples in (A). The cyan curve shows the F-actin histogram, normalized so the maximum frequency is 1.5. The dark-red curve shows the initial distribution of microtubule orientation, which is always random. The light-red curve shows the final microtubule orientation distribution after ~1 h of growth, showing that longer microtubules can be better organized by the F-actin bundles. In all graphs, the full width at half maximum (estimated from fitting Gaussian functions to the distributions) is shown for the F-actin bundle array as well as the resulting microtubule (MT) organization (light-red curves). Increasing the F-actin bundle density leads to better actin–microtubule coalignment through steric interactions. Scale bars: 10 μm. (See the color plate.)

prepolymerized F-actin containing a small fraction of G-actin labeled with Alexa Fluor® 488, 594, or 647 (Molecular Probes, Life Technologies, Carlsbad, CA, USA) (Gentry et al., 2012).

1. In G-buffer, prepare a G-actin mix (15% labeled protein) at a concentration such that the final G-actin concentration will be 1.5 μ*M* (for experiments with F-actin bundles, Section 2.4) or 20 μ*M* (for experiments with centrosomes, Section 2.5). Keep on ice.

2. In a separate tube, prepare an actin polymerization mix in MRB80 such that the final concentrations will be 5 m*M* DTT, phalloidin at a 1:1 molar ratio with G-actin, 50 m*M* KCl, and 0.2 m*M* ATP.
3. Add the actin polymerization mix to the G-actin mix and incubate for 1 h at RT in the dark, to avoid photobleaching.
4. The F-actin solution can be used for 1 week.

2.2.2 GMPCPP-stabilized microtubule seeds

We nucleate microtubules from surface-immobilized microtubule "seeds". These seeds are biotinylated and stabilized against depolymerization with the slowly hydrolyzable GTP analog guanylyl-(α,β)-methylene-diphosphonate (GMPCPP; Hyman, Salser, Drechsel, Unwin, & Mitchison, 1992). For labeling, we use TRITC Rhodamine, HiLyte® 488, or HiLyte® 635 and biotin-labeled tubulins from Cytoskeleton Inc. (Denver, CO, USA). The following protocol gives short microtubule seeds (~5 μm in length).

1. Prepare a 20-μ*M* tubulin mix (70% unlabeled, 12% fluorescently labeled, and 18% biotinylated) in MRB80 on ice.
2. Spin down for 5 min at RT at 149,000 × *g*.
3. Add GMPCPP to a final concentration of 1 m*M*.
4. Incubate at 37 °C for 30 min to induce polymerization.
5. Spin down as in step 2.
6. Resuspend the pellet in MRB80 to a final volume that corresponds to 20 μ*M* tubulin, assuming that 80% of the tubulin is recovered.
7. Keep the solution of microtubule seeds in the dark at RT and use within 2 days of preparation.
   Tip: The most convenient way to spin down small volumes (5–200 μl) at high speeds is with an Airfuge® ultracentrifuge (Beckman Coulter, Brea, CA, USA). Depending on the rotor, the maximum speed (30 PSI) corresponds to ~149,000 × *g*.

More stable seeds that can be frozen and stored at −80 °C can be obtained by cycling the microtubule seeds twice (Gell et al., 2010) to replace more of the GDP–tubulin with GMPCPP–tubulin. After thawing, we use these microtubule seeds for up to 3 days, since they anneal and get longer with time.

2.3. Microscopy and imaging

The best signal-to-noise ratio for *in vitro* surface assays with fluorescently labeled proteins is achieved with total-internal reflection fluorescence (TIRF) microscopy. Since we must visualize two or even three proteins within one experiment, we use up to three different wavelengths: 488,

561, and 635 nm. We obtain optimal fluorophore/laser combinations using TRITC Rhodamine-labeled tubulin, Alexa 647-labeled actin, and GFP-labeled EBs. If unlabeled EBs are used, Alexa 488-labeled G-actin works equally well.

We typically image microtubules at 2 s per frame with 100–200 ms exposure times. To ensure consistent microtubule dynamics and control the microtubule growth speed and average length (Fygenson, Braun, & Libchaber, 1994), we regulate the sample temperature with a home-built objective temperature controller with a range of 15–40 ± 1 °C. For all assays presented here, we typically work at 25–28 °C, to ensure long enough microtubules and hence many interaction events with F-actin.

2.4. Assay with randomly oriented microtubule seeds and aligned F-actin bundles

With this assay, one can probe how growing microtubules that are initially randomly oriented respond to a linear array of rigid F-actin bundles (organized by fascin, Gentry et al., 2012). We use flow-cells with entry/exit points on the glass slide side (Fig. 17.2B), which allow us to sequentially flow in different solutions while monitoring the sample by TIRF microscopy (Fig. 17.2D).

2.4.1 Surface functionalization

We immobilize microtubule seeds on the coverslip surface via biotin–streptavidin (Fig. 17.2C, top). We first coat the clean and negatively charged glass (Section 2.1.1) with the block copolymer poly(L-lysine)-graft-poly (ethylene glycol) functionalized with a biotin tag (PLL-PEG-biotin). We then bind streptavidin and coat hydrophobic patches with κ-casein and Pluronic® F-127. All solutions and rinsing steps are based on MRB80.

1. Prepare a flow-cell from a clean coverslip and a glass slide with holes (Fig. 17.2B).
2. Flow in 0.2 mg/ml PLL-PEG-biotin and incubate for 30–45 min.
3. Rinse with 4–5 channel volumes (CVs).
4. Flow in 50–100 μg/ml streptavidin and incubate for 10 min. Rinse.
5. Flow in 0.5 mg/ml κ-casein and incubate for 10 min. Rinse.
6. Flow in 1% (w/v) Pluronic® F-127 and incubate for 10 min. Rinse. Now the flow-cell is ready for use.

   Tip: To avoid drying of the channels during incubation steps, place the flow-cell in a Petri-dish along with tissue paper soaked in water.

   Tip: Streptavidin can be replaced with NeutrAvidin™.

2.4.2 Surface deposition of microtubule seeds and F-actin bundles

Before adding the microtubule polymerization mix to the flow-cell we prepare the surface with randomly oriented microtubule seeds and aligned F-actin bundles (Fig. 17.2E). During each step, F-actin bundle formation is monitored by TIRF microscopy (Fig. 17.2D). All rinsing steps are with MRB80.

1. Flow in microtubule seeds diluted to an equivalent of 50–100 n*M* tubulin in MRB80 with 0.1% (w/v) methylcellulose, which exerts an entropic force pushing the seeds toward the surface. Incubate for 2 min.
2. Gently rinse with two CVs.
3. Flow in phalloidin-stabilized actin filaments at the equivalent of 1 μ*M* G-actin in MRB80 containing 0.2% (w/v) methylcellulose, which is sufficient to push the filaments toward the surface without inducing bundle formation (Popp, Yamamoto, Iwasa, & Maéda, 2006).
4. Place a drop of 0.5% (w/v) methylcellulose at one end of the channel and let it diffuse in for 5 min; this will induce F-actin bundling and cause a pressure-driven flow that aligns the bundles.
5. Slowly rinse the channel with three CVs of a solution containing 200 n*M* fascin, an actin bundling protein, and 0.1% (w/v) methylcellulose. This will stabilize the F-actin bundles and remove excess methylcellulose. Since fascin exchanges quickly (Courson & Rock, 2010), always keep its concentration constant in the channel to maintain the F-actin bundles.

   Tip: In step 2, rinse slowly, to prevent attachment and alignment of freely floating microtubule seeds on the surface.

   Tip: Pipette F-actin solutions with a pipette tip whose tip is cut off, to avoid filament breakage.

   Tip: Letting the sample equilibrate for ~10–20 min after fascin addition leads to more needle-like F-actin bundles (Fig. 17.2E).

   Tip: Note that the density and degree of alignment of the F-actin bundles typically vary across the flow-cell channel, so regions of interest should be picked with care.

2.4.3 Dynamic microtubules growing within F-actin bundle arrays

A minimal mixture for microtubule polymerization should contain tubulin, GTP, a blocking agent to avoid non-specific protein interactions, an oxygen scavenging system to prevent photodamage, and a reducing agent. Additionally, we add EB3 (EB family member 3, purified as in Komarova et al., 2009; Montenegro Gouveia et al., 2010) and use KCl to tune

EB3's binding at microtubule tip and lattice. Methylcellulose is included to confine out-of-plane microtubule fluctuations, as needed for TIRF microscopy.

1. Make the following mixture in MRB80, keep on ice:
   - 0.1% methylcellulose
   - 0.2 mg/ml κ-casein
   - Oxygen scavenging and reducing cocktail (0.2 mg/ml catalase, 0.4 mg/ml glucose oxidase, 4 m*M* DTT)
   - 50 m*M* glucose
   - 50 m*M* KCl
   - 1 m*M* GTP
   - 100 n*M* EB3
   - 0.5 μ*M* fascin
   - 23 μ*M* unlabeled tubulin
   - 2 μ*M* TRITC Rhodamine-labeled tubulin
2. Clarify the mixture at ~ 149,000 × *g* for 5 min and immediately add to the flow-cell.
3. Seal the channel with vacuum grease or wax to prevent solvent evaporation.

   Tip: An alternative oxygen scavenging system consists of protocatechuate 3,4-dioxygenase and protocatechuic acid, which has the advantage that it does not acidify over time (Shi, Lim, & Ha, 2010).

   Tip: Higher tubulin concentrations will yield a larger microtubule growth speed and average length (Fygenson et al., 1994). Since longer microtubules align better with the F-actin bundles, tubulin concentrations ≥25 μ*M* work best.

   Tip: The concentration of EB proteins is chosen to enhance the ratio of microtubule tip over lattice binding (Bieling et al., 2007).

   Tip: Additional proteins such as +TIPs, actin-binding proteins, or F-actin–microtubule cross-linkers can be included in the microtubule polymerization mix.

2.4.4 F-actin bundle and microtubule orientation analysis

We quantify the degree of F-actin–microtubule coalignment by analyzing the local anisotropy in fluorescence intensity for all image pixels using the OrientationJ plugin (Rezakhaniha et al., 2012) developed for ImageJ (Rasband, 2012). This plugin calculates the spatial gradients of the fluorescence intensity within a user-defined window centered on each pixel. The dominant orientation of a pixel is given by the direction of minimum change in intensity. In addition, the plugin returns the coherency of each pixel, that is, the degree of coalignment of a pixel with its neighbors. We use the Gaussian gradient analysis method with a window size of two pixels; this combination reduces artifacts and the contribution of noise to the orientation measurements (Alvarado, Mulder, & Koenderink, 2013).

Finally, we construct orientation histograms for the microtubules and F-actin bundles by weighting the orientation values by the coherency. Figure 17.3C shows an example of a time-lapse experiment, where we automatically track microtubule coalignment with F-actin bundle arrays of three different densities. Figure 17.3D shows the resulting orientation histograms for the microtubules at the start and the end (~1.5 h) of the experiment. At all F-actin densities, the orientation distributions are initially rather random. However, only at the higher bundle densities, the microtubule orientation distributions develop a clear peak at an angle close to the average direction of the actin bundles. The higher the density of the actin bundle array, the stronger the degree of coalignment. This finding suggests that steric interactions can drive actin–microtubule coalignment, which may act in concert with specific interactions mediated by F-actin–microtubule cross-linkers such as the MACF/EB complex, which is thought to mediate microtubule alignment with actin stress fibers (Wu et al., 2008).

2.5. Radial arrays of microtubules growing within isotropic F-actin networks

This assay probes how a radial array of dynamic microtubules (nucleated by a centrosome) interacts with an entangled meshwork of actin filaments (Fig. 17.2C, bottom). Although we focus here on how the F-actin network responds to microtubules growing within it, we also use this assay to study how microtubule growth responds to a decreasing mesh size of the actin network (to be published elsewhere).

2.5.1 Surface preparation

1. Prepare a flow cell (Section 2.1 and Fig. 17.2A).
2. Prewarm the centrosomes to 37 °C for 10 min and gently dilute them with MRB80 at RT. We purify centrosomes from KE37 cells and store them at −80 °C in a 30% sucrose solution (Laan et al., 2012).
3. Add centrosomes to the flow cell. Wait 5 min to allow for surface adhesion.
4. Rinse with 5–10 CVs of MRB80.
5. Flow in a 0.2 mg/ml PLL-PEG solution. Incubate 10 min.

6. Rinse with 5–10 CVs of a 1 mg/ml κ-casein and 1% Pluronic® F-127 solution. Incubate 15 min.
7. Rinse with a 1 mg/ml κ-casein solution.
   Tip: Since centrosomes are nonspecifically bound to the glass, any contaminating proteins will bind as well. To prevent protein aggregates on the surface, substantial rinsing in step 4 is essential to achieve satisfactory passivation in later steps.
   Tip: After step 6, samples can be stored on ice, but immediate use is recommended since the ability of centrosomes to nucleate microtubules deteriorates.

2.5.2 Dynamic microtubule assay within entangled actin filament networks

We use a similar polymerization mixture as in Section 2.4, but in this case, phalloidin-stabilized actin filaments are included, and fascin is left out.

1. Prepare the following mix in MRB80:
   - 0.1% (w/v) methylcellulose
   - 1 mg/ml κ-casein
   - Oxygen scavenging and reducing cocktail
   - 50 m*M* glucose
   - 50 m*M* KCl
   - 1 m*M* GTP
   - 16–30 μ*M* tubulin (1:15 labeled to unlabeled subunits)
   - 100–200 n*M* EB3
   - 1–10 μ*M* phalloidin-stabilized F-actin
2. After mixing, immediately add to the flow-cell channel and seal to prevent solvent evaporation.
   Tip: Higher tubulin concentrations facilitate microtubule nucleation and growth at lower temperatures (e.g., 30 μ*M* tubulin at 25 °C). For lower tubulin concentrations, higher temperatures may be needed (e.g., 15–20 μ*M* tubulin at 30 °C).
   Tip: We use actin concentrations between 1 and 10 μ*M*, which are sufficient for filaments to entangle. At concentrations above 48 μ*M*, the filaments will form a nematic phase (Käs et al., 1996).

2.5.3 Analysis of local changes in F-actin network density

At the start of these experiments, we typically observe a depletion of F-actin in close vicinity to the centrosome (Fig. 17.4A). This depletion zone closes over time as the actin filaments diffuse inward. With the use of a MATLAB

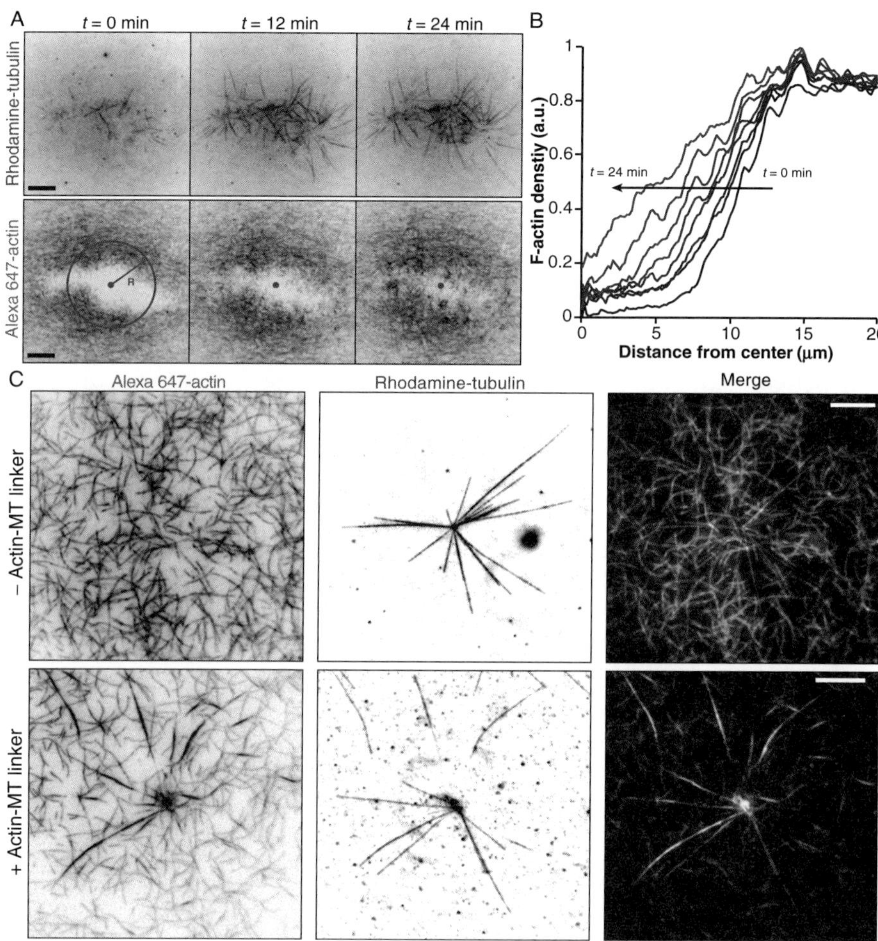

Figure 17.4 TIRF images of radial microtubule growth from centrosomes within isotropic solutions of actin filaments. (A) Initially, we observe a zone depleted of F-actin around the centrosomes, which typically closes within 15–30 min (in this example, multiple centrosomes are bound to the surface). (B) Radial fluorescence intensity profiles calculated at different time points demonstrate inward diffusion of actin filaments toward the centrosome cluster. (C) Once equilibrated, the actin filaments form a homogeneous entangled network (left) that fully encloses the dynamic centrosome-based microtubule aster (middle). (Right) Merge. (D) In the presence of an engineered F-actin-binding protein that binds to microtubule (MT) plus tips via EB, the actin filaments become radially organized as microtubules grow. Scale bars: 10 μm. (See the color plate.)

(MathWorks Inc., Natick, MA, USA) script, we can quantify the time scale of "F-actin invasion" by measuring the average F-actin fluorescence intensity profile along circles of increasing radius centered about the centrosome position. The resulting radial intensity profiles show that the actin filaments reach a fairly homogeneous distribution within 15–30 min (Fig. 17.4B). This time scale is consistent with the typical time scale of actin filament diffusion by reptation observed previously at comparable F-actin concentrations (Käs et al., 1996). Figure 17.4C shows a typical example of the steady-state arrangement of microtubules and actin filaments after equilibration. The actin filaments appear unperturbed by the microtubules growing into the network, and conversely, the microtubules appear to grow unhampered.

3. OUTLOOK

In this chapter, we introduced methods that allow for the simultaneous reconstitution of dynamic microtubules and actin networks. By preorganizing the actin network in different ways, either in parallel bundles or in loosely connected networks, we show that microtubules and actin can influence each other's organization and dynamics in different ways, purely as a result of steric effects. In our lab, we are using these assays as a starting point for mechanistic studies of the effect of specific cross-linking proteins on the coorganization of actin and microtubule networks (to be published elsewhere). As a preliminary example, we show in Fig. 17.4D how individual filaments in a loose actin network can become radially organized by a dynamic aster of microtubules in the presence of an F-actin-binding microtubule +TIP.

4. REAGENTS AND STOCKS

Here, we provide a list of key reagents and stock solutions in alphabetical order. Unless stated otherwise, chemicals were obtained from Sigma and stock solutions are prepared in MRB80. For stock solutions, information is provided as follows: name (catalog number, company), concentration, optional details, and storage temperature.

- ATP (disodium salt hydrate, A2383), 50 m*M*, −80 °C
- D-Glucose (G7528), 1 *M*, filter, −80 °C
- G-actin (labeled and unlabeled), for long-term storage keep concentration $\geq$48 μ*M* in G-buffer, −80 °C

- Alexa Fluor® 488, Alexa Fluor® 594, and Alexa Fluor® 647 succinimidyl ester dyes (A-20000, A-20004, A-20006, Molecular Probes, Life Technologies, Carlsbad, CA, USA), −20 °C
- GMPCPP (10 m*M* solution, NU-405, Jena Biosciences, Jena, Germany), −80 °C
- GTP (G8877), 50 m*M*, −80 °C
- H_2O_2 (30% in water, non-stabilized, 95313), 4 °C
- κ-Casein (C0406), 5 mg/ml, filter, −80 °C
- Methylcellulose (M0512), 1% (w/v), −80 °C
- NeutrAvidin™ (31055, Thermo Scientific, Pierce Protein Biology Products, Rockford, IL, USA), 5 mg/ml, −80 °C
- NH_4OH (30% in water, 320145), RT
- Oxygen scavenging and reducing cocktail (50×): 200 m*M* DTT, 10 mg/ml catalase, and 20 mg/ml glucose oxidase in MRB80, filter, −80 °C
- DTT (D0632), 1 *M*, −80 °C
- Catalase (C9322), powder, −20 °C
- Glucose oxidase (G7016), powder, −20 °C
- Phalloidin (P2141), 1 m*M* in methanol, −20 °C
- PLL-PEG (PLL(20)-g[3.5]-PEG(2), Susos AG, Dübendorf, Switzerland), 2 mg/ml, −80 °C
- PLL-PEG-biotin (PLL(20)-g[3.5]-PEG(2)/PEG(3.4)-biotin (50%), Susos AG), 2 mg/ml, −80 °C
- Pluronic® F-127 (P2443), 10% (w/v) in DMSO, RT
- Streptavidin (85878), 5 mg/ml, −80 °C
- Tubulin (T240, Cytoskeleton Inc., Denver, CO, USA), 50 μ*M*, −80 °C
- Tubulin HiLyte 488-labeled, HiLyte 647-labeled, TRITC Rhodamine-labeled, and biotin-labeled (TL488M, TL670M, TL590M, T333P Cytoskeleton Inc.), 50 μ*M*, −80 °C

ACKNOWLEDGMENTS

We are grateful to K.L. Yu, B. van der Vaart, and K. Jiang for help with cloning, C. Manatschal for help with protein purification and D. Mullins for providing us with the fascin plasmid. We thank B. Mulder and J. Alvarado for advice with the orientation analysis and N. Kurniawan for helpful comments on the manuscript. This work is part of the research program of the Foundation for Fundamental Research on Matter (FOM), which is part of the Netherlands Organization for Scientific Research (NWO). F. H. was supported by a Marie Curie fellowship.

REFERENCES

Akhmanova, A., & Steinmetz, M. O. (2008). Tracking the ends: A dynamic protein network controls the fate of microtubule tips. *Nature Reviews Molecular Cell Biology*, *9*, 309–322.

Ali, M. Y., Krementsova, E. B., Kennedy, G. G., Mahaffy, R., Pollard, T. D., Trybus, K. M., et al. (2007). Myosin Va maneuvers through actin intersections and diffuses along microtubules. *Proceedings of the National Academy of Sciences United States of America*, *104*(11), 4332–4336.

Alvarado, J., Mulder, B. M., & Koenderink, G. H. (2013). Alignment of nematic and bundled semiflexible polymers in cell-sized confinement. *Soft Matter*, http://dx.doi.org/10.1039/C3SM52421C. Epub ahead of print.

Applewhite, D. A., Grode, K. D., Keller, D., Zadeh, A. D., Slep, K. C., & Rogers, S. L. (2010). The spectraplakin Short stop is an actin-microtubule cross-linker that contributes to organization of the microtubule network. *Molecular Biology of the Cell*, *21*(10), 1714–1724.

Bieling, P., Laan, L., Schek, H., Munteanu, L., Sandblad, L., Dogterom, M., et al. (2007). Reconstitution of a microtubule plus-end tracking system *in vitro*. *Nature*, *450*, 1100–1105.

Brangwynne, C., Koenderink, G., MacKintosh, F., & Weitz, D. (2008). Nonequilibrium microtubule fluctuations in a model cytoskeleton. *Physical Review Letters*, *100*, 118104.

Breitsprecher, D., Jaiswal, R., Bombardier, J. P., Gould, C. J., Gelles, J., & Goode, B. L. (2012). Rocket launcher mechanism of collaborative actin assembly defined by single-molecule imaging. *Science*, *336*, 1164–1168.

Courson, D. S., & Rock, R. S. (2010). Actin cross-link assembly and disassembly mechanics for apha-actinin and fascin. *Journal of Biological Chemistry*, *285*, 26350–26357.

Deery, W. J., & Brinkley, B. R. (1983). Cytoplasmic microtubule assembly-disassembly from endogenous tubulin in a Brij-lysed cell model. *The Journal of Cell Biology*, *96*, 1631–1641.

Dogterom, M., & Surrey, T. (2013). Microtubule organization *in vitro*. *Current Opinion in Cell Biology*, *25*(1), 23–29.

Etienne-Manneville, S., & Hall, A. (2002). Rho GTPases in cell biology. *Nature*, *420*, 629–635.

Frey, N., Klotz, J., & Nick, P. (2009). Dynamic bridges—A calponin-domain kinesin from rice links actin filaments and microtubules in both cycling and non-cycling cells. *Plant and Cell Physiology*, *50*, 1493–1506.

Fygenson, D. K., Braun, E., & Libchaber, A. (1994). Phase diagram of microtubules. *Physical Review E*, *50*(2), 1579–1588.

Gell, C., Bormuth, V., Brouhard, G. J., Cohen, D. N., Diez, S., Friel, C. T., et al. (2010). Microtubule dynamics reconstituted *in vitro* and imaged by single-molecule fluorescence microscopy. In L. Wilson & J. J. Correia (Eds.), *Methods in cell biology*: *Vol. 95*. (pp. 221–245). Amsterdam: Elsevier, Chapter 13.

Gentry, B. S., van der Meulen, S., Noguera, P., Alonso-Latorre, B., Plastino, J., & Koenderink, G. H. (2012). Multiple actin binding domains of Ena/VASP proteins determine actin network stiffening. *European Biophysics Journal*, *41*, 979–990.

Geraldo, S., Khanzada, U. K., Parsons, M., Chilton, J. K., & Gordon-Weeks, P. R. (2008). Targeting of the F-actin-binding protein drebrin by the microtubule plus-tip protein EB3 is required for neuritogenesis. *Nature Cell Biology*, *10*, 1181–1189.

Huang, S. F., Jin, L. F., Du, J. Z., Li, H., Zha, Q., Ou, G. S., et al. (2007). SB401, a pollen-specific protein from *Solanum berthaultii*, binds to and bundles microtubules and F-actin. *Plant Journal*, *51*, 406–418.

Huber, F., Schnauß, J., Rönicke, S., Rauch, P., Müller, K., Fütterer, C., et al. (2013). Emergent complexity of the cytoskeleton: From single filaments to tissue. *Advances in Physics*, *62*(1), 1–112.

Hyman, A. A., Salser, S., Drechsel, D. N., Unwin, N., & Mitchison, T. J. (1992). Role of GTP hydrolysis in microtubule dynamics: Information from a slowly hydrolysable analogue, GMPCPP. *Molecular Biology of the Cell*, *3*(10), 1155–1167.

Janson, M. E., de Dood, M. E., & Dogterom, M. (2003). Dynamic instability of microtubules is regulated by force. *The Journal of Cell Biology*, *161*(6), 1029–1034.

Jiang, K., Toedt, G., Montenegro Gouveia, S., Davey, N. E., Hua, S., van der Vaart, B., et al. (2012). A proteome-wide screen for mammalian SxIP motif-containing microtubule plus-end tracking proteins. *Current Biology*, *22*(19), 1800–1807.

Käs, J., Strey, H., Tang, J. X., Finger, D., Ezzell, R., Sackmann, E., et al. (1996). F-actin, a model polymer for semiflexible chains in dilute, semidilute, and liquid crystalline solutions. *Biophysical Journal*, *70*(2), 609–625.

Komarova, Y., De Groot, C. O., Grigoriev, I., Montenegro Gouveia, S., Munteanu, E. L., Schober, J. M., et al. (2009). Mammalian end binding proteins control persistent microtubule growth. *The Journal of Cell Biology*, *184*(5), 691–706.

Laan, L., Pavin, N., Husson, J., Romet-Lemonne, G., van Duijn, M., Preciado López, M., et al. (2012). Cortical dynein controls microtubule dynamics to generate pulling forces that position microtubule asters. *Cell*, *148*, 502–514.

Lansbergen, G., & Akhmanova, A. (2006). Microtubule plus end: A hub of cellular activities. *Traffic*, 7, 499–507.

Lee, L., Tirnauer, J. S., Li, J., Schuyler, S. C., Liu, J. Y., & Pellman, D. (2000). Positioning of the mitotic spindle by a cortical-microtubule capture mechanism. *Science*, *287*, 2260–2262.

Lin, Y.-C., Koenderink, G. H., MacKintosch, F. C., & Weitz, D. A. (2010). Control of non-linear elasticity in F-actin networks with microtubules. *Soft Matter*, 7, 902–906.

Mao, Y. (2011). FORMIN a link between kinetochores and microtubule ends. *Trends in Cell Biology*, *21*(11), 625–629.

Montenegro Gouveia, S., Leslie, K., Kapitein, L. C., Buey, R. M., Grigoriev, I., Wagenbach, M., et al. (2010). In vitro reconstitution of the functional interplay between MCAK and EB3 at microtubule plus ends. *Current Biology*, *20*, 1717–1722.

Moseley, J. B., Bartolini, F., Okada, K., Wen, Y., Gundersen, G. G., & Goode, B. L. (2007). Regulated binding of adenomatous polyopsis coli protein to actin. *Journal of Biological Chemistry*, *282*(17), 12661–12668.

Mullins, R. D., & Hansen, S. D. (2013). *In vitro* studies of actin filament network dynamics. *Current Opinion in Cell Biology*, *25*(1), 6–13.

Nadar, V. C., Lin, S., & Baas, P. W. (2012). Microtubule redistribution in growth cones elicited by focal inactivation of kinesin-5. *Journal of Neuroscience*, *32*(17), 5783–5794.

Olmsted, J. B., & Borisy, G. G. (1975). Ionic and nucleotide requirements for microtubule polymerization *in vitro*. *Biochemistry*, *14*(13), 2996–3005.

Pardee, J. D., & Spudich, J. A. (1982). Mechanism of K^+-induced actin assembly. *Journal of Cell Biology*, *93*, 648–654.

Popp, D., Yamamoto, A., Iwasa, M., & Maéda, Y. (2006). Direct visualization of actin nematic network formation and dynamics. *Biochemical and Biophysical Research Communications*, *351*(2), 348–353.

Rasband, W. S. (1997–2012). *ImageJ*, U.S. National Institutes of Health, Bethesda, Maryland, USA, http://imagej.nih.gov/ij.

Rezakhaniha, R., Agianniotis, A., Schrauwen, J. T., Griffa, A., Sage, D., Bouten, C. V., et al. (2012). Experimental investigation of collagen waviness and orientation in the arterial adventitia using confocal laser scanning microscopy. *Biomechnics and Modeling in Mechanobiology*, *11*, 461–473.

Rodriguez, O. C., Schaefer, A. W., Mandato, C. A., Forscher, P., Bement, W. M., & Waterman-Storer, C. M. (2003). Conserved microtubule-actin interactions in cell movement and morphogenesis. *Nature Reviews Molecular Cell Biology*, *5*, 599–609.

Schroeder, H. W., III, Mitchell, C., Shuman, H., Holzbaur, E. L., & Goldman, Y. E. (2010). Motor number controls cargo switching at actin-microtubule intersections *in vitro*. *Current Biology*, *20*(8), 687–696.

Selden, S. C., & Pollard, T. D. (1986). Interaction of actin filaments with microtubules is mediated by microtubule-associated proteins and regulated by phosphorylation. *Annals of the New York Academy of Sciences*, *466*, 803–812.

Shen, Q. T., Hsiue, P. P., Sindelar, C. V., Welch, M. D., Campellone, K. G., & Wang, H. W. (2012). Structural insights into WHAM-mediated cytoskeletal coordination during membrane remodeling. *Journal of Cell Biology*, *199*(1), 111–124.

Shi, X., Lim, J., & Ha, T. (2010). Acidification of the oxygen scavenging system in single-molecule fluorescence studies: In situ sensing with ratiometric dual-emission probe. *Analytical Chemistry*, *82*(14), 6132–6138.

Sider, J. R., Mandato, C. A., Weber, K. L., Zandy, A. J., Beach, D., Finst, R. J., et al. (1999). Direct observation of microtubule-f-actin interaction in cell free lysates. *Journal of Cell Science*, *112*, 1947–1956.

Siegrist, S. E., & Doe, C. Q. (2007). Microtubule-induced cortical cell polarity. *Genes and Development*, *21*, 483–496.

Solinet, S., Mahmud, K., Stewman, S. F., Ben El Kadhi, K., Decelle, B., Talje, L., et al. (2013). The actin-binding ERM protein Moesin binds to and stabilizes microtubules at the cell cortex. *Journal of Cell Biology*, *202*(2), 251–260.

Tamura, N., & Draviam, V. M. (2012). Microtubule plus-ends within a mitotic cell are 'moving platforms' with anchoring, signaling and force-coupling roles. *Open Biology*, *2*, 120132.

Vasiliev, M., Gelfand, I. M., Domnina, L. V., Ivanova, O. Y., Komm, S. G., & Olshevskaja, L. V. (1970). Effect of colcemid on the locomotory behavior of fibroblasts. *Journal of Embryology & Experimental Morphology*, *24*(3), 625–640.

Vitre, B., Coquelle, F. M., Heichette, C., Garnier, C., Chretien, D., & Arnal, I. (2008). EB1 regulates microtubule dynamics and tubulin sheet closure *in vitro*. *Nature Cell Biology*, *10*, 415–421.

Waterman-Storer, C., Duey, D. Y., Weber, K. L., Keech, J., Cheney, R. E., Salmon, E. D., et al. (2000). Microtubules remodel actomyosin networks in *Xenopus* egg extracts via two mechanisms of F-actin transport. *Journal of Cell Biology*, *150*(2), 361–376.

Waterman-Storer, C. M., & Salmon, E. D. (1999). Positive feedback between microtubule and actin dynamics during cell motility. *Current Opinion in Cell Biology*, *11*, 61–67.

Wu, X., Kodama, A., & Fuchs, E. (2008). ACF7 regulates cytoskeletal-focal adhesion dynamics and migration and has ATPase activity. *Cell*, *135*(1), 137–148.

Yarm, F., Sagot, I., & Pellman, D. (2001). The social life of actin and microtubules: Interaction versus cooperation. *Current Opinion in Microbiology*, *4*(6), 696–702.

Yin, H., Pruyne, D., Huffaker, T. C., & Brescher, A. (2000). Myosin V orientates the mitotic spindle in yeast. *Nature*, *406*, 1013–1015.

CHAPTER EIGHTEEN

Measuring Kinetochore–Microtubule Interaction *In Vitro*

Jonathan W. Driver*, Andrew F. Powers*, Krishna K. Sarangapani*, Sue Biggins[†,1], Charles L. Asbury*

*Department of Physiology and Biophysics, University of Washington School of Medicine, Seattle, Washington, USA

†Division of Basic Sciences, Fred Hutchinson Cancer Research Center, Seattle, Washington, USA

[1]Corresponding author: e-mail address: sbiggins@fhcrc.org

Contents

Abstract

Many proteins and protein complexes perform sophisticated, regulated functions *in vivo*. Many of these functions can be recapitulated using *in vitro* reconstitution, which serves as a means to establish unambiguous cause–effect relationships, for example, between a protein and its phosphorylating kinase. Here, we describe a protocol to purify kinetochores, the protein complexes that attach chromosomes to microtubules during mitosis, and quantitatively assay their microtubule-binding characteristics. Our assays, based on DIC imaging and laser trapping microscopy, are used to measure the attachment of microtubules to kinetochores and the load-bearing capabilities of those attachments. These assays provide a platform for studying kinase disruption of kinetochore–microtubule attachments, which is believed to be critical for correcting erroneous kinetochore–spindle attachments and thereby avoiding chromosome mis-segregation. The principles of our approach should be extensible to studies of a wide range of force-bearing interactions in biology.

Methods in Enzymology, Volume 540
ISSN 0076-6879
http://dx.doi.org/10.1016/B978-0-12-397924-7.00018-2

1. INTRODUCTION

Kinetochores are protein complexes that attach chromosomes to microtubules of the mitotic spindle (Fig. 18.1). These remarkable microtubule-binding machines are capable of tracking the dynamic tips of growing and shortening microtubules for many minutes without detaching (Nicklas & Koch, 1969). They are also regulatory hubs that can break improper kinetochore–microtubule attachments using kinase reactions to prevent chromosomal missegregation (Biggins et al., 1999; Cheeseman et al., 2002; Hauf et al., 2003; Lampson, Renduchitala, Khodjakov, & Kapoor, 2004; Pinsky, Kung, Shokat, & Biggins, 2006; Tanaka et al., 2002). On the other hand, kinetochore–microtubule attachments are paradoxically *stabilized* when tension is applied to them (Akiyoshi et al., 2010; Nicklas & Koch, 1969; Nicklas & Ward, 1994). *In vitro* biophysical studies provide powerful tools for dissecting these regulatory mechanisms.

Our *in vitro* assays have already revealed that tension directly stabilizes kinetochore–microtubule attachments through effects on microtubule dynamics (Akiyoshi et al., 2010). Tension reduces the frequency of "catastrophes," or switches from polymerization-driven microtubule growth to depolymerization/shortening, during which a kinetochore–microtubule attachment is particularly susceptible to spontaneous rupture. This stabilization mechanism is completely independent of the aforementioned kinase activity and would have been nearly impossible to detect *in vivo* using other existing technologies, as tension cannot be measured inside a cell. The *in vitro* approach also makes

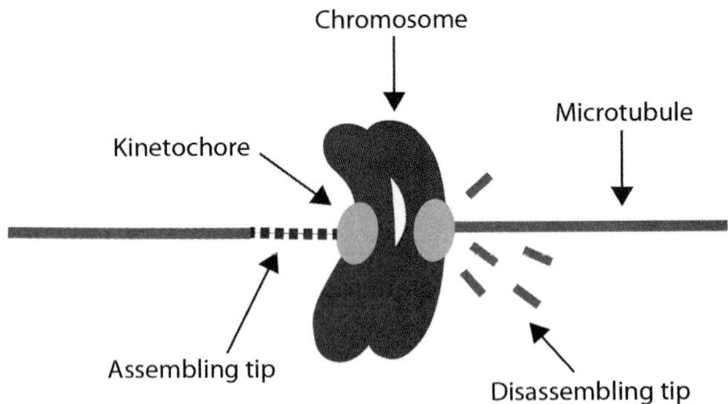

Figure 18.1 A metaphase chromosome is coupled to microtubules from opposite sides of the mitotic spindle through its kinetochores. The kinetochores remain attached as the microtubules undergo assembly and disassembly. (For the color version of this figure, the reader is referred to the online version of this chapter.)

quantitative comparison of mutant or modified kinetochores straightforward, sidestepping the problems presented by subtle phenotypes (where small but measureable attachment strength changes do not lead to observable changes in spindle structure) and pleiotropic effects (where multiple changes occur at once, producing a confusing result). We have shown that the Ipl1/Aurora B kinase very likely uses a small set of phosphorylation sites on the Ndc80 and Dam1 kinetochore subcomplexes to disrupt attachments, and that phosphorylation on these two subcomplexes contributes additively to attachment weakening (Sarangapani, Akiyoshi, Duggan, Biggins, & Asbury, 2013).

In this chapter, we present methods for studying kinase disruption of kinetochore–microtubule attachments *in vitro*, which involve modifications to the assays used in our previous work (Akiyoshi et al., 2010; Sarangapani et al., 2013). These experiments directly test how kinases affect the strength of kinetochore-microtubule attachments, which is important for understanding the error correction process of improper chromosome attachment (Maure, Kitamura, & Tanaka, 2007). These experiments also enable more advanced tests of mechanoregulation of the kinases, which is believed (but not proved) to help the kinases identify improper attachments (Maresca & Salmon, 2010). A deeper understanding of the error correction process is of wide interest because chromosomal missegregation is associated with birth defects and the development of cancers (Ganem, Godinho, & Pellman, 2009; Rajagopalan & Lengauer, 2004).

2. THE LEAP TO *IN VITRO*: PURIFICATION OF INTACT, FUNCTIONAL KINETOCHORES

2.1. An antibody-based approach

Purifying native protein complexes can be challenging. Antibody-based strategies are an attractive option since dissociation of the complex from the antibody can be achieved by introducing an excess amount of the protein antigen, as opposed to altering solution pH or ionic strength or introducing reactive small molecules which might disrupt the complex. We use antibodies raised against the 3 × Flag peptide, a 24 (3 × DYKDDDDK) amino acid sequence that we fused to the Dsn1 protein, which is part of the Mtw1 complex and a central structural subunit of the kinetochore. Dsn1 is the kinetochore subunit bound by the monopolin complex to promote sister kinetochore coorientation during meiosis I, so it can be thought of as a handle for the kinetochore (Corbett & Harrison, 2012; Corbett et al., 2010). By replacing the endogenous Dsn1 gene with a gene coding for

Dsn1–6 × His–3 × Flag, we can effectively immunoprecipitate whole kinetochores (Akiyoshi et al., 2010), using a protocol similar to that previously used to purify minichromosomes (Akiyoshi, Nelson, Ranish, & Biggins, 2009). The histidine tag is used in the assay protocols and is not involved in the purification. Yeast with this new, tagged Dsn1 gene have no phenotype relative to wild-type strains. We find magnetic beads to be a superior purification substrate, because they produce very clean kinetochore preparations with few contaminants.

2.1.1 Antibody conjugation to magnetic beads

Immobilization of anti-Flag antibodies (M2 anti-Flag, Sigma F3165) to magnetic beads (protein G Dynabeads, Life Technologies 10003D) is achieved through protein G. To begin, rinse 300 μL Dynabeads twice with 1 mL sodium phosphate buffer (0.1 *M* Na-PO_4, pH 7.0) using a magnetic particle concentrator (MPC, Life Technologies). After removing the second milliliter of buffer, resuspend the beads in 60 μL of 2.5 μg/μL M2 antibodies (in sodium phosphate buffer) with gentle rotation for 30 min. Rinse the beads twice with 1 mL PBST (0.1 *M* Na-PO_4, 0.01% Tween 20, pH 7.0) and twice with triethanol amine buffer (0.2 *M* triethanol amine, 20 m*M* dimethyl pimelimidate, pH 8.2) using the MPC. Initiate covalent coupling through DMP chemistry by resuspending the beads in 1 mL of 20 m*M* DMP in triethanol amine buffer. Quench the reaction after 30 min (do not overcouple) by resuspending in 1 mL Tris buffer (50 m*M* Tris–HCl, pH 7.5) and then rinse three times with 1 mL PBST. Resuspend in 300 μL PBST. M2 antibody-coated Dynabeads can be stored at 4 °C for use within 1 month.

2.1.2 Yeast lysate preparation

Grow a 2-L culture of a Dsn1–6 × His–3 × Flag epitope-tagged yeast strain (SBY8253) to late log phase ($OD_{600}=1.5$–2.0) and harvest the yeast by centrifugation. All subsequent steps should be performed at 4 °C by keeping buffers and yeast on ice. Rinse (resuspend with 25 mL pipette + re-pellet) the pellet with cold H_2O + 0.2 m*M* PMSF, and then with BH0.15 (25 m*M* Hepes, 2 m*M* $MgCl_2$, 0.1 m*M* EDTA, 0.5 m*M* EGTA, 15% glycerol, 0.1% NP-40, 150 m*M* KCl) + protease inhibitors (0.2 m*M* PMSF, 10 μg/mL leupeptin, 10 μg/mL pepstatin, 10 μg/mL chymostatin) + phosphatase inhibitors (100 ng/mL Microcystin-LR, 1 m*M* sodium pyrophosphate, 2 m*M* sodium-beta-glycerophosphate, 100 n*M* sodium orthovanadate, 5 m*M* sodium fluoride). Freeze the pellet in liquid nitrogen and weigh (usually ~9 g) in preparation for mechanical disruption. The pellet can be stored at −80 °C if necessary. We use a small blender (Oster brand) to lyse the yeast, but other methods may be effective as well (Aitchison, Blobel, & Rout, 1996; Schultz, Hockman, Harkness, Garinther, & Altheim, 1997). Regardless of the method used, care must be taken to keep the pellet frozen during the grinding process, which invariably creates heat. Shake the pellet into the blender cup and add dry ice to the blender at around five times the volume of the pellet, grind for 10–12 min, replace the evaporated dry ice, and repeat once more. Allow the dry ice to evaporate at room temperature, occasionally stirring the powder. Once the dry ice evaporates and the ground cell mass begins to melt, lyse the pellet in 16.6 mL of BH0.15 + protease inhibitors + phosphatase inhibitors (same as pellet rinse buffer except double the amount of leupeptin, pepstatin, chymostatin, and microcystin-LR). Use a 25-mL pipette to completely resuspend the cell mass and then incubate on ice for an additional 15 min with occasional mixing. Clarify by ultracentrifugation (24,000 rpm in a Beckman-Coulter SW-41 Ti or 71,124 × *g* for 90 min at 4 °C). Three layers are visible in the centrifuge tube after clarification: (1) an insoluble cell debris mass at the bottom, (2) an aqueous layer containing the protein, (3) a thin film of lipid at the top. Extract only the aqueous layer using a needle and syringe to puncture the wall of the plastic centrifuge tube, being very cautious not to disturb the lipid layer which would contaminate the lysate. Use a Bradford assay to estimate protein concentration and total protein yield. If needed, the lysate can be flash-frozen in liquid nitrogen and stored at −80 °C for future use.

2.1.3 Purification step

The final step is immunoprecipitation. Incubate M2 anti-Flag Dynabeads (15 μL per 12.6 mg protein in lysate) with lysate for 3 h at 4 °C with gentle rotation. Rinse the beads four times with 1 mL BH0.15 + protease inhibitors + phosphatase inhibitors + 2 m*M* DTT and then three times with 1 mL BH0.15 + protease inhibitors. Incubate with elution buffer (BH0.15 + protease inhibitors + 0.5 mg/mL 3 × Flag peptide) in the original volume of Dynabeads for 30 min with constant gentle vortexing to elute the kinetochores from the beads. We use a custom-synthesized 3 × Flag peptide, but it is also commercially available (Sigma F4799). Freeze the product in liquid nitrogen in small (~2–5 μL) aliquots.

2.1.4 Analysis of kinetochore particles

To analyze the kinetochore purification, run an aliquot on a 4–12% NuPage Bis–Tris gel (Life Technologies) and analyze the stoichiometry of kinetochore components after silver staining the gel. After a successful purification, the stoichiometry of the Ndc80, Spc105, and Dsn1 components are similar. On the same gel, run BSA protein standards to estimate the Dsn1 concentration of the preparation, which is usually between 1.5 and 6.0 ng/μL. Particles can also be analyzed by immunoblotting for specific candidate proteins that are present at substoichiometric amounts (Akiyoshi et al., 2010).

2.2. Microbeads for visualization and manipulation of single kinetochores

Polystyrene microbeads provide a handle for manipulating protein complexes such as microtubule motors, DNA enzymes, and kinetochore particles (Akiyoshi et al., 2010; Vale, Reese, & Sheetz, 1985; Yin et al., 1995). The complex of interest can be targeted to the microbead surface through an antibody recognizing a tag on the complex. We cannot use anti-Flag antibodies for this since our kinetochore preparations contain excess 3 × Flag peptide; instead, we use antibodies that bind the His epitope on Dsn1–6 × His–3 × Flag. We purchase streptavidin-conjugated microbeads and biotin-conjugated anti-His antibodies that self-associate. Alternatively, antibodies can be covalently linked directly to microbeads using established protocols (Bangs Laboratories, 2002; Hermanson, 1996).

To prepare microbeads coated in anti-His antibody, combine 112.5 μL H_2O, 45 μL 5 × BRB80 (400 m*M* pipes, 62.5 m*M* NaOH, 5 m*M* $MgCl_2$, 5 m*M* EGTA, pH 6.9), and 22.5 μL PSS4 (Spherotech, 0.44 μm) beads in a 500-μL Eppendorf tube. Sonicate with a tip sonicator in an ice bath for 1 min. Add 45 μL biotin-conjugated anti-His antibody (Qiagen #34440) to the mixture and incubate at 4 °C for 1 h. Add 275 μL bead storage buffer (1 × BRB80 (80 m*M* pipes, 12.5 m*M* NaOH, 1 m*M* $MgCl_2$, 1 m*M* EGTA, pH 6.9) + 8 mg/mL BSA + 1 m*M* DTT) and mix well, sonicate again for 1 min and centrifuge at 16.1 × *g* for 9 min at 4 °C. Aspirate the supernatant gently. Repeat the resuspension and centrifugation steps five times, resuspending in 500 μL bead storage buffer each time. After the last centrifuge spin and aspiration step, resuspend the bead pellet in 225 μL bead storage buffer, sonicate for 1 min, and store at 4 °C with inversion mixing.

Microbeads prepared this way (which we refer to as PSS4/10) can be stored and used for 2–3 months. Over time, the beads will clump/deposit on the sides of the tube. Sonication should correct this. Clumping is more pronounced in samples that have been insufficiently washed (too few resuspensions/centrifugations).

3. THE ASSAYS

3.1. Bead-binding assay

Sometimes, a simple measure of a complex's ability to bind its target (microtubule, DNA, etc.) is useful. Polystyrene microbeads are plainly visible in DIC, so they can be used as a marker instead of a fluorescent tag on the complex. The principle behind the assay is to densely coat a coverslip with microtubules so that all microbeads with active kinetochores have an opportunity to bind (Biggins et al., 1999; Sassoon et al., 1999). Measure the dimensions of the field only if you wish to compare measurements between different instruments. Of course, bead binding is a function of the amount of kinetochore material used to coat the beads, so this is a design parameter that may need to be optimized for different experiments to maximize sensitivity of the assay to differences between the intrinsic binding capabilities of the different kinetochores in the comparison.

3.1.1 Materials

All coverslips should be cleaned with KOH solution (80 g KOH in ethanol, 300 mL total volume in a 1-L beaker). This solution is corrosive and flammable and should be handled with care in a fume hood. Load coverslips into a Teflon rack (this rack needs hooks so that a metal positioning rod can be used to pull it in and out of the KOH) and immerse them in the KOH solution. Sonicate the beaker with the coverslips for 10 min using a bath sonicator. Prepare two beakers with 1.5 and 0.8 L deionized water. Using a positioning rod, remove the coverslip rack from the KOH and dunk it twenty times in the beaker with 1.5 L of water, and then remove and place in the smaller water beaker. Sonicate the small beaker for 10 min. Take the coverslip rack out and spray the coverslips forcefully with deionized water from a spray bottle. Use about 250 mL of water. Last, spray the coverslips with around 250 mL of ethanol. Place the rack in an oven at around 50 °C for a few hours to dry and then transfer the rack to a covered container for storage.

To prepare taxol-stabilized microtubules, mix 48.8 μL of polymerization buffer (1 × BRB80 + 5.7% DMSO + 4.1 m*M* $MgCl_2$ + 9.8 m*M* GTP) with 12.8 μL tubulin (100 μM). Incubate 45 min at 37 °C. Add 200 μL warm BTAX (1 × BRB80 + 10 μ*M* paclitaxel) to the polymerization mix to stabilize the microtubules. Transfer to a TLA 100 centrifuge tube and spin at 58,000 rpm (129,797 × *g*) for 10 min at 37 °C. Discard the supernatant and resuspend the pellet in 200 μL BTAX with a cut P200 tip to avoid excessive shearing of the microtubules.

Prepare a flow chamber using a glass slide, coverslip, and double-sided tape. Place two pieces of tape on the slide parallel with its short dimension with a 1-mm gap between them centered on the slide. The width of the gap is important; it influences the speed of the flow through the channel as well as the total channel area over which the bead binding is distributed. The best practice is to measure the width with calipers and to only use channels that fall within ±0.1 mm of the target width. Place the coverslip over the channel so that it covers it and overhangs each edge, standardizing the length of the channel to exactly the width of the slide.

For the assay, you will need rigor kinesin, which serves as a linker between the microtubules and the surface (we use the G234A mutant: Rice et al., 1999). Typically, a fivefold dilution of BL-21 expressed, Ni^{2+}-NTA purified motor will do well, but optimization may be needed to achieve high microtubule density. In this protocol, a 6 × His-tagged GFP (0.5 mg/mL working concentration) is used to saturate the antibodies on the microbeads that remain unoccupied after the kinetochore incubation so that they do not bind directly to the 6 × His-tagged kinesin on the surface (Fig. 18.2A).

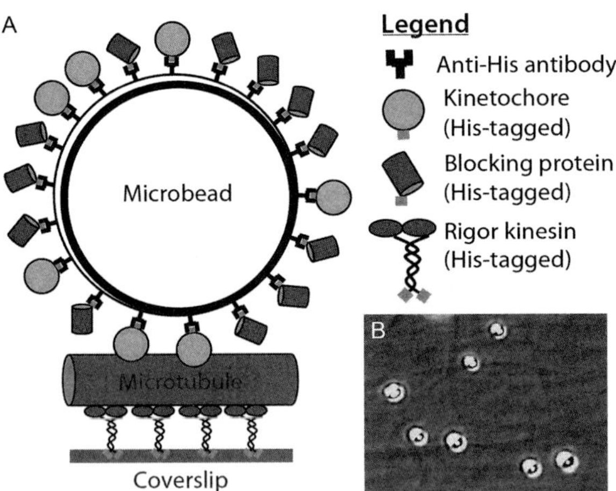

Figure 18.2 (A) Diagram of a kinetochore-coated bead binding to a microtubule immobilized to a coverslip through rigor kinesins. The blocking protein described in the text is 6 × His-GFP, but any protein that has a 6 × His tag and that does not stick to the surface will work. (B) DIC image of kinetochore-beads bound to a microtubule mat. (For the color version of this figure, the reader is referred to the online version of this chapter.)

3.1.2 Basic binding assay

To test the ability of the kinetochores to bind microtubules, mix 20 μL of BTAX with 5 μL of PSS4/10 to make PSS4/50. Sonicate the PSS4/50 under a tip sonicator in an ice bath for 1 min. Dilute a kinetochore aliquot with BTAX so that the concentration of Dsn1 is 5 nM (0.33 ng/μL), mix 5 μL of PSS4/50 with 5 μL of diluted kinetochores, and incubate at 4 °C for 45 min, with rotational mixing. Add 10 μL of GFP and incubate 15 min more at 4 °C. Five minutes into the incubation, fill the chamber with rigor kinesin and incubate for 5 min. Using an aspirator or a filter paper to draw the liquid through the channel, rinse the channel with 30 μL blocking buffer (1 × BRB80 + 2 mg/mL κ-casein + 10 μM paclitaxel), add 20 μL of the taxol-stabilized microtubules, and incubate for 5 min. Rinse the channel once more with 30 μL blocking buffer. When the GFP incubation completes, add 20 μL blocking buffer to the mixture and sonicate it for 45 s using the tip sonicator. Flow 20 μL of this final mixture through the slide and begin counting the number of surface-associated beads in each field 5 min later. Count as many fields as you can in 10 min. For a control measurement, substitute pure BTAX for diluted kinetochores.

While you are counting, qualitatively assess the microtubule coverage of the coverslip. The microtubules should densely cover the coverslip in a "mat" (Fig. 18.2B). If the microtubules become sparse, non-specific binding will increase and specific binding will decrease. High concentrations of casein and 6 × His-GFP are necessary to prevent non-specific binding of the beads to the surface (binding not mediated by kinetochore–microtubule attachments). Do not attempt to differentiate beads that are associated with microtubules on the coverslip versus beads associated with the coverslip directly. Be sure to count beads attached to microtubules protruding from the surface; they are valid attachments. Develop a system for counting beads that are on the edge of a field and be consistent (e.g., all beads visible on the right edge count, all beads on the left edge do not). Make sure that you traverse the channel in such a way as to sample across its width while avoiding the very edges. This will average out spatial variations in bead deposition. Stick to the time limits. Beads continue to accumulate slowly throughout the experiment.

3.1.3 Timecourse experiment for assaying kinase activity

A protein kinase called Mps1 copurifies with the kinetochores from Section 2.1 (London, Ceto, Ranish, & Biggins, 2012). It phosphorylates the kinetochores *in vitro* in the presence of ATP, and this phosphorylation may interfere with the formation of attachments to microtubules (Maure et al., 2007). The progress of this reaction can be monitored in the binding assay, with minor modifications. Instead of diluting the PSS4/10 bead stock in BTAX, dilute it in bead buffer that contains the phosphatase inhibitors described in Section 2.1.2 for rinsing the cell pellet. The phosphatase inhibitors stop any phosphatases in the prep from undoing the phosphorylation by Mps1 so that its effects can be measured. One may also deactivate the Mps1 kinase using a temperature- (Schutz & Winey, 1998; Winey, Goetsch, Baum, & Byers, 1991) or nucleotide analogue-sensitive mutant (Jones et al., 2005; Tighe, Staples, & Taylor, 2008) and then add recombinant Ipl1/Sli15 to the reaction mixture below before adding ATP.

Before starting, prepare four slides as described earlier. Follow the previous instructions, but use 10 μL of PSS4/50 in bead buffer (1× BRB80 + phosphatase inhibitors), 10 μL of diluted kinetochores, and 20 μL of GFP. After 15 min GFP incubation, add 4 μL of 100 m*M* ATP and move the reaction to a 37 °C incubator. This creates a one-pot reaction that will be sampled at four different time points. Preparation for each slide requires ~15 min, and they should be prepared immediately before use. Two people must be involved: one to prepare the samples and another to count the beads per field in each sample. For time points of 5, 25, 45, and 65 min after ATP introduction, start the preparation of the first slide at the same time as in the original protocol. Extract 10 μL of material from the pot at each time point, dilute with 10 μL blocking buffer, and sonicate for 45 s. Formally, the person preparing the slides has 5 min to begin preparing the next sample, but practically the 5 min of extra time will be taken by the various aspiration and sonication operations required at each step. Use a master timer that monitors the overall time in the 65-min experiment rather than counting off 5 min for each step. This will help with comparisons between runs. If three people and two microscopes are available, the kinetochore-bead–GFP mixture volume can be doubled and then split in half into experimental (ATP) and control (H_2O or ADP) reactions to be counted simultaneously.

3.2. Rupture force

With an optical trap, it is possible to measure the force that a single complex attachment can withstand (Fig. 18.3A). In essence, optical traps are simply microscopes with powerful, focused laser beams in the sample plane, but the optics and instrumentation required to focus, image, and detect the laser require expertise to set up (Block, 1990). Like most, our traps are feedback

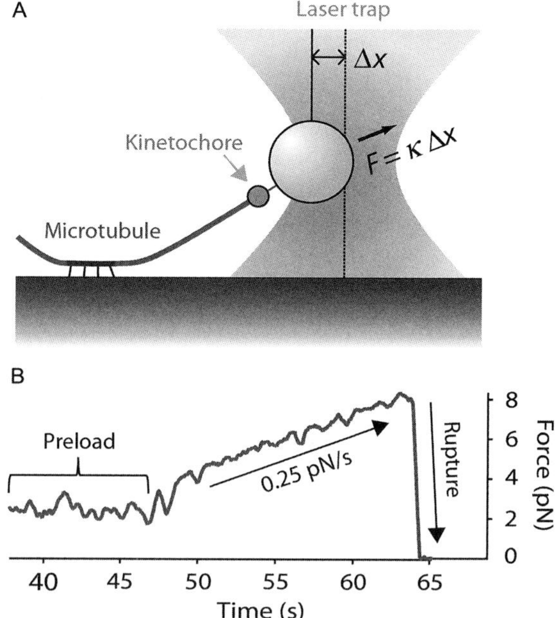

Figure 18.3 (A) Diagram of a kinetochore–microtubule attachment under tension in an optical trap. (B) Example rupture force measurement. The microbead returns to the center of the trap (zero force) after the attachment ruptures. The peak force reached during the force ramp is the "rupture force" that is recorded for that attachment. (For the color version of this figure, the reader is referred to the online version of this chapter.)

controlled such that the separation between the laser focus and a trapped polystyrene bead can be maintained or changed in a preprogrammed manner (Visscher & Block, 1998). We translate a piezoelectric sample stage (to which the polystyrene bead is attached) relative to the trap laser, which is fixed (Frank, Powers, Gestaut, Davis, & Asbury, 2010), but one may alternatively steer the trap beam using an acousto- or electrooptic deflector. Our custom instrument control software is written in LabView and is freely available upon request. In these experiments, the control software is used to steadily increase or "ramp" the separation between the bead and the trap laser, consequently increasing the restoring force applied by the trap on the bead (Fig. 18.3B). Trap control software can be easily written to facilitate other measurements, such as attachment lifetime under oscillating loads (as occurs during metaphase; Dumont, Salmon, & Mitchison, 2012) to test whether attachment strength is affected by the pattern or structure of the applied load in addition to its magnitude. Since single complexes detach from their targets stochastically under any load (or none), the result of the force ramp experiment is a rupture force distribution that is dependent on both the intrinsic properties of the attachment and the rate at which load on the attachment is increased (Evans, 2001). With this measurement, one can cope with some amount of heterogeneity in a sample and distinguish relatively small differences between species.

3.2.1 Microtubules for optical trapping

The microtubules are anchored to the coverslip surface so that they remain fixed while force is applied to the kinetochores through the microbeads. Through a serial polymerization strategy, microtubules with two compositionally distinct ends are synthesized. One end binds the surface through biotin/streptavidin interactions, while the other remains unbound and projects into solution so that it can interact with beads held away from the surface with the optical trap. This second end can be either dynamic (spontaneously polymerizing and depolymerizing throughout the experiment) or static (stabilized using paclitaxel). Kinetochores couple to the ends of dynamic microtubules *in vivo*, so reconstituting this behavior *in vitro* is desirable. However, for consistent polymerization behavior, buffer conditions must be strictly controlled.

Before starting, prepare biotinylated microtubule seeds. These can be stored in small (~5 μL) aliquots at −80 °C. Mix 10 μL biotin tubulin (2 μg/μL in 1× BRB80, cytoskeleton #T333P) with 33 μL unlabeled tubulin (100 μM in 1× BRB80). Withdraw 40 μL of this mixture and add to it 250 μL of seeds buffer (2× BRB80 + 20.8% glycerol + 2 m*M* GMPCPP + 1.6 m*M* DTT). Incubate for 1 h at 37 °C, aliquot, and flash freeze in liquid nitrogen.

3.2.2 Dynamic microtubule rupture force assay

To perform our dynamic microtubule rupture force assay, use chambers that are identical to those described in Section 3.1.1, though the channel width does not need to be stringently controlled. The solutions required for this assay are assay buffer (1× BRB80 + 1 mg/mL κ-casein), GB2× (1× BRB80 + 2 mg/mL κ-casein + 2 m*M* GTP), GB1× (50% GB2× in H_2O), and GBTB (50% GB2× in H_2O + 12% KTB + 16 μM tubulin + 0.8 m*M* DTT + 50 m*M* glucose + 0.5 mg/mL glucose oxidase + 60 μg/mL catalase). The glucose oxidase, catalase, and glucose consume oxygen in the GBTB solution so that kinetochores are protected from oxidative damage when they are in the trapping laser. One must wait to add the glucose oxidase and catalase until just before the GBTB is used so that the reaction does not run its course before the measurement. Also wait until the end to add the tubulin to avoid unnecessary polymerization. Prepare PSS4/50 as before using assay buffer to dilute. Dilute the kinetochores to 1 ng/μL Dsn1 with assay buffer. Combine 5 μL of each to make KTB and incubate at 4 °C. Fifteen minutes into the KTB incubation, add biotin BSA (Vector Labs #B-2007, 10 mg/mL) to the flow chamber and incubate for 15 min. Rinse the chamber with 100 μL warm 1 × BRB80, introduce avidin (Vector Labs #A-3100, 0.33 mg/mL), and then incubate for 3 min. Rinse again with 100 μL 1 × BRB80, dilute a 5-μL aliquot of microtubule seeds with 15–35 μL of warm 1 × BRB80, and then flow the entire dilution through the chamber and incubate for 3 min. The appropriate seeds concentration can vary based on properties of the coverslip. After the seeds incubation, rinse the channel with 100 μL GB1×. Finish preparation of the GBTB by adding KTB, glucose oxidase, and catalase, sonicating under the tip sonicator for 45 s, and then adding the tubulin. Warm up the GBTB to approximately room temperature just before introduction to the chamber to avoid depolymerization of the seeds without risking polymerization of the free tubulin in the GBTB. Seal the ends of the chamber with nail polish and take the sample to the optical trap.

To test the attachment strength of the kinetochores, you may capture beads floating in solution and attach them to microtubules (only a fraction will bind), or find preattached beads. With each bead, apply a load in the direction of the microtubule's free end. Start with a very low load (e.g., 1 pN) and increase it only as necessary to pull the bead to the microtubule tip. Kinetochores can slide along the lattice of a microtubule without detaching, but some amount of force is needed to overcome protein friction. Once at the tip, the bead will advance at a rate of ~10 nm/s during microtubule assembly (which is far more common than disassembly). Increase the load to a predetermined level (e.g., 4 pN) and hold for 30–60 s before initiating a force ramp to increase the force at a constant rate (e.g., 0.25 pN/s) until the attachment ruptures. Repeat this for as many attachments as you wish within a 2-h time window.

3.2.3 Taxol microtubule rupture force assay for kinase activity

In the cell, kinases must recognize erroneous attachments that have already formed and disrupt them; we have modified the previous assay to match this order of events. In this last protocol, beads that do not spontaneously bind are rinsed away with a reaction buffer so that subsequent enzyme reactions are only carried out on attached kinetochores (which may be acted upon differently than unattached kinetochores). This rinsing operation is sensitive because the flow driven through the channel can easily be forceful enough to remove attached kinetochore-beads. The most ideal solution to the problem is to use a syringe pump to drive a slow, smooth, consistent rinse (Fig. 18.4, diagram). If this is not possible, wicking with wet filter paper may be a viable alternative. Note that the use of a syringe pump may require custom-machined adaptors to interface the tubing to the slide/chamber (Gestaut, Cooper, Asbury, Davis, & Wordeman, 2010). Channel width must again be closely controlled (Fig. 18.4).

Be sure to rinse the channel with H_2O thoroughly before you begin. Refer to the dynamic microtubule assay protocol (Fig. 18.4). New solutions are 2× BTAX+PIs (2× BRB80+2× phosphatase inhibitors+20 μ*M*

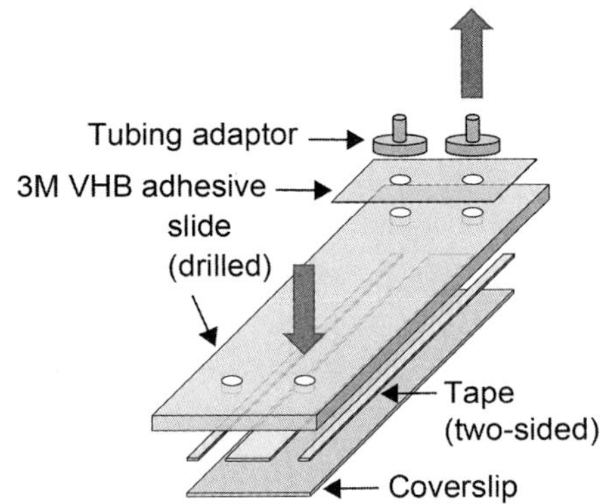

Solution	**Volume (μL)**	**Rate (μL/min)**	**Wait (min)**
Biotin BSA	25	50	15
1xBRB80	100	50	0
Avidin	75	50	3
1xBRB80	100	50	0
Seeds	25	50	3
GB1x	100	50	3
GB1x+tub.	50	50	10
BTAX+PIs	200	50	0
KTB mix	50	10	5
Rxn buff	50	25	0

Figure 18.4 (Diagram) Illustration of a flow chamber for syringe-pump-driven buffer exchange. (Table) Flow protocol for creating a chamber with kinetochore-beads bound to surface-immobilized microtubules. (For the color version of this figure, the reader is referred to the online version of this chapter.)

paclitaxel), BTAX + PIs (50% 2× BTAX + PIs in H_2O), GB1× + tubulin (50% GB2× + 18 μ*M* tubulin in H_2O; reserve tubulin until just before use), KTB mix (50% 2× BTAX + PIs + 20% KTB in H_2O; KTB will be added after incubation), and reaction buffer (50% 2× BTAX + PIs in H_2O + 0.8 m*M* DTT + 50 m*M* glucose + 0.5 mg/mL glucose oxidase + 60 μg/mL catalase + 2 m*M* ATP; reserve glucose oxidase, catalase, and ATP until just before use). During the KTB incubation, construct the slide (Fig. 18.4, table). The GB1× + tubulin polymerizes nonbiotinylated extensions from the seeds, and the BTAX + PIs stops the polymerization and stabilizes the extensions. Kinetochore-beads (KTB mix) are then introduced at low velocity so that they can bind the extensions, and finally, the free kinetochore-beads are rinsed away with reaction buffer containing ATP, which activates the kinase. Use a razor blade to cut the tubing away from the tubing adaptor to avoid disturbing the chamber. Take the sample to the optical trap. There should be some beads attached to microtubules "wagging" just off the coverslip surface. The strength of those attachments can be tested in a manner similar to that in the previous assay, but if kinase activity has a strong disruptive effect on the attachments, a lower initial load may be required.

4. SUMMARY

Kinetochores form load-bearing attachments with microtubules to link chromosomes to microtubules during cell division. Mechanoregulation of these attachments is very difficult to study in living cells since load on the attachments cannot be precisely controlled or measured. As a solution, we have recently shown that kinetochores can be purified from yeast by immunoprecipitation using an epitope tag on the Dsn1 protein. This one-step strategy gives high yield sufficient for quantitative biophysical assays. The bead-binding assay measures the propensity of kinetochores to form attachments, while the rupture force assay measures their strength. Kinase disruption of these attachments can be studied using (1) genetic manipulation to introduce phosphorylation-mimicking aspartic acid mutations at kinase target sites and (2) by phosphorylation *in vitro* by active kinase. These strategies provide the most direct test of the effects of phosphorylation at specific sites by specific kinases and do so with instrumentation and expertise that is becoming increasingly accessible to biochemists. It should be possible to adapt these assays to study a wide range of protein-binding interactions regulated by enzyme activity.

ACKNOWLEDGMENTS

We would like to thank Nitobe London for his input on the manuscript and for his assistance in providing purified kinetochores for the development of the assays described. This work was supported by NIH Grants RO1GM064386 and RO1GM078079 to S. B., NIH Grants RO1GM079373 and S10RR026406 to C. L. A., and a Packard Fellowship to C. L. A. J. W. D. is supported by a Fellow Award from the Leukemia and Lymphoma Society.

REFERENCES

Aitchison, J. D., Blobel, G., & Rout, M. P. (1996). Kap104p: A karyopherin involved in the nuclear transport of messenger RNA binding proteins. *Science, 25*, 624–627.

Akiyoshi, B., Nelson, C. R., Ranish, J. A., & Biggins, S. (2009). Quantitative proteomic analysis of purified yeast kinetochores identifies a PP1 regulatory subunit. *Genes and Development, 23*, 2778–2899.

Akiyoshi, B., Sarangapani, K. K., Powers, A. F., Nelson, C. R., Reichow, S. L., Arellano-Santoyo, H., et al. (2010). Tension directly stabilizes reconstituted kinetochore-microtubule attachments. *Nature, 468*(7323), 576–579.

Bangs Laboratories, Inc. (2002). Tech note 205: Covalent coupling, http://www.bangslabs.com/support/technotes.

Biggins, S., Severin, F., Bhalla, N., Sassoon, I., Hyman, A. A., & Murray, A. W. (1999). The conserved protein kinase Ipl1 regulates microtubule binding to kinetochores in budding yeast. *Genes and Development, 13*(5), 532–544.

Block, S. M. (1990). Optical tweezers: A new tool for biophysics. In S. Grinstein & K. Foskett (Eds.), *Modern biology series: 9. Noninvasive techniques in cell biology*. (pp. 375–401). NewYork: Wiley-Liss.

Cheeseman, I. M., Anderson, S., Jwa, M., Green, E. M., Kang, J., Yates, J. R., et al. (2002). Phospho-regulation of kinetochore-microtubule attachments by the Aurora kinase Ipl1p. *Cell, 111*(2), 163–172.

Corbett, K. D., & Harrison, S. C. (2012). Molecular architecture of the yeast monopolin complex. *Cell Reports, 1*(6), 583–589.

Corbett, K. D., Yip, C. K., Ee, L. S., Walz, T., Amon, A., & Harrison, S. C. (2010). The monopolin complex crosslinks kinetochore components to regulate chromosome-microtubule attachments. *Cell, 142*(4), 556–567.

Dumont, S., Salmon, E. D., & Mitchison, T. J. (2012). Deformations within moving kinetochores reveal different sites of active and passive force generation. *Science, 337*, 355–358.

Evans, E. (2001). Probing the relation between force, lifetime, and chemistry in single molecular bonds. *Annual Reviews of Biophysics and Biomolecular Structure, 30*, 105–128.

Frank, A. D., Powers, A. F., Gestaut, D. R., Davis, T. N., & Asbury, C. L. (2010). Direct physical study of kinetochore-microtubule interactions by reconstitution and interrogation with an optical force clamp. *Methods, 51*(2), 242–250.

Ganem, N. J., Godinho, S. A., & Pellman, D. (2009). A mechanism linking extra centrosomes to chromosomal instability. *Nature, 460*, 278–282.

Gestaut, D. R., Cooper, J., Asbury, C. L., Davis, T. N., & Wordeman, L. (2010). Reconstitution and analysis of kinetochore subcomplexes. *Methods in Cell Biology, 95*, 641–656.

Hauf, S., Cole, R. W., LaTerra, S., Zimmer, C., Schnapp, G., Walter, R., et al. (2003). The small molecule Heperadin reveals a role for Aurora B in correcting kinetochore-microtubule attachment and in maintaining the spindle assembly checkpoint. *The Journal of Cell Biology, 161*(2), 281–294.

Hermanson, G. T. (1996). *Bioconjugate techniques*. San Diego: Academic Press.

Jones, M., Huneycutt, B., Pearson, C., Zhang, C., Morgan, G., Shokat, K., et al. (2005). Chemical genetics reveals a role for Mps1 kinase in kinetochore attachment during mitosis. *Current Biology*, *15*(2), 160–165.

Lampson, M. A., Renduchitala, K., Khodjakov, A., & Kapoor, T. M. (2004). Correcting improper chromosome-spindle attachments during cell division. *Nature Cell Biology*, *6*(3), 232–237.

London, N., Ceto, S., Ranish, J. A., & Biggins, S. (2012). Phosphoregulation of Spc105 and Mps1 and PP1 regulates Bub1 localization to kinetochores. *Current Biology*, *22*(10), 900–906.

Maresca, T. J., & Salmon, E. D. (2010). Welcome to a new kind of tension: Translating kinetochore mechanics into a wait-anaphase signal. *Journal of Cell Science*, *123*(6), 825–835.

Maure, J.-F., Kitamura, E., & Tanaka, T. U. (2007). Mps1 kinase promotes sister-kinetochore bi-orientation by a tension-dependent mechanism. *Current Biology*, *17*(24), 2175–2182.

Nicklas, R. B., & Koch, C. A. (1969). Chromosome micromanipulation III. Spindle fiber tension and the reorientation of mal-oriented chromosomes. *The Journal of Cell Biology*, *43*(1), 40–50.

Nicklas, R. B., & Ward, S. C. (1994). Elements of error correction in mitosis: Microtubule capture, release, and tension. *The Journal of Cell Biology*, *126*(5), 1241–1253.

Pinsky, B. A., Kung, C., Shokat, K. M., & Biggins, S. (2006). The Ipl1-aurora protein kinase activates the spindle checkpoint by creating unattached kinetochores. *Nature Cell Biology*, *8*(1), 78–83.

Rajagopalan, H., & Lengauer, C. (2004). Aneuploidy and cancer. *Nature*, *432*, 338–341.

Rice, S., Lin, A. W., Safer, D., Hart, C. L., Naber, N., Carragher, B. O., et al. (1999). A structural change in the kinesin motor protein that drives motility. *Nature*, *402*(6763), 778–784.

Sarangapani, K. K., Akiyoshi, B., Duggan, N. M., Biggins, S., & Asbury, C. L. (2013). Phosphoregulation promotes release of kinetochores from dynamic microtubules via multiple mechanisms. *Proceedings of the National Academy of Sciences of the United States of America*, *100*(18), 7282–7287.

Sassoon, I., Severin, F. F., Andrews, P. D., Taba, M.-R., Kaplan, K. B., Ashford, A. J., et al. (1999). Regulation of Saccharomyces cerevisiae kinetochores by the type 1 phosphatase Glc7p. *Genes and Development*, *13*(5), 545–555.

Schultz, M. C., Hockman, D. J., Harkness, T. A. A., Garinther, W. I., & Altheim, B. A. (1997). Chromatin assembly in a yeast whole-cell extract. *Proceedings of the National Academy of Sciences of the United States of America*, *94*(17), 9034–9039.

Schutz, A. R., & Winey, M. (1998). New alleles of the yeast Mps1 gene reveal multiple requirements in spindle pole body duplication. *Molecular Biology of the Cell*, *9*(4), 759–774.

Tanaka, T. U., Rachidi, N., Janke, C., Pereira, G., Galova, M., Schiebel, E., et al. (2002). Evidence that the Ipl1-Sli15 (Aurora kinase-INCENP) complex promotes chromosome bi-orientation by altering kinetochore-spindle pole connections. *Cell*, *108*(3), 317–329.

Tighe, A., Staples, O., & Taylor, S. (2008). Mps1 kinase activity restrains anaphase during an unperturbed mitosis and target Mad2 to kinetochores. *The Journal of Cell Biology*, *181*(6), 893–901.

Vale, R. D., Reese, T. S., & Sheetz, M. P. (1985). Identification of a novel force generating protein, kinesin, involved in microtubule-based motility. *Cell*, *42*, 39–50.

Visscher, K., & Block, S. M. (1998). Versatile optical traps with feedback control. *Methods in Enzymology*, *298*, 460–489.

Winey, M., Goetsch, L., Baum, P., & Byers, B. (1991). Mps1 and Mps2: Novel yeast genes defining distinct steps of spindle pole body duplication. *The Journal of Cell Biology*, *114*, 745–754.

Yin, H., Wang, M. D., Svoboda, K., Landick, R., Block, S. M., & Gelles, J. (1995). Transcription against an applied force. *Science*, *270*(5242), 1653–1657.

CHAPTER NINETEEN

Micropattern-Guided Assembly of Overlapping Pairs of Dynamic Microtubules

Franck J. Fourniol[*,1], **Tai-De Li**[†,1], **Peter Bieling**[†,‡,1], **R. Dyche Mullins**[‡,§], **Daniel A. Fletcher**[†,2], **Thomas Surrey**[*,2]

[*]London Research Institute, Cancer Research UK, London, United Kingdom
[†]Department of Bioengineering and Biophysics Group, University of California-Berkeley, Berkeley, California, USA
[‡]Department of Cellular and Molecular Pharmacology, University of California-San Francisco, San Francisco, California, USA
[§]Howard Hughes Medical Institute, Chevy Chase, Maryland, USA
[1]These authors contributed equally
[2]Corresponding authors: e-mail address: fletch@berkeley.edu; thomas.surrey@cancer.org.uk

Contents

Abstract

Interactions between antiparallel microtubules are essential for the organization of spindles in dividing cells. The ability to form immobilized antiparallel microtubule pairs *in vitro*, combined with the ability to image them via TIRF microscopy, permits detailed biochemical characterization of microtubule cross-linking proteins and their effects on microtubule dynamics. Here, we describe methods for chemical micropatterning of microtubule seeds on glass surfaces in configurations that specifically promote the

Methods in Enzymology, Volume 540
ISSN 0076-6879
http://dx.doi.org/10.1016/B978-0-12-397924-7.00019-4

formation of antiparallel microtubule overlaps *in vitro*. We demonstrate that this assay is especially well suited for reconstitution of minimal midzone overlaps stabilized by the antiparallel microtubule cross-linking protein PRC1 and its binding partners. The micropatterning method is suitable for use with a broad range of proteins, and the assay is generally applicable to any microtubule cross-linking protein.

1. INTRODUCTION

Microtubule-cross-linking proteins are important for microtubule organization in living cells. During cell division, cross-linkers help to connect the two halves of the spindle apparatus by mediating antiparallel microtubule contacts (Duellberg, Fourniol, Maurer, Roostalu, & Surrey, 2013; Glotzer, 2009). A crucial cross-linker in anaphase is PRC1 (protein required for cytokinesis 1) (Janson et al., 2007; Mollinari et al., 2002; Verni et al., 2004). PRC1, a homodimeric molecule with microtubule-binding sites at opposite ends of the dimer (Subramanian, Ti, Tan, Darst, & Kapoor, 2013), cross-links antiparallel microtubules with considerably higher affinity than parallel microtubules (Bieling, Telley, & Surrey, 2010; Gaillard et al., 2008; Janson et al., 2007; Subramanian et al., 2010). During anaphase, it localizes selectively to the central spindle where it contributes to spindle stability and recruits several other proteins with critical functions for the central spindle (Duellberg et al., 2013). The PRC1 homologs in plants also play important roles in microtubule bundling in interphase (Gaillard et al., 2008).

Microtubule cross-linking activity has often been assayed *in vitro* by mixing purified cross-linkers and microtubules to form microtubule bundles. The large and variable number of microtubules and their mixed orientation can make it difficult to determine the properties of the proteins bound to the bundle. Therefore, *in vitro* experiments with pairs of microtubules with known orientation have been developed. For the study of microtubule-cross-linking and sliding motors (such as kinesin-5 or kinesin-14), microtubule pairs are formed from two stabilized microtubules, one of which is surface immobilized and the other of which is tethered to the immobilized microtubule by the cross-linking motors, either in the absence or presence of other cross-linkers (Braun et al., 2011; Hentrich & Surrey, 2010; Kapitein et al., 2005; Roostalu et al., 2011; van den Wildenberg et al., 2008). The goal of this assay is to determine how motors slide two microtubules with respect to each other.

A complementary assay was designed to form microtubule pairs that exhibit dynamic polymerization and depolymerization behavior (Bieling et al., 2010). This assay is a variation of a commonly used microtubule dynamics assay in which short, stabilized microtubules are bound to a glass surface and then extended by the addition of free tubulin (Telley, Bieling, & Surrey, 2011). When two growing microtubules oppose each other in this assay, antiparallel encounters lead to the formation of antiparallel microtubule overlaps that can be used to study PRC1 binding, PRC1-dependent recruitment of other proteins to these overlaps, and their effect on the dynamic properties of the microtubules themselves. This assay has been used to study the combined effects of Xenopus PRC1 and kinesin-4 Xklp1 on setting the length of antiparallel microtubule overlaps (Bieling et al., 2010; Nunes Bastos et al., 2013).

A technical challenge in this type of assay is how to orient microtubules and control the density of immobilized seeds so that the chance of antiparallel overlap is relatively high. Recent developments in techniques for micropatterning glass surfaces now enable more spatially controlled microtubule nucleation or seed immobilization (Aoyama, Shimoike, & Hiratsuka, 2013; Ghosh, Hentrich, & Surrey, 2013; Portran, Gaillard, Vantard, & Thery, 2013; Waichman, You, Beutel, Bhagawati, & Piehler, 2011). Patterning enables the growth of microtubules from distinct foci with well-defined positions and dimensions, and it has recently been used to reconstitute bipolar microtubule bundles (consisting of several microtubules) *in vitro* (Portran et al., 2013; Su et al., 2013).

We describe here a high contrast micropatterning method and demonstrate its use to chemically micropattern microtubule seeds on glass surfaces to guide formation of antiparallel microtubule pairs with defined seed-to-seed distance (Fig. 19.1). We produce micropatterns of maleimide functionalization on polyethylene glycol (PEG) brushes covalently linked to glass (Waichman et al., 2011). Maleimide is then used to covalently link either thiol-biotin or cystein-tagged streptavidin to the maleimide-functionalized areas, generating biotin-PEG or streptavidin-PEG micropatterned glass. Both methods achieve selective immobilization of biotinylated microtubule seeds via a Cys-streptavidin or a biotin–neutravidin sandwich. In combination with using fluid flow for seed orientation, this method allows the generation of pairwise antiparallel microtubule overlaps with controlled seed-to-seed distance.

2. REAGENTS AND EQUIPMENT

The rationale and practical details for TIRF microscopy have been described elsewhere (Gell et al., 2010). Here, we focus on the glass

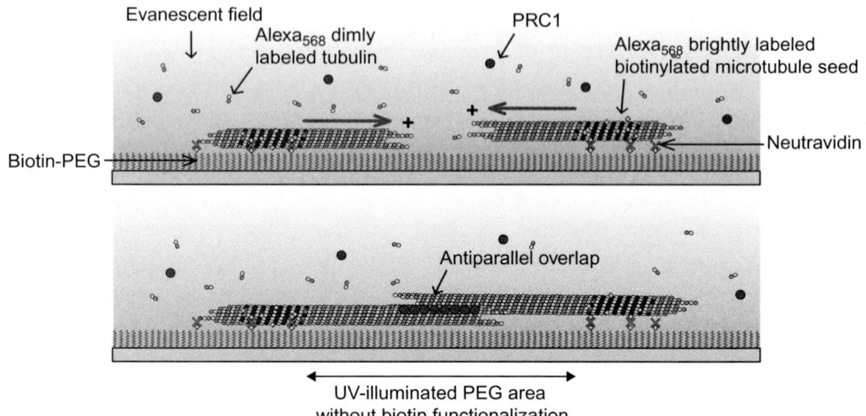

Figure 19.1 Schematic overview of the dynamic antiparallel microtubule assay on micropatterned coverslips. Brightly labeled microtubule seeds are immobilized on the functionalized areas of the coverslip surface by a biotin–neutravidin sandwich (or alternatively by a streptavidin-Cys directly coupled to the glass—not shown in this scheme) and are visualized using TIRF microscopy. From these seeds, dynamic microtubule extensions are grown with dimly labeled tubulin. When antiparallel microtubules meet, PRC1 accumulates in the overlap region. *(Adapted from Bieling et al. (2010)).* (See the color plate.)

treatment, patterning process, and the sample preparation for the dynamic microtubule overlap assay.

2.1. Reagents for glass treatment

- NaOH (3 *M* solution).
- Hydrogen peroxide (30% stabilized).
- Sulfuric acid (Sigma; concentrated, 95–97%).
- (3-Glycidyloxypropyl)-trimethoxysilane (GOPTS; Sigma #440167-100ML).
- H_2N-PEG-NH_2 2000 (DAPEG; Rapp-Polymere #11 2000-2).
- *N*-[β-maleimidopropyloxy] succinimide ester (BMPS; Thermo Scientific, 50 mg #22298).
- Acetone (Sigma, Chromasolv Plus).
- Dimethylformamide (DMF, Sigma, Chromasolv).
- Dimethylsulfoxide (DMSO, Sigma, ACS grade).
- Coverslips (Zeiss, High Precision, 22 × 22 mm, 170 μm thick, No. 1.5H).
- Microscope slides (Thermo Scientific, 76 × 26 mm).
- Poly-L-lysin-PEG (PLL-PEG; SuSoS, PLL-g-PEG).

- Double-sided tape (Tesa).
- Inverted tweezers (Dumont, N2A Inox.).
- Coverslip washing racks (ceramic or Teflon).
- Diamond pen.
- Kimwipes (lint free).
- Coverslip spinner or nitrogen gas (for drying).
- Oven (75 °C).
- Hamilton syringe #710 (100 μL). Used for GOPTS exclusively.
- Hamilton syringe #1001 (1 mL). Used for acetone and DMF.
- Beakers 1 L, 600 mL, 100 mL.
- Closed weighing jars.
- Disposable culture tubes, borosilicate glass grade 3.3 (VWR #212-0028).
- Kimtech gloves grade 3 (residue free, #99237).

  Note: all glassware should be cleaned with 3 *M* NaOH.

2.2. Material and reagents for patterning and functionalization

- 0.5 mL desalting columns (Thermo Scientific, Zebaspin 7K MWCO).
- Carbon-fiber tipped tweezers (Agar Scientific).
- Dry seal desiccator (Wheaton #02-913-360).
- Quartz masks with positive chromium stripe pattern (NB Technologies). Custom design, with stripes spaced 30, 20, or 10 μm apart and 2, 5, or 10 μm wide.
- Newport Solar Simulator with 300 W xenon arc lamp.

  Note 1: Make sure all optical components transmit UV wavelengths down to 300 nm, as these short wavelengths are most efficient for photodestruction of maleimide groups.

  Note 2: A 150-W lamp can be used as an alternative to the 300 W lamp, though illumination time must be doubled.
- Ethanol (Sigma, Chromasolv).
- Ammonium hydroxide (Sigma, ACS grade).
- Microscope slides (Fisher Scientific) with two plasma-bonded 3 mm, wide PDMS stripes aligned along the long axis edges. PDMS stripes are manually cut from a thin PDMS sheet (SSP-M823, Specialty Silicone Products).
- Vacuum-sealed illumination chamber made with quartz microscope slide, acrylic plates, sponge gasket, UV glue, and acrylic cement (Fig. 19.2A).

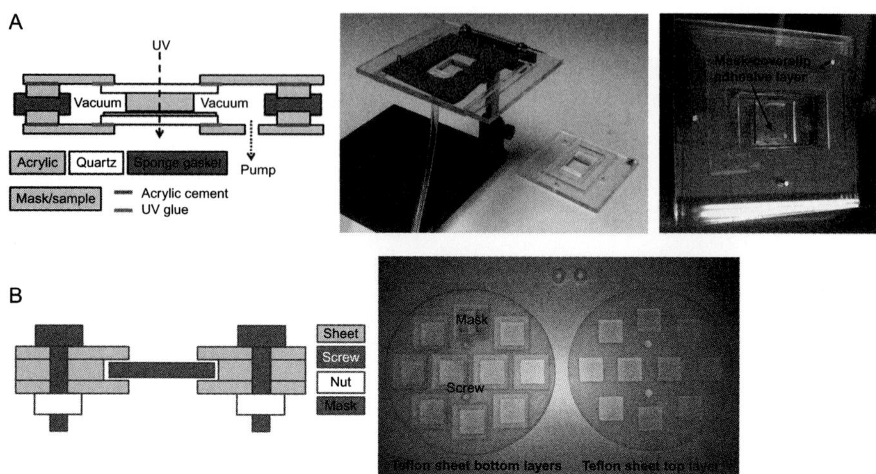

Figure 19.2 Custom-built sample preparation devices for UV micropatterning. (A) Schematic diagram and photographs of the UV illumination vacuum-sealed chamber; Photos: left: empty chamber with lid off; right: sample-loaded and vacuum-sealed chamber. Dark areas seen at the coverslip-mask interface in the loaded chamber correspond to regions of tight adhesion. (B) Schematic diagram and photo of the Teflon holder for piranha cleaning of the masks. Photo: *quartz and chromium mask loaded in the open Teflon holder.* (For color version of this figure, the reader is referred to the online version of this chapter.)

- Teflon holder for masks. Custom made from Teflon sheets, screws, and nuts (Fig. 19.2B).
- Vacuum pump (2511B-01, Gardner Denver Welch Vacuum Technology).
- Humidifier.

  Note: Alternatively, a hot water beaker and air pump (Aquarium 212, Petco) can be used.
- Sonicator bath with heating function.
- Humid storage container for slide incubation (e.g., an empty pipet tip container filled with MilliQ (MQ) water at the bottom, sealed with parafilm).
- *Maleimide coupling buffer*: 20 m*M* HEPES pH 7.4, 300 m*M* NaCl, 1 m*M* EDTA.
- Thiol-biotin (Prochimia, #FT 005-0.2).

2.3. Proteins

- Pig brain tubulin in BRB80, purified as described (Castoldi & Popov, 2003), flash-frozen, and stored in single-use aliquots at a concentration of 150–200 μ*M* at liquid nitrogen temperature.

- Biotinylated-tubulin in BRB80, labeled as described (Hyman et al., 1991), stored in 2 μL aliquots at 150 μ*M*, in liquid nitrogen.
- Alexa568- and/or Alexa647-tubulin, labeled as described (Hyman et al., 1991), stored in 2 μL aliquots at 150 μ*M*, in liquid nitrogen.
- PRC1 from *Xenopus laevis* (Bieling et al., 2010) stored in 2 μL aliquots at 18 μ*M*, in liquid nitrogen.
- Streptavidin-Cys. We introduced a single cysteine by site-directed mutagenesis into a previously described construct (Sørensen, Sperling-Petersen, & Mortensen, 2003), containing domain I of the *Escherichia coli* translation initiation factor 2 (IF2) as an N-terminal solubility tag. The IF2(I)-streptavidin-His6-Cys protein was expressed in BL21 pRIL cells overnight at 18 °C after induction with 0.5 m*M* IPTG. Bacterial pellets were lysed in 50 m*M* Tris–HCl pH 8.0, 150 m*M* NaCl, 4 m*M* imidazole, 1 m*M* EDTA, 1 m*M* $MgCl_2$, 0.2 m*M* PMSF, 10 m*M* β-mercaptoethanol, supplemented with Complete Protease Inhibitors and DNase I (Roche). The lysate was ultracentrifuged 30 min at 50,000 × *g*, at 4 °C, and the clear lysate then incubated with 2 g protino Ni-TED resin (Macherey-Nagel). The resin was loaded on a gravity column and washed with 80 mL wash buffer (50 m*M* Tris–HCl pH 8.0, 150 m*M* NaCl, 4 m*M* imidazole, 1 m*M* EDTA, 10 m*M* β-mercaptoethanol), before elution with 50 m*M* Tris–HCl pH 8.0, 150 m*M* NaCl, 500 m*M* imidazole, 1 m*M* EDTA, 10 m*M* β-mercaptoethanol. The imidazole was removed using desalting PD10 columns (GE Healthcare) equilibrated with 50 m*M* Tris–HCl pH 8.0, 150 m*M* NaCl, 1 m*M* EDTA, 10 m*M* β-mercaptoethanol. The protein was aliquoted (50 μL, 72 μ*M* monomeric concentration) and flash-frozen in liquid nitrogen.
- Constructs were made for the expression of wild-type mCherry (Shaner et al., 2004) and a nonfluorescent variant (Tyr72Ser) fused to an N-terminal His10-Lys-Cys-Lys-tag using a modified pETMz vector. The protein was expressed in Rosetta cells overnight at 18 °C and purified by IMAC over a 5-mL HiTrap Chelating column (GE Healthcare) followed by TEV cleavage, desalting over a HiPrep XK26/10 Desalting column (GE Healthcare) to remove imidazole and a second round of IMAC to remove the z-tag. The eluate after the second round of IMAC containing His10- and Cys-tagged mCherry was gel filtered over a Superdex 200 XK26/60 PG column (GE Healthcare). The protein was flash-frozen in aliquots in liquid nitrogen.

  Note: In contrast to most other fluorescent proteins, mCherry does not contain a cysteine in its wild-type sequence.

2.4. Microtubule polymerization and overlap assay

- BRB80 buffer: 80 m*M* K-PIPES pH 6.8 at RT, 1 m*M* $MgCl_2$, 1 m*M* EGTA. Kept maximum 6 weeks at 4 °C.
- Guanosine-5′-[(α,β)-methylene]triphosphate (GMPCCP; Jena Biosciences). 10 m*M*, 5 μL aliquots stored in liquid nitrogen.
- Guanosine-5′-triphosphate (GTP; Sigma). 200 m*M*, 10 μL aliquots stored in liquid nitrogen.
- Kappa casein (Sigma, #C0406-100MG). 5 mg/mL in BRB80, 90 μL aliquots stored in liquid nitrogen. Before use, ultracentrifuge at 80,000 rpm for 10 min, at 4 °C.
- Glucose 20% stock solution, kept at 4 °C.
- Glucose oxidase (Serva, #22778.01). 40 mg/mL, 30 μL aliquots stored in liquid nitrogen. Before use, ultracentrifuge at 279,000 × *g* for 10 min, at 4 °C.
- Pluronic F-127, 5% solution (in water), kept at 4 °C.
- Catalase (Sigma, #C40-100MG). 6.9 mg/mL, 30 μL aliquots stored in liquid nitrogen. Before use, ultracentrifuge at 279,000 × *g* for 10 min, at 4 °C.
- *Midzone buffer*: 80 m*M* K-PIPES pH 6.8 at RT, 85 m*M* KCl, 85 m*M* K-acetate, 4 m*M* $MgCl_2$, 1 m*M* EGTA, 0.25% (v/v) Brij-35. Kept for maximum 6 weeks at 4 °C.
- Methylcellulose 2% solution (viscosity 4,000 cP; Sigma). Kept maximum about 6 weeks at 4 °C.

3. GLASS TREATMENT AND SURFACE CHEMISTRY

To generate a micropatterned streptavidin or neutravidin surface for the patterned immobilization of biotinylated microtubule seeds, we first generate a maleimide-PEG micropattern on glass. Diamino-PEG is covalently coupled to glass using silane chemistry followed by coupling maleimide to the PEG. UV-irradiation through a photolithographic mask is used to destroy the maleimide locally in the irradiated areas, generating a micropattern of areas with thiol-reactive maleimide-PEG separated by PEG areas. We describe two methods that can be used to produce either neutravidin-PEG or streptavidin-PEG micropatterns (Fig. 19.3). In the first method, maleimide is allowed to react with thiol-biotin, producing a biotin-PEG pattern to which commercially available neutravidin is bound. In the second method, a cysteine-tagged recombinant streptavidin is allowed to react with the maleimide, producing a streptavidin-PEG pattern. Both

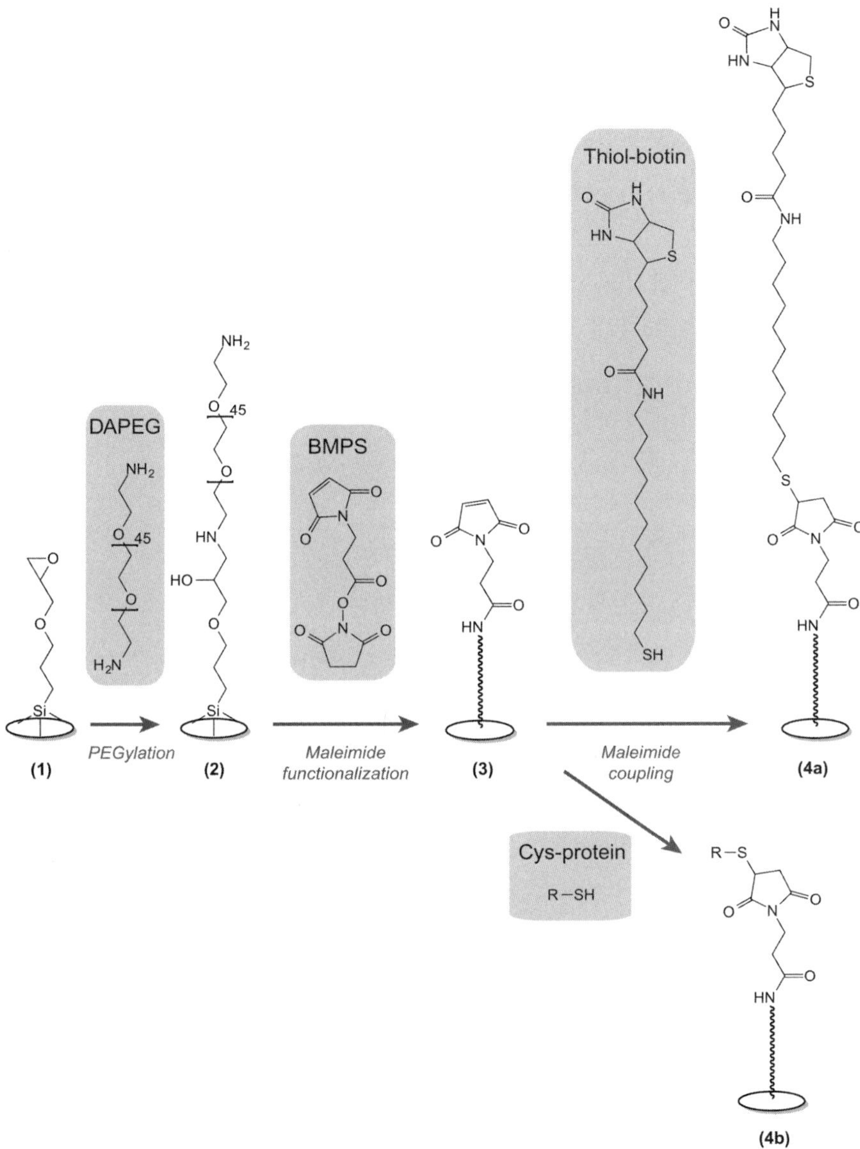

Figure 19.3 Surface chemistry for coverslip functionalization. The glass is silanized (1), PEGylated (2), and reacted with the maleimide derivative BMPS (3). At that stage, the coverslip, functionalized with a uniform layer of maleimide-PEG, can be patterned using a UV source and a quartz/chromium mask. UV illumination destroys the maleimide groups. In contrast, the maleimide-PEG groups remain intact in the photoprotected areas and will react with thiols (here: thiol-biotin (4a), or cysteine-containing proteins (4b)) to generate covalently linked biotin-PEG or protein-PEG chains. (For color version of this figure, the reader is referred to the online version of this chapter.)

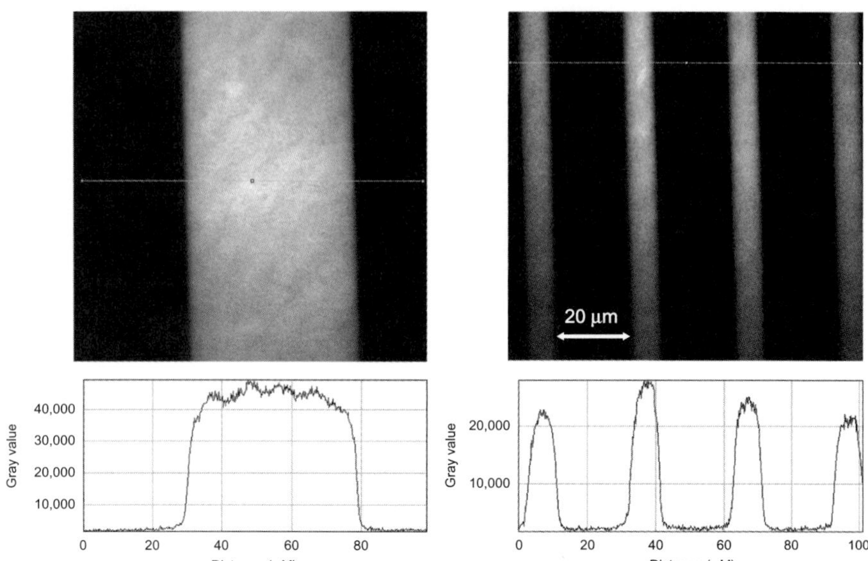

Figure 19.4 High contrast micropatterning of cross-linked Cys-mCherry. Micropatterned coverslips coupled to a cysteine-containing mCherry and imaged by TIRF microscopy. Two fields of view and fluorescent intensity line profiles are shown. Comparison of fluorescence intensity of the photoprotected and illuminated areas reveals the highly selective patterning of proteins that can be achieved with this method.

methods produce high contrast patterns (see Fig. 19.4) of biotinylated microtubule seeds in geometries defined by the photolithographic mask. The highly specific interactions between the protein and surface-bound PEG result in very good control of seed binding. The uniform PEG coating of the glass surface greatly reduces non-specific protein adsorption, thus limiting protein depletion and improving the imaging quality. To ensure that protein was not depleted by other parts of the chamber, the unpatterned glass surface opposite of the patterned surface was passivated using poly-L-lysine-PEG (PLL-PEG). Careful attention to surface passivation is important for the correct quantitative determination of biochemical parameters from the analysis of fluorescence signals.

3.1. Coverslip treatment and functionalization

3.1.1 PEG-maleimide coating

Glass cleaning

- Using a diamond pen, mark each coverslip with an asymmetric triangular cut in the top right corner. Arrange the coverslips in racks and place the racks in a large beaker.

Note: We typically process 36 coverslips at a time.

- Cover the coverslips with 3 *M* NaOH and sonicate for 15 min. Rinse them with plenty of MQ—purified water. At this stage, rinse the weighing jars with 3 *M* NaOH and MQ water and dry them off in the oven.

  Note: The coverslips can stand in NaOH several hours.
- Drain the coverslips and, under a fume hood, gently pour in two parts of hydrogen peroxide and three parts of sulfuric acid ("piranha" solution). Sonicate the glass in piranha for 30 min.

  Note: This piranha solution is extremely corrosive and volatile—it is essential to keep it under the hood; wear a labcoat and protective goggles. Discard the piranha in the appropriate acid waste (do not screw the cap tightly as gas accumulation might result in an explosion).
- Rinse the glass in two successive baths of 1 L MQ water.

Silanization

- Spin-dry the coverslips and lay them on dust-free Kimwipes. Place half of them functional side up in weighing jars.
- Take GOPTS out with the 100 μL Hamilton syringe. Put 2.5 drops of GOPTS onto each slide. Form a sandwich by putting another slide face-to-face onto the GOPTS. Repeat until all coverslips are sandwiched. Incubate the closed weighing jars 30 min at 75 °C.

  Note: Once opened, a bottle of GOPTS can be kept for years at RT; however, the concentration of unreacted molecules will slowly decrease and you may have to adjust the incubation time accordingly, in order to get similar surface densities of silane groups.
- Get the weighing jars out of the oven and let them cool down for 15 min on the bench.
- Open the jars and quickly transfer the closed coverslip sandwiches into a 100-mL beaker filled with dry acetone (80 mL) and a Teflon rack. Separate the coverslips in acetone. Dip-wash the rack into successive acetone baths (80, 400, and 200 mL in a 1-L beaker). Repeat with as many racks as necessary to hold the coverslips.
- Spin-dry the acetone off from each coverslip, as quickly as possible. Store them functionalized side up on Kimwipes.

  Note: The GOPTS reaction with the PEG is the most sensitive step of the procedure. GOPTS is extremely reactive, and it has to be kept water-free at all times to remain active. Use a freshly opened bottle of acetone. Dry the coverslips quickly; otherwise, acetone evaporation will cool down the glass and cause water condensation.

PEGylation

- Prepare a DAPEG mix in acetone in a glass tube. Use 150 mg of PEG in 500 μL acetone. Vortex at medium speed for 2 min, while warming up the tube between your fingers (PEG should dissolve fully if vortexed long enough).
- Arrange half of the coverslips functionalized side up into clean weighing jars.
- Pipet 25 μL of the DAPEG solution on each coverslip and form sandwiches. Repeat until all coverslips are sandwiched.
- Close the jars and incubate for 4 h.

  Note: Longer incubation can result in oxidation of the PEG (brown material), which ultimately will reduce the surface density of functional groups.
- On a hot plate, separate the sandwiches using a razor blade and tweezers. Dip-wash each coverslip in two successive 1-L beakers of MQ water and store in racks in another beaker with MQ water. Sonicate for 3 min. Spin-dry each coverslip and store in a box in between layers of lens paper, at 4–8 °C.

  Note: This is a possible break point as the DAPEG-functionalized coverslips are chemically stable for several days/weeks.

Maleimide functionalization

- Arrange the slides in the weighing jars and form a sandwich with 7.5 μL of the 1 *M* β-mercaptoethanol in DMF solution in between two slides.
- Incubate for 10 min at RT.
- In the hood, transfer the slide sandwiches in a fresh beaker containing dry DMF (80 mL) and a Teflon rack. Separate the coverslips in DMF, dip wash the racks in two more baths of DMF (80, 400 mL), and store the slides in racks in a beaker containing dry DMF.

  Note: Make sure to remove β-mercaptoethanol fully. Use fresh gloves, etc.
- Prepare a solution of BMPS in dry DMF (use a fresh 50 mg aliquote, weigh in a glass tube, and add 500 μL of DMF per 50 mg BMPS). Vortex until fully dissolved.
- Spin-dry the slides in the hood and collect them on Kimwipes outside of the hood. Then form a sandwich with 25 μL of the BMPS solution in between two glass pieces in clean weighing jars.
- Incubate for 30 min at RT.
- In the hood, transfer the slide sandwiches in a fresh beaker containing dry DMF and a Teflon rack. Separate the slides in DMF, dip wash once more in DMF, and store the slides in racks in a beaker containing dry DMSO.

- Immediately prior to UV illumination, spin-dry each coverslip to remove DMSO. Dip in EtOH 2 × and spin-dry.

3.1.2 UV micropatterning

Preparation of chrome-on-glass photomasks

- Mount the individual photomasks in the custom-built Teflon holder using carbon-fiber tipped tweezers (all mask handling must be done using these tweezer types to avoid mechanical damage to the mask). The chrome sides of the masks should face the same direction. Put the Teflon holder containing the masks into a 1-L beaker with the chrome layers facing upward.
- Under a fume hood, gently pour in two parts of hydrogen peroxide and three parts of sulfuric acid. Sonicate the masks in piranha for 30 min at RT.
- Carefully decant the piranha solution and wash the Teflon holder containing the masks in the beaker under running MQ water for 5 min. Disassemble the Teflon holder and dip wash the masks individually in a clean 1-L beaker containing MQ water. Individually dry the masks under a clean nitrogen stream and keep them in vacuum in the desiccator.
- Directly before illumination, immerse the individual photomask into a clean beaker containing heated (50 °C), diluted SC1 solution (diluted Standard Clean 1 solution: 1 part ammonium hydroxide (28% solution): four parts hydrogen peroxide (30% solution): 50 parts MQ water). Sonicate for 12 min, wash in a clean beaker under running MQ water for 2 min, dip wash in two subsequent beakers containing EtOH, and dry under a stream of clean nitrogen.

UV illumination

- Place a spin-dried, PEG-maleimide coverslip with the functionalized side up in the custom-built vacuum illumination device.
- Briefly pass a freshly SC1-cleaned quartz photomask through a stream of water vapor generated from a humidifier with the chrome layer facing the vapor. The water vapor should leave a minimum uniform layer of condensed water on the mask.
- Immediately, right before the vapor layer has completely gone, put the mask with the chrome side down on the PEG-maleimide coverslip forming a sandwich.
- Close the custom-built vacuum illumination device by mounting the top assembly and apply vacuum with the vacuum pump. The residual water on the photomask should form a visible adhesive layer between the mask and coverslip surface covering at least 10% of the overall contact

area (see Fig. 19.2A). Failure to form this adhesive layer will result in patterns with less sharp edges and less homogenous ligand densities in the photoprotected area.

Note: The extent of the adhesive layer depends on the cleanliness of the glass and mask, and on the surface properties conferred by the successive cleaning and functionalization steps.

- Mark the center position of the UV beam emitted from the Newport Solar Simulator. Close the illumination shutter and position the vacuum illumination device with the chrome patterns of the photomask in the beam center. Open the illumination shutter and illuminate with UV light for 6 min.

3.1.3 Maleimide coupling

3.1.3.1 Coverslip preparation before coupling

- After UV illumination, store coverslips in a rack covered with DMSO.
- Transfer the rack into EtOH and sonicate for 3 min.
- Transfer into water and sonicate for 3 min.

3.1.3.2 Cys-mCherry and streptavidin-Cys coupling

- Prereduce a sufficient amount (40 μL of a 10 μ*M* solution per individual sample) of Cys-containing protein by adding fresh β-mercaptoethanol to 1 m*M* final concentration. Incubate for 30 min on ice.
- Desalt the protein into maleimide coupling buffer not containing reducing agents.
- Determine protein concentration by UV absorbance at 280 nm. Adjust protein concentration to 10 μ*M* with maleimide coupling buffer.
- Spin-dry coverslips and mount them on the microscope slides with plasma-bonded PDMS stripes with the functionalized slide down. This should form a central flow chamber of approximately 40 μL volume covering the whole patterned area of the coverslip. The adhesion between the PDMS and the coverslip should be strong enough to keep the coverslip in place but sufficiently weak to allow later removal of the coverslip with tweezers.
- Add 40 μL of the 10 μ*M* protein solution to the flow chamber. Incubate for 25 min at RT in a humid storage container.
- Wash with 250 μL maleimide coupling buffer containing 5 m*M* β-mercaptoethanol. Incubate for 5 min to quench the residual maleimide moieties.

- Wash with 250 μL maleimide coupling buffer containing 2 m*M* TCEP. Store in a cooled humid storage container for up to 2 weeks.

  Note: This method can achieve a high contrast of mCherry bound on photoprotected versus UV-illuminated areas (Fig. 19.4).

3.1.3.3 Thiol-biotin coupling

- Spin-dry coverslips and arrange half of them in weighing jar, store the rest functionalized side up on Kimwipes.
- Prepare 6 mL of 1:1 mix of DMF and 50 m*M* HEPES pH 7.4 aqueous solution.
- Weigh the thiol-biotin in a glass tube and make a 5-m*M* suspension in the previous mix (e.g., 5 mg in 2.5 mL). Vortex well. Then by serial dilution prepare a 10 μ*M* solution—the thiol-biotin should fully dissolve.
- Use 25 μL of the 10 μ*M* thiol-biotin solution per coverslip sandwich and incubate for 30 min at RT.
- Arrange the sandwiches and separate the coverslips in a rack covered in EtOH. Transfer the rack into a fresh beaker of EtOH and sonicate for 3 min.
- Spin-dry and store the coverslips in between layers of lens paper, in the fridge.

  Note: The biotin-PEG coverslips can keep for several months.

3.2. PLL-PEG counterglass preparation

- For each glass slide, fix two parallel strips of double-sided tape (15 × 3 mm each, 5 mm apart). Apply pressure and let stand for 5 min to allow bonding.
- Pipet 3 μL PLL-PEG 2 mg/mL aqueous solution and spread in between the strips of tape, by pressing the pipet tip on the glass. Let it dry for 20 min.
- Rinse with plenty of MQ water and let the slides fully dry.

4. MICROTUBULE OVERLAP ASSAY ON MICROPATTERN

In the assay described here, immobilized seeds are aligned perpendicular to patterned functionalized stripes by flow. This increases the likelihood that microtubules extending from these seeds will make head-on encounters so that antiparallel overlaps can form. To promote oriented seed immobilization to the neutravidin or streptavidin micropatterns, long GMPCPP seeds are bound to the surface under strong laminar flow.

4.1. Microtubule seeds preparation

- On ice, prepare a 40-μL mix containing 1 m*M* GMPCPP, 20 μ*M* biotinylated-tubulin (30%), 15 μ*M* Alexa568- or Alexa647-tubulin (25%), and 25 μ*M* unlabelled tubulin, in BRB80.
- Incubate 5 min on ice, to allow nucleotide exchange.
- Transfer to a water bath at 37 °C for 30 min.
- Dilute with 400 μL warm BRB80. Pipet up and down. Spin for 7 min at 17,000 × *g* in a benchtop centrifuge at RT.
- Discard the supernatant. Pipet gently 400 μL warm BRB80 into the tube, making sure not to disturb the pellet (pink dot). Discard the solution and resuspend the pellet in 50 μL warm BRB80—pipet up and down 20 times slowly, using a cut pipet tip to limit the shearing of the seeds.

  Note 1: GMPCPP microtubule seeds keep for 1–2 days at RT (they are cold-sensitive).

  Note 2: Tubulin labeling can affect differently seed nucleation and elongation. We found that Cy5-tubulin tends to cause more nucleation and therefore much shorter seeds than, for example, Alexa647 or Alexa568-tubulin. A low labeling ratio is advisable (e.g., we used a 0.4 ratio Alexa647/tubulin).

  Note 3: You may need to adjust tubulin concentrations given here. Keep in mind that lowering tubulin concentration will result in fewer microtubule nuclei and longer seeds. To double the amount of seeds without changing their length, double the volume of the polymerization mix.

  Note 4: GMPCPP seeds fuse over time. You can take advantage of this to obtain longer seeds.

4.2. Flow chamber assembly and seed attachment

4.2.1 Protein-coupled patterns

- Flow 250 μL water into the glass-PDMS-coverslip flow chamber. Carefully lift the coverslip with tweezers, briefly spin-dry it, and fix it on the counter glass. Flow BRB80 immediately.

  Note: The protein-coupled patterns can tolerate brief dehydration, but it is advisable to be quick during this step.

- Flow 100 μL BRB80 supplemented with glycerol (we used 7–21%).

  Note: Seeds will align in a laminar flow, which can be readily obtained using a viscous buffer.

  Note: When using 21% glycerol, you might want to build a chamber with a double layer of tape to increase its volume and thus facilitate flowing the viscous buffer.
- Prepare 50 μL diluted seeds (e.g., 10 μL seed mastermix in 40 μL BRB80 + 21% glycerol). Mix by pipetting up and down with a cut pipet tip.
- In a P200, pipet 150 μL BRB80. Set the P200 to 200 μL and pipet in the seeds so that the pipet tip contains 50 μL seeds + 150 μL BRB80 wash solution stacked on top of each other.
- Flow steadily the 200 μL through the chamber.

  Note: To flow steadily, one can wick out the solution using a long filter paper strip that is steadily swiped at the outlet of the chamber. One can alternatively use a vacuum pump to suck the liquid out of the chamber at constant speed.
- *Optional step*: flow 50 μL BRB80 + oxygen scavengers (1% glucose, glucose oxidase 0.32 mg/mL, catalase 0.055 mg/mL) to visualize the seeds in the TIRF microscope (Fig. 19.5A).

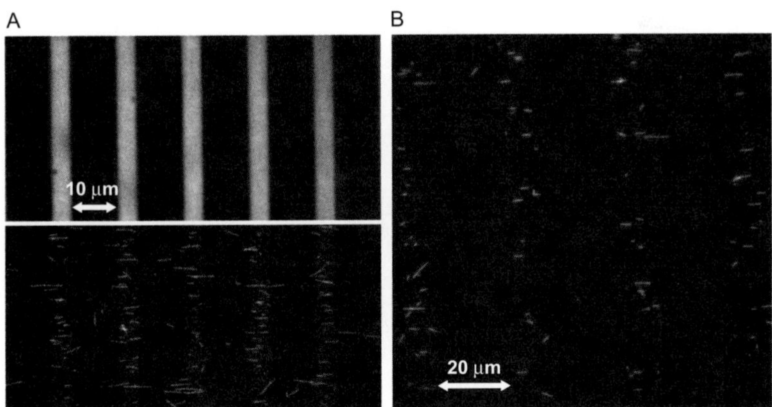

Figure 19.5 Alignment of microtubule seeds on micropatterned surfaces under flow. (A) Alexa647-labeled GMPCPP microtubule seeds containing 30% biotinylated-tubulin in BRB80 + 21% glycerol were perfused into a chamber formed with a patterned coverslip coupled to 500 n*M* Cys-mCherry and 250 n*M* streptavidin-Cys (concentration expressed for the tetramer). (B) Alexa568-labeled GMPCPP microtubule seeds containing 30% biotinylated-tubulin in BRB80 + 7% glycerol were perfused into a chamber formed with a patterned coverslip coupled to thiol-biotin and previously incubated with 50 μg/mL neutravidin.

4.2.2 PEG-biotin pattern

- Using a diamond pen, cut a coverslip into two pieces (22 × 11 mm).
- For each counter glass, fix a coverslip piece across the two stripes of tape so that its functionalized side is inside the chamber. Apply pressure and let stand for 5 min to ensure the coverslip is tightly bound.
- Flow 50 μL pluronic 5% and kappa casein 100 μg/mL in MQ water, at RT. The amphiphilic molecule pluronic and the blocking protein kappa casein will, in addition to the PEG coating, further prevent adsorption of the proteins of interest to the chamber surface.
- Wash with 100 μL kappa casein 50 μg/mL in midzone buffer with 10 m*M* β-mercaptoethanol, on ice.
- Flow 50 μL neutravidin 50 μg/mL and kappa casein 50 μg/mL in midzone buffer with 10 m*M* β-mercaptoethanol, on ice. Incubate for 4 min.

  Note: neutravidin concentration can be varied to achieve different PEG-biotin–neutravidin densities. We found that the maximal density was achieved with 50 μg/mL and incubated for 4 min on ice.
- Wash with 100 μL BRB80 + 7% glycerol.
- Prepare and flow a seed mix as in Section 4.2.1.
- *Optional step*: flow 50 μL BRB80 + oxygen scavengers (1% glucose, glucose oxidase 0.32 mg/mL, catalase 0.055 mg/mL) to visualize the seeds in the TIRF microscope (Fig. 19.5B).

4.3. TIRF microscopy

- For the dynamic assay, prepare the following 50 μL mix on ice: tubulin 17 μ*M* (from a pre-mix of freshly thawed 10 μL unlabelled tubulin and 1 μL Alexa647- or Alexa568-tubulin), and PRC1-SNAP-Alexa647 5 n*M* (concentration expressed for the monomer) in midzone buffer supplemented with 10 m*M* β-mercaptoethanol, 1 m*M* GTP, 0.1% methylcellulose, and oxygen scavengers (1% glucose, glucose oxidase 0.32 mg/mL, catalase 0.055 mg/mL).
- Pre-warm the mix and flow it in a glass chamber loaded with seeds.
- Seal the chamber with nail varnish or Valap (1:1:1 mix of Vaseline, lanolin, and paraffin wax heated to 50 °C).
- Image immediately in the TIRF microscope. To capture the formation of antiparallel overlaps, timelapse movies are typically recorded with a rate of 1 frame per 10 s (Fig. 19.6A and B).

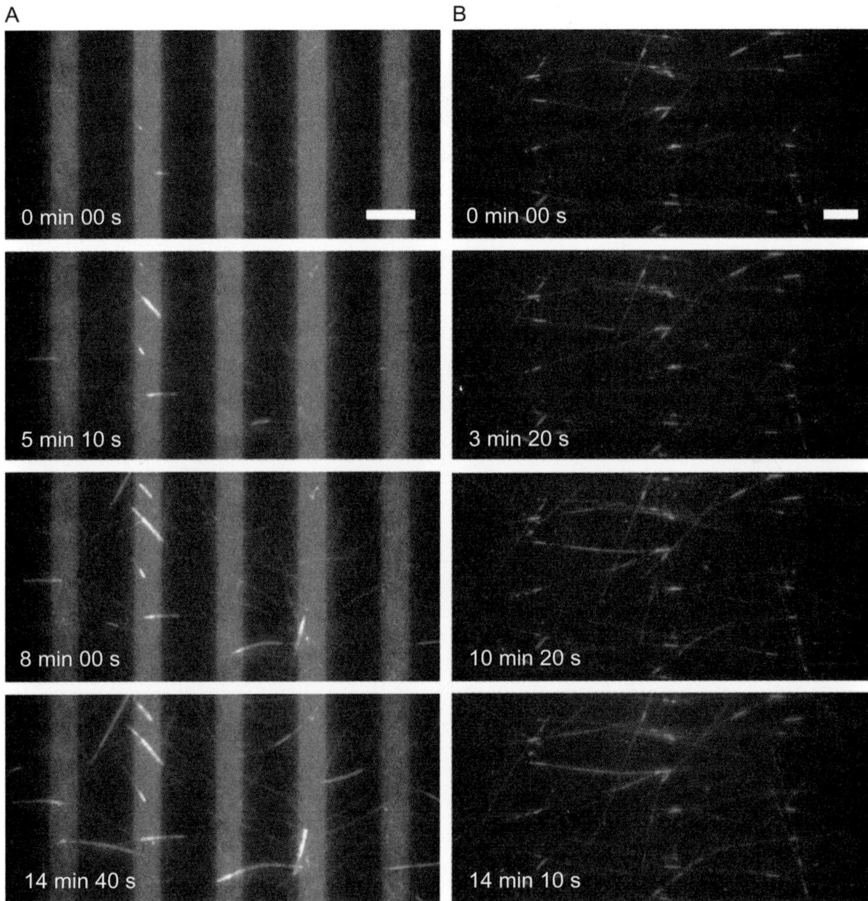

Figure 19.6 TIRF imaging of antiparallel microtubule nucleated from micropatterned surfaces. (A) Time-course of dynamic microtubules grown from Alexa647-labeled microtubule seeds (dim magenta) aligned on Cys-mCherry (green) and streptavidin-Cys patterns, in the presence of 5 n*M* PRC1-SNAP-Alexa647 (bright magenta) and 22.5 μM Alexa568-tubulin (green). Also see Video 1 (http://dx.doi.org/10.1016/B978-0-12-397924-7.00019-4). (B) Time-course of dynamic microtubules grown from Alexa568 brightly labeled microtubule seeds (bright green) aligned on PEG-biotin patterns, in the presence of 5 n*M* PRC1-SNAP-Alexa647 (magenta) and 17 μM Alexa568-tubulin (green). Scale bars are 10 μm. Also see Video 2 (http://dx.doi.org/10.1016/B978-0-12-397924-7.00019-4). (See the color plate.)

5. DISCUSSION

The assay described here is a modification of a previously reported assay where microtubules were grown from surface-immobilized seeds in order to form dynamic antiparallel microtubule overlaps (Bieling et al., 2010). The novel aspect introduced here is the controlled immobilization of biotinylated microtubule seeds in precisely defined areas of the glass surface, using micropatterning, to promote the formation of antiparallel microtubule interactions with controlled seed spacing. This is achieved by photolithographic surface patterning that produces a thiol-reactive maleimide-pattern on a PEG-passivated glass surface (Waichman et al., 2011). This is a versatile method for the covalent immobilization of thiol-containing small functional groups or proteins to which we have added detailed sample preparation steps. Here, we use this method to produce biotin-PEG or streptavidin-PEG patterns with high contrast. These functionalized patterns allow the localized and specific immobilization of biotinylated microtubule seeds via biotin–streptavidin or biotin–neutravidin bonds on a PEG brush-passivated glass. This method is designed to create reproducible protein immobilization and preserve protein activities on micropatterned surfaces, because it relies on defined and oriented protein immobilization via a PEG-linker rather than non-specific adsorption to a glass surface.

The advantage of micropatterned seed attachment for the antiparallel overlap assay described here is that the distance between microtubule seeds can be well controlled by the geometry of the chosen micropattern, improving the efficiency and reliability of the assay. Biotinylated seeds are attached to functionalized stripes with perpendicular orientation using flow. This causes microtubules to grow over PEG-passivated areas of the glass where they meet "head-on" and start forming antiparallel overlaps in the presence of the antiparallel cross-linker PRC1. This assay can easily be extended to study other homologs of PRC1, such as the various MAP65 homologs in plants that are responsible for interphase microtubule bundling (Gaillard et al., 2008). Another natural extension of the assay is the study of the role of the other anaphase midzone proteins in microtubule overlap formation that are known to be recruited by PRC1 (Duellberg et al., 2013; Glotzer, 2009), leading toward the reconstitution of a more complex "minimal midzone." Since the biotin patterns are quite versatile, further applications with other proteins can also be envisaged.

ACKNOWLEDGMENTS

We thank Jacob Piehler for helpful discussions, Surajit Ghosh and Nicholas Cade for help setting up the UV patterning devices in London, and Iris Lueke for protein expression. F. J. F. and T. S. acknowledge support from the ERC. D. A. F. acknowledges support from the NIH.

REFERENCES

Aoyama, S., Shimoike, M., & Hiratsuka, Y. (2013). Self-organized optical device driven by motor proteins. *Proceedings of the National Academy of Sciences of the United States of America, 110*(41), 16408–16413.

Bieling, P., Telley, I. A., & Surrey, T. (2010). A minimal midzone protein module controls formation and length of antiparallel microtubule overlaps. *Cell, 142*(3), 420–432. http://dx.doi.org/10.1016/j.cell.2010.06.033, S0092-8674(10)00723-3 [pii].

Braun, M., Lansky, Z., Fink, G., Ruhnow, F., Diez, S., & Janson, M. E. (2011). Adaptive braking by Ase1 prevents overlapping microtubules from sliding completely apart. *Nature Cell Biology, 13*(10), 1259–1264. http://dx.doi.org/10.1038/ncb2323, ncb2323 [pii].

Castoldi, M., & Popov, A. V. (2003). Purification of brain tubulin through two cycles of polymerization-depolymerization in a high-molarity buffer. *Protein Expression and Purification, 32*, 83–88.

Duellberg, C., Fourniol, F. J., Maurer, S. P., Roostalu, J., & Surrey, T. (2013). End-binding proteins and Ase1/PRC1 define local functionality of structurally distinct parts of the microtubule cytoskeleton. [Review]. *Trends in Cell Biology, 23*(2), 54–63.

Gaillard, J., Neumann, E., Van Damme, D., Stoppin-Mellet, V., Ebel, C., Barbier, E., et al. (2008). Two microtubule-associated proteins of Arabidopsis MAP65s promote antiparallel microtubule bundling. *Molecular Biology of the Cell, 19*(10), 4534–4544. http://dx.doi.org/10.1091/mbc.E08-04-0341, E08-04-0341 [pii].

Gell, C., Bormuth, V., Brouhard, G. J., Cohen, D. N., Diez, S., Friel, C. T., et al. (2010). Microtubule dynamics reconstituted in vitro and imaged by single-molecule fluorescence microscopy. *Methods in Cell Biology, 95*, 221–245.

Ghosh, S., Hentrich, C., & Surrey, T. (2013). Micropattern-controlled local microtubule nucleation, transport, and mesoscale organization. *ACS Chemical Biology, 8*(4), 673–678.

Glotzer, M. (2009). The 3Ms of central spindle assembly: Microtubules, motors and MAPs. *Nature Reviews Molecular Cell Biology, 10*(1), 9–20. http://dx.doi.org/10.1038/nrm2609, nrm2609 [pii].

Hentrich, C., & Surrey, T. (2010). Microtubule organization by the antagonistic mitotic motors kinesin-5 and kinesin-14. *Journal of Cell Biology, 189*(3), 465–480.

Hyman, A., Drechsel, D., Kellogg, D., Salser, S., Sawin, K., Steffen, P., et al. (1991). Preparation of modified tubulins. *Methods in Enzymology, 196*, 478–485.

Janson, M. E., Loughlin, R., Loiodice, I., Fu, C., Brunner, D., Nedelec, F. J., et al. (2007). Crosslinkers and motors organize dynamic microtubules to form stable bipolar arrays in fission yeast. *Cell, 128*(2), 357–368. http://dx.doi.org/10.1016/j.cell.2006.12.030, S0092-8674(07)00048-7 [pii].

Kapitein, L. C., Peterman, E. J., Kwok, B. H., Kim, J. H., Kapoor, T. M., & Schmidt, C. F. (2005). The bipolar mitotic kinesin Eg5 moves on both microtubules that it crosslinks. *Nature, 435*(7038), 114–118.

Mollinari, C., Kleman, J. P., Jiang, W., Schoehn, G., Hunter, T., & Margolis, R. L. (2002). PRC1 is a microtubule binding and bundling protein essential to maintain the mitotic spindle midzone. *Journal of Cell Biology, 157*(7), 1175–1186. http://dx.doi.org/10.1083/jcb.200111052, jcb.200111052 [pii].

Nunes Bastos, R., Gandhi, S. R., Baron, R. D., Gruneberg, U., Nigg, E. A., & Barr, F. A. (2013). Aurora B suppresses microtubule dynamics and limits central spindle size by locally activating KIF4A. *Journal of Cell Biology*, *202*(4), 605–621.

Portran, D., Gaillard, J., Vantard, M., & Thery, M. (2013). Quantification of MAP and molecular motor activities on geometrically controlled microtubule networks. *Cytoskeleton*, *70*(1), 12–23.

Roostalu, J., Hentrich, C., Bieling, P., Telley, I. A., Schiebel, E., & Surrey, T. (2011). Directional switching of the kinesin Cin8 through motor coupling. *Science*, *332*(6025), 94–99.

Shaner, N. C., Campbell, R. E., Steinbach, P. A., Giepmans, B. N. G., Palmer, A. E., & Tsien, R. Y. (2004). Improved monomeric red, orange and yellow fluorescent proteins derived from Discosoma sp. red fluorescent protein. *Nature Biotechnology*, *22*, 1567–1572.

Sørensen, H. P., Sperling-Petersen, H. U., & Mortensen, K. K. (2003). A favorable solubility partner for the recombinant expression of streptavidin. *Protein Expression and Purification*, *32*(2), 252–259.

Su, X., Arellano-Santoyo, H., Portran, D., Gaillard, J., Vantard, M., Thery, M., et al. (2013). Microtubule-sliding activity of a kinesin-8 promotes spindle assembly and spindle-length control. *Nature Cell Biology*, *15*(8), 948–957.

Subramanian, R., Ti, S. C., Tan, L., Darst, S. A., & Kapoor, T. M. (2013). Marking and measuring single microtubules by PRC1 and kinesin-4. *Cell*, *154*(2), 377–390.

Subramanian, R., Wilson-Kubalek, E. M., Arthur, C. P., Bick, M. J., Campbell, E. A., Darst, S. A., et al. (2010). Insights into antiparallel microtubule crosslinking by PRC1, a conserved nonmotor microtubule binding protein. *Cell*, *142*(3), 433–443. http://dx.doi.org/10.1016/j.cell.2010.07.012, S0092-8674(10)00781-6 [pii].

Telley, I. A., Bieling, P., & Surrey, T. (2011). Reconstitution and quantification of dynamic microtubule end tracking in vitro using TIRF. *Methods in Molecular Biology*, *777*, 127–145.

van den Wildenberg, S. M., Tao, L., Kapitein, L. C., Schmidt, C. F., Scholey, J. M., & Peterman, E. J. (2008). The homotetrameric kinesin-5 KLP61F preferentially crosslinks microtubules into antiparallel orientations. *Current Biology*, *18*(23), 1860–1864.

Verni, F., Somma, M. P., Gunsalus, K. C., Bonaccorsi, S., Belloni, G., Goldberg, M. L., et al. (2004). Feo, the Drosophila homolog of PRC1, is required for central-spindle formation and cytokinesis. *Current Biology*, *14*(17), 1569–1575. http://dx.doi.org/10.1016/j.cub.2004.08.054, S0960982204006578 [pii].

Waichman, S., You, C., Beutel, O., Bhagawati, M., & Piehler, J. (2011). Maleimide photolithography for single-molecule protein-protein interaction analysis in micropatterns. *Analytical Chemistry*, *83*(2), 501–508.

SECTION V

Cell Extract Systems

CHAPTER TWENTY

WAVE Regulatory Complex Activation

Peter J. Hume, Daniel Humphreys, Vassilis Koronakis[1]
Department of Pathology, University of Cambridge, Cambridge, United Kingdom
[1]Corresponding author: e-mail address: vk103@cam.ac.uk

Contents

Abstract

The WAVE regulatory complex (WRC) is critical to control of actin polymerization at the eukaryotic cell membrane. By reconstituting WAVE-dependent actin assembly on silica microspheres coated with phospholipid bilayers in mammalian brain extracts, we discovered that membrane recruitment and activation of WRC require the cooperative action of two mammalian GTPases, Arf and Rac. Here, we describe detailed methods to generate phospholipid-coated microspheres and porcine brain extract and outline conditions necessary to reconstitute WRC-dependent motility. In addition, we describe how to generate acylated recombinant GTPases, anchor them to lipid-coated microspheres, and reconstitute GTPase activation of WRC.

Methods in Enzymology, Volume 540
ISSN 0076-6879
http://dx.doi.org/10.1016/B978-0-12-397924-7.00020-0

1. INTRODUCTION

The ubiquitous Arp2/3 complex is the key nucleator of branched networks of actin filaments, which drive the formation of membrane ruffles and lamellipodia (Rotty, Wu, & Bear, 2013). Arp2/3 must be activated by nucleation-promoting factors (NPFs), the best characterized being N-WASP (neural Wiskott–Aldrich syndrome protein) and the WAVE (WASP family veroprolin homologue) regulatory complex (WRC) (Campellone & Welch, 2010). N-WASP was the first NPF whose regulation was understood at the molecular level (Miki, Sasaki, Takai, & Takenawa, 1998). N-WASP exists in an autoinhibited state in the cell, in which the Arp2/3-activating VCA (*V*eroprolin-homology, *C*ofilin-homology, *A*cidic) domain is tightly bound to an intramolecular domain known as the GTPase-binding domain (GBD). Direct binding of the Rho GTPase Cdc42 to the GBD domain can induce a conformational change that exposes the VCA domain and thus relieves autoinhibition (Padrick & Rosen, 2010). The regulation of WAVE, which helps to generate actin filaments at the leading edges of motile cells, has proved more complicated (Bisi et al., 2013). WAVE also contains a VCA domain, but in contrast to N-WASP, isolated WAVE is not autoinhibited. In the cell, however, WAVE is part of an inactive heteropentameric complex (Ismail, Padrick, Chen, Umetani, & Rosen, 2009), comprising WAVE, Cyfip, Nap1, Abi1, and HSPC300 (or their homologues). This complex integrates multiple signals to regulate the nucleation-promoting activity of WAVE, including direct binding by the GTPase Rac1, interaction with certain phospholipids and phosphorylation (Chen et al., 2010; Lebensohn & Kirschner, 2009; Pocha & Cory, 2009; Suetsugu et al., 2006; Ura et al., 2012). Much recent progress has been made in understanding the structure of WRC and its interaction with these various ligands, largely from biochemical/biophysical studies using recombinant WRC, which can be purified in milligram quantities (see Chapter 4 in this volume by Chen et al).

Here, we describe the *in vitro* reconstitution of membrane-associated WRC-dependent actin assembly. Briefly, solid silica microspheres are coated with a stable, defined, single phospholipid bilayer, and incubated in an optimized cell-free brain extract, from which protein complexes are recruited to the bilayer. This system has several advantages. The dense lipid-coated silica microspheres can be isolated by simple centrifugation and recruited protein complexes analyzed by proteomics, while in parallel, actin filament assembly induced by recruited NPFs can be readily tracked by fluorescence microscopy and measured as microsphere motility. The presence of the lipid bilayer mimics the cellular environment in which the WRC operates (i.e., the plasma membrane), allowing lateral diffusion of recruited signaling complexes and also enabling the direct role of signaling phospholipids in WRC activation to be assessed. Performing these experiments in extract, as opposed to in intact cells or whole organisms, facilitates easy biochemical manipulation of the system, for example, removing proteins by immunodepletion, adding exogenous proteins and derivatives, pharmacological inhibition, etc. We recently used this reconstitution approach to demonstrate that Rac1 is necessary, but not sufficient, for WRC activation (Koronakis et al., 2011). We identified an unexpected role for Arf family GTPases, which synergize with Rac1 to efficiently recruit WRC to the membrane and trigger WAVE-dependent actin assembly.

Here, we describe how to generate phospholipid-coated microspheres and porcine brain extract, and then two approaches to reconstitute WRC-dependent actin assembly. First, through phosphatidylinositol-3,4,5-trisphosphate (PIP_3)-containing bilayers that recruit the signaling machinery required for WRC activation from the extract. This system is especially useful for identifying signaling networks upstream of WRC. Second, we describe how individual pathways can be recapitulated by incorporating recombinant GTPases into the bilayer prior to incubation in the brain extract.

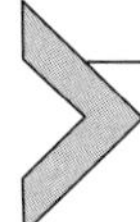

2. GENERATING PHOSPHOLIPID BILAYER-COATED SILICA MICROSPHERES

Phospholipid bilayers can be assembled onto silica microspheres by fusion of small unilamellar vesicles (SUVs) (Bayerl & Bloom, 1990). The phospholipid bilayer is destabilized by heating, and when vortexed, the unstable highly curved SUVs collapse onto the surface of the microspheres, generating a stable, continuous single bilayer (Fig. 20.1). A thin water layer (1–2 nm thick) is present between the bilayer and the surface of the bead, and the lateral mobility of the phospholipids and any incorporated proteins is equivalent to that seen in typical fluid membranes.

The precise phospholipid composition of the bilayer can be varied as desired. To reconstitute PIP_3-dependent WRC activity, we routinely use either a mixture of phosphatidylcholine (PC), phosphatidylinositol (PI), and PIP_3 (at a molar ratio of 48:48:4) or a mixture of phosphatidylethanolamine, PC, phosphatidylserine, cholesterol, and PIP_3 (28:28:30:10:4),

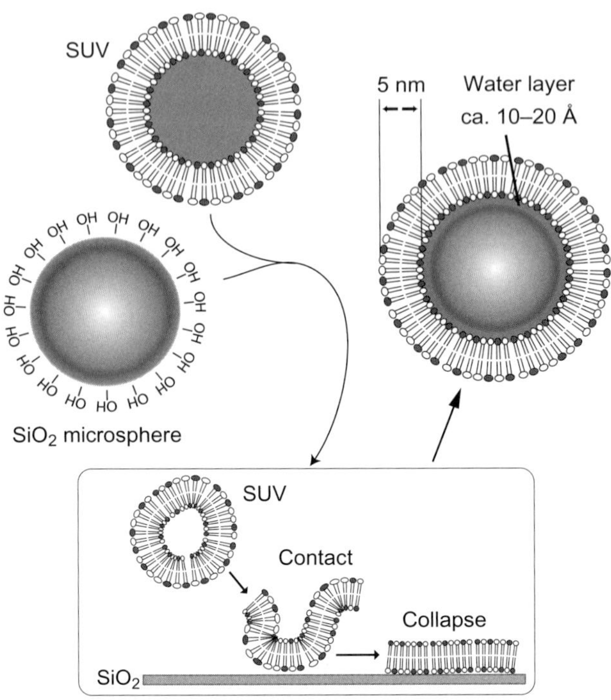

Figure 20.1 Schematic outlining the coating of silica microspheres with defined phspholipid bilayers. Highly curved small unilamellar vesicles collapse upon contact with the surface of the microspheres and fuse together to form a stable single bilayer. A thin (10–20 Å) water layer remains between the bilayer and the surface of the microspheres. (For color version of this figure, the reader is referred to the online version of this chapter.)

which more accurately reflects the typical phospholipid composition of the plasma membrane. To assess the role of individual proteins (e.g., GTPases) in WRC activation, we use bilayers lacking PIP_3, for example, PC:PI (50:50).
Materials

All phospholipids are purchased as powders from Avanti Phospholipids. We generally use 1 μm diameter acid-washed silica microspheres (Bangs Laboratories) but have successfully reconstituted WRC-dependent motility using diameters ranging from 0.1 to 10 μm.

Buffers

Resuspension buffer	10 m*M* HEPES, pH 7.4, 100 m*M* NaCl, 50 m*M* sucrose, 2 m*M* EGTA
HKS	20 m*M* HEPES, pH 7.4, 100 m*M* KCl

Method

1. The phospholipids are dissolved in chloroform:methanol (2:1, v/v) in glass tubes. Individual phospholipids can be mixed at this stage to obtain the required composition. It is important that phosphoinositide lipids are protonated to allow efficient incorporation into liposomes. To achieve this, phosphoinositides are first dissolved in chloroform:methanol:1 N HCl (2:1:0.01), incubated (15 min), then dried under nitrogen (1 h), washed by dissolving in chloroform:methanol (2:1), and again dried under nitrogen (1 h). The protonated phosphoinositides are then dissolved in choloroform:methanol ready for use.
2. To generate multilamellar vesicles, the chloroform:nitrogen is evaporated under nitrogen and the resultant film dried extensively in a rotary vacuum evaporator (8 h) to remove any residual solvent. The phospholipid film is then hydrated in resuspension (2 h) followed by brief vortexing (10 min) to give a final concentration of 2 mg/ml. The suspension after vortexing should appear turbid.

   To generate SUVs, we pass the phospholipid suspension three times through a French pressure cell (82,800 kPa), until the suspension clears, followed by clarification by centrifugation (100,000 × *g*, 10 min). To avoid possible contaminating phosphatases, it is essential to ensure that the French pressure cell is sterilized thoroughly, especially if the same cell is also used for lysing bacteria. SUVs can also be produced by other methods, for example, extrusion through a microfilter (Mui, Chow, & Hope, 2003) or sonication (Duzgunes, 2003). SUVs can be stored at 4 °C for several weeks, but we usually use them immediately for coating microspheres.
3. The approximate amount of phospholipid (M_{pl}) required to coat the required mass of microspheres with a single bilayer is calculated as follows: $M_{pl} = 2(A_m/A_{pl})/N_A$. The surface average area per gram of microspheres (A_m) is available via the Bangs Laboratories website (www.bangslabs.com; e.g., 5.7×10^{12} μm^2/g for 1 μm diameter microspheres). The surface area of a single phospholipid headgroup (A_{pl}) can be approximated as 5.5×10^{-7} μm^2. N_A is Avogadro's number. Therefore, approximately 34 μmol of phospholipid is required to coat 1 g of 1 μm diameter microspheres with a single bilayer. If we assume an average molecular weight of 800 Da for the phospholipids, this equates to around 28 mg.
4. Acid-washed silica microspheres are washed in lipid suspension buffer by centrifugation (1000 × *g*, 1 min) prior to resuspension in the same buffer containing a fivefold excess of the SUV titer required to coat the bead surface with a single lipid bilayer (calculated in step 4—e.g., for 1 g of

1 μm diameter microspheres we use 140 mg). The suspension is incubated (5 min) at 55 °C (to ensure that the lipid bilayers are in the liquid phase) and then vortexed (30 min). The suspension is subjected to one freeze–thaw cycle in liquid nitrogen, washed in HKS (5 ×), and then resuspended at a concentration of 1 g microspheres in 30 ml HKS. The suspension is aliquoted and stored at −80 °C.

3. Preparation of Brain Extract

We have reconstituted WAVE-dependent motility in diverse extracts, for example, porcine brains, *Xenopus laevis* oocytes, and cultured HeLa cells. Here, we provide a protocol for generating porcine brain extract optimized for WRC reconstitution. It is critical that the brains are from freshly slaughtered animals, are chilled and kept on ice immediately after collection, and then processed within 2–3 h. Frozen brains are available from commercial sources (e.g., Pel-Freez Biologicals), but we find that results are far more reproducible using brains obtained from a local abattoir.

Buffers

Extraction buffer	20 m*M* HEPES, pH 7.4, 0.5 m*M* ATP, 0.5 m*M* DTT, 100 m*M* KCl, 2 m*M* $MgCl_2$, 1 m*M* EGTA, 0.1 m*M* EDTA, protease inhibitors (leupeptin, pepstatin, chymostatin, all at final concentration of 10 μg/ml), Complete™ EDTA-free protease inhibitor tablets (Roche, 10 tablets per liter)

Method

1. All procedures are performed using prechilled buffers in a 4 °C cold room. Three brains will produce a final volume of approximately 50 ml of concentrated extract, enough for ~5000 motility assays, but the following procedures can be scaled up or down as required.
2. The meninges and any blood vessels are carefully removed from three fresh porcine brains and discarded. The brains are then chopped into 2–3 cm pieces.
3. The volume of brain tissue is estimated and an equal volume of extraction buffer added. For three brains, this will be approximately 400 ml. Any less than an equal volume can cause problems when collecting supernatants following centrifugation, so it is better to err on the side of too much buffer.
4. The brains are homogenized in a chilled Waring blender, using 2 × 15 s pulses, with a 60-s gap in between to prevent sample heating.

5. The homogenate is centrifuged at 12,000 × g for 30 min at 4 °C. The supernatant is carefully decanted, taking care not to dislodge the large, loose pellets. The supernatant is poured through cheese cloth to remove any contaminating pellet material. The pellets can be discarded. From three brains, around 250 ml of supernatant (brain extract) is obtained.
6. For WRC reconstitution, the brain extract is concentrated approximately fivefold. This can be achieved using a variety of methods, but we find that centrifugation ultrafiltration concentrators clog very easily. Instead, we use tangential flow filtration, for example, using Vivaflow 50 disposable modular concentrators (Sartorius), with a 10 kDa molecular weight cutoff. Figure 20.2 demonstrates the setup of such a device.

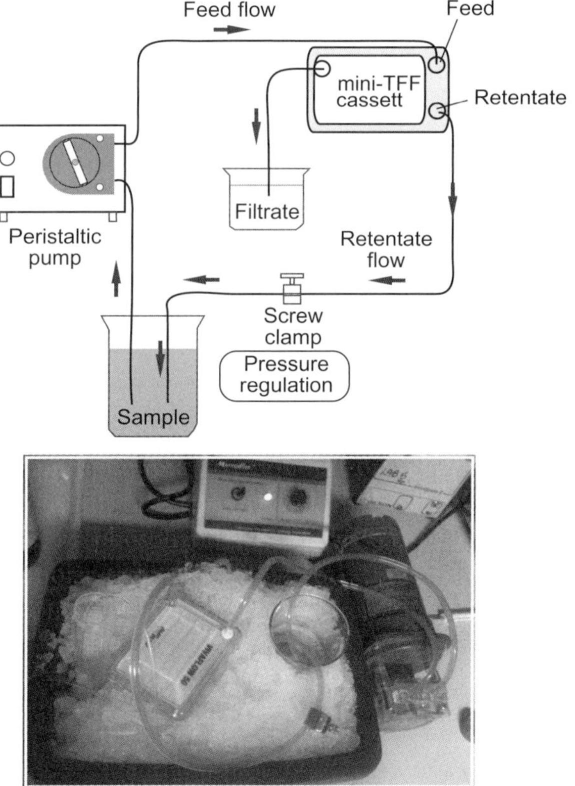

Figure 20.2 Setup of tangential flow filtration (TFF) cassette to concentrate brain extract. It is essential to use some form of flow restrictor, for example, a screw clamp as shown here, in order to achieve the necessary pressure. The TFF cassette and all containers are kept on ice, in a 4 °C cold room. (For color version of this figure, the reader is referred to the online version of this chapter.)

The concentrator and sample reservoir are kept on ice at all times. We use one Vivaflow cassette per 100 ml and achieve fivefold concentration in approximately 1–2 h.

7. The concentrated brain extract is aliquoted and stored at −80 °C. The extract can be stored for up to 6 months. Before use, the concentrated extract is clarified by centrifugation at 100,000 × *g* for 5 min (TLA-120.2 rotor, Beckman-Coulter).

4. Assay of WAVE-Dependent Motility by Phospholipid-Coated Microspheres

The dynamic assembly of actin filaments requires a significant source of energy, that is, ATP. Indeed, as much as 50% of a cell's energy expenditure can be due to actin polymerization and molecular motor dynamics (Daniel, Molish, Robkin, & Holmsen, 1986). To reconstitute actin assembly in a cell-free extract, it is therefore critical to provide a sustained energy source. We supply two energy regeneration sources. First, we add creatine phosphate, which can be utilized by endogenous creatine kinase in the brain extract to generate ATP. Second, we add the glycolytic intermediate 3-phosphoglycerate, which again is used by endogenous enzymes to regenerate ATP. Sufficient GTP levels are maintained in the extract by the action of endogenous nucleoside diphosphate kinase, which can generate GTP from GDP and ATP. WRC-dependent actin-based bead motility can thus be sustained for over 1 h in our extracts (Koronakis et al., 2011).

4.1. Basic bead motility assay

Buffers

All buffers are freshly prepared.

20 × energy mix	300 m*M* creatine phosphate, 40 m*M* $MgCl_2$, 40 m*M* ATP
10 × salt buffer	600 m*M* KCl, 200 m*M* 3-phosphoglycerate
Actin solution	140 μ*M* G-actin (ATP-bound form) containing 1% rhodamine-labeled actin, prepared as described in a previous volume (Theriot & Fung, 1998)
BAPTA	50 m*M* 1,2-bis(o-aminophenoxy)ethane-*N*,*N*,*N*′,*N*′-tetraacetic acid, pH 7.0
DTT	300 m*M* dithiothreitol
VALAP	1:1:1 mix of petroleum jelly (e.g., Vaseline), lanolin and paraffin, heated in a glass beaker and mixed. Can be stored at room temperature and reheated when needed

Method

1. A 60 μl motility mix is assembled as follows—40 μl brain extract, 3 μl 20× energy mix, 3 μl actin solution, 6 μl salt buffer, 6 μl BAPTA, and 1 μl DTT. 1 μl 30 m*M* GTPγS is added to the motility mix to ensure extract GTPases are activated. 60 μl of motility mix is sufficient for six motility assays and can be held on ice until needed.
2. For each motility assay, 1 μl of lipid-coated microspheres is added to 10 μl of motility mix and gently mixed by flicking the tube.
3. 1 μl of the motility reaction is then applied to the center of a microscope slide. A glass cover slip is inverted onto the slide. Melted VALAP is used to seal all four sides of the cover slip, using a fine brush. The slide can be imaged immediately using a fluorescence microscope.
4. Long actin comet tails should be immediately visible associated with each lipid-coated bead, which are consequently propelled through the extract. The microspheres themselves can be easily identified by imaging using phase contrast microscopy.

4.2. Isolating the WRC pathway

To identify the signaling pathways driving actin assembly at the membrane, we scaled up the PIP_3 motility assay and subjected the lipid-coated microspheres to proteomic analysis (Koronakis et al., 2011). This confirmed that two NPFs were recruited, WRC and N-WASP, and that no Arp2/3-independent actin nucleators were present. Consequently, to study the WRC pathway, it is necessary to establish conditions in which the N-WASP pathway is inhibited. We achieve this using the dominant-negative derivative N-WASP$^{\Delta VCA}$. N-WASP is activated by the GTPase Cdc42 (Padrick & Rosen, 2010). N-WASP$^{\Delta VCA}$ lacks the Arp2/3-activating VCA domain but still binds Cdc42, and can consequently be used out-titrate Cdc42 in the extract and block the activation of wild type N-WASP.

Materials and buffers

V-lysis buffer	10 m*M* Tris, pH 8.3, 50 m*M* NaCl, 1 m*M* dithiothreitol, 1× Complete™ EDTA-free protease inhibitor tablet per 50 ml

Method

1. A 5-l culture of *Escherichia coli* BL21 transformed with a plasmid encoding residues 1–420 from *Rattus norvegicus* N-WASP is grown at 30 °C until $OD_{600} \approx 0.4$. The temperature is then reduced to 16 °C, and after 30 min 0.1 m*M* Isopropyl β-D-1-thiogalactopyranoside

(IPTG) added to induce protein expression and incubation continued at 16 °C for 16 h. The cells are harvested by centrifugation (10 min, 6000 × *g*), resuspended in 40 ml lysis buffer, and lysed by passage through a cell disruptor (Constant Systems; 30,000 psi). The lysate is then clarified by centrifugation (60 min, 100,000 × *g*).

2. The clarified lysate is applied to a 5-ml HiTrap Q HP column, and the flow-through collected. The N-WASP$^{\Delta VCA}$ should not bind at this pH, but several common breakdown products will. The flow-through is then applied to a 5-ml HiTrap Heparin HP column. Bound proteins are eluted with a continuous NaCl gradient, and peak fractions are analyzed by SDS-PAGE.
3. Fractions containing N-WASP$^{\Delta VCA}$ are pooled, dialyzed extensively against V-lysis buffer, before aliquots are flash frozen in liquid nitrogen and stored at −80 °C.
4. The WRC-dependent motility assay is performed as described above, except that prior to the addition of the lipid-coated microspheres, N-WASP$^{\Delta VCA}$ is added to the extract at a final concentration of 10 μ*M* and incubated for 10 min. The PIP_3-containing lipid-coated microspheres are then added, and the assay continued exactly as described above. Figure 20.3A shows an example of typical PIP_3-induced, WRC-dependent comet tails.

5. RECONSTITUTING ACTIVATION OF WRC BY COOPERATING Arf AND Rac1 GTPases

Section 4 described how to reconstitute WAVE-dependent motility of microspheres coated with PIP_3-containing bilayers. This assay is well suited to determine the requirements for motility, for example, using proteomics, inhibitors, immunodepletion, etc. Indeed, we used such approaches to determine that both Arf and Rac1 GTPases are necessary to activate WRC (Koronakis et al., 2011). We describe here how this individual Arf–Rac–WAVE pathway can be studied.

To reconstitute WRC activation by cooperating GTPases, it is necessary to anchor both Arf and Rac1 to the same lipid-coated microspheres. Small GTPases require posttranslational acylation to interact with membranes. Correctly acylated recombinant GTPases can be produced using various eukaryotic expression systems, for example, expression in yeast, baculovirus-mediated expression in *Spodoptera frugiperda* cells, and these have been described in detail in previous volumes of *Methods in Enzymology* (e.g.,

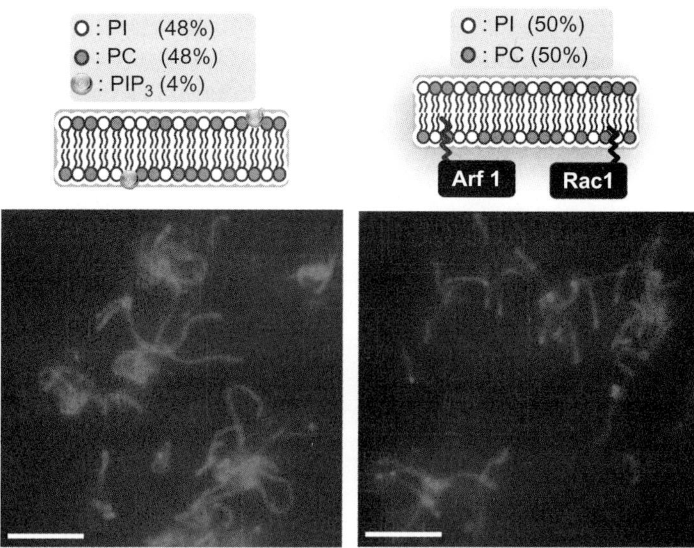

Figure 20.3 WAVE-dependent actin-based motility of silica microspheres coated with lipid bilayers composed of PC:PI and either PIP_3 (left) or Arf1 and Rac1 (right). Extract contains rhodamine-labeled G-actin, allowing visualization of comet tails (red). Comet tail formation and consequent microsphere motility triggered by Arf1 and Rac1 (right) are indistinguishable from that induced by PIP_3 (left). Scale bars, 15 μm. (See the color plate.)

Read & Nakamoto, 2000; Seifert, Snyder, Sondek, & Harden, 2006). However, to avoid the possibility of any contaminating eukaryotic proteins interfering with our *in vitro* reconstitutions, we prefer to isolate GTPases following expression in bacteria. The lack of acylation machinery in bacteria can be overcome in a number of different ways for each family of GTPases.

5.1. Preparation of recombinant Arf GTPases

Arf family GTPases are cotranslationally myristoylated at a glycine residue at position 2 in the amino acid sequence. As bacteria lack any myristoylation activity, we coexpress recombinant Arfs with the yeast *N*-myristoyltransferase in standard *E. coli* strains (e.g., BL21), followed by selective purification of the acylated GTPases. This protocol has been extensively described in a previous volume (Ha, Thomas, Stauffer, & Randazzo, 2005). A typical yield from a 2-l culture will be around 10 mg at a concentration of around 2 mg/ml. The purified myristoylated Arf can be flash frozen in liquid nitrogen and stored at −80 °C until required.

5.2. Purification and *in vitro* prenylation of Rac1

The Rho family GTPase Rac1 is geranylgeranylated at a C-terminal cysteine residue by geranylgeranyltransferase type I (GGTase I). We purify non-acylated Rac1 following expression in *E. coli* using standard procedures and subsequently geranylgeranylate the GTPase *in vitro* using GGTase I (which is itself also expressed and purified from *E. coli*; Fu, Moomaw, Moomaw, & Casey, 1996). We obtained bacterial expression constructs for the two subunits of GGTase I from Patrick Casey (Duke University) and carry out the *in vitro* prenylation reaction using a protocol from the Casey lab (Fu et al., 1996), with minor modifications.

Buffers

H-lysis buffer	10 m*M* Tris, pH 8.0, 100 m*M* NaCl, 1 m*M* $MgCl_2$, 1 m*M* dithiothreitol (DTT), 5 m*M* imidazole, 1 × Complete™ EDTA-free protease inhibitor tablet per 50 ml
5 × prenylation buffer	250 m*M* Tris, pH 7.4, 100 m*M* KCl, 25 m*M* $MgCl_2$, 25 μ*M* $ZnCl_2$, 1 m*M* *tris* (2-carboxyethyl) phosphine (TCEP)
GGpp solution	100 μ*M* geranylgeranyl pyrophosphate (Sigma), 20 m*M* Tris pH 7.4, 0.02% Zwittergent 3-14

Method

1. A 2-l culture of *E. coli* BL21 transformed with a plasmid encoding Rac1 fused to an N-terminal 6 × histidine tag is grown at 30 °C until $OD_{600} \approx 0.8$. Expression is induced by addition of 0.1 m*M* IPTG, and growth continued at 30 °C for 3 h. The cells are harvested by centrifugation (10 min, 6000 × *g*), resuspended in 40 ml H-lysis buffer, and lysed by passage through a cell disruptor (Constant Systems; 30,000 psi). The lysate is then clarified by centrifugation (30 min, 100,000 × *g*) and the supernatant incubated with 3 ml NiNTA sepharose (Qiagen) for 1 h at room temperature. The slurry is then poured into an empty column and the flow-through collected. The remaining resin is washed with 50 ml of H-lysis buffer, and the bound Rac1 eluted with lysis buffer containing 400 m*M* imidazole, collecting 10 × 1 ml fractions. Fractions are analyzed by SDS-PAGE, and the eluted protein is then extensively dialyzed against H-lysis buffer before aliquots are flash frozen in liquid nitrogen prior to storage at −80 °C.
2. *E. coli* BL21 are cotransformed with plasmids encoding the α and β subunits of GGTase I, and grown and induced as above for Rac1. At the time of induction, 0.5 m*M* $ZnSO_4$ is added to the culture medium to

ensure the availability of zinc, a crucial cofactor for GGTase I activity. Following a further 3 h at 30 °C, the bacteria are harvested and lysed as above, and the GGTase complex purified identically to Rac1. The purified GGTase is then dialyzed against lysis buffer containing 0.5 μ*M* $ZnCl_2$, and aliquots flash frozen in liquid nitrogen and stored at −80 °C until required.

3. A 20 ml prenylation reaction is assembled as follows—4 ml of 5 × prenylation buffer, 50 μg GGTase I, 1 mg Rac1, and 1.2 ml GGpp solution. The reaction is gently mixed by inverting several times and then incubated at 30 °C for 1 h. The resulting geranylgeranylated Rac1 is then immediately anchored to lipid-coated microspheres (see below). The final concentration of Zwittergent 3-14 in this reaction (0.0012%) is well below the cmc (~0.01%) and will therefore not disrupt the lipid bilayer.

5.3. Anchoring Arf and Rac1GTPases to phospholipid-coated microspheres

Following the above procedures, both the acylated Arf and Rac1 GTPases will be in the inactive, GDP-bound form. For WRC reconstitution, both GTPases need to be activated, that is, bound to GTP. Nucleotide binding is stabilized by magnesium, so exchange of GDP for GTP can be achieved by incubating with EDTA (to destabilize the bound GDP) and an excess of GTP. To ensure that the GTPases remain in the active state, the non-hydrolyzable analog GTPγS is used in place of GTP.

Binding of Rac1 to the membrane is independent of activation status, so it is possible to activate Rac1 either before or after binding to lipid-coated microspheres. However, as GTPγS is expensive, we prefer to carry out nucleotide exchange in as small a volume as possible, and therefore first anchor Rac1 to the microspheres, which can then be resuspended in a small volume of buffer for activation. Arf GTPases, however, can only efficiently associate with membranes when GTP-bound. We therefore perform Arf activation concomitantly with lipid binding. As myristoylated Arfs can be purified at relatively high concentrations (see above), the volume can still be kept to a minimum.

Buffers

HKS	20 m*M* HEPES, pH 7.4, 100 m*M* KCl
HKSM	20 m*M* HEPES, pH 7.4, 100 m*M* KCl, 1 m*M* $MgCl_2$

Method

1. 100 mg of thawed PC:PI-coated microspheres are washed twice in HKS by centrifugation (1000 × *g*, 1 min).
2. The pelleted microspheres are gently resuspended in 20 ml of Rac1 prenylation reaction (Section 5.2) and incubated at room temperature for 2 h.
3. The microspheres washed twice in HKS and then once in HKS + 1 *M* NaCl (to remove any Rac1 not anchored to the bilayer) by centrifugation (1000 × *g*, 1 min).
4. The washed Rac-anchored microspheres are then gently resuspended in 1 mg of purified myristoylated Arf (this will be approximately 500 μl). To this suspension, 5 μl of 500 m*M* EDTA and 5 μl of 100 mM GTPγS are added to activate both GTPases. The reaction is then incubated at room temperature for 2 h.
5. The nucleotide exchange reaction is stopped by the addition of 2.5 μl of 2 *M* $MgCl_2$, and the microspheres washed twice in HKSM, followed by once in HKSM + 1 *M* NaCl, by centrifugation (1000 × *g*, 1 min). Finally, the microspheres are gently resuspended in 500 μl of HKSM. 1 μl of the suspension can be analyzed by SDS-PAGE to confirm the association of both GTPases. Lipid-coated microspheres containing anchored GTPases are used immediately for WRC reconstitution assays.

5.4. Assay of Arf-Rac1-dependent bead motility

Once Arf and Rac1 are anchored to lipid-coated microspheres, a motility assay is carried out in brain extract essentially as described above for microspheres coated with PIP_3-containing bilayers. As the lipid-anchored GTPases are activated prior to the motility assay, no additional GTPγS is required in the extract. To ensure that only the WRC pathway is activated, the assay is performed in extract preincubated with N-WASP$^{\Delta VCA}$.

Buffers

20 × energy	mix 300 m*M* creatine phosphate, 40 m*M* $MgCl_2$, 40 m*M* ATP
10 × salt buffer	600 m*M* KCl, 200 m*M* 3-phosphoglycerate
Actin solution	140 μ*M* G-actin containing 1% rhodamine-labeled actin, prepared as described in a previous volume (Theriot & Fung, 1998)
BAPTA	50 m*M* 1,2-bis(*o*-aminophenoxy)ethane-*N*,*N*,*N*′,*N*′-tetraacetic acid

DTT	300 m*M* dithiothreitol
VALAP	1:1:1 mix of petroleum jelly (e.g., Vaseline), lanolin and paraffin, heated in a glass beaker and mixed. Can be stored at room temperature and reheated when needed

Method

1. A 60 μl motility mix is assembled as follows—40 μl brain extract, 3 μl 20× energy mix, 3 μl actin solution, 6 μl salt buffer, 6 μl BAPTA, and 1 μl DTT. 60 μl of motility mix is sufficient for six motility assays and can be held on ice until needed.
2. A final concentration of 10 μ*M* N-WASP$^{\Delta VCA}$ is added to the motility mix and incubated on ice for 10 min.
3. For each motility assay, 1 μl of lipid-coated microspheres is added to 10 μl of motility mix and gently mixed by flicking the tube.
4. 1 μl of the motility reaction is then applied to the center of a microscope slide. A glass cover slip is inverted onto the slide. Melted VALAP is used to seal all four sides of the cover slip, using a fine brush. The slide can be imaged immediately using a fluorescence microscope.
5. Long actin comet tails should be immediately visible associated with each lipid-coated bead, which are consequently propelled through the extract. The comet tails and motility of the microspheres are essentially identical to PIP_3-coated microspheres. Typical results are shown in Fig. 20.3B.

6. CONCLUDING REMARKS

In this chapter, we have described methods for reconstituting WRC-dependent actin assembly at phospholipid membranes. These powerful approaches are ideally suited for not only exploring the molecular details of WRC signaling but also reconstituting other signal transduction pathways. They are especially suited to studying small GTPase signaling, as the membrane plays such a key role, but could be applied to any membrane signaling platforms. We have used these assays to study cytoskeleton regulation and subversion by bacterial pathogens (Humphreys, Davidson, Hume, & Koronakis, 2012; Humphreys, Davidson, Hume, Makin, & Koronakis, 2013), and consequently use bead motility as a read out for actin assembly, but they could easily be modified to examine diverse cellular functions controlled by membrane signaling platforms, for example, the role of Rab GTPases in vesicle trafficking.

ACKNOWLEDGMENTS

This work was supported by the Wellcome Trust and the Cambridge Isaac Newton Trust.

REFERENCES

Bayerl, T. M., & Bloom, M. (1990). Physical properties of single phospholipid bilayers adsorbed to micro glass beads. A new vesicular model system studied by 2H-nuclear magnetic resonance. *Biophysical Journal*, *58*(2), 357–362.

Bisi, S., Disanza, A., Malinverno, C., Frittoli, E., Palamidessi, A., & Scita, G. (2013). Membrane and actin dynamics interplay at lamellipodia leading edge. *Current Opinion in Cell Biology*, *25*(5), 565–573.

Campellone, K. G., & Welch, M. D. (2010). A nucleator arms race: Cellular control of actin assembly. *Nature Reviews Molecular Cell Biology*, *11*(4), 237–251.

Chen, Z., Borek, D., Padrick, S. B., Gomez, T. S., Metlagel, Z., Ismail, A. M., et al. (2010). Structure and control of the actin regulatory WAVE complex. *Nature*, *468*(7323), 533–538.

Daniel, J. L., Molish, I. R., Robkin, L., & Holmsen, H. (1986). Nucleotide exchange between cytosolic ATP and F-actin-bound ADP may be a major energy-utilizing process in unstimulated platelets. *European Journal of Biochemistry*, *156*(3), 677–684.

Duzgunes, N. (2003). Preparation and quantitation of small unilamellar liposomes and large unilamellar reverse-phase evaporation liposomes. *Methods in Enzymology*, *367*, 23–27.

Fu, H. W., Moomaw, J. F., Moomaw, C. R., & Casey, P. J. (1996). Identification of a cysteine residue essential for activity of protein farnesyltransferase. Cys299 is exposed only upon removal of zinc from the enzyme. *The Journal of Biological Chemistry*, *271*(45), 28541–28548.

Ha, V. L., Thomas, G. M., Stauffer, S., & Randazzo, P. A. (2005). Preparation of myristoylated Arf1 and Arf6. *Methods in Enzymology*, *404*, 164–174.

Humphreys, D., Davidson, A., Hume, P. J., & Koronakis, V. (2012). Salmonella virulence effector SopE and Host GEF ARNO cooperate to recruit and activate WAVE to trigger bacterial invasion. *Cell Host & Microbe*, *11*(2), 129–139.

Humphreys, D., Davidson, A. C., Hume, P. J., Makin, L. E., & Koronakis, V. (2013). Arf6 coordinates actin assembly through the WAVE complex, a mechanism usurped by Salmonella to invade host cells. *Proceedings of the National Academy of Sciences of the United States of America*, *110*(42), 16880–16885.

Ismail, A. M., Padrick, S. B., Chen, B., Umetani, J., & Rosen, M. K. (2009). The WAVE regulatory complex is inhibited. *Nature Structural & Molecular Biology*, *16*(5), 561–563.

Koronakis, V., Hume, P. J., Humphreys, D., Liu, T., Horning, O., Jensen, O. N., et al. (2011). WAVE regulatory complex activation by cooperating GTPases Arf and Rac1. *Proceedings of the National Academy of Sciences of the United States of America*, *108*(35), 14449–14454.

Lebensohn, A. M., & Kirschner, M. W. (2009). Activation of the WAVE complex by coincident signals controls actin assembly. *Molecular Cell*, *36*(3), 512–524.

Miki, H., Sasaki, T., Takai, Y., & Takenawa, T. (1998). Induction of filopodium formation by a WASP-related actin-depolymerizing protein N-WASP. *Nature*, *391*(6662), 93–96.

Mui, B., Chow, L., & Hope, M. J. (2003). Extrusion technique to generate liposomes of defined size. *Methods in Enzymology*, *367*, 3–14.

Padrick, S. B., & Rosen, M. K. (2010). Physical mechanisms of signal integration by WASP family proteins. *Annual Review of Biochemistry*, *79*, 707–735.

Pocha, S. M., & Cory, G. O. (2009). WAVE2 is regulated by multiple phosphorylation events within its VCA domain. *Cell Motility and the Cytoskeleton*, *66*(1), 36–47.

Read, P. W., & Nakamoto, R. K. (2000). Expression and purification of Rho/RhoGDI complexes. *Methods in Enzymology*, *325*, 15–25.

Rotty, J. D., Wu, C., & Bear, J. E. (2013). New insights into the regulation and cellular functions of the ARP2/3 complex. *Nature Reviews Molecular Cell Biology*, *14*(1), 7–12.

Seifert, J. P., Snyder, J. T., Sondek, J., & Harden, T. K. (2006). Direct activation of purified phospholipase C epsilon by RhoA studied in reconstituted phospholipid vesicles. *Methods in Enzymology*, *406*, 260–271.

Suetsugu, S., Kurisu, S., Oikawa, T., Yamazaki, D., Oda, A., & Takenawa, T. (2006). Optimization of WAVE2 complex-induced actin polymerization by membrane-bound IRSp53, PIP(3), and Rac. *The Journal of Cell Biology*, *173*(4), 571–585.

Theriot, J. A., & Fung, D. C. (1998). Listeria monocytogenes-based assays for actin assembly factors. *Methods in Enzymology*, *298*, 114–122.

Ura, S., Pollitt, A. Y., Veltman, D. M., Morrice, N. A., Machesky, L. M., & Insall, R. H. (2012). Pseudopod growth and evolution during cell movement is controlled through SCAR/WAVE dephosphorylation. *Current Biology*, *22*(7), 553–561.

CHAPTER TWENTY-ONE

Dissecting Principles Governing Actin Assembly Using Yeast Extracts

Alphée Michelot[*,1], **David G. Drubin**[†,1]

[*]Physics of the Cytoskeleton and Morphogenesis Group, institut de Recherches en Technologies et Sciences pour le Vivant, iRTSV, LPCV/CNRS/CEA/INRA/UJF, Grenoble, France
[†]Department of Molecular and Cell Biology, University of California, Berkeley, California, USA
[1]Corresponding authors: e-mail address: alphee.michelot@cea.fr; drubin@berkeley.edu

Contents

Abstract

In this chapter, we describe recent protocols that we have developed to trigger actin assembly and actin-based motility in yeast cell extracts. Our method allows for the fast

Methods in Enzymology, Volume 540
ISSN 0076-6879
http://dx.doi.org/10.1016/B978-0-12-397924-7.00021-2

preparation of yeast extracts that are competent in dynamic assembly of distinct actin filament structures of biologically appropriate protein composition. Compared to previous extract-based systems using other eukaryotic cell types, yeast provides a unique advantage for combining reconstituted assays with the preparation of extracts from genetically modified yeast strains. We present a global strategy for dissecting the functions of individual proteins, where the activities of the proteins are analyzed in systems of variable complexity, ranging from simple mixtures of pure proteins to the full complexity of a cell's cytoplasm.

1. INTRODUCTION

Analysis of cytoskeletal-based cellular activities presents a daunting challenge: How to determine the precise functions of each of the multitudes of proteins that collaborate for proper function within complex networks of interactions? In this context, the objective of reconstituting cytoskeletal events using a classical reductionist approach (i.e., with purified components) can be difficult to attain for several reasons. First, such a reconstitution requires that all of the essential constituents that mediate this process are known and can be purified. Second, addition of multiple constituents to an experimental system demands that many experimental parameters such as each protein's concentration be controlled. These considerations have motivated us and other investigators to use reconstituted systems based on cellular protein extracts. In these extracts, one expects the molecular complexity to be nearly equivalent to what is present in the physiological cellular environment, with most important protein–protein interactions represented.

In the case of actin-based assembly and motility, a number of studies have used cell extracts produced from a range of cell types. Originally, *Xenopus laevis* egg extracts (Theriot, Rosenblatt, Portnoy, Goldschmidt-Clermont, & Mitchison, 1994) were used successfully because it was possible and easy to prepare large amounts of concentrated protein extracts with a minimal dilution. Later, fractionated human platelets and mouse brain (Carlier et al., 1997; Laurent et al., 1999; Welch, Iwamatsu, & Mitchison, 1997), and HeLa cells extracts (Fradelizi et al., 2001) were also used successfully to reconstitute actin structures *in vitro*, demonstrating that similar approaches could be used with a wide variety of cell types. While assays performed with cell extracts reveal emergent properties of ensembles of proteins, they have been somewhat limited in their ability to reveal the functions of individual proteins. Indeed, defining systems of increasing complexity with a "bottom-up" approach allows for the precise control of the protein composition (e.g., it is straightforward to titrate the levels of a single protein). In contrast, modification of protein composition in extracts is difficult, although proteins have been depleted in some cases (Laurent et al., 1999).

Ideally one could overcome the difficulty associated with testing functions of individual proteins in complex cell extracts by the choice of an appropriate model organism. The organism of choice should satisfy several requirements. It should be easy to grow and well studied, and should exhibit actin-dependent cellular functions conserved in other eukaryotes. Most importantly, it should be an organism with advanced genetics so that mutants can be readily isolated. The budding yeast *Saccharomyces cerevisiae* meets all of these criteria, and mutants of each of its ~6000 genes are available. Moreover, the proteins required for actin-based motility appear to be present in budding yeast at sufficiently high levels for biochemical studies, and these proteins are soluble when cells are lysed (Goode, 2002). Lastly, actin filaments from *S. cerevisiae* are able to assemble in extracts (Geli, Lombardi, Schmelzl, & Riezman, 2000; Goode, 2002; Idrissi, Wolf, & Geli, 2002).

Here, we describe the protocols that we developed around the idea of combining the advantages of biochemical reconstitution and genetics (Michelot et al., 2010, 2013). We will describe how we prepare yeast extracts and analyze actin networks that assemble in them. We will also describe how these protocols can be integrated with yeast genetic screens to dissect the functions of individual proteins in complex processes.

2. YEAST CULTURE AND EXTRACT PREPARATION

This section describes a method to prepare yeast cell extracts that are functional for actin assembly. Because the actin concentration in yeast cells is relatively low (i.e., in the 10 μM range) compared to at least an order of magnitude higher in more complex eukaryotes (Moseley & Goode, 2006; Pollard, Blanchoin, & Mullins, 2000), we developed an extract preparation protocol that minimizes dilution as much as possible. Protocols presented in this section ensure that the actin concentration remains sufficiently high to support efficient actin-based motility. In addition, the protocol endeavors to maintain ionic and protein concentration conditions that are as close as possible to the cytosolic environment.

Required materials, chemicals, and equipment

- Autoclaved YPD medium (1% yeast extract; 2% peptone; 2% dextrose).
- Beckman Coulter Avanti J-26 XP Centrifuge and JLA 8.1 rotor.

- Beckman Coulter Optima Max Ultracentrifuge and TLA 100.3 rotor.
- Liquid nitrogen.
- Cryomill grinder from Retsch, or steel blender 7011S from Waring.
- Thermo Scientific Nalgene Versi-dry Lab Soaker.
- H buffer: 10 m*M* Hepes pH 7.5.
- HK50 buffer: H buffer + 50 m*M* KCl.
- Protease Inhibitors Set IV, Calbiochem, Merck4Biosciences.

2.1. Cell culture and harvesting

Because a small part of the cell culture will be lost when processing the cell pellets, and because we preferentially harvest cells when growing in log phase (up to an $1.5 < OD_{600} < 2$), we usually consider that a minimum volume of 2 L is necessary to obtain a sufficient amount of extract for our experiments. Yeast cells are usually cultured in the rich medium YPD. However, specific experiments might require the use of different media, or the use of additional chemicals, for example, if cells need to be arrested at a specific cell-cycle stage before harvesting (Miao et al., 2013).

1. Start an overnight culture of yeast cells in 100 ml of YPD medium (Fig. 21.1A). Cells must be maintained at 30 °C with good aeration. Starting with an $OD_{600} = 0.1$, cells grown from a healthy strain will reach stationary phase in the morning.
2. In the morning, resuspend cells at OD 0.1 in > 2 l of YPD and culture in the same conditions.
3. When cells reach an $1.5 < OD_{600} < 2$, centrifuge at 3000 rpm for 5 min at 4 °C. Remove supernatant. Maintain cell pellets at 4 °C or on ice (Fig. 21.1B).
4. Resuspend cell pellets in ice-cold ddH_2O (100 ml per liter of cell culture). Pool cell suspensions and centrifuge again at 3000 rpm for 5 min.
5. Remove supernatant. Using a clean spatula, collect hazelnut-sized cell pellets and drop them into liquid nitrogen (Fig. 21.1C).
6. Isolate the yeast with a metal strainer and store at −80 °C. The frozen yeast appears as small balls.
7. Cool a Waring blender (up to the top of the rotor part) with small volumes of liquid nitrogen (Fig. 21.1E). Drop frozen yeast into the blender (Fig. 21.1D), fill with liquid nitrogen up to the same level, and blend for 30 s (Fig. 21.1E). We usually place a square-piece of fabric Versi-dry between the cap and the grinder to allow for gas pressure release during grinding, and to prevent escape of liquid nitrogen. Alternatively, a

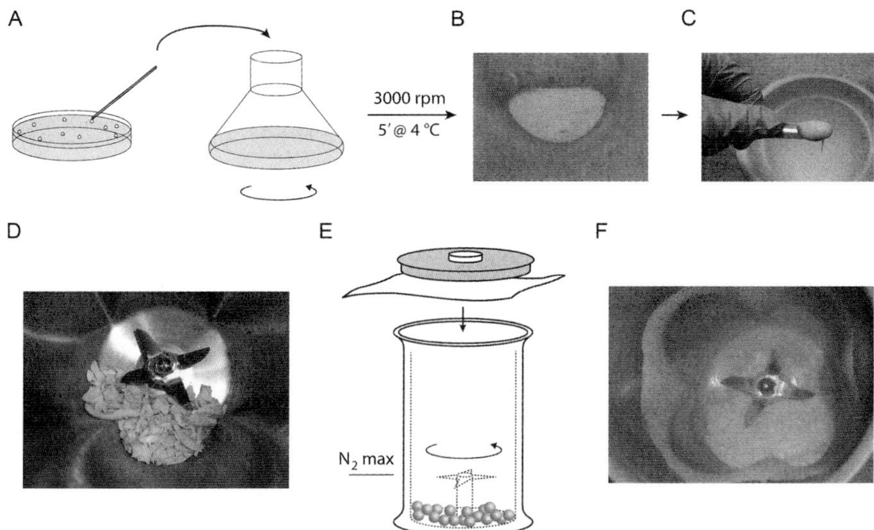

Figure 21.1 Cell culture and harvesting. (A) Illustration of yeast cell liquid culture inoculation from a single colony. (B) Photograph of a yeast cell pellet after centrifugation. (C) Photograph of a hazelnut-sized fraction of the pellet before being dropped into a pool of liquid nitrogen. (D) Photograph of the bottom of a grinder's stator with frozen yeast balls before grinding. (E) Illustration of grinder. At the time of grinding, we recommend filling the chamber with liquid nitrogen to a level not higher than the top of the rotor. (F) Photograph of the bottom of a grinder's stator with frozen yeast powder after grinding. (For color version of this figure, the reader is referred to the online version of this chapter.)

Cryomill grinding system can be used for efficient breakage. Use of liquid nitrogen with blenders can be hazardous. We recommend that users of this protocol wear appropriate eye protection. The resulting yeast powder can be stored for months at −80 °C (Fig. 21.1F).

2.2. Extract preparation

To minimize dilution during the preparation of the extracts, buffers, salts, and chemicals are added as concentrated solutions before thawing the frozen yeast powder.

1. Pour yeast powder into an ice-cold beaker. Do not allow the powder to thaw.
2. For each gram of yeast powder, add 100 μl of 10× HK buffer and 20 μl of 10× H buffer supplemented with 4 μl of protease inhibitors (Fig. 21.2A). The yeast powder will absorb the drops of liquid.

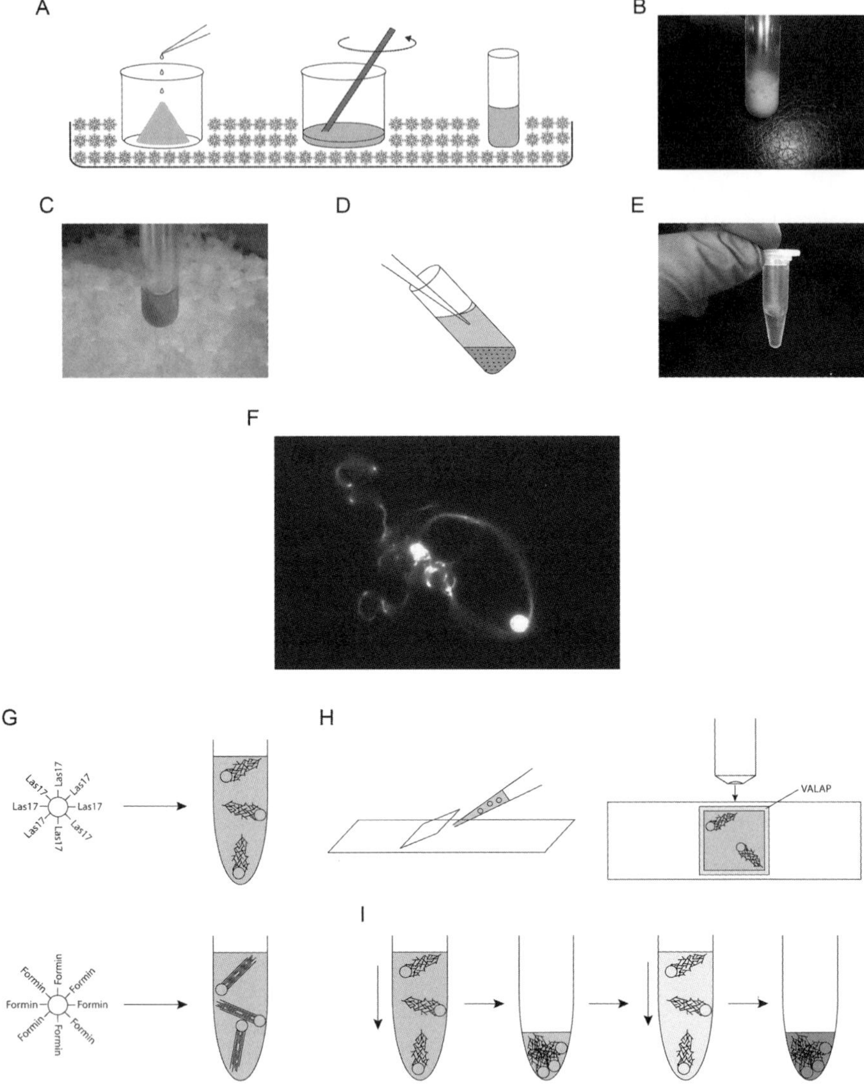

Figure 21.2 Preparation and use of yeast extracts for actin network assembly. (A) Illustration of yeast lysate solubilization. Left drawing represents addition of buffers to the yeast powder. Central panel represents the solubilization step. Right panel represents the transfer of the solution to an ultracentrifuge tube. (B) Photograph of solubilized yeast lysates before high-speed centrifugation. (C) Photograph of yeast extract with cell debris pellet after high-speed centrifugation. (D) Illustration of yeast extract collection. Nonsoluble pelleted fraction (dark orange area) and floating lipid layer (yellow area) are to be avoided. (E) Photograph of a yeast extract ready to be used for actin assembly assays. (F) Example of Abp140-3GFP extract observed in fluorescence microscopy 5 h after preparation (right panel; cable-like actin filament structures of actin filaments

3. With a spatula, mix the liquid with the powder. When the mixture is homogenous, slowly thaw it on ice (Fig. 21.2B).
4. Centrifuge twice at 80,000 × rpm for 20 min in a TLA 100.3 rotor (Fig. 21.2C). After each spin, collect the clear supernatant, avoiding as much as possible the thin layer of lipids floating at the surface. Also, do not pipet near the surface of the pellet to avoid insoluble particles (Fig. 21.2D and E). Keep the extract on ice.

2.3. Protein solubility in extracts: Choice of buffer

Having soluble and active proteins is essential for successful assays. The buffer indicated in the previous paragraph maintains activity of proteins essential for actin assembly assays. Studies indicate that yeast contain high potassium (in the 50–250 m*M* range) and magnesium concentrations (in the 0.1–1 m*M* range), and low calcium concentrations (Rodriguez-Navarro, 2000; van Eunen et al., 2010). Even with the dilution factor resulting from preparation of the extracts, the ionic contribution of the yeast is favorable for actin assembly. Addition of extra chemicals is a possibility, but one should be careful not to add anything that would interfere with actin assembly. For example, addition of EDTA should be avoided because magnesium is necessary for actin polymerization in physiological conditions.

2.4. Extract stability

We usually use yeast extracts for up to 4 h. Extracts must appear clear and homogenous when observed by microscopy and should be free from any visible filamentous actin structures before use in an actin assembly assay (Fig. 21.2F). After 4 h, structures such as actin cables sometimes appear (Fig. 21.2F). For unknown reasons, flash freezing of extracts in liquid nitrogen or storing extracts as glycerol stocks at −20 °C does not preserve actin assembly activity.

may appear). (G) Illustration of the protocol for the assembly of patch-like structures of actin filaments from Las17-coated microbeads and cable-like structures from formin-coated microbeads. (H) Illustration of the protocol for microscopy observation of actin networks. Samples are dropped between a slide and a coverslip, and the sealed with VALAP (in yellow). (I) Illustration of the protocol for the partial purification of actin structures bound to beads for protein composition analysis. Beads are pelleted at low speed, washed carefully multiple times, and quickly pelleted again before resuspension in the buffer of choice. (See the color plate.)

3. SURFACE PROTEIN ACTIVATION

This section describes the procedure to functionalize objects such as microspheres with proteins of interest in order to trigger actin assembly in the yeast extracts. All the steps of the protocols presented in this section must be done on ice or at 4 °C.

Required materials and chemicals

- *HK150 buffer*: 20 m*M* Hepes pH 7.5; 150 m*M* KCl.
- *HKG buffer*: HK150 buffer + 5% glycerol.
- Polybead® Microspheres 2 μm, Polysciences, Inc.
- BSA Sigma A7030. Prepare a 10% filtered stock solution in ddH_2O. Store at −20 °C.
- Slide-A-lyzer Dialysis Cassettes with 10 kDa MWCO, Thermo Scientific.
- Sypro® Ruby Protein Gel Stain, Invitrogen.

3.1. Yeast WASP (Las17) and formin (Bni1) expression and purification

Yeast are useful for efficient expression and purification of recombinant proteins with high yields (Rodal, Duncan, & Drubin, 2002). Isolation of yeast proteins from yeast makes it more likely that proteins will have the appropriate posttranslational modifications. A detailed description of the expression and purification procedure that we use can be found in a previous publication (Rodal et al., 2002). After purification, proteins are dialyzed for 3 h in HKG buffer at 4 °C with a Fastdialyzer, aliquoted, flash-frozen, and stored at −80 °C. For the reasons discussed in Section 3.3, the presence of EDTA should be avoided in the storage buffer. Protein concentrations are determined on a SDS-PAGE Gel stained with Sypro, using a range of actin concentrations as a standard.

We generally express full-length proteins and cleave off the affinity tag using TEV protease after purification to work with proteins that are as similar as possible to the ones present in cells. For example, full-length Las17 is fully functional for Arp2/3 complex activation, and beads coated with full-length Las17 generate patch-like actin networks on beads. However, in specific cases, truncated versions of proteins need to be generated. For example, full-length formins are not functional for actin nucleation because they are autoinhibited (Goode & Eck, 2007). In this case, truncations containing only the FH1 and FH2 C-terminal domains can be used to trigger actin cable assembly (Miao et al., 2013).

3.2. Functionalizing microbeads with actin nucleators

For the functionalization of proteins on polystyrene microbeads, we use the simple method of non-specific adsorption to the bead surface. This protocol has been reliable in our hands, although we cannot rule out the possibility that a fraction of the proteins lose their activity when binding to the surface. Full-length Las17 appears to bind to beads with minimal activity loss. However, formin FH1–FH2 domains alone lose their activity when bound to beads, probably because the flexibility between the two halves of the formin dimer is lost (Xu et al., 2004). The presence of an N-terminal Gst-tag on the protein allows for preservation of formin activity on functionalized beads (Michelot et al., 2007).

1. Dilute Las17 to 100 n*M* in HK150 buffer.
2. Wash 5 μl stock solution of polystyrene microbeads with 100 μl of HK150 buffer. Spin for 1 min at 13,000 rpm in a standard laboratory microfuge. Remove the supernatant, being careful to not disturb the bead pellet.
3. Add 50 μl of the diluted Las17 solution to the top of the beads and pipet up and down several times quickly.
4. Leave on ice for 30 min to allow Las 17 to bind to the beads.
5. Add 5 μl of a 1% BSA solution to saturate the bead surfaces. Pipet up and down thoroughly. Leave on ice for an additional 10 min.
6. Spin beads for 20 s at 5000 rpm in a standard laboratory microfuge. Remove supernatant carefully and immediately resuspend the beads in 100 μl of HK150 buffer supplemented with 0.1% BSA. This will wash off unbound Las17.
7. Spin the bead stock again for 20 s at 5000 rpm. Resuspend the beads in 50 μl of HK150 buffer supplemented with 0.1% BSA.
8. Vortex the beads for 5 s before use. Beads can be used for 12 h without any significant loss of activity if kept at 4 °C.

4. ACTIN ASSEMBLY FROM YEAST EXTRACTS

This section describes the procedure to stimulate actin network assembly in yeast extracts, how to prepare the sample for imaging, and how to handle the beads after network assembly for determining protein composition of actin structures.

Required materials, chemicals, and equipment

- HK150 buffer.

- *T buffer*: 50 m*M* Tris pH 8.8.
- *TU buffer*: T buffer + 8 *M* urea.
- Thermo Scientific Superfrost 76 × 26 mm Slides.
- Agar Scientific L46S24-1 24 × 24 mm Coverglasses No. 1.
- VALAP (1/3 Vaseline; 1/3 lanolin; 1/3 paraffin). Place VALAP on a warm plate (~75 °C) to maintain in liquid state.
- Vortex.
- Olympus IX71 Fluorescence Microscope equipped with a Plan-APOCHROMAT 63 ×/1.4 Oil Ph3 objective. Ideally, the microscope is equipped with a motorized stage to image multiple fields of the same slide.
- Acetone, EM grade.

4.1. Assembly of actin networks from yeast extracts

1. Mix yeast extract with the protein-activated bead stock at a 20:1 volume: volume ratio (Fig. 21.2G).
2. Vortex thoroughly for 15 s.

4.2. Observation of actin networks using phase-contrast or fluorescence microscopy

1. Spread 2.7 μl of reaction between a slide and coverslip (Fig. 21.2H). The small drop must spread well and fill the glass–glass interface without trapping any air bubbles.
2. Seal the coverslip to the slide with melted VALAP. Any defect in the preparation of the sample might lead to an annoying flow of the solution.
3. Once the VALAP has solidified, place the sample on the microscope stage at room temperature.
4. Using bright-field illumination, focus on the beads.

With bright-field illumination, only the polystyrene beads will be seen. After several minutes, if the objective allows for phase-contrast optics, thick actin structures such as cables or Arp2/3-branched actin tails can also be detected. Fluorescence will allow for more sensitive detection of small actin structures if the yeast extract is prepared from a strain expressing an actin-binding protein tagged with a fluorescent protein.

4.3. Actin network protein composition analysis

1. Assemble actin networks around functionalized microbeads for 30–45 min in at least 500 μl of yeast extract at room temperature.

2. Pellet beads with a low-speed centrifugation (1 min at $1000 \times g$). Carefully remove the supernatant.
3. Wash the pellet with 1 ml of HK150 buffer and centrifuge the beads (1 min at $1000 \times g$—Fig. 21.2I).
4. Repeat step 3 once or twice to obtain a more pure sample. Steps 3 and 4 must be performed as rapidly as possible to avoid as much as possible the disassembly of actin structures and dissociation of actin-binding proteins from the beads and networks. The final pellet volume should be less than 25 μl.

Resuspend the purified sample in the buffer that is appropriate to the detection technique being used. For Western blot analysis, dilute the sample in SDS sample buffer. For mass spectrometry analysis, it is necessary to precipitate the proteins and the beads need to be eliminated from the samples. We use the following procedure:

1. Denature the protein mixture by the addition of 250 μl of TU buffer. Incubate for 20 min. Denatured proteins will be detached from the beads.
2. In the presence of urea, polystyrene microbeads do not pellet with centrifugation because the density of the medium is too high. We observed that a fourfold dilution of urea allows the beads to pellet. Add 750 μl of T buffer to lower the density of the medium. Centrifuge 1 min at $1000 \times g$. Beads will pellet in these conditions. Keep the supernatant and discard the pellet.
3. Precipitate the protein mixture with addition of 20% TCA. Leave on ice overnight.
4. Pellet the proteins by centrifugation at 4 °C for 15 min at 13,000 rpm in a standard laboratory microfuge.
5. Wash pellets three times with cold acetone (stored at −20 °C). Spin down pellets in the same conditions for each wash.
6. Dry the pellets for 3 min with a speed-vacuum pump and perform mass spectrometry analysis. Otherwise, samples can be stored for an indefinite time at −80 °C.

As detailed in Michelot et al., (2010), it is important to note that the procedure presented in this section allows only for an enrichment of the proteins bound to actin networks, not a complete purification from the extracts. Therefore, it is important to keep in mind during Western blot or mass spectrometry analysis that not all proteins present in the sample interact specifically with the actin networks. Rather, it is important to employ reference proteins that are not likely to interact with the network to deduce enrichment levels.

4.4. Critical parameters for the success of the assay

Performing successful experiments relies on the fine-tuning of critical experimental parameters. This paragraph aims to provide a few tips for the troubleshooting of yeast extract-based actin assembly assays. We have found that assembling actin networks at the surface of polystyrene microbeads can be relatively straightforward, but optimizing conditions to obtain efficient bead motility requires careful, systematic parameter optimization. We have identified three parameters that are critical for optimal bead motility.

- *Concentration of Las17 on the beads for Arp2/3-based motility.* Studies have shown that efficient Arp2/3-based motility requires precisely tuned protein concentrations and protein–protein ratios (Loisel, Boujemaa, Pantaloni, & Carlier, 1999). Because most parameters are fixed in yeast extracts, the experimenter needs only to systematically vary the surface density of Las17 on the beads. On one hand, if Las17 densities are too low, bead motility will not be achieved because the high concentrations of capping protein present in the extracts (Michelot et al., 2013) will be detrimental to filament elongation after nucleation. On the other hand, Las17 densities that are too high will only result in the formation of dense Arp2/3-rich shells of actin filaments around the beads, but not in motility because symmetry breaking will not occur.
- *Photodamage of the sample.* Fluorophore excitation can have negative effects on biological samples. Interaction of light with fluorescent species induces formation of oxidative species, leading to the photodamage of nearby molecules. In the case of actin polymers, we have observed that excitation of fluorescently labeled samples with too much intensity induces actin filament severing and destroys actin networks. In the case of our yeast extract-based bead assay, too intense or too frequent illumination of the samples arrests bead motility rapidly. There are several ways to limit these unwanted effects. For example, oxygen scavengers might be employed (Aitken, Marshall, & Puglisi, 2008). It is also essential to limit light exposure of the sample as much as possible, and to verify that the actin networks have similar properties in the light exposed versus nonexposed areas of the sample.
- *Number of beads in solution.* It is also important to limit the number of beads present in solution. If the bead concentration is too high, the available pool of actin monomers may be depleted and bead movement affected.

5. DEDUCING PROTEIN FUNCTIONS WITH MUTANT STRAINS

This section explores general strategies for using the yeast system for determining individual protein functions within complex networks. For a specific example of this approach, we refer the readers to one of our recent publications (Michelot et al., 2013).

5.1. Effects of mutations on actin network properties

Mutants of essentially all *S. cerevisiae* proteins are available. Genetic crosses allow for the production of strains with new mutant combinations. For the corresponding protocols, we refer our readers to a book of protocols published on the subject (Amberg, Burke, Strathern, & Cold Spring Harbor Laboratory, 2005).

For any mutation analyzed, wild-type and mutant strains are grown and harvested in identical conditions. Typical effects observed between wild-type and mutated conditions are actin network dynamic property differences, detected, for example, when comparing bead motility rates. Also, we have successfully used these assays to determine how the absence of a given protein affects the recruitment of others (Michelot & Drubin, 2011; Michelot et al., 2013). In order to verify that the absence of the protein of interest is responsible for any observed phenotypes, we recommend a rescue experiment, in which the purified protein is added back to the mutant extract (Michelot et al., 2013).

5.2. Coupling yeast genetics and actin assembly assays to determine protein functions

One problem intrinsic to extract-based assays is that molecular complexity makes dissection of a process into clear functional pathways difficult. Moreover, it is common that several proteins have overlapping functions, thus reducing the phenotypic impact of single mutants. For example, although capping is an essential activity for Arp2/3-derived motility, loss of capping protein function did not cause major motility defects in extracts because other factors have partially overlapping functions (Michelot et al., 2013).

To circumvent this complication, a genetic strategy can be used to dissect complex molecular pathways (Fig. 21.3). Not only is yeast a well-studied organism amenable to advanced genetic analysis, but recent genome-wide

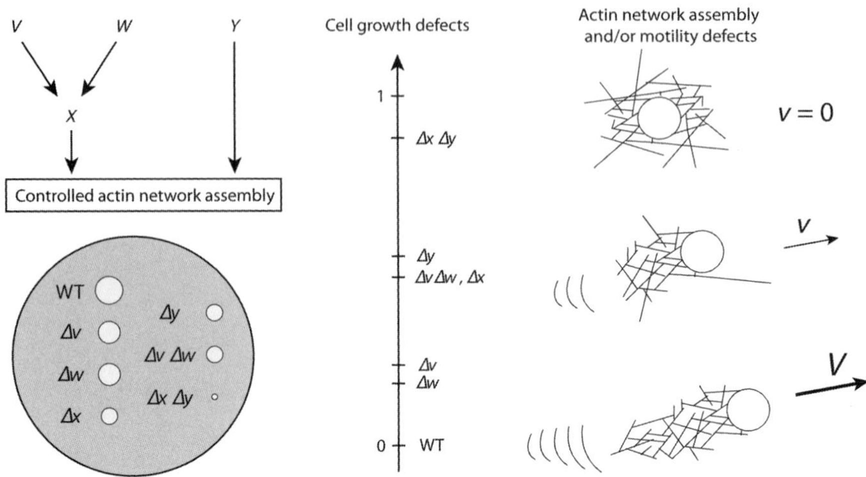

Figure 21.3 Combing yeast genetics with actin network reconstitution assays. The top left hypothetical network represents an example in which four proteins, V, W, X, and Y, function in parallel pathways for assembly of a functional actin network. V, W, and X provide an activity, but Y can provide the same function. The bottom left illustration represents the growth on an agar plate of single yeast colonies from several mutants. The growth rate for each single or double mutant is quantified, as represented on the central scale. Mutants defective in X and Y, which function in parallel "genetically redundant" pathways, have pronounced growth defects. The right drawings indicate that genetic interactions are reflected in defects in actin network assembly, motility, and/or protein composition. (For color version of this figure, the reader is referred to the online version of this chapter.)

genetic interaction screens also provide a wealth of information (Costanzo et al., 2010; Ooi et al., 2006; Schuldiner, Collins, Weissman, & Krogan, 2006). These screens are powerful approaches for unveiling gene functions, functionally grouping genes, and organizing them into pathways.

5.3. Strategy for investigating the function of a protein: From single molecule to cell biology

In this section, we discuss how it can be difficult to interpret phenotypes observed in the bead assays if the biochemical activities of the proteins are not well defined. For example, detection of a decrease in the rate of actin assembly can be due to multiple factors, such as a lower actin filament nucleation frequency, a defective barbed end capping activity, or a drop in the pool of monomeric actin available in solution. Therefore, additional information is often necessary to link protein activities with their contributions to global actin network behavior. In this context, we believe that use of yeast extracts

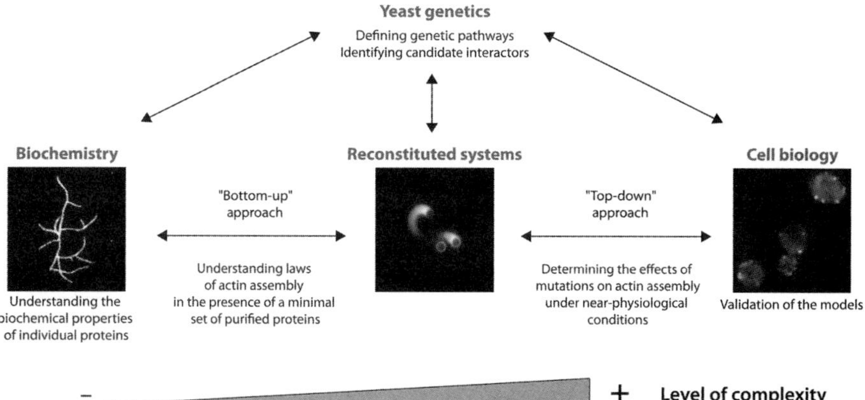

Figure 21.4 An integrated approach for studying the function of proteins with systems of progressive complexity. (For color version of this figure, the reader is referred to the online version of this chapter.)

in our studies needs to be coupled with other adapted experimental setups allowing for the characterization of individual biochemical activities (Michelot et al., 2013). These assays include, for example, pyrene-labeled actin assembly kinetics or total internal reflection microscopy assays to visualize the effect of proteins on the polymerization of individual actin filaments.

More generally, yeast affords an opportunity for an integrated strategy to determine the functions of proteins, using systems of increasing complexity, from simple biochemistry to reconstituted systems and cells (Fig. 21.4). In this context, yeast appears to be a particularly well-suited organism for integrating these types of studies. One particular advantage of studies in yeast is that there is no need to study mixtures of proteins from different species; studies at all levels of complexity can and should be performed on a homologous set of proteins from the same organism, budding yeast.

6. PERSPECTIVES

In this chapter, we have presented protocols and a strategy for determining the activities and functions of proteins in actin-derived cellular processes. We believe that such an approach will also be very useful in the future for the study of other types of cellular processes.

ACKNOWLEDGMENTS

The authors wish to acknowledge support from National Institutes of Health Grant RO1 GM42759.

REFERENCES

Aitken, C. E., Marshall, R. A., & Puglisi, J. D. (2008). An oxygen scavenging system for improvement of dye stability in single-molecule fluorescence experiments. *Biophysical Journal, 94*(5), 1826–1835.

Amberg, D. C., Burke, D., Strathern, J. N., & Cold Spring Harbor Laboratory, (2005). *Methods in yeast genetics: A Cold Spring Harbor Laboratory course manual* (2005 ed.). Cold Spring Harbor, NY: Cold Spring Harbor Laboratory Press.

Carlier, M. F., Laurent, V., Santolini, J., Melki, R., Didry, D., Xia, G. X., et al. (1997). Actin depolymerizing factor (ADF/cofilin) enhances the rate of filament turnover: Implication in actin-based motility. *The Journal of Cell Biology, 136*(6), 1307–1322.

Costanzo, M., Baryshnikova, A., Bellay, J., Kim, Y., Spear, E. D., Sevier, C. S., et al. (2010). The genetic landscape of a cell. *Science, 327*(5964), 425–431.

Fradelizi, J., Noireaux, V., Plastino, J., Menichi, B., Louvard, D., Sykes, C., et al. (2001). ActA and human zyxin harbour Arp2/3-independent actin-polymerization activity. *Nature Cell Biology, 3*(8), 699–707.

Geli, M. I., Lombardi, R., Schmelzl, B., & Riezman, H. (2000). An intact SH3 domain is required for myosin I-induced actin polymerization. *The Embo Journal, 19*(16), 4281–4291.

Goode, B. L. (2002). Purification of yeast actin and actin-associated proteins. *Methods in Enzymology, 351*, 433–441.

Goode, B. L., & Eck, M. J. (2007). Mechanism and function of formins in the control of actin assembly. *Annual Review of Biochemistry, 76*, 593–627.

Idrissi, F. Z., Wolf, B. L., & Geli, M. I. (2002). Cofilin, but not profilin, is required for myosin-I-induced actin polymerization and the endocytic uptake in yeast. *Molecular Biology of the Cell, 13*(11), 4074–4087.

Laurent, V., Loisel, T. P., Harbeck, B., Wehman, A., Grobe, L., Jockusch, B. M., et al. (1999). Role of proteins of the Ena/VASP family in actin-based motility of Listeria monocytogenes. *The Journal of Cell Biology, 144*(6), 1245–1258.

Loisel, T. P., Boujemaa, R., Pantaloni, D., & Carlier, M. F. (1999). Reconstitution of actin-based motility of Listeria and Shigella using pure proteins. *Nature, 401*(6753), 613–616.

Miao, Y., Wong, C. C., Mennella, V., Michelot, A., Agard, D. A., Holt, L. J., et al. (2013). Cell-cycle regulation of formin-mediated actin cable assembly. *Proceedings of the National Academy of Sciences of the United States of America, 110*(47), E4446–E4455.

Michelot, A., Berro, J., Guerin, C., Boujemaa-Paterski, R., Staiger, C. J., Martiel, J. L., et al. (2007). Actin-filament stochastic dynamics mediated by ADF/cofilin. *Current Biology, 17*(10), 825–833.

Michelot, A., Costanzo, M., Sarkeshik, A., Boone, C., Yates, J. R., 3rd, & Drubin, D. G. (2010). Reconstitution and protein composition analysis of endocytic actin patches. *Current Biology, 20*(21), 1890–1899.

Michelot, A., & Drubin, D. G. (2011). Building distinct actin filament networks in a common cytoplasm. *Current Biology, 21*(14), R560–R569.

Michelot, A., Grassart, A., Okreglak, V., Costanzo, M., Boone, C., & Drubin, D. G. (2013). Actin filament elongation in Arp2/3-derived networks is controlled by three distinct mechanisms. *Developmental Cell, 24*(2), 182–195.

Moseley, J. B., & Goode, B. L. (2006). The yeast actin cytoskeleton: From cellular function to biochemical mechanism. *Microbiology and Molecular Biology Reviews, 70*(3), 605–645.

Ooi, S. L., Pan, X., Peyser, B. D., Ye, P., Meluh, P. B., Yuan, D. S., et al. (2006). Global synthetic-lethality analysis and yeast functional profiling. *Trends in Genetics, 22*(1), 56–63.

Pollard, T. D., Blanchoin, L., & Mullins, R. D. (2000). Molecular mechanisms controlling actin filament dynamics in nonmuscle cells. *Annual Review of Biophysics and Biomolecular Structure, 29*, 545–576.

Rodal, A. A., Duncan, M., & Drubin, D. (2002). Purification of glutathione S-transferase fusion proteins from yeast. *Methods in Enzymology, 351*, 168–172.

Rodriguez-Navarro, A. (2000). Potassium transport in fungi and plants. *Biochimica et Biophysica Acta, 1469*(1), 1–30.

Schuldiner, M., Collins, S. R., Weissman, J. S., & Krogan, N. J. (2006). Quantitative genetic analysis in Saccharomyces cerevisiae using epistatic miniarray profiles (E-MAPs) and its application to chromatin functions. *Methods, 40*(4), 344–352.

Theriot, J. A., Rosenblatt, J., Portnoy, D. A., Goldschmidt-Clermont, P. J., & Mitchison, T. J. (1994). Involvement of profilin in the actin-based motility of L. monocytogenes in cells and in cell-free extracts. *Cell, 76*(3), 505–517.

van Eunen, K., Bouwman, J., Daran-Lapujade, P., Postmus, J., Canelas, A. B., Mensonides, F. I., et al. (2010). Measuring enzyme activities under standardized in vivo-like conditions for systems biology. *The FEBS Journal, 277*(3), 749–760.

Welch, M. D., Iwamatsu, A., & Mitchison, T. J. (1997). Actin polymerization is induced by Arp2/3 protein complex at the surface of Listeria monocytogenes. *Nature, 385*(6613), 265–269.

Xu, Y., Moseley, J. B., Sagot, I., Poy, F., Pellman, D., Goode, B. L., et al. (2004). Crystal structures of a Formin Homology-2 domain reveal a tethered dimer architecture. *Cell, 116*(5), 711–723.

CHAPTER TWENTY-TWO

Xenopus Egg Cytoplasm with Intact Actin

Christine M. Field[*,†,1], **Phuong A. Nguyen**[*,†], **Keisuke Ishihara**[*,†], **Aaron C. Groen**[*,†], **Timothy J. Mitchison**[*,†]

[*]Department of Systems Biology, Harvard Medical School, Boston, Massachusetts, USA
[†]The Marine Biological Laboratory, Woods Hole, Massachusetts, USA
[1]Corresponding author: e-mail address: Christine_field@hms.harvard.edu

Contents

Abstract

We report optimized methods for preparing Xenopus egg extracts without cytochalasin D, that we term "actin-intact egg extract." These are undiluted egg cytoplasm that contains abundant organelles, and glycogen which supplies energy, and represents the least perturbed cell-free cytoplasm preparation we know of. We used this system to probe cell cycle regulation of actin and myosin-II dynamics (Field et al., 2011), and to reconstitute the large, interphase asters that organize early Xenopus embryos (Mitchison et al., 2012; Wühr, Tan, Parker, Detrich, & Mitchison, 2010). Actin-intact Xenopus egg extracts are useful for analysis of actin dynamics, and interaction of actin with other cytoplasmic systems, in a cell-free system that closely mimics egg physiology, and more generally for probing the biochemistry and biophysics of the egg, zygote, and early embryo. Detailed protocols are provided along with assays used to check cell cycle state and tips for handling and storing undiluted egg extracts.

Methods in Enzymology, Volume 540
ISSN 0076-6879
http://dx.doi.org/10.1016/B978-0-12-397924-7.00022-4

1. INTRODUCTION

Here, we report methods for preparing undiluted cytoplasmic extracts with intact actin dynamics from unfertilized Xenopus eggs. Our system builds on "cytostatic factor (CSF) extract," which is prepared by centrifugal crushing of packed, unfertilized eggs, where the meiotic state is stabilized by CSF. CSF extract maintains physiological energy homeostasis from endogenous stores (Niethammer, Kueh, & Mitchison, 2008) and has long provided an excellent system for reconstitution of cell cycle oscillation, meiotic spindle assembly, and many other processes (Chang & Ferrell, 2013; Desai, Murray, Mitchison, & Walczak, 1999; Maresca & Heald, 2006; Murray & Kirschner, 1989). CSF extract is normally prepared from cytochalasin D-treated eggs to block actin polymerization, and additional cytochalasin D is added after extract preparation. This prevents actomyosin-driven gelation-contraction that tends to interfere with spindle assembly and other reactions, but precludes reconstitution of actin-dependent processes. Previous reports described egg extracts prepared in the absence of cytochalasin (Pinota et al., 2012; Valentine, Perlman, Mitchison, & Weitz, 2005; Waterman-Storer et al., 2000). Our approach is similar, but we have systematically investigated preparation methods and potential technical problems.

Actin-intact egg extract stabilized in a state of high Cdk1 activity by CSF, which we will term "M-phase extract," and its low Cdk1 activity counterpart "I-phase extract" reconstitute endogenous actin dynamics and represent a step forward in terms of physiological relevance. We used these preparations to probe cell cycle regulation of bulk actomyosin behavior (Field et al., 2011), which is relevant to the mechanical behavior of the interior of the egg and early embryo, and to reconstitute growth and interaction of the large, interphase microtubule asters that organize early embryos (Mitchison et al., 2012). Actin-intact extract should be adaptable to microfluidic, water-in-oil droplet, and giant unilamellar vesicle formats that are increasingly being used to generate cell-like enclosures with controlled size and shape (Good, Vahey, Skandarajah, Fletcher, & Heald, 2013; Hazel et al., 2013; Pinota et al., 2012; Richmond et al., 2011).

2. Actin-Intact Extract Preparation

2.1. Requirements for extract preparation

2.1.1 Buffer and reagent stocks

10× MMR: 50 m*M* Na-HEPES, pH to 7.8 with NaOH; 1 m*M* EDTA; 1 *M* NaCl; 20 m*M* KCl; 10 m*M* $MgCl_2$; 20 m*M* $CaCl_2$. Autoclave and store at RT; MMR can also be prepared as a 25× stock.

20× XB salts: 2 *M* KCl; 20 m*M* $MgCl_2$; 2 m*M* $CaCl_2$.

2 M sucrose: sterile filter and store at 4 °C.

1 M K-HEPES: pH to 7.7 with KOH, sterile filter, and store at 4 °C.

0.5 M K-EGTA: pH to 7.7 with KOH.

1 M $MgCl_2$: sterile filter and store at RT.

Protease inhibitors (LPC):

10 mg/ml each of leupeptin, pepstatin A, and chymostatin (Sigma #L2884, P5318, C7268, respectively) dissolved in DMSO; store in 100 μl aliquots at −20 °C.

Cysteine, free base (Sigma #C7352).

Silicone oil AP 100 (Sigma #10838).

100× energy mix: 750 m*M* creatine phosphate; 100 m*M* ATP; 100 m*M* $MgCl_2$, pH to 7.0.

Store in 100 μl aliquots at −20 °C.

Hormone stocks for priming frogs and inducing ovulation:

Pregnant mare serum gonadotropin (PMSG) (Calbiochem #367222): 1000 U/ml dissolved in water and stored at −20 °C. Note: Do not freeze/thaw.

Human chorionic gonadotropin (HCG) (Sigma #CG-10): 1000 U/ml dissolved in water and stored at 4 °C.

Hormones are injected into the dorsal lymph sac using a 27-gage needle.

2.1.2 Equipment

Glass crystallizing dishes (VWR #89000-294): 150 mm diameter × 75 mm height or equivalent. For washing eggs.

13 × 51 mm UltraClear tubes (Beckman #344057).

Clinical centrifuge: Low-speed spin to pack eggs.

Beckman ultracentrifuge and SW 55 Ti or MLS-50 rotor.

Note: A swinging-bucket rotor is required to achieve proper separation of cell components.

2.1.3 Primed frogs

Frogs for extract preparation are primed using progesterone (present in PMSG) that induces maturation of oocytes. Inject 100 U (0.1 ml) of PMSG per frog. Primed frogs are stored in dechlorinated water containing 2 g/L rock salt (frog water) in a cool room (ambient temperature of 16–20 °C). We wait for 2 days after priming before inducing them to lay eggs (see below). The priming lasts for up to 2 weeks.

2.1.4 Day before extract prep

Sixteen to eighteen hours before extract preparation, two to three primed frogs are induced to ovulate by injection with 500 U (0.5 ml) of HCG. After injection, frogs are left in frog water for at least 3 h. (Occasionally, frogs vomit after injection, which can be deleterious for egg quality.) Frogs are then moved to separate containers (Fisher #03-484-21) with 1 l of 1 × MMR for egg laying. Containers are placed in a dark and quiet 16 °C incubator overnight.

Eggs sitting in MMR longer than 18 h produce inferior extract. Egg quality must take precedence over egg quantity. A good frog usually produces about 0.8 ml of extract. It is best not to use any batches of eggs containing >10% of lysed eggs, activated[1] eggs, or "puffballs" (swollen eggs that are often white and puffy). Strings of eggs that do not have fully separated jelly coats can be used as a last resort. In general, we prepare separate extracts from each frog.

2.1.5 Setup for extract preparation (morning of the prep)

For a two- to three-frog prep:

a. *Buffers:* Prepare just before use unless indicated; otherwise, all glassware should be rinsed well with double distilled water (ddH_2O) prior to use. All buffers should be at 16–18 °C.

600 ml Dejellying solution: 2% (w/v) cysteine in 1 × XB Salts; pH to 7.8 with 10 N KOH. Use within 1–2 h.

[1] If Xenopus eggs are activated, they lose their metaphase arrest. The black pigment in the animal pole of the egg will contract if activation occurs. If significant numbers of activated eggs are included in the preparation, the M-phase arrest of the extract will be lost.

1–1.5 l CSF-XB: 10 mM K-HEPES, pH 7.7; 100 mM KCl; 1 mM $MgCl_2$; 0.1 mM $CaCl_2$; 50 mM sucrose; 1 mM $MgCl_2$; and 5 mM K-EGTA.[2] Prepare using 20× XB salts, 2 M sucrose, 1 M $MgCl_2$, 0.5 M K-EGTA, and 1 M K-HEPES, pH 7.7 stocks; to maintain a pH of 7.7 at 10 mM HEPES, add 11 μl 10 N KOH per 100 ml buffer.

100 ml CSF-XB + PIs: CSF-XB + 10 μg/ml LPC; use 100 ml of CSF-XB (of the liter prepared above), add LPC to 10 μg/ml, and mix immediately.

b. *Equipment:* The centrifuge and rotor should both be at 4 °C.

Wide bore Pasteur pipets: Glass Pasteur pipets with tips broken and fire-polished to ~3–4 mm diameter; or polypropylene transfer pipets with tips cut off.

UltraClear tubes (13 × 51 mm).

Glass crystallizing dish: rinse well with ddH_2O and then with MMR, one per frog.

2.2. Procedure for M-phase extract preparation

All steps should be performed at 16–18 °C unless noted otherwise.

Always keep eggs under buffer.

2.2.1 Washing eggs

(1) Pour off half of MMR (0.5 l) from each frog container, and transfer eggs in MMR to a glass crystallizing dish ("egg dish"). Remove as much MMR as possible, while keeping egg dish tilted so as to keep eggs under buffer.

(2) Wash eggs in fresh MMR (3 × 100–200 ml) until all of the debris is removed. As eggs settle quickly, pouring out the MMR immediately after the eggs have settled allows easy removal of debris. "Gardening" of eggs (removing puffballs, activated and defective eggs with a wide bore Pasteur pipet) should start during this step and continue during dejellying and subsequent wash steps.

2.2.2 Removing the jelly coat

(3) Remove as much MMR as possible and add dejellying solution (100–200 ml). Swirl eggs gently in the dejellying solution. After

[2] Unfertilized Xenopus eggs are arrested at metaphase of meiosis II by a biochemical activity—partly composed of the APC/C inhibitor, XErp1/Emi2-termed cytostatic factor (CSF) (Wu & Kornbluth, 2008). Upon fertilization, sperm entry induces a calcium rise, which results in degradation of XErp1/Emi2 and release of M-phase arrest (Rauh, Schmidt, Bormann, NIgg & Mayer, 2005). The CSF-XB buffer contains the calcium chelator EGTA and thus metaphase arrest is maintained during extract preparation.

approximately 3 min, pour off the dejelly solution and add more (100–200 ml).

(4) While eggs are dejellying, pipet 1 ml of CSF-XB + PIs to each UltraClear centrifuge tube.

(5) After eggs are dejellied (5–7 min total in the dejelly solution), they will pack tightly and orient with their animal poles upward. The volume of settled eggs will decrease approximately fivefold. Remove as much dejellying solution as possible.

2.2.3 Washing and collecting eggs

(6) Rinse eggs thoroughly with CSF-XB (3–4 × 100–200 ml, removing as much buffer as possible each time). Dejellied eggs are quite fragile. Pour solutions down the side of the wash vessel and gently swirl the eggs ensuring maximal buffer exchange between washes. During the washes, the eggs should be "gardened," see Step 2 and large pieces of debris should be removed using a wide bore Pasteur pipet.

(7) Rinse eggs in CSF-XB + PIs (2 × 25–50 ml). Leave eggs in a small volume of CSF-XB + PIs after the second wash.

(8) Draw eggs up into the wide bore Pasteur pipet and slowly deposit them into the UltraClear tube with 1 ml CSF-XB + PIs. Insert pipet below the meniscus to break surface tension and gently release eggs into the solution. Minimize transfer of buffer along with the eggs.

(9) Fill the tube three quarters full—note there is a packing spin (below), so the egg height in the tube will be reduced. Aspirate excess buffer from the top of the eggs.

(10) Gently layer 0.5 ml of silicone oil[3] on top of the eggs.

2.2.4 Packing and lysis of eggs, and stratification of cellular components

(11) *Packing spin*: Pack the eggs by spinning at setting #5 (2000 rpm) for 30 s and full speed (#7; 2500–3000 rpm) for 15 s in a clinical centrifuge.[4] Aspirate all buffer and silicone oil from the top of packed eggs. A successful packing spin is critical for a good extract. Eggs must be still intact but tightly packed so that most buffer has been displaced and the resultant extract is not diluted. Signs of good packing: eggs are no

[3] Use of silicone oil minimizes dilution during extract preparation. The density of silicone oil AP 100 is intermediate between that of buffer and cytoplasm; thus, the oil displaces buffer between the eggs during the packing spin but floats to the top of the cytoplasmic layer during the crushing spin.

[4] All clinical centrifuges are not equivalent and you might have to adjust the speed or timing.

longer spherical, and they do not move within the tube if you tilt the tube with open end downward to remove buffer (see Fig. 22.1A). Time is another important variable. Generally, Steps 1–12 (from collecting eggs to the crushing spin) should take no more than 45–60 min.

(12) *Crushing spin*: Transfer the tubes containing packed eggs to the SW 55 Ti rotor in an ultracentrifuge. Crush eggs by centrifugation at 11,100 rpm ($15{,}000 \times g$) for 15 min at 4 °C (full brake). Figure 22.1B compares tubes of eggs crushed in the presence (left) or absence (right) of cytochalasin D. The top light yellow layer contains lipid droplets and the bottom dark layer contains the yolk, pigment, and nuclei. The middle slightly muddy, straw-colored layer is the desired cytoplasmic extract. Without cytochalasin, there is an additional black layer directly under the lipid.

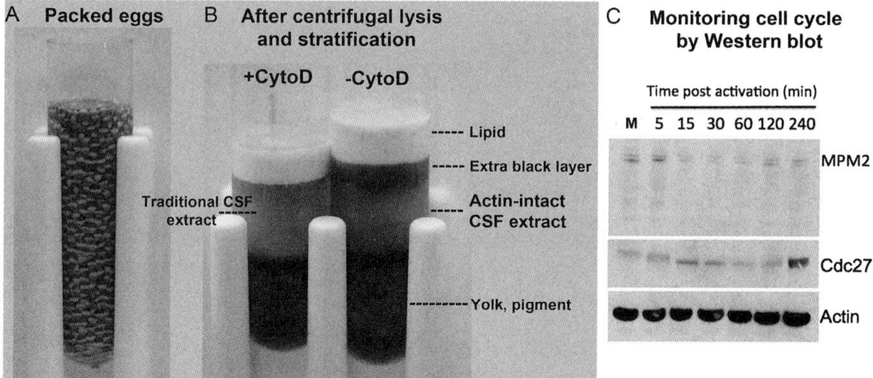

Figure 22.1 Extract preparation and monitoring cell cycle progression. (A) A centrifuge tube full of unfertilized eggs after the packing spin. Note that the eggs have become slightly deformed as they packed together, forcing out excess buffer. (B) Centrifuge tubes after successful crushing spins. On the left, eggs were pretreated with cytochalasin D to yield traditional M-phase (CSF) extract. The cytoplasmic layer always separates well. On the right, cytochalasin D was omitted to yield actin-intact M-phase extract. Note the presence of an additional band of dark material in the actin-intact system. In this prep, it is well separated from the cytoplasmic layer. This layer is a mixture of mitochondria, smooth ER and lipid droplets as shown by electron microscopy (data not shown). (C) Western blot characterization of cell cycle state of extract. M-phase extract (M) is compared to extract activated by calcium addition. Calcium is added, the extract incubated at 20 °C, and samples taken at times shown. Note the lower intensity of MPM2 starting at 15 min, indicating progression from M-phase arrest and the beginning of I-phase. At 120 min, increased intensity indicates reentry into M-phase. The same samples were also probed with an antibody against Cdc27. Note the shift in electrophoretic mobility following M-phase exit (see 5 min). (See the color plate.)

2.2.5 Harvesting and freezing egg extract

(13) *Harvesting extract*: Puncture tube near the bottom of the cytoplasmic layer with an 18-gage needle (on an 1-cc syringe) and gently draw out extract with the needle bevel upward. For the cytochalasin-free extract, make sure to avoid collecting any of the upper black layer.[5] Gently deposit extract in an Eppendorf tube on ice. Immediately add 1/1000 volume of protease inhibitors, 1/40 volume of 2 *M* sucrose, and 1/100 volume of energy mix. A good M-phase extract can be stored on ice for 4–9 h without loss of any activity.

2.3. Converting M-phase extract to I-phase

Sperm entry upon fertilization induces a transient rise in cytosolic calcium, triggering degradation of CSF, and release of M-phase arrest. This natural response to calcium is exploited in the preparation of I-phase extracts. $CaCl_2$ from a 50 × or 100 × stock is added to M-phase extract to a final concentration of 0.4 m*M*, and the extract is immediately mixed. This causes a transient increase in Ca^{2+} that mimics fertilization. Calcium concentration decreases rapidly due to uptake into ER, and failure to mix rapidly and completely is a likely source of failure to release the CSF arrest. We use 1.5-ml Eppendorf tubes with 25–100 μl of extract. After calcium is added, gently pipet the entire solution up and down three to five times to insure mixing, or flick the tube ~8 ×. To prevent M-phase reentry, add protein synthesis inhibitors (cycloheximide 100 μg/ml or puromycin 50 μ*M*). This blocks synthesis of cyclin B and cell cycle progression.

Two alternative methods for making I-phase extract with intact actin have previously been described:

(1) Treat dejellied eggs with 0.2 μg/ml calcium ionophore (Sigma A23187) for 5–7.5 min to mimic fertilization, followed by extensive washing in XB (CSF-XB without $MgCl_2$ and EGTA) before the crushing spin (Leno & Laskey, 1991; Valentine et al., 2005). Here, the CSF arrest is released before the packing spin so the cell cycle clock is already ticking before the extract is made. Electrical activation of eggs (Murray, 1991) can be substituted for ionophore. Activated eggs have

[5] If the crushing spin has not been successful, this black material is scattered throughout the cytoplasmic layer and the extract will be inferior. It can be filtered through a 0.9- to 1.2-μm syringe filter (at a loss of approximately 50% of the volume). This filtered extract maintains its cell cycle state. Alternatively, extracts can be spun for 30 s at 7000 × *g* in a microfuge at room temperature, placed on ice for 1 min, and then spun for 30 s at 7000 × *g* at room temperature again. The black material will float to the top; isolate the lower straw-colored cytoplasm.

different mechanical properties from CSF-arrested eggs (Elinson, 1983). We found that preactivated eggs pack less tightly during the packing spin, which leads to diluted extracts. This method tends to generate "cycling" extracts that naturally undergo multiple rounds of cell cycle.

(2) No deliberate activation of eggs, but packing and crushing spins performed in a buffer without EGTA (Waterman-Storer et al., 2000). As the eggs are packed and crushed, the CSF arrest is released, presumably by Ca^{2+} entry.

2.4. Monitoring cell cycle state

2.4.1 Phosphoprotein analysis (Fig. 22.1C)

The mitotic phosphoprotein monoclonal antibody (MPM2; Millipore #05-368) recognizes many phosphorylated Cdk1 substrates. Cdc27 protein, recognized by antibody AF 3.1 (Upstate #05-906), undergoes a large mobility shift by SDS page due to loss of M-phase phosphorylation (Rankin & Kirschner, 1997).

2.4.2 Gelation-contraction

Actin-intact M-phase extract undergoes bulk gelation-contraction within a few minutes of warming to RT (Fig. 22.2B). We use this routinely as a spot-check of extract quality, and also to monitor progress to I-phase following Ca^{2+} addition.

2.4.3 Microtubule aster morphology

Add fluorescent tubulin probe, then 7.5% DMSO to induce polymerization. Incubate at 20 °C for 20 min and image. In M-phase, the microtubules assemble into asters under the influence of dynein and NUMA (Verde, Berrez, Antony, & Karsenti, 1991), while in I-phase they form long, interconnected bundles.

2.4.4 Chromatin morphology

Add permeabilized Xenopus sperm nuclei (Desai et al., 1999) to extract at a final concentration of ~100/μl. Visualize by adding a DNA dye to the extract (e.g., Hoechst 1 μg/ml). In M-phase, chromatin is condensed and slowly resolves into mitotic chromosomes. In I-phase, it gradually decondenses, acquires a nuclear envelope, and forms spherical nuclei (Murray, 1991).

2.5. General considerations when working with extract

2.5.1 Storage

Best results are obtained with freshly made extract. This can be stored on ice for up to ~9 h, but best results are obtained in the first 4 h. Extract can be flash-frozen in liquid N_2 in PCR tubes and stored at −80 °C for up to 1 month. We suspect freezing mainly damages organelles, as organelle-free cytosol can be frozen without loss of activity (Groen, Nguyen, Field, Ishihara, & Mitchison, 2014). The sucrose (50 m*M* final) and energy mix added to the extract immediately after it is collected (see above) are important during freezing. Sucrose protects the organelles and the energy mix compensates for possible mitochondrial damage. On thawing, it behaves similarly to fresh extract in many assays, for example, gelation-contraction, but is never quite as good as fresh for reconstituting complex microtubule arrays.

2.5.2 Mixing and dilution

Egg extract is quite viscous, and it is important to mix well when adding reagents and probes. See Section 2.4 about I-phase extract preparation for advice on mixing. It is important to keep the extract concentrated. The total volume of all aqueous additions should be <10%. If the volume added is kept small, the buffer usually does not matter, but high salt concentrations should be avoided. The total volume of DMSO added must be <0.5% if microtubule dynamics are important. Higher concentrations of DMSO induce microtubule polymerization.

2.5.3 Small molecule drugs

We often use these to perturb cytoskeletal proteins and kinases. We find that the EC_{50} for drug action is often 10- to 1000-fold higher in extract than in tissue culture cells, although mechanism and specificity are maintained. We believe that there are two reasons: (1) tissue culture cells are surrounded by a large reservoir of drug in the medium and can accumulate drugs to a higher internal concentration; (2) hydrophobic drugs partition into lipid reservoirs and/or acidic compartments present in extract. For these reasons, EC_{50} for drug action in extract must be determined empirically, and we aim to use them at ~5 × the EC_{50} to maximize specificity. If drugs are added from a DMSO stock, it is important to keep the total percent DMSO at <0.5% (see above).

2.6. Imaging experiments

2.6.1 Passivation of glass surfaces

We typically image thin squashes between two cover slips, or a slide and a cover slip. Glass tends to bind dynein and other proteins that influence microtubule organization. To prevent this, we passivate glass surfaces with a polyethylene glycol (PEG) layer. The most rigorous approach involves chemically coupling the PEG to the glass surface following a method described in Bieling, Telley, Hentrich, Piehler, and Surrey (2010). We made the following modifications:

(a) Instead of cleaning cover slips with "Piranha" solution, we use oxygen plasma (Technics plasma etcher 500-II, oxygen gas at 100 mTorr, glow discharge at 200 W for 4 min).

(b) Instead of a two-step silanization followed by PEG coupling procedure, we couple a silane–PEG directly (methoxy PEG silane, MW 5000, JenKem Technology USA #A3037) to the glass. Silane–PEG is dissolved in 100% ethanol to 10% (w/v) (needs heating to fully dissolve); cover slip sandwiches are made with this solution in between and incubated at 75 °C for 2 h. These cover slips can be stored with a desiccant at −20 °C for up to 1 year. A simpler PEG passivation that is good for exploratory studies is to adsorb poly-(L-lysine)–PEG to a cleaned glass surface (Bieling et al., 2010). This works, but tends to generate inhomogeneously passivated surfaces.

We also found passivation of cover slips by either whole casein or kappa-casein (see Groen et al., 2014) minimized the absorption of many microtubule-interacting proteins on glass cover slips—including large amounts of dynein, as judged by microtubule motility and its inhibition by dynein inhibitors CC1 or ciliobrevin (Firestone et al., 2012; Quintyne et al., 1999). An advantage of casein coating is that (unlike PEG passivation) growing microtubules track along the coated surface, within a Total Internal Reflection Fluorescence (TIRF) illumination field. Specifically, cover slips passivated with whole casein (Sigma C3400) bind less dynein, but enrich for plus end-directed motors (Maloney, Herskowitz, & Koch, 2011) that can sequester kinesin cargoes to the glass surface. Whereas cover slips passivated with kappa-casein (Sigma C0406) bind much less kinesin but more dynein, resulting in significant dynein activity on the glass surface. We obtained good TIRF images of astral microtubules while minimizing motor-generated forces from the substrate by coating the imaging cover slip with kappa-casein and the other (or top) cover slip with PEG, and using a dynein inhibitor. We routinely use 1 mg/ml CC1, an easily expressed fragment of p150Glued that blocks dynein function (Quintyne et al., 1999). Vanadate usually has similar effects in extract systems, but may interfere with phosphatase activity.

2.6.2 Microscopy

For bulk gelation-contraction assays (Fig. 22.2), we used darkfield illumination to visualize organelles, or widefield fluorescence to image actin probes. For interphase aster assays, we used widefield (Fig. 22.3) and spinning disc confocal fluorescence microscopy. For widefield, we used mainly 10 × Plan/APO NA 0.45 and 20 × Plan/APO NA 1.4 objectives on an upright Nikon Eclipse 90i microscope. With a mercury-halide light source and cooled CCD camera, typical exposures were 300–1000 ms at 10 × and 50–600 ms and 20 ×.

2.6.3 Fluorescent probes

Antibody and tubulin probes are discussed in another article in this volume (Groen et al., 2014). For imaging F-actin, we used actin- and phalloidin-based probes (Field et al., 2011), fluorescein–Lifeact peptide (Riedl et al., 2008), Lifeact–GFP (Riedl et al., 2008), and GFP–Utrophin (Burkel, von Dassow, & Bement, 2007). Directly labeled actin gives very low contrast images, presumably due to the large monomer pool (Rosenblatt, Agnew, Abe, Bamburg, & Mitchison, 1997), and is not usable for most applications. Phalloidin is strongly perturbing of dynamics. Lifeact–GFP and GFP–Utrophin give similar images. We suspect that Utrophin–GFP is slightly more perturbing, as it tends to induce bundles in a concentration-dependent manner. All probes have the potential to perturb and should be added at the minimum concentrations needed for imaging. We often aliquot fluorescent probes as 10–100 × stocks in actin-intact extract and then freeze in small aliquots. This preserves the activity of protein probes and allows efficient use. Freezing tends to compromise the activity of whole extract, but adding 10% (v/v) of frozen extract to fresh has no discernable negative effect.

2.6.4 Specimen preparation

For interphase aster imaging experiments, we add fluorescent probes and mix well before adding Ca^{2+}. After calcium addition, extract is preincubated for 2–20 min at 20 °C with occasional flicking of the tube to allow progression to interphase and prevent gelation-contraction after making a squash. We then add permeabilized sperm (Desai et al., 1999), centrosomes, or artificial centrosomes made by coating Dynabeads with activating antibody to AurkA (Tsai & Zheng, 2005). The sample is mixed well and squashed (4–5 μl) between two passivated cover slips (bottom: 22 × 22 mm, top: 18 × 18 mm) for imaging.

Gelation-contraction makes imaging M-phase extract difficult. One approach is to make a thin squash (1–2 μl) between uncoated cover slips, which physically blocks contraction.

3. CELL CYCLE-REGULATED ACTOMYOSIN CONTRACTILITY

Figure 22.2 shows examples of cell cycle regulation of bulk actomyosin contractility in actin-intact egg extract. In M-phase, where Cdk1 activity is high, we observed dramatic gelation-contraction that transported light-scattering

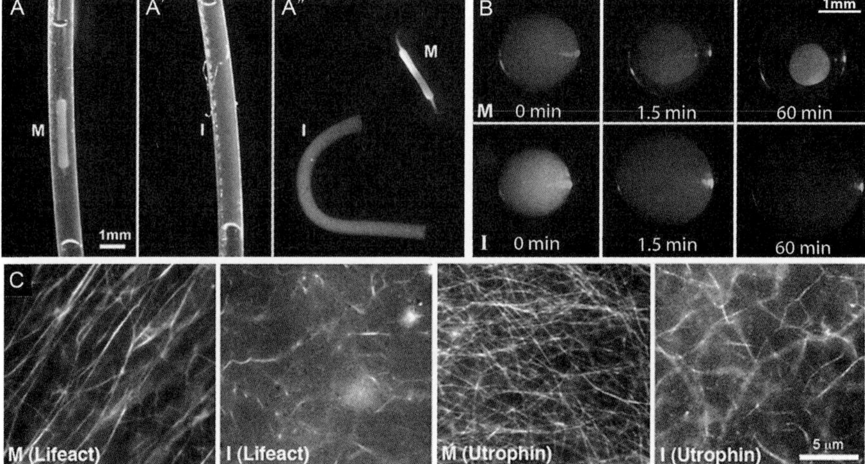

Figure 22.2 Cell cycle regulated bulk actin gelation-contraction in actin-intact egg extract. (A, A′) M-phase (M) and I-phase (I) actin-intact extract were incubated at 20 °C in Teflon tubing. Note the gelation-contraction in the M-phase, but not in the I-phase tubing. Darkfield microscopy through a dissecting microscope. (A″) After 32 min at 20 °C, the same extracts as A, A′ were extruded into a physiological buffer (CSF-XB) and imaged ~1 min later. The organelle-rich, contracted part of the M-phase cylinder was gelled, while the remainder disperses as a liquid. The entire I-phase cylinder remained gelled for many minutes. (B) 1 μl droplets of M-and I-phase extracts were pipetted onto a cover slip under mineral oil and imaged by darkfield microscopy. M-phase extract initiated gelation-contraction 1–3 min after warming to 20 °C, I-phase did not, and the droplet spread over time. This simple assay was used routinely to check extract quality and test cell cycle state. (C) Comparison of actin filament morphology in M- and I-phase extract using two probes, Lifeact peptide coupled to fluorescein and GFP–Utrophin. 100× objective, widefield fluorescence. Note more abundant F-actin bundles in M-phase. In this experiment, 2 μl of extract was squashed between a slide and a 22-mm^2 cover slip, which blocked gelation-contraction and allowed imaging. Note that the portions of this figure A, A′, A″, and C are from figure 3 in Field et al. (2011).

organelles to the center of the contracted gel (M in Fig. 22.2A and B). In I-phase, there was no bulk gelation-contraction (I in Fig. 22.2A and B). Although I-phase extract did not contract, it did gel; when extract that had been incubated in a piece of tubing was extruded into buffer, the cylindrical shape of the gelled cytoplasm was retained (I in Fig. 22.2A″). When M-phase cytoplasm was extruded, only the contracted gel region remained solid. Cytochalasin treatment blocked gelation-contraction, showing that actin polymerization was required (not shown). F-actin imaging showed that actin bundles were more numerous in M- than in I-phase extract (Fig. 22.2C), which may account for enhanced contractility. Myosin-II activity was similar in M- and I-phase (Field et al., 2011).

The physiological significance of bulk gelation-contraction in M- but not in I-phase extract is not clear. We believe that it reflects physiological regulation, but in the whole egg, the actin cytoskeleton is anchored to the cortex, which prevents bulk contraction. Actin-intact egg extract presumably mimics the interior cytoplasm of eggs. M-phase contractility in bulk cytoplasm may be important for (i) concentration of chromosomes during meiosis-I prophase, (ii) movement of meiotic spindles to the cortex, and (iii) maintaining the position of the spindle during mitosis (Field & Lenart, 2011).

4. CELL-FREE RECONSTITUTION OF I-PHASE ASTER GROWTH AND INTERACTION

Figure 22.3 shows examples of large I-phase asters assembled in actin-intact egg extract. We are using this system to investigate the dynamic organization of the I-phase asters that position centrosomes, nuclei, and cleavage furrows in early embryos (Mitchison et al., 2012; Wühr, Tan, Parker, Detrich, & Mitchison, 2010). For these experiments, the extract was supplemented with various fluorescent probes and permeabilized Xenopus sperm, whose centrosomes serve at microtubule-organizing centers. Note that F-actin and intermediate filaments colocalize with microtubule asters. Addition of nocodazole to block microtubule polymerization showed that this global organization of actin and keratin filaments into islands depended on microtubules (not shown). ER networks (Wang, Romano, Field, Mitchison, & Rapoport, 2013), mitochondria, and endosomes + lysosomes visualized using Lysotracker dyes are also organized by microtubules in this system, which opens new approaches to probing the biochemistry and biophysics of global cytoplasmic organization in Xenopus zygotes and early embryos.

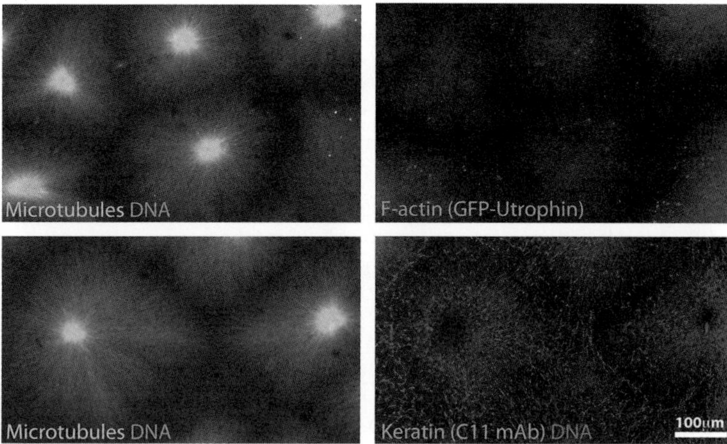

Figure 22.3 Actin and keratin filaments associate with I-phase microtubule asters. Actin-intact M-phase extract was supplemented with fluorescently labeled tubulin and either GFP–Utrophin to visualize F-actin or directly labeled monoclonal antibody to keratin to visualize keratin filaments. Permeabilized Xenopus sperm (Desai et al., 1999) was added to supply centrosomes, and Ca^{2+} added to initiate I-phase. A volume of 8 μl extract was spread between a whole casein-passivated slide and a 18-mm^2 cover slip. Aster assembly was imaged live with a 10 × objective and widefield fluorescence. These images were taken ~40 min after initiating aster assembly. Note that both F-actin and keratin filaments coorganize into islands associated with each microtubule aster. (See the color plate.)

Actin filaments are not required for robust growth of interphase asters in egg extract, and cytochalasin B does not block aster growth or positioning in Zebrafish (Wühr et al., 2008) or Xenopus embryos (M. Wühr, personal communication). We nevertheless prefer actin-intact over cytochalasin-treated extract for aster reconstitution because (i) it is intrinsically more physiological; (ii) it gels (see Fig. 22.2), which reduces convective flow under the cover slip and facilitates imaging; and (iii) it allows probing of actin organization in asters (Fig. 22.3). Actin-intact egg cytoplasm will likely be useful for reconstitution of many other aspects of cytoplasmic organization, including signaling from microtubules to the cortex to position cleavage furrows, which we are actively investigating.

ACKNOWLEDGMENTS

Supported by GM39565. We thank Nikon Inc. for support of microscopy at MBL and the Nikon Imaging Center at Harvard Medical School. MBL work was supported by fellowships from the Evans Foundation, MBL Associates, and the Laura and Arthur Colwin Summer Research Fund. We thank Margaret Coughlin for investigation of the "black matter" in Fig. 22.1B.

REFERENCES

Bieling, P., Telley, I. A., Hentrich, C., Piehler, J. J., & Surrey, T. (2010). Fluorescence microscopy assays on chemically functionalized surfaces for quantitative imaging of microtubule, motor, and +TIP dynamics. *Methods in Cell Biology*, *95*, 555–580.

Burkel, B. M., von Dassow, G., & Bement, W. M. (2007). Versatile fluorescent probes for actin filaments based on the actin-binding domain of utrophin. *Cell Motility and the Cytoskeleton*, *64*, 822–832.

Chang, J. B., & Ferrell, J. E., Jr. (2013). Mitotic trigger waves and the spatial coordination of the Xenopus cell cycle. *Nature*, *500*, 603–607.

Desai, A., Murray, A., Mitchison, T. J., & Walczak, C. E. (1999). The use of Xenopus egg extracts to study mitotic spindle assembly and function in vitro. *Methods in Cell Biology*, *61*, 385–412.

Elinson, R. P. (1983). Cytoplasmic phases in the first cell cycle of the activated frog egg. *Developmental Biology*, *100*, 440–451.

Field, C. M., & Lenart, P. (2011). Bulk cytoplasmic actin and its functions in meiosis and mitosis. *Current Biology*, *21*, R825–R830.

Field, C. M., Wuhr, M., Anderson, G. A., Kueh, H. Y., Strickland, D., & Mitchison, T. J. (2011). Actin behavior in bulk cytoplasm is cell cycle regulated in early vertebrate embryos. *Journal of Cell Science*, *124*, 2086–2095.

Firestone, A. J., Weinger, J. S., Maldonado, M., Barlan, K., Langston, L. D., O'Donnell, M., et al. (2012). Small-molecule inhibitors of the AAA+ ATPase motor cytoplasmic dynein. *Nature*, *484*, 125–129.

Good, M. C., Vahey, M. D., Skandarajah, A., Fletcher, D. A., & Heald, R. (2013). Cytoplasmic volume modulates spindle size during embryogenesis. *Science*, *342*, 856–860.

Groen, A. C., Nguyen, P., Field, C., Ishihara, K., & Mitchison, T. J. (2014). Glycogen-supplemented mitotic cytosol for analyzing Xenopus egg microtubule organization. *Methods in Enzymology*, *540*, 417–433.

Hazel, J., Krutkramelis, K., Mooney, P., Tomschik, M., Gerow, K., Oakey, J., et al. (2013). Changes in cytoplasmic volume are sufficient to drive spindle scaling. *Science*, *342*, 853–856.

Leno, G. H., & Laskey, R. A. (1991). DNA replication in cell-free extracts from Xenopus laevis. *Methods in Cell Biology*, *36*, 561–579.

Maloney, A., Herskowitz, L. J., & Koch, S. J. (2011). Effects of surface passivation on gliding motility assays. *PLoS One*, *6*, e19522. http://dx.doi.org/10.1371/journal.pone.0019522.

Maresca, T. J., & Heald, R. (2006). Methods for studying spindle assembly and chromosome condensation in Xenopus egg extracts. *Methods in Molecular Biology*, *322*, 459–474.

Mitchison, T. J., Wühr, M., Nguyen, P., Ishihara, K., Groen, A., & Field, C. M. (2012). Growth, interaction, and positioning of microtubule asters in extremely large vertebrate embryo cells. *Cytoskeleton*, *69*, 738–750.

Murray, A. W. (1991). Cell cycle extracts. *Methods in Cell Biology*, *36*, 581–605.

Murray, A. W., & Kirschner, M. W. (1989). Cyclin synthesis drives the early embryonic cell cycle. *Nature*, *25*, 275–280.

Niethammer, P., Kueh, H. Y., & Mitchison, T. J. (2008). Spatial patterning of metabolism by mitochondria, oxygen, and energy sinks in a model cytoplasm. *Current Biology*, *18*, 586–591.

Pinota, M., Steinerb, V., Dehapiotc, B., Yoob, B.-K., Chesnelc, F., Blanchoind, L., et al. (2012). Confinement induces actin flow in a meiotic cytoplasm. *PNAS*, *109*, 11705–11710.

Quintyne, N. J., Gill, S. R., Eckley, D. M., Crego, C. L., Compton, D. A., & Schroer, T. A. (1999). Dynactin is required for microtubule anchoring at centrosomes. *The Journal of Cell Biology*, *147*, 321–334.

Rankin, S., & Kirschner, M. W. (1997). The surface contraction waves of Xenopus eggs reflect the metachronous cell-cycle state of the cytoplasm. *Current Biology*, 7, 451–454.

Rauh, N. R., Schmidt, A., Bormann, J., NIgg, E. A., & Mayer, T. U. (2005). Calcium triggers exit from meiosis II by targeting the APC/C inhibitor XErp1 for degradation. *Nature, 437*, 1049–1052.

Richmond, D. L., Schmid, E. M., Martens, S., Stachowiak, J. C., Liska, N., & Fletcher, D. A. (2011). Forming giant vesicles with controlled membrane composition, asymmetry, and contents. *PNAS, 108*, 9431–9436.

Riedl, J., Crevenna, A. H., Kessenbrock, K., Yu, J. H., Neukirchen, D., Bista, M., et al. (2008). Lifeact: A versatile marker to visualize F-actin. *Nature Methods, 5*, 605–607.

Rosenblatt, J., Agnew, B. J., Abe, H., Bamburg, J. H., & Mitchison, T. J. (1997). Xenopus actin depolymerizing factor/cofilin (XAC) is responsible for the turnover of actin filaments in Listeria monocytogenes tails. *The Journal of Cell Biology, 136*, 1323–1332.

Tsai, M. Y., & Zheng, Y. (2005). Aurora A kinase-coated beads function as microtubule-organizing centers and enhance RanGTP-induced spindle assembly. *Current Biology, 15*, 2156–2163.

Valentine, M. T., Perlman, Z. E., Mitchison, T. J., & Weitz, D. A. (2005). Mechanical properties of Xenopus egg cytoplasmic extracts. *Biophysical Journal, 88*, 680–689.

Verde, F., Berrez, J. M., Antony, C., & Karsenti, E. (1991). Taxol-induced microtubule asters in mitotic extracts of Xenopus eggs: Requirement for phosphorylated factors and cytoplasmic dynein. *The Journal of Cell Biology, 112*, 1177–1187.

Wang, S., Romano, F. B., Field, C. M., Mitchison, T. J., & Rapoport, T. A. (2013). Mechanisms determining ER network morphology during the cell cycle in Xenopus egg extracts. *The Journal of Cell Biology, 203*, 801–814.

Waterman-Storer, C., Duey, D. Y., Weber, K. L., Keech, J., Cheney, R. E., Salmon, E. D., et al. (2000). Microtubules remodel actomyosin networks in *Xenopus* egg extracts via two mechanisms of F-actin transport. *The Journal of Cell Biology, 150*, 361–376.

Wu, J. Q., & Kornbluth, S. (2008). Across the meiotic divide-CSF activity in the post-Emi2/XErp1 era. *Journal of Cell Science, 121*, 3509–3514.

Wühr, M., Chen, Y., Dumont, S., Groen, A. C., Needleman, D. J., Salic, A., et al. (2008). Evidence for an upper limit to mitotic spindle length. *Current Biology, 18*, 1256–1261.

Wühr, M., Tan, E. S., Parker, S. K., Detrich, H. W., III, & Mitchison, T. J. (2010). A model for cleavage plane determination in early amphibian and fish embryos. *Current Biology, 20*, 2040–2045.

CHAPTER TWENTY-THREE

Glycogen-Supplemented Mitotic Cytosol for Analyzing Xenopus Egg Microtubule Organization

Aaron C. Groen[*,†], **Phuong A. Ngyuen**[*,†], **Christine M. Field**[*,†], **Keisuke Ishihara**[*,†], **Timothy J. Mitchison**[*,†,1]

[*]Department of Systems Biology, Harvard Medical School, Boston, Massachusetts, USA
[†]Marine Biological Laboratory, Woods Hole, Massachusetts, USA
[1]Corresponding author: e-mail address: timothy_mitchison@hms.harvard.edu

Contents

Abstract

Undiluted cytoplasmic extract prepared from unfertilized *Xenopus laevis* eggs by low-speed centrifugation (CSF extracts) is useful for reconstitution of egg microtubule dynamics and meiosis-II spindle organization, but it suffers limitations for biochemical analysis due to abundant particulates. Here, we describe preparation and the use of fully clarified, undiluted mitotic cytosol derived from CSF extract. Addition of glycogen

Methods in Enzymology, Volume 540
ISSN 0076-6879
http://dx.doi.org/10.1016/B978-0-12-397924-7.00023-6

improves the ability of this cytosol to reconstitute microtubule organization, in part through improved energy metabolism. Using fully clarified, glycogen-supplemented mitotic cytosol, we reconstituted (i) stimulation of microtubule polymerization by Ran.GTP (Groen, Coughlin, & Mitchison, 2011; Ohba, Nakamura, Nishitani, & Nishimoto, 1999) and (ii) self-organization of highly regular bipolar arrays of taxol-stabilized microtubules that we termed "pineapples" (Mitchison, Nguyen, Coughlin, & Groen, 2013). Both systems will be useful for biochemical dissection of spindle assembly mechanisms. We also describe reliable small-scale methods for preparing fluorescent antibody probes that can be used for live imaging in egg extract systems as well as standard immunofluorescence.

1. INTRODUCTION

Assembly of the Xenopus egg meiosis-II spindle, as reconstituted in egg extract (Sawin & Mitchison, 1991), provides a paradigm for dynamic organization of the microtubule cytoskeleton. Spindle assembly can be conceptually separated into processes that regulate microtubule polymerization, such as nucleation, stabilization, and depolymerization, and processes that act on preformed microtubules, such as motor-promoted sliding, minus-end clustering, bundling, cross-linking, and friction that opposes sliding forces. These processes can also be separated experimentally. Nucleation, polymerization, and depolymerization can be assayed with motors inhibited. The actions of motors and microtubule-binding proteins can be assayed in the absence of polymerization dynamics by adding taxol to force nucleation of stable microtubules and assaying subsequent spatial organization (Verde, Berrez, Antony, & Karsenti, 1991). Here, we discuss recent technical advances for analyzing the actions of an important polymerization enhancer, Ran.GTP, and for analyzing how motors and binding proteins collectively promote self-organization of taxol-stabilized microtubules into bipolar arrays. Key to both approaches was the development of stable mitotic cytosol that was fully clarified to remove organelles and supplemented with glycogen to improve energy metabolism.

The standard system for investigating Xenopus egg meiosis-II spindle assembly and dynamics has been an egg extract made by gentle centrifugal crushing with minimal dilution. We will refer to this type of extract as "CSF extract" when it is prepared in a naturally mitosis-arrested state stabilized by cytostatic factor. CSF extract has provided a versatile system for mechanistic analysis of spindles, DNA replication, nuclear assembly and transport, and many other processes (Desai, Murray, Mitchison, & Walczak, 1999;

Maresca & Heald, 2006). It is essentially undiluted egg cytoplasm arrested in a high CDK1 activity state characteristic of meiosis-II metaphase and contains abundant ER, mitochondria, ribosomes, and glycogen. Its ability to reconstitute complex processes depends in part on its vigorous energy metabolism. The endogenous ATP pool (~2 m*M*) in CSF extract turns over with a half-life of ~1 min at 20 °C under both aerobic and anaerobic conditions (Niethammer, Kueh, & Mitchison, 2008). When drops of extracts are sealed from the air, as in typical imaging experiments, oxygen is rapidly depleted by mitochondria (ibid). This stabilizes chemical fluorochromes against photobleaching and likely accounts for the outstanding imaging characteristics of the extract.

While unparalleled for its ability to reconstitute complex processes, CSF extract suffers significant limitations for biochemical analysis. Its abundant particulate components preclude many biochemical manipulations such as spinning out microtubules or loading directly onto chromatography columns. Freezing and thawing tend to reduce its spindle assembly activity, probably by damaging mitochondria and energy metabolism. We also need to separate cytosol from particulates to test potential roles of organelles. Many groups have prepared partially clarified cytosol from egg extract simply by centrifuging sufficiently hard to pellet ribosomes and most organelles (Maresca & Heald, 2006). We will call such preparations high-speed supernatant (HSS). We found several limitations with simple HSS preps: (1) their ability to retain a mitotic state with high CDK1 activity after freeze–thaw was variable; (2) they still contain light membrane fractions that are buoyant at the high protein concentration; and (3) because they lack endogenous energy sources, their metabolic state is inferior to crude egg extracts. Supplementing with an energy-regenerating system such as creatine phosphate may help, but this leads to buildup of phosphate and ADP over minutes. To address these limitations, we developed an optimized cytosol preparation where the mitotic state was stabilized with a stable form of Cyclin B (Cyclin B-Δ90; Murray, Solomon, & Kirschner, 1989); HSS was further clarified by dilution, centrifugation, and reconcentration; and glycogen was added to restore physiological pathways for making ATP and reducing equivalents (NADPH; Kruger & von Schaewen, 2003; Schaftingen, 1993). The resulting system has proved superior to HSS for reconstitution of (i) Ran. GTP stimulation of microtubule assembly and (ii) bipolar microtubule arrays by mitotic motor protein activity. In unpublished work, the system is also proving ideal for mass-spec analysis of how microtubules binding proteins are regulated by Ran.GTP and mitotic kinases.

2. DEVELOPMENT OF THE CYTOSOL SYSTEM

Assembly of meiosis-II spindles in Xenopus eggs is triggered in part by Ran.GTP generated at chromatin by the GEF RCC1 (Askjaer, Galy, Hannak, & Mattaj, 2002; Carazo-Salas et al., 1999; Karsenti & Vernos, 2001). Addition of Ran.GTP, provided as a hydrolysis-defective Q69L mutant, to CSF extract triggers assembly of microtubule asters and spindle-like assemblies (Nachury et al., 2001; Ohba et al., 1999). We took a fractionation approach to investigating this pathway and asked if any particulate material was required for induction of microtubule polymerization by Ran.GTP (Groen et al., 2011). We first established methods for making HSS from egg extract that was reliably and stably mitotic and could be frozen and thawed without loss of activity. Addition of Cyclin B-Δ90, a nondegradable mutant of sea urchin Cyclin B, before centrifugation helped stabilize the mitotic state of the cytosol. Mitotic HSS did not exhibit microtubule polymerization when Ran.GTP was added, revealing a particulate requirement. Fractionation of the pellet revealed that only glycogen was required. Commercial, purified glycogen could completely replace egg-derived glycogen. Glycogen is naturally present in eggs at high concentrations (~30 mg/ml) as an energy store. To confirm that glycogen was the only particulate required, we had to completely remove membranes from HSS. We diluted it 10-fold with dilution buffer, centrifuged to pellet light membranes, and then reconcentrated back to the original volume (see Fig. 23.1). Imaging and biochemical analysis showed that the resulting clarified-HSS (C-HSS) is free of organelles. While optimizing the dilution buffer, we found that phosphate (~10 m*M*) was required for Ran-promoted microtubule assembly. ATP and GTP were also required. The role of phosphate is unclear. It probably promotes phosphorolysis of glycogen but might have other interesting effects.

Investigation of the glycogen add-back effect revealed several potential benefits: (i) regeneration of ATP without the buildup of free phosphate that occurs with typical regenerating systems such as creatine phosphate; (ii) generation of reducing equivalents, presumably as NADPH, which promote thiol homeostasis (Jackson, Herbert, Campbell, & Hunt, 1983); and (iii) a mild crowding effect that promotes microtubule assembly, though to a lesser extent, and probably in a more physiological manner, than classic crowding agents such as poly(ethylene glycol). Whether these activities together account for all the beneficial effects of glycogen add-back is

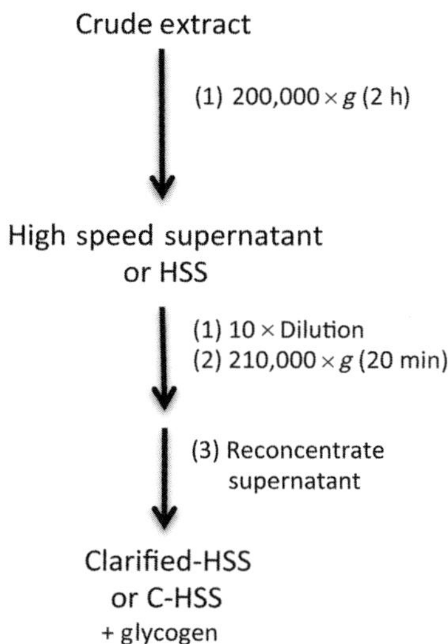

Figure 23.1 Isolation of clarified high-speed supernatant (C-HSS).

currently unclear. Previous studies reported beneficial effects of glycogen for the biochemistry of egg-derived HSS (Hartl, Olson, Dang, & Forbes, 1994), and we found it was also useful for reconstituting motor-dependent self-organization of microtubules (Mitchison et al., 2013). Glycogen may be a beneficial additive to any cytosol system that retains the enzymes necessary for glycolysis and NADPH production.

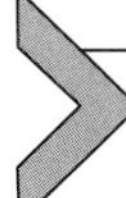

3. PREPARATION OF GLYCOGEN-SUPPLEMENTED CYTOSOL

3.1. Purification of recombinant proteins

GST-Ran(Q69L) and MBP-Cyclin B-Δ90 (a kind gift of A. Salic, Harvard Medical School, Boston, MA, USA) were conventionally purified in bacteria. Briefly, expressed cells were lysed by sonication in lysis buffer (50 m*M* sodium phosphate, pH 7.7, 500 m*M* NaCl, 100 μ*M* $MgCl_2$, 10% Triton [Sigma], 1 m*M* DTT [Sigma], 1 m*M* phenylmethylsulfonyl fluoride [Sigma], plus 100 μ*M* GTP for Ran only), using ~100 ml buffer per 1 l of bacteria. After sonication (six 30-s pulses), clarification (150,000 × *g* for 1 h), and purification by chromatography using either glutathione or amylose agarose (2 ml/l of culture) and being washed in 10 column volumes of wash buffer (lysis buffer with 10% glycerol instead of Triton), purified proteins were dialyzed overnight in extract buffer and concentrated if required in a 10-kDa molecular-weight-cutoff Centricon Centrifugal Concentration Device (Millipore, Billerica, MA, USA). Aliquots were flash-frozen in liquid nitrogen for storage at −80 °C. Typical stock concentrations were ~10 m*M* (Ran) and 100 μg/ml (Cyclin B-Δ90). Ran activity was scored using the microtubule polymerization assay described below. Cyclin B-Δ90 activity was routinely scored by morphology of DMSO-induced microtubule asters. If needed, it can also be scored by CDK1 activity assays or the morphology of chromatin.

3.2. Preparation of stably mitotic, partially clarified cytosol (HSS)

Mitotic egg extract (CSF extract) was prepared from cytochalasin-treated Xenopus eggs and supplemented with cytochalasin D and protease inhibitors as described (Desai et al., 1999; Maresca & Heald, 2006). This is the same extract we routinely use for studies of meiosis-II spindle assembly and dynamics. Chapter 22 describes preparation of CSF extract with intact actin for actin-dependent assays. To stabilize the mitotic state, CSF was supplemented with MBP-Cyclin B-Δ90 derived from sea urchin (100 μg/ml final), energy mix (1 m*M* ATP, 7.5 m*M* creatine phosphate final), and sucrose to stabilize organelles (100 m*M* final). It was incubated for 15 min at 20 °C, cooled to 0 °C, and centrifuged at 200,000 × *g* for 2 h in a swinging bucket rotor at 4 °C (see Fig. 23.1). The clear, straw-colored cytosolic layer was collected with a syringe, and either used immediately for further clarification or frozen in aliquots using liquid N_2 and stored at −80 °C. This HSS is free of ribosomes, glycogen, and most organelles, but it still contains considerable amounts of light membranes, presumably mitotic smooth ER.

3.3. Preparation of fully clarified cytosol (C-HSS)

See Fig. 23.1 for outline of the procedure. HSS, freshly prepared or thawed, was diluted 10-fold in dilution buffer-containing phosphate, ATP and GTP (100 m*M* KCl, 10 m*M* K-phosphate, pH 7.2, 5 m*M* ethylene glycol tetra-acetic acid, 1 m*M* $MgCl_2$, 1 m*M* ATP, 1 m*M* GTP, 1 m*M* DTT) and centrifuged at 210,000 × *g* for 20 min in a fixed-angle rotor at 4 °C (see Fig. 23.1). The supernatant was collected and concentrated back to the original volume using a 10-kDa molecular-weight-cutoff Centricon Centrifugal Concentration Device (Millipore, Billerica, MA). C-HSS was stored frozen as per HSS.

3.4. Preparation of glycogen

Glycogen from oyster was purchased from Sigma-Aldrich. Some batches were impure, as judged by smell, yellowish color, and inhibitory effects on extract. We routinely purified this commercial glycogen by one round of ethanol precipitation. Glycogen was dissolved in water at ~50 mg/ml, precipitated with three volumes of ethanol, and collected by centrifugation (5000 × *g*, 5 min 20 °C). The supernatant was discarded and the pellet dried in air overnight. The glycogen was dissolved in dilution buffer at 200–300 mg/ml and stored in aliquots at −20 °C. It took some time and agitation to completely dissolve the glycogen; the final solution should be clear and viscous with no color or smell. Some batches of commercial glycogen might benefit from a second round of ethanol precipitation.

4. ASSAY OF RAN-PROMOTED MICROTUBULE ASSEMBLY

4.1. Assembly in CSF extract

Freshly made CSF extract was supplemented with fluorochrome-labeled tubulin (~100 n*M* final). Labeled tubulin was prepared as described (Hyman et al., 1991), except we now use the NHS ester forms of Alexa 488, 568, and 646 for labeling (http://www.lifetechnologies.com). High-quality-labeled tubulin can also be purchased from Cytoskeleton Inc. (www.cytoskeleton.com). RanQ69L was added between 1/20 and 1/100 final. The extract was incubated 20 °C for 30 min and then squashed between a slide and coverslip and imaged directly. We used uncoated glass for these experiments. Dynein activity caused the Ran-induced microtubules to aggregate into asters, and we typically scored the number of asters per microscope field with a 20 × objective.

4.2. Assembly in HSS or C-HSS + glycogen

HSS or C-HSS was thawed and supplemented with glycogen (30–40 mg/ml final) and labeled tubulin (~100 n*M* final). RanQ69L was added and incubated as above. Aster assembly was completely dependent on glycogen. Typical Ran-induced asters assembled in C-HSS + glycogen are shown in Fig. 23.2. Tests with other polymers and metabolites in place of glycogen are discussed in Groen et al. (2011). Briefly, a combination of a mild crowing agent and glucose-6-phosphate could partially replace glycogen. Figure 23.2 shows examples of asters induced by RanQ69L in CSF extract and C-HSS.

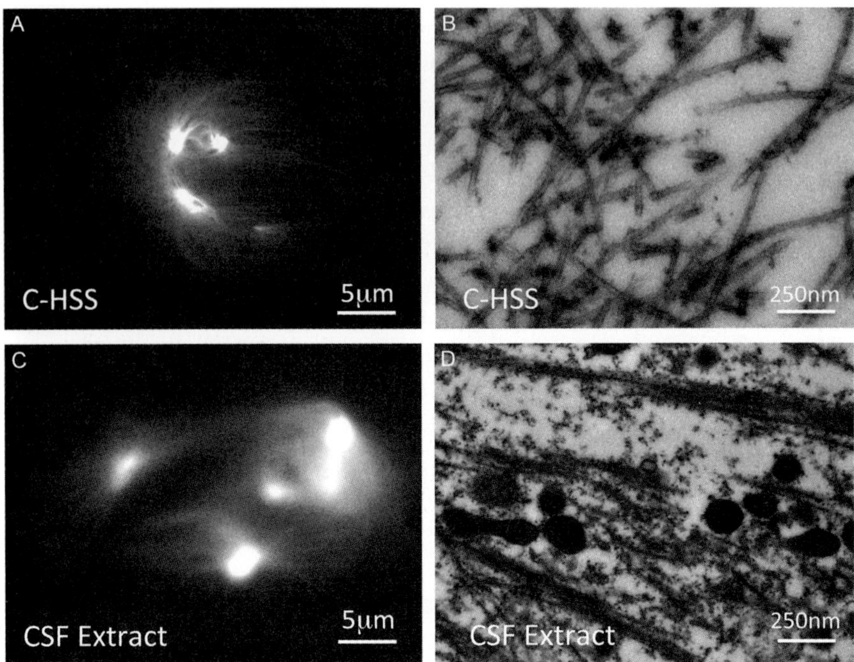

Figure 23.2 Ran aster assembly in C-HSS+glycogen. Ran asters assembled in C-HSS+glycogen or crude extract. The constitutively active RanQ69L (16 μ*M* final) mutant was added to either C-HSS+glycogen or crude extract supplemented with Alexa 568 tubulin and imaged by either wide-field microscopy (A and C) or electron microscopy (B and D). Scale bars = 5 μm (A and C) and 250 nm (B and D). Note similar aster morphology, with abundant vesicles in CSF extract and no vesicles in C-HSS.

5. SELF-ORGANIZATION OF SPINDLE-LIKE MICROTUBULE ARRAYS

5.1. Introduction: The "pineapple" system

Addition of taxol to mitotic extracts allows analysis of the collective action of motors and binding proteins on microtubule organization independent of polymerization regulators. Addition of taxol to CSF extract first revealed the powerful minus-end clustering activity of mitotic dynein, which is now recognized as a major driver of spindle morphogenesis (Verde et al., 1991). We sought to extend this approach to more spindle-like bipolar arrays and explore possible roles of membranes in self-organization. We added various combinations of taxol and DMSO to mitotic HSS and C-HSS supplemented with fluorescent probes for tubulin and proteins that

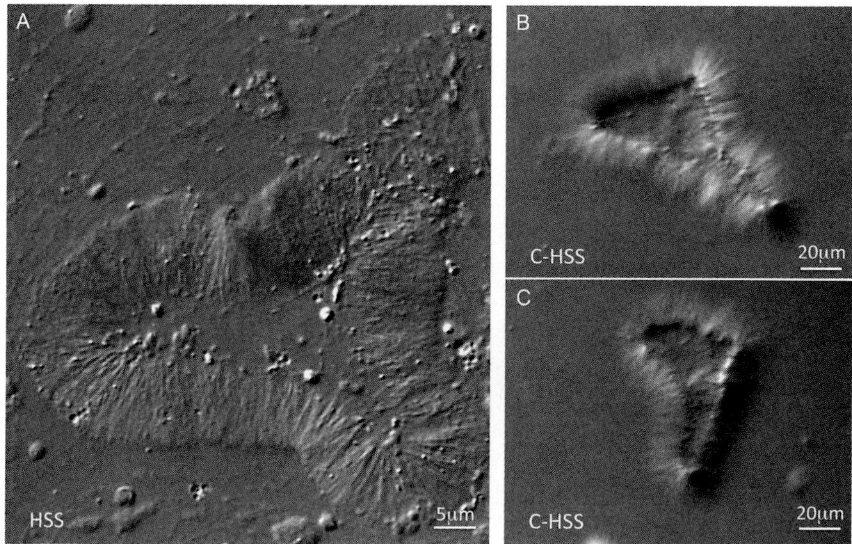

Figure 23.3 Pineapples assembled in HSS or C-HSS (DIC). DIC images of pineapples structures assembled in HSS+glycogen (A) or C-HSS+glycogen (B and C). Scale bars = 5 μm (A) and 20 μm (B and C). Note the presence of some vesicles and debris in the HSS reaction, and much less in the fully C-HSS reaction.

accumulate at microtubule plus and minus ends. This approach culminated in a system that self-organized highly regular bipolar arrays (Figs. 23.3 and 23.4) (Mitchison et al., 2013). We termed these assemblies "pineapples" because round examples resembled a slice of canned pineapple (see Figs. 23.3 and 23.4).

We investigated the role of different motor systems in pineapple assembly. Microtubule aggregation and minus-end clustering required dynein, presumably working in combination with NUMA (Gaglio et al., 1996; Merdes, Ramyar, Vechio, & Cleveland, 1996). Bipolarization required Kinesin-5/Eg5; when this motor was inhibited pineapples assembled as asters with minus ends on the inside. These findings were expected given the role of dynein and Kinesin-5 in spindle assembly. Plus ends also aligned dramatically in pineapples (Figs. 23.3 and 23.4). This is not seen in spindles, where dynamic instability randomizes plus-end positions, except in kinetochore fibers. We found that aligned plus ends in pineapples accumulated proteins that are normally found at overlapping plus ends in cytokinesis midzones. These included the chromosomal passenger complex or CPC (e.g., AurkB kinase and Dasra; see Fig. 23.4), Kif4, and Kif23, part of Centralspindlin (Bieling, Telley, Hentrich, Piehler, & Surrey, 2010; Carmena, Wheelock,

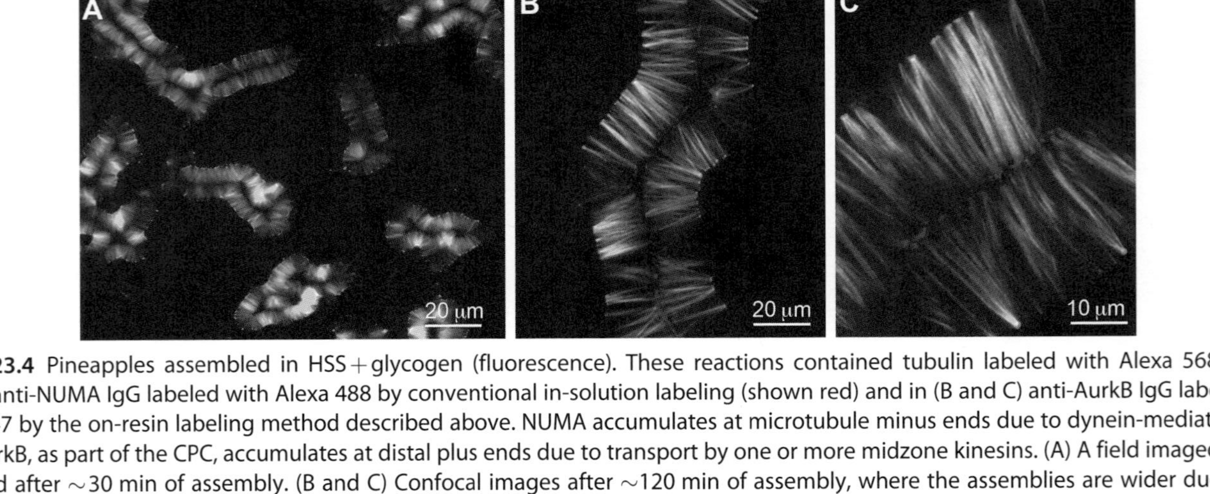

Figure 23.4 Pineapples assembled in HSS + glycogen (fluorescence). These reactions contained tubulin labeled with Alexa 568 (shown green), anti-NUMA IgG labeled with Alexa 488 by conventional in-solution labeling (shown red) and in (B and C) anti-AurkB IgG labeled with Alexa 647 by the on-resin labeling method described above. NUMA accumulates at microtubule minus ends due to dynein-mediated transport. AurkB, as part of the CPC, accumulates at distal plus ends due to transport by one or more midzone kinesins. (A) A field imaged by 20 × widefield after ∼30 min of assembly. (B and C) Confocal images after ∼120 min of assembly, where the assemblies are wider due to slow plus-end growth. (See the color plate.)

Funabiki, & Earnshaw, 2012; Glotzer, 2005). Plus-end alignment in pineapples depended on the activity of both AurkB and Kif4, as does plus-end alignment in midzones in cells (Hu, Coughlin, Field, & Mitchison, 2011). Thus, pineapples provide a system for investigation of the morphogenetic activity of both spindle and midzone assembly factors. It is unclear why midzone factors, which normally act after anaphase onset, operate so strongly in a system where the Cdk1 activity is high. We suspect that stabilization of microtubules with taxol and DMSO leads to high AurkB activity, and this induces activity of midzone proteins. Misregulation of Plk1 kinase may also be involved, since midzone-like arrays form prematurely during metaphase in living cells when Plk1 is inhibited (Hu, Ozlü, Coughlin, Steen, & Mitchison, 2012).

5.2. Pineapple assembly reactions

Slides and coverslips were passivated to reduce protein binding. Typically, we took clean coverslips straight out of the box, incubated them with casein (0.2% in water, 25 °C, 10–60 min), and then rinsed with water several times, dried in air, and used within 60 min. In some cases, we used more complex poly(ethylene glycol) coatings (Bieling, Telley, & Surrey, 2010). The results were similar, though pineapples tended to adhere more to casein, than to PEG-coated surfaces. Mitotic HSS or C-HSS, prepared as above and stored frozen, was thawed and supplemented with glycogen (20 mg/ml final) and fluorescent probes. We made extensive use of labeled antibody probes for analyzing pineapples (see labeling methods below), as well as fluorescent fusion proteins. Pineapple assembly was initiated by the addition of 1/20th volume of 100 μM taxol in DMSO. Taxol is easily hydrolyzed and can be unstable in impure DMSO, so it is important to use fresh, dry DMSO to make stock solutions. Immediately after mixing, a drop of the assembly reaction (typically, 5 μl) was spread between a slide and coverslip, or two cover slips (typically, 18 × 18 mm). Optionally, the edges were sealed with VALAP (33% paraffin wax, 33% Vaseline, 33% lanolin) to prevent evaporation. Reactions were imaged live at 20 °C, or at fixed time points. Microtubule polymerization was complete by ~5 min, and aggregation was complete by ~15 min. Parallel bundling and plus- and minus-end alignment were completed by ~30 min. After ~40 min pineapple morphology did not change, but the width of the structures slowly increased over ~2 h. Pulse labeling suggested that this occurs by slow growth on plus ends. Subunits for growth are provided by disassembly of smaller pineapples at the expense of larger ones, which we believe is due to Ostwald ripening (Yao, Elder, Guo, & Grant, 1993).

To test the roles of specific proteins in pineapple assembly, we either added inhibitors to the reaction before assembly or depleted proteins with antibodies. In a typical depletion experiment, 2-μm magnetic beads coated with Protein A (Dynabeads, www.lifetechnologies.com) were saturated with an affinity-purified polyclonal antibody (~20 μg per 50 μl of bead slurry) to the protein of interest and washed. HSS or C-HSS was incubated with beads at a 1:20 dilution, rotating at 4 °C for 30 min. The beads were removed with a magnet and the depletion repeated. After the second depletion, the depleted C-HSS was either used immediately for pineapple assembly or frozen. Examples of depleting Kif4A and AurkB/CPC are shown in Mitchison et al. (2013).

6. ANTIBODY-BASED FLUORESCENT PROBES

6.1. Introduction

To localize endogenous proteins in Xenopus extract systems, we often used labeled antibody probes (e.g., Fig. 23.4). Such probes must be used with caution; antibody binding can readily activate or inactive the protein target, so testing the effect of added antibody on the process of interest is essential. One advantage of antibody probes is to prove a rapid readout of possible location, before going to the trouble of expressing and purifying a fusion protein. Another is avoiding potential overexpression or dominant-negative artifacts of expressed protein. We tend to trust localization data when both labeled antibody and fluorescent fusion protein probes reveal the same localization. When abundant antibody is available, we use traditional in-solution labeling with fluorochrome–NHS esters followed by gel filtration. However, we often work with small amounts of antibody where this method can be problematic. Below are two microscale labeling methods that work for as little as a few micrograms of antibody. The on-resin labeling procedure has proved particularly reliable, and we now use it routinely, even for larger amounts of antibody. It does not require calculating how much dye to add, and it separates IgG-bound and free dye very effectively. Rabbit IgG that binds to a protein of interest was purified by affinity chromatography using published methods (Field, Oegema, Zheng, Mitchison, & Walczak, 1998). We have had good luck with both antibodies to whole proteins and to C-terminal peptides if they are highly charged.

6.2. IgG labeling with anti-IgG FABs

This is a rapid method we use for testing antibody batches and other prototyping experiments. Rabbit IgG was mixed with FAB fragments of goat anti-rabbit IgG labeled with a fluor, typically DyeLight 568 or 647 (Jackson Labs, www.jax.org). We mixed equal mass of IgG and FAB, corresponding to a molar excess of FAB. After 30 min at 0 °C, the mixture is diluted into CSF extract, HSS or C-HSS for imaging. Typically, we would aim for a final concentration of labeled IgG of ~0.5–5 μ*M*, using imaging to find the optimum. The IgG–FAB mixture can be stored for months at 4 °C.

6.3. On-resin IgG labeling

This reliable, small-scale method was used to make probes for live imaging in the pineapple system and CSF extract. It should be useful for any application requiring small amounts of directly labeled antibody, including multicolor immunofluorescence. It was adapted from a commercial kit (APEX® Antibody Labeling Kits, Life Technologies, www.lifetechnologies.com) optimized for small amounts of antibody and has given reliable results for many antibodies and several fluors. Labeled antibodies always had a useful dye:protein ratios and very little unbound dye, as assayed by silica gel TLC with a methanol solvent. On-resin labeling works well using Protein G ultralink resin (ThermoFisher Scientific, www.thermofisher.com) (rabbit and mouse IgG) and Protein A Affiprep resin (Biorad, www.bio-rad.com) (rabbit IgG). To hold the resin, a 200-μl polypropylene pipette tip was sealed at the pointed end by brief passage through the flame of a Bunsen burner. Aim to just seal the tip, avoid a big blob of molten plastic. To allow liquid egress while retaining resin a 1- to 2-mm slit was made parallel to the long axis of the sealed tip by cutting with a thin razor blade. Fluid was forced through the tip with a pipetting device, taking ~30 s for passage of 100 μl of liquid through the resin during loading and washes. All subsequent procedures were performed at room temp (20–25 °C). Forty microliters of resin slurry was loaded into the tip and washed 3× with 0.15 *M* NaCl, 10 m*M* K-HEPES (pH 7.7) to give a packed resin bed of ~20 μl. Purified antibody to the protein of interest, typically an affinity-purified rabbit IgG or a commercial monoclonal antibody, was passed through the resin bed five times sequentially to bind. The resin can hold up to 400 μg of IgG, we typically labeled 20–100 μg per tip. The resin was washed 3× with 200 m*M* K-HEPES (pH 7.7). Alexa dyes as the NHS ester (Life Technologies, NY, USA) were dissolved in DMSO at ~50 m*M* and stored as frozen aliquots. 0.5 μl of dye stock was diluted into 25 μl of 200 m*M* K-HEPES (pH 7.7) and immediately loaded onto the resin bed, making sure the resin was saturated with dye and leaving a small amount of dye solution on top of the bed. NHS esters hydrolyze rapidly in alkaline solution, with a half-life of minutes at pH 7.7, so it is important to get the diluted dye onto the resin rapidly. We used the same amount of NHS ester independent of the amount of labeled IgG on the column. It is a large molar excess over the IgG, but most reacts with the resin or gets hydrolyzed. The tip was then incubated 25 °C for 10–60 min. A second aliquot of dye was diluted, added to the resin bed, and incubated in the same way. After the second labeling reaction, the bed was washed 5× with 0.15 *M* NaCl, 10 m*M* K-HEPES (pH 7.7) to remove unbound dye. The wash buffer should be free of color by the third wash. Labeled IgG was then eluted with 5 × 15 μl aliquots of 200 m*M* acetic acid. Each aliquot of eluate was immediately mixed with 5 μl of 1 *M* Tris–Cl, pH 9, to neutralize and cooled to 0 °C. The colored fractions were pooled. Typically, these were fractions 2 and 3 only. The resin is still highly colored after elution, presumably due to labeling of the Protein A or G. Labeled antibody is stable for several months at 4 °C in the Tris–Cl acetate buffer, which has a near neutral pH after neutralization. If desired, the labeled antibody can be exchanged into a different buffer and concentrated by dilution and reconcentration using a centrifugal concentrating device, but some loss is likely using low antibody amounts unless carrier protein is added. Dye:protein ratios can be measured by spectrophotometry, but we usually omitted this step to conserve scarce antibody and used localization by immunofluorescence or live imaging as our quality metric. The on-resin method gave useful labeling reagents over a large range of IgG concentrations on the resin for all NHS ester dyes tested (Alexa 488, 568, 647) (Fig. 23.5).

7. FUTURE DIRECTIONS

With a clarified mitotic cytosol system in hand that reconstitutes Ran-promoted microtubule polymerization and assembly of bipolar microtubule arrays, it should be possible to make further progress on the biochemistry of mitotic spindle assembly. Cytosol approaches, where all the relevant proteins are present in a cell-free system at their physiological concentrations and modification states, will complement pure protein reconstitutions and genetic manipulation of living cells. We are especially excited about application of multiplexed, quantitative mass spectrometry to the systems

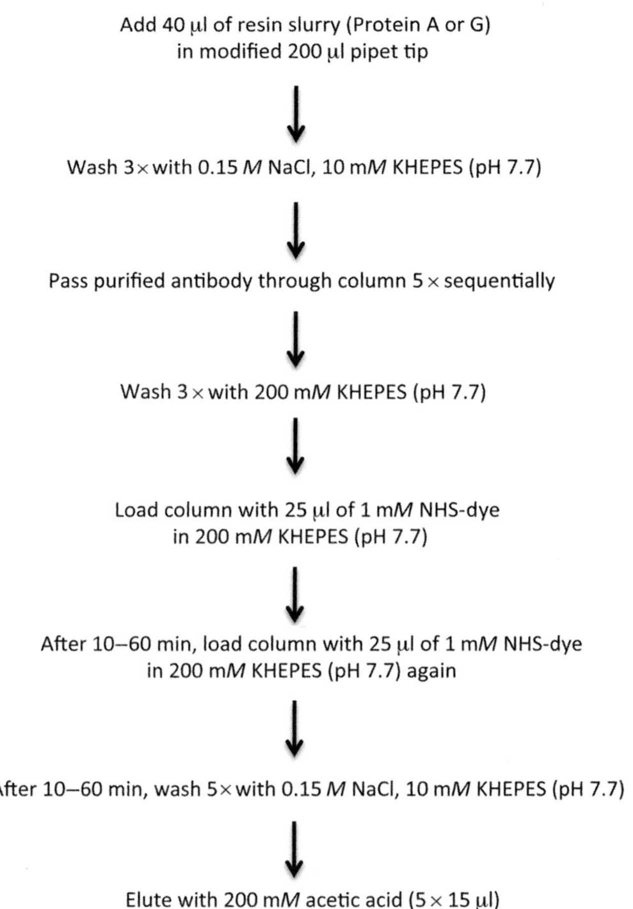

Figure 23.5 On-column antibody labeling protocol.

described above. By diluting pineapples, collecting them by centrifugation, and washing once, we have been able to obtain highly enriched preparations of mitotic microtubule-binding proteins. Mass spectrometry reveals essentially all the binding proteins and motors that have been implicated in spindle and midzone assembly. By comparing the amount of each binding proteins that cosediments under different conditions—for example, plus/minus RanQ69L, plus/minus kinase inhibitors—we can systematically identify the most Ran- or kinase-regulated-binding proteins. This approach can be extended by immunodepletion of specific motors or MAPs and will enable a systematic analysis of how the microtubule-binding proteome is regulated during mitosis and cytokinesis.

ACKNOWLEDGMENTS

This work was supported by NIH Grant GM39565 and by MBL summer fellowships. Microscopy support was provided by the NIC at HMS and by Nikon Inc. at MBL. Key experiments were initiated at MBL as part of the Physiology Course in 2007 and 2008, and we thank the students involved.

REFERENCES

Askjaer, P., Galy, V., Hannak, E., & Mattaj, I. W. (2002). Ran GTPase cycle and importins alpha and beta are essential for spindle formation and nuclear envelope assembly in living Caenorhabditis elegans embryos. *Molecular Biology of the Cell, 13*, 4355–4370.

Bieling, P., Telley, I. A., Hentrich, C., Piehler, J., & Surrey, T. (2010). Fluorescence microscopy assays on chemically functionalized surfaces for quantitative imaging of microtubule, motor, and +TIP dynamics. *Methods in Cell Biology, 95*, 555–580.

Bieling, P., Telley, I. A., & Surrey, T. (2010). A minimal midzone protein module controls formation and length of antiparallel microtubule overlaps. *Cell, 142*, 420–432.

Carazo-Salas, R. E., Guarguaglini, G., Gruss, O. J., Segref, A., Karsenti, E., & Mattaj, I. W. (1999). Generation of GTP-bound Ran by RCC1 is required for chromatin-induced mitotic spindle formation. *Nature, 400*, 178–181.

Carmena, M., Wheelock, M., Funabiki, H., & Earnshaw, W. C. (2012). The chromosomal passenger complex (CPC): From easy rider to the godfather of mitosis. *Nature Reviews Molecular Cell Biology, 13*, 789–803.

Desai, A., Murray, A., Mitchison, T. J., & Walczak, C. E. (1999). The use of Xenopus egg extracts to study mitotic spindle assembly and function in vitro. *Methods in Cell Biology, 61*, 385–412.

Field, C. M., Oegema, K., Zheng, Y., Mitchison, T. J., & Walczak, C. E. (1998). Purification of cytoskeletal proteins using peptide antibodies. *Methods in Enzymology, 298*, 525–541.

Gaglio, T., Saredi, A., Bingham, J. B., Hasbani, M. J., Gill, S. R., Schroer, T. A., et al. (1996). Opposing motor activities are required for the organization of the mammalian mitotic spindle pole. *Journal of Cell Biology, 135*, 399–414.

Glotzer, M. (2005). The molecular requirements for cytokinesis. *Science, 307*, 1735–1739.

Groen, A. C., Coughlin, M., & Mitchison, T. J. (2011). Microtubule assembly in meiotic extract requires glycogen. *Molecular Biology of the Cell, 22*, 3139–3151.

Hartl, P., Olson, E., Dang, T., & Forbes, D. J. (1994). Nuclear assembly with lambda DNA in fractionated Xenopus egg extracts: An unexpected role for glycogen in formation of a higher order chromatin intermediate. *Journal of Cell Biology, 124*, 235–248.

Hu, C.-K., Coughlin, M., Field, C. M., & Mitchison, T. J. (2011). KIF4 regulates midzone length during cytokinesis. *Current Biology: CB, 21*, 815–824.

Hu, C.-K., Ozlü, N., Coughlin, M., Steen, J. J., & Mitchison, T. J. (2012). Plk1 negatively regulates PRC1 to prevent premature midzone formation before cytokinesis. *Molecular Biology of the Cell, 23*, 2702–2711.

Hyman, A., Drechsel, D., Kellogg, D., Salser, S., Sawin, K., Steffen, P., et al. (1991). Preparation of modified tubulins. *Methods in Enzymology, 196*, 478–485.

Jackson, R. J., Herbert, P., Campbell, E. A., & Hunt, T. (1983). The roles of sugar phosphates and thiol-reducing systems in the control of reticulocyte protein synthesis. *European Journal of Biochemistry FEBS, 131*, 313–324.

Karsenti, E., & Vernos, I. (2001). The mitotic spindle: A self-made machine. *Science, 294*, 543–547.

Kruger, N. J., & von Schaewen, A. (2003). The oxidative pentose phosphate pathway: Structure and organisation. *Current Opinion in Plant Biology, 6*, 236–246.

Maresca, T. J., & Heald, R. (2006). Methods for studying spindle assembly and chromosome condensation in Xenopus egg extracts. *Methods in Molecular Biology (Clifton, NJ), 322*, 459–474.

Merdes, A., Ramyar, K., Vechio, J. D., & Cleveland, D. W. (1996). A complex of NuMA and cytoplasmic dynein is essential for mitotic spindle assembly. *Cell, 87*, 447–458.

Mitchison, T. J., Nguyen, P., Coughlin, M., & Groen, A. C. (2013). Self-organization of stabilized microtubules by both spindle and midzone mechanisms in Xenopus egg cytosol. *Molecular Biology of the Cell, 24*, 1559–1573.

Murray, A. W., Solomon, M. J., & Kirschner, M. W. (1989). The role of cyclin synthesis and degradation in the control of maturation promoting factor activity. *Nature, 339*, 280–286.

Nachury, M. V., Maresca, T. J., Salmon, W. C., Waterman-Storer, C. M., Heald, R., & Weis, K. (2001). Importin beta is a mitotic target of the small GTPase Ran in spindle assembly. *Cell, 104*, 95–106.

Niethammer, P., Kueh, H. Y., & Mitchison, T. J. (2008). Spatial patterning of metabolism by mitochondria, oxygen, and energy sinks in a model cytoplasm. *Current Biology: CB, 18*, 586–591.

Ohba, T., Nakamura, M., Nishitani, H., & Nishimoto, T. (1999). Self-organization of microtubule asters induced in Xenopus egg extracts by GTP-bound Ran. *Science, 284*, 1356–1358.

Sawin, K. E., & Mitchison, T. J. (1991). Mitotic spindle assembly by two different pathways in vitro. *Journal of Cell Biology, 112*, 925–940.

Schaftingen, E. V. (1993). Glycolysis revisited. *Diabetologia, 36*, 581–588.

Verde, F., Berrez, J. M., Antony, C., & Karsenti, E. (1991). Taxol-induced microtubule asters in mitotic extracts of Xenopus eggs: Requirement for phosphorylated factors and cytoplasmic dynein. *Journal of Cell Biology, 112*, 1177–1187.

Yao, J. H., Elder, K. R., Guo, H., & Grant, M. (1993). Theory and simulation of Ostwald ripening. *Physical Review B, 47*, 14110–14125.

CHAPTER TWENTY-FOUR

Spindle Assembly on Immobilized Chromatin Micropatterns

Céline Pugieux[1], Serge Dmitrieff[1], Katarzyna Tarnawska, François Nédélec[2]

Cell Biology and Biophysics Unit, European Molecular Biology Laboratory, Heidelberg, Germany
[1]These authors contributed equally
[2]Corresponding author: e-mail address: nedelec@embl.de

Contents

Abstract

We describe a method to assemble meiotic spindles on immobilized micropatterns of chromatin built on a first layer of biotinylated BSA deposited by microcontact printing. Such chromatin patterns routinely produce bipolar spindles with a yield of 60%, and offer the possibility to follow spindle assembly dynamics, from the onset of nucleation

Methods in Enzymology, Volume 540
ISSN 0076-6879
http://dx.doi.org/10.1016/B978-0-12-397924-7.00024-8

to the establishment of a quasi steady state. Hundreds of spindles can be recorded in parallel for different experimental conditions. We also describe the semi-automated image analysis pipeline, which is used to analyze the assembly kinetics of spindle arrays, or the final morphological diversity of the spindles.

1. INTRODUCTION

Mitotic or meiotic spindles, which are made from microtubules (MT), are fascinating structures that are able to segregate the genetic material to the future daughter cells. Cell division must be regulated in time and space to ensure functional tissue development and ultimately yield a viable organism. Aneuploidy is a common feature of cancer (Noatynska, Gotta, & Meraldi, 2012) and deciphering the mechanisms of chromosome segregation, and more generally understanding spindle assembly is of high interest. *In vitro*, spindle assembly can be studied in extracts prepared from unfertilized *Xenopus laevis* eggs that are naturally arrested in metaphase of meiosis II (Murray, 1991). The endogenous DNA is lost during preparation, but spindle will spontaneously form if exogenous chromatin is added to a fresh extract. Such chromatin can be obtained from *Xenopus* sperm, but clumps of 2.8-μm diameter chromatin-coated beads (Heald et al., 1996) or a single 10-μm chromatin bead (Halpin, Kalab, Wang, Weis, & Heald, 2011) are also able to induce spindle assembly.

Using microcontact printing, we could immobilize 2.8-μm chromatin beads, and develop arrays of chromatin micropatterns on which meiotic spindles assemble with a yield that is comparable to that of older methods also based on *Xenopus* egg extracts. The new method offers a way to control the number of beads per clump and the geometry of the clump around which spindles assemble, which is not possible if the beads are in solution. The amount and the geometry of the chromatin directly affect the formation of the spindles (Dinarina et al., 2009) and it is important to control these parameters to correctly measure spindle assembly. Another important advantage of *spindle arrays* over older methods is that kinetic studies can be performed for hundreds of spindles simultaneously. Because the spots on which spindles may form are predefined, image analysis of the spindle arrays is simplified compared to previous methods. Moreover, different conditions can be imaged side-by-side to compare the effects of altering the same *Xenopus* egg extract, for example, by drug addition.

So far, this technique has allowed us to measure the influence of chromatin on the polymerization of asymmetric asters from centrosomes (Athale et al., 2008), to quantify the influence of chromatin shape and amount on spindle geometry (Dinarina et al., 2009), and to study the role of Augmin in spindle organization (Petry, Pugieux, Nedelec, & Vale, 2011) in *Xenopus* egg extract. We describe in detail the different steps that are necessary to build spindle arrays, and the software used to analyze the acquired microscopy images using MATLAB(R).

2. OVERVIEW OF THE METHOD

Microcontact printing (μCP) is a soft lithography method (Kumar & Whitesides, 1994) whereby a polydimethylsiloxane (PDMS) stamp with a desired motif is used to deposit molecules of interest onto a surface. In our assay (Fig. 24.1), we print biotinylated bovine serum albumin (BSA) on PDMS-coated glass coverslips. The PDMS stamp and the target surface are treated to maximize the transfer efficiency of biotinylated BSA and to ensure its covalent immobilization. To limit the unspecific adsorption of proteins present in *X. laevis* egg extract, two passivation layers are applied. The surface is then incubated with 1-μm streptavidin-coated paramagnetic beads that bind to the biotinylated BSA on the pattern. The unbound beads are washed away revealing the pattern. The surface is incubated with 2.8-μm chromatin-coated beads, which dock to the 1-μm streptavidin-coated beads. Finally incubating in *Xenopus* egg extract is then sufficient to induce the formation of meiotic spindles. We routinely used a pattern of circular patches, separated by 52 or 62 μm, center to center, and with a diameter of 12 μm, nine chromatin beads dock on average on each patch. With 0.5 pg of DNA per bead, the chromatin content of each patch is close to that of the *X. laevis* haploid genome of 3.2 pg (Thiébaud & Fischberg, 1977). The images are analyzed with custom MATLAB(R) programs.

3. PREPARATIVE STEPS

Some standard steps are only briefly described here, as more detailed protocols are available elsewhere (Qin, Xia, & Whitesides, 2010; Théry & Piel, 2009).

3.1. Mask design and master fabrication

Conventional photolithography techniques are followed to produce the silicon masters. Purchase a bright field 4 × 4 × 0.09 in. mask with an antireflective chrome pattern on soda lime (Delta Mask, NL, www.deltamask.

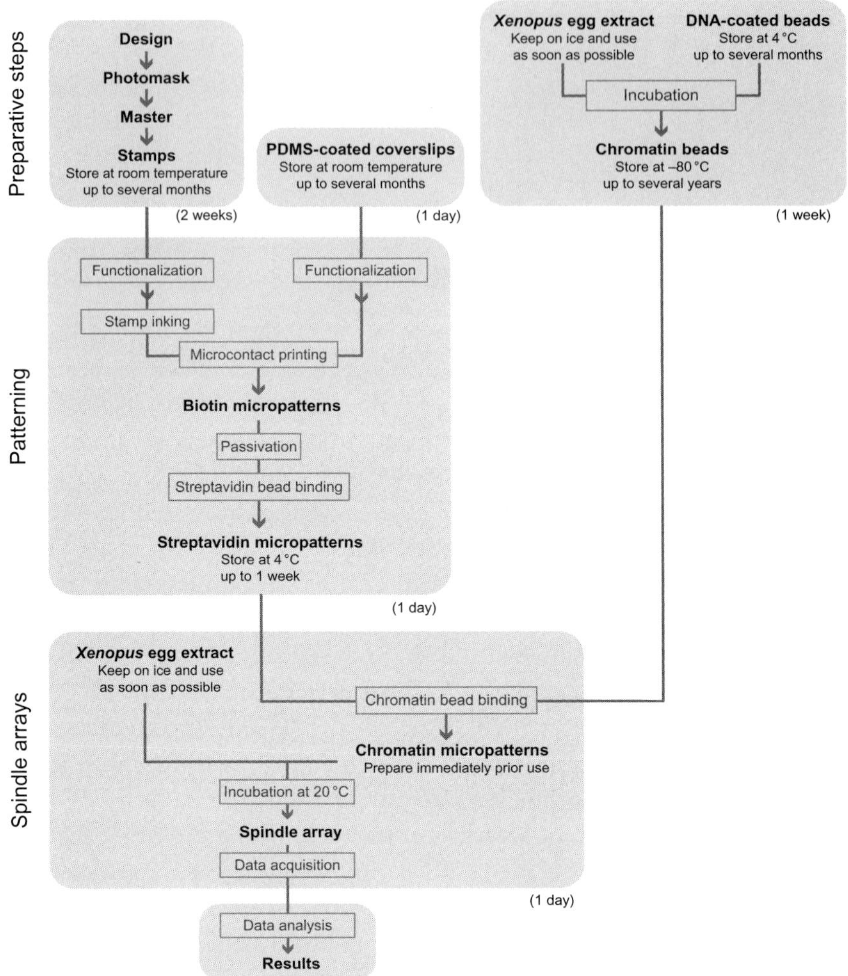

Figure 24.1 Flow chart of the experimental and data analysis steps. The approximate time necessary to perform the operations is indicated below each box.

nl) following a custom design. Spin-coat negative photoresist onto a 10 cm diameter silicon wafer (Silicon Materials, Inc.) down to a thickness of 8–10 μm. Expose coated wafer to UV light through the mask, bake, and develop. The resist master is stored with its patterned surface facing up in a dust-free place such as a plastic Petri dish.

3.2. PDMS mixture preparation

Mix prepolymer and curing agent (Dow Corning, Sylgard 184 kit) in a ratio of 10:1 for around 5 min.

3.3. Stamp fabrication

Place the resist master in a 150-mm Petri dish, cover it with 55 g of PDMS mixture and degas the mixture under vacuum for 1 h. Cure the PDMS at 65 °C overnight (o/n). Once the PDMS has cooled down to RT, carefully peel it off from the master and place it in a Petri dish with its patterned surface facing up. It can be stored in a dust-free place up to several months. Finally, cut-out stamps of the desired size. An area much larger than strictly needed should be printed to later select the best-printed area, and also because larger stamps are easier to handle. We always cut off one corner of the stamps on their nonpatterned side to mark their orientation.

3.4. Coverslip coating with PDMS

We coat microscopy coverslips with PDMS for three reasons: (1) coated coverslips are less prone to break while being manipulated, (2) the activated side is easily identified, and (3) the area around the pattern can be delimited to reduce the amount of 1-μm streptavidin bead solution used in the procedure. Clean bare glass was, however, also successfully used.

Procedure: place 30 mm round #1 glass coverslips (Menzel-Glaser) in a slide holder; immerse in a 1:1 ethanol:water solution and sonicate in a water bath (Bandelin Sonorex Super RK255H) for 1 h. Remove slides from the ethanol solution, and transfer them to a dust-free place to dry and be stored. Place a clean coverslip in the spin coater (EMBL CB-151D). Deposit about 100 μl of PDMS mixture to the center of the coverslip and spin at 8500 rpm (12 V) for 15 s. Cure PDMS-coated coverslips at 65 °C o/n and store in a dust-free environment up to several months.

3.5. PDMS cover fabrication

PDMS covers are fabricated to enclose the chromatin patterns and create incubations chambers later filled with *Xenopus* egg extract. For incubation chambers of approximately 30 μl, use elongated hexagonal stainless steel metallic pieces of size 15 × 5 × 0.5 mm ($L \times W \times H$). Pour the PDMS mixture over the metal pieces placed in a Petri dish. Degas under vacuum for 1 h and cure at 65 °C o/n. Cut covers individually and punch inlet and outlet using a 1.5 mm diameter biopsy puncher (Harris Uni-Core™, Ted Pella, Inc.). Store the PDMS covers in a dust-free place.

3.6. Chromatin bead preparation

The DNA sequence used to coat the beads does not seem to matter while a minimum length of 5 kb is recommended. We use a 10 kb plasmid with three unique restriction sites (one blunt site between two sticky end sites) in the polylinker. The plasmid is linearized using two restrictions enzymes to produce one blunt end and one sticky end. The sticky end is biotinylated by filling-in the overhangs using Klenow DNA polymerase and biotin-14-dATP, biotin-16-dUTP, thio-dCTP, and thio-dGTP. This biotinylated DNA is then bound to 2.8-μm paramagnetic streptavidin-coated beads. The ratio of DNA to beads will determine the final quantity of DNA per bead, and we recommend using 0.5 pg DNA per bead. The DNA bound to the beads is then further digested using the third restriction site and again biotinylated as previously described. This second DNA biotinylation step is needed to immobilize the beads (once organized as chromatin beads) on the 1-μm streptavidin bead patterns. If the DNA bead stock is clumpy after the second biotinylation step, it should be drawn up several times into a syringe fitted with a 27-gauge needle (Hannak & Heald, 2006). If the problem persists, we recommend preparing a new DNA bead stock as the 0.5 *M* KCl solution will not be efficient at individualizing the chromatin beads before binding them to the pattern. The DNA beads can be stored in bead storage buffer (PBS pH 7.4, 0.01% Tween®-20, 0.09% NaN_3) at 4 °C up to several months. Chromatin beads are made by incubation of DNA-coated beads in *Xenopus* extracts, which contain nucleosomes and chromatin factors. Chromatin beads are aliquoted, flash frozen in liquid nitrogen, and stored at −80 °C. The ability of each chromatin bead stock to assemble spindles should be tested either in solution or on a pattern. It seems critical to only use very potent chromatin bead stocks to obtain reliable spindle arrays. A detailed protocol to prepare chromatin beads can be found elsewhere (Hannak & Heald, 2006; Tarnawska, Pugieux, & Nedelec, 2014).

4. PATTERNING

4.1. Stamp and surface functionalization

Stamps and coverslips are activated in parallel. Place stamps (with their patterned side facing up) in a plastic Petri dish. Using a sharp blade, gently delimit a region on each PDMS-coated coverslip corresponding to the stamp size. Expose stamps and PDMS-coated coverslips to an air plasma (GaLa Instrumente, Germany) for 1 min (at 100 W) to make their surface hydrophilic. Incubate stamps in a freshly prepared 2% (3-aminopropyl) triethoxysilane (Sigma, cat. no. A3648) ethanolic solution for 40 min. Place activated coverslips on a piece of Parafilm® M (Bemis), and cover each of them with 150 μl of (3-glycidyloxypropyl)trimethoxysilane (Sigma, cat.

no. 440167). Incubate coverslips until the stamps are ready for the printing step.

4.2. Stamp inking

Rinse stamps once with ethanol, twice with MilliQ water and keep them immersed in water until needed. Dry stamps briefly using compressed nitrogen, and incubate their patterned side for 20 min with 10 mg/ml biotinylated BSA in PBS (Sigma, cat. no. A8549).

4.3. Microcontact printing

Handling the stamps is easier with self-closing tweezers (Dumont N2A). Rinse a coverslip in acetone and carefully dry it using compressed nitrogen. Rinse a stamp in PBS to remove unbound biotinylated BSA and dry it using compressed nitrogen. Carefully deposit the stamp, patterned side down, on the coverslip and leave in contact for 10 min. Optionally, placing a weight on top of the stamp improves the contact, and thus the quality of printing (we used 40 g weights for stamps of 20×5 mm). Carefully remove the stamp. It can be stored (patterned face up) and reused.

4.4. Passivation

Prepare a fresh solution of 2.5 mg/ml NHS-PEG (Nanocs, USA, cat. no. PG1-SC-5k) in PBS. Cover the surface of the coverslip with 200 μl of NHS-PEG solution and incubate for 1 h at 4 °C. Prepare 10 mg/ml BSA (Sigma, cat. no. A2153) in PBS. Rinse coverslip with BSA solution and further incubate in BSA solution for 1 h at 4 °C.

4.5. Streptavidin bead binding

The first bead layer on the pattern consists of Dynabeads® MyOne™ Streptavidin C1 beads (Invitrogen, cat. no. 650-01). To maximize the number of beads binding to the pattern, it is important to use a concentrated bead solution, and a volume proportional to the area covered by the pattern. Twenty microliters of streptavidin bead stock solution is sufficient to cover a pattern of few mm^2, providing a large excess of beads.

Procedure: Recover the streptavidin beads from the storage buffer, following the manufacturer's instructions and resuspend them in cold PBS with a dilution ratio of 5. Wash printed coverslips in PBS to remove the excess BSA. Dry the area surrounding the printed area using Whatman paper, without touching the printed area itself, which must remain wet.

Gently deposit 100 μl of PBS-containing beads on the printed area. Cover the coverslips with a Petri dish lid to protect the bead solution from drying, and incubate at 4 °C for 45 min. The beads will sediment by gravity onto the surface and some of them will bind to the biotinylated BSA patterns.

4.6. Pattern visualization and storage

Immerse the patterned coverlip in a Petri dish filled with PBS and gently pipette some PBS above the pattern to remove unbound beads. At this stage, rinsing the unbound streptavidin beads will reveal the pattern, and the quality of the printing can be assessed. Only well-printed portions of the surface are used, where beads densely cover the features of the design. The patterned coverslips are stored in PBS at 4 °C and should be used within a week.

5. SPINDLE ARRAYS

Prepare a *Xenopus* egg extract and assess its quality (Hannak & Heald, 2006) (see also Chapter 22) in order to decide whether to proceed. Thaw a frozen aliquot of chromatin beads on ice. Collect beads on the side of the tube using a magnetic particle concentrator (Dynabeads® MPC®-S). Remove supernatant and immediately replace with 20 μl of cold 0.5 *M* KCl (prepared in CSF-XB). The salt is used to dissociate any clumps of chromatin beads that may have formed during the procedure. Remove the tube from the MPC, keep it on ice and resuspend the beads by pipetting. Transfer the patterned coverslip, stored in PBS, into a Petri dish filled with CSF-XB. Gently pipette CSF-XB above the pattern to clean it. If the size of the pattern is larger than desired, scrape the excess area to prevent the chromatin beads from binding outside the region of interest.

The following steps should be done without interruption, on ice whenever possible. Take the coverslip out of the CSF-XB and place it on the lid of a Petri dish, which can be more comfortably handled. Dry around the pattern using Whatman paper strips without drying the pattern itself. Immediately deposit the chromatin beads onto the pattern. Place an office magnet below the coverslip for around 10 s to attract the chromatin beads to the pattern. To increase the amount of chromatin beads on the pattern, the unbound chromatin beads can be resuspended very gently and once again attracted down with the office magnet for few seconds. We recommend using a bench-top microscope to verify the immobilization of the chromatin beads on the pattern. If the chromatin beads that have fallen outside the regions covered by 1-μm streptavidin beads cannot be removed, this may indicate a problem related to passivation, and our advice is to discard this slide and to prepare a new one using another printed coverslip and a second chromatin bead aliquot. Immediately place the coverslip back into cold CSF-XB. Remove unbound chromatin beads by pipetting gently while sweeping the pipette stream over the pattern. This step is crucial, since leftover beads immobilized outside the original pattern can affect the experiment outcome. Take the coverslip out of the CSF-XB solution and dry the region around the chromatin pattern with strips of Whatman paper. It is essential to dry the coverslip as close as possible to the pattern otherwise the PDMS cover will not adhere properly. Place a PDMS cover, previously prepared with an inlet and outlet, above the chromatin pattern and press gently to ensure adhesion to the coverslip. Immediately fill the chamber with 80 μl of cold CSF-XB, and subsequently flow in *Xenopus* egg extract supplemented with a Hoechst DNA dye and fluorescently labeled tubulin. Transfer the coverslip onto the microscope stage. Record the time and start observations.

We monitor such experiments using a fast confocal microscope, which scans a slit instead of a pinhole (Zeiss LSM 5Live controlled by the software Zen 2009), with a 40 × oil immersion lens. This microscope is equipped with a Plexiglas chamber and a temperature controller (EMBL workshop) to maintain a temperature of 20 °C during the imaging. We define six positions to be imaged every minute. Each position covers 600 × 600 μm^2 (3 × 3 fields of view). The total area will contain nearly 500 chromatin dots when the dots are separated by 62 μm (Fig. 24.2). Altogether this large amount of collected data makes an automated image analysis very desirable.

6. Analysis

Spindle arrays are typically imaged every minute over 1–2 h, but DNA is imaged only once. To analyze such set of images, we developed semiautomated analysis software in MATLAB (The Mathworks Inc., USA), available from the authors on request.

6.1. Program architecture and usage

The user is provided an ordered panel of buttons to call the underlying functions (Fig. 24.3B). Most functions are independent and buttons can be added, allowing great flexibility at low maintenance costs. One function

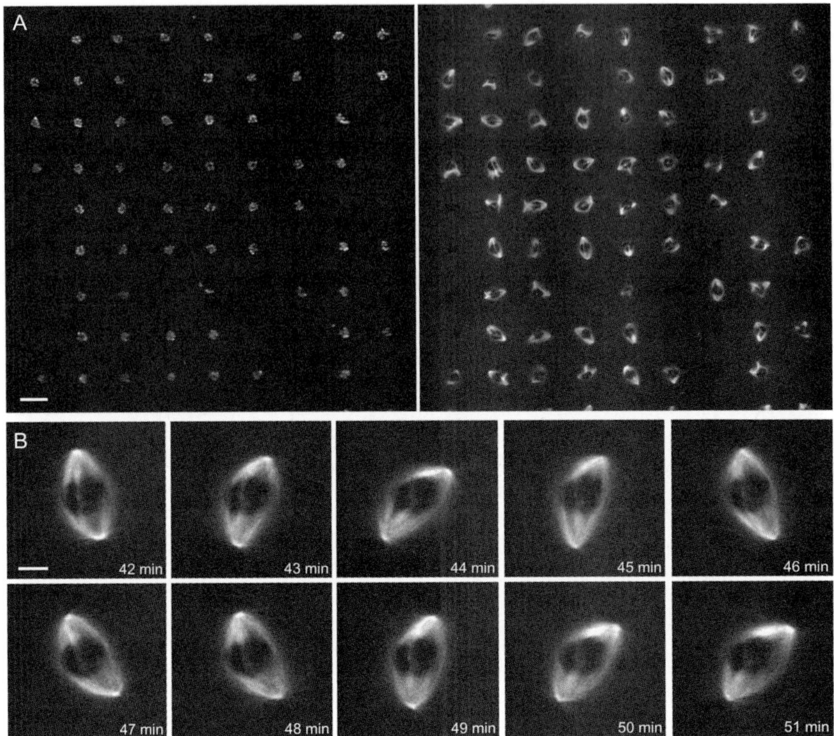

Figure 24.2 Meiotic spindles organized on chromatin pattern incubated with IgG mock-depleted *Xenopus* egg extract. (A) 3×3 field of view of spindles organized on chromatin patches separated by 62 μm, center to center. (Left) Chromatin pattern visualized with Hoechst dyes for DNA. (Right) Spindles formed after 42 min of incubation. The tubulin is labeled with Cy3. Scale bar 50 μm. (B) Rotation and oscillation of a bipolar spindle over 10 min. Scale bar 10 μm.

creates the list of images to analyze. The user then defines which image shows the DNA and may customize different parameters. Regions of interest are then automatically created to match the DNA pattern, but the user has the possibility to edit them (Fig. 24.3A). Other functions allow the clicking of chromatin patches and spindle poles (Fig. 24.3C). From then on, extraction of quantities such as total fluorescence (Fig. 24.3D) and classification based on the number of poles (Fig. 24.3E) is automated. The program also extracts trajectories, that is, all the consecutive positions of the poles of each spindle, automatically reordering the poles that may have been clicked in a different order by the user.

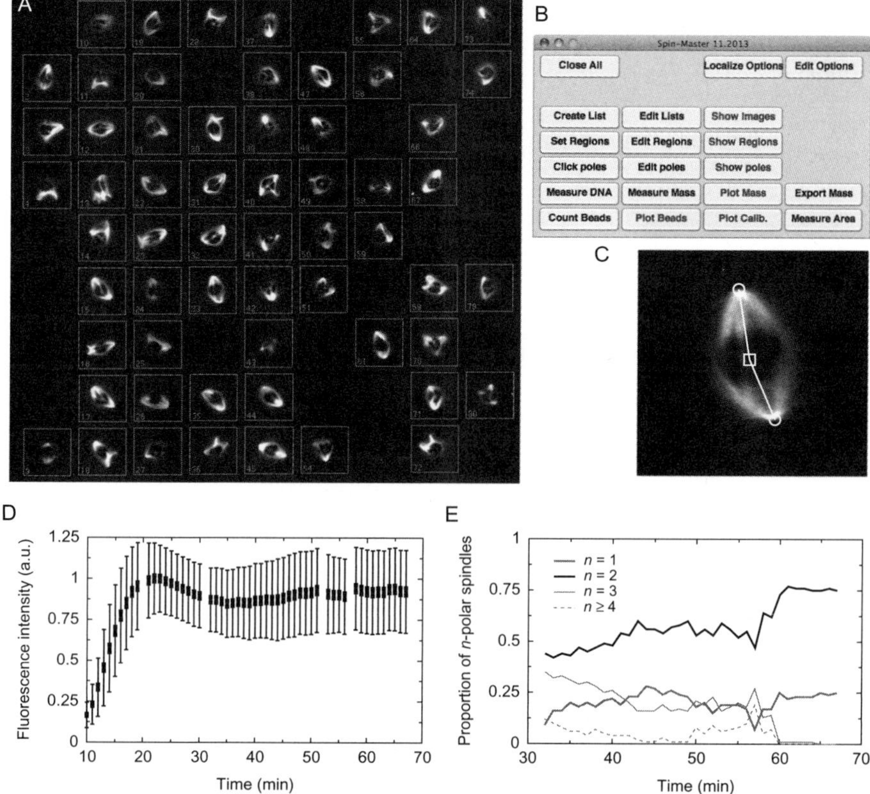

Figure 24.3 Data analysis pipeline. (A) Regions of interest (ROIs) are used to isolate the chromatin patches on which microtubules appear. ROIs (green) are initially generated automatically but the user has the possibility to delete or edit them. (B) Front end to the different MATLAB(R) functions that drive the analysis. (C) Example of a manual analysis. Each spindle is automatically presented to the user who defines the pole location by mouse clicks. (D) The total fluorescence in the selected regions of interest as a function of time (avg. and std. dev.). Regions containing less than five chromatin beads were discarded. (E) Proportion of monopolar ($n=1$), bipolar ($n=2$), tripolar ($n=3$), and multipolar ($n \geq 4$) spindles, as a function of time. (For interpretation of the references to color in this figure legend, the reader is referred to the online version of this chapter.)

6.2. Sample analysis

To illustrate the versatility of the method, we quantified two spindle movements: oscillation and rotation (Fig. 24.4A). We analyzed eight movies covering 10 min each of structures assembled in IgG mock-depleted *Xenopus* egg extracts. First, the user clicks the center C of each chromatin patch on the DNA image. Second, the user clicks all spindle poles (points P_1,

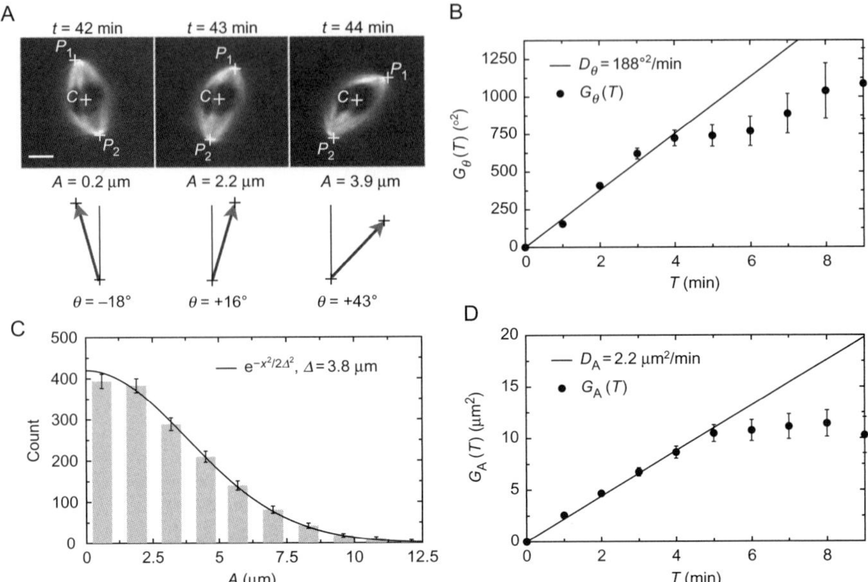

Figure 24.4 Spindle motion on immobilized DNA. (A) Spindle orientation θ is the angle of P_1P_2 with an arbitrary axis, and spindle acentricity is defined as $A = |P_1C - P_2C|$. These quantities show variability, here at times 42, 43, and 44 min from the beginning of incubation. Scale bar: 10 μm. Rotation is defined as the change of θ with time while spindle oscillations would result in A being a periodic function of time. (B) The mean squared displacement (MSD) of θ is computed as $G_\theta(T) = <(\theta(t+T) - \theta(t))^2>$. G_θ is linear at short time intervals ($D_\theta = 188^{\circ 2}$/min) and then saturates, indicating a constrained rotational diffusion. Here and thereafter, error bars represent standard error. (C) Histogram of spindle acentricity is nearly Gaussian ($\sigma = 3.8$ μm). (D) The MSD of spindle acentricity shows constrained diffusion ($D_A = 2.2$ μm^2/min), saturating at 10.9 μm^2. Results were obtained with IgG mock-depleted egg extract (i.e., they reflect wild-type conditions).

P_2, P_3, etc.) on each tubulin image (Fig. 24.4A). We analyzed all the structures that had two poles, for which the angle formed by P_1, C, and P_2 was greater than 150°. Such bipolar spindles represented approximately 40% of the structures, and provided 403 trajectories.

The resulting angular mean squared displacement (MSD) G_θ, exhibits a linear increase at short time intervals characterized by a rotational diffusion coefficient of $D_\theta = 188^{\circ 2}$/min, and saturates to a value of $\sim 1000^{\circ 2}$ after 10 min (Fig. 24.4B). This means that spindles rotate in both directions, with an average net rotation of 13.7° in a minute, but do not rotate more than ~30° overall. This is the signature of a constrained rotational diffusion.

The values of acentricity are normally distributed with a width of 3.8 μm (Fig. 24.4C). This means that the chromatin dot is precisely centered between the two poles, with an average offset of around 10% compared to spindle length, or in other words that all spindles were symmetrically positioned around the DNA. The MSD G_A shows a linear behavior at short time intervals and plateaus at 10.9 μm^2 at time intervals larger than 5 min (Fig. 24.4D). Once again, this is the signature of a constrained diffusion, with a diffusion coefficient $D_A = 2.2\ \mu m^2/min$. Our analysis also shows that there is no oscillation at a preferred frequency, as seen before for spindles formed around single chromatin-coated 10-μm beads (Halpin et al., 2011).

7. CONCLUSIONS

Spindles formed on chromatin micropatterns are geometrically similar to spindles formed in *Xenopus* extract around chromatin beads in solution (Heald et al., 1996), but their motions are different. If spindles formed in solution are put between slides and coverslip, as needed for observation, dynein motors adsorbed on the surfaces move the spindles and eventually destroy them. In contrast, on micropatterns only rotation is observed, as the spindles always remain symmetric around the immobilized spot of chromatin on which they form. Spindle arrays are useful to measure spindle behavior and they greatly facilitate automation of the analysis. For example, in Section 6.2, we used this approach to extract statistically significant results on dynamical properties such as spindle rotation and spindle oscillation. Further observables can be extracted by adapting the analysis tools that we provide on request.

The quality of the chromatin beads, together with the quality of the *Xenopus* egg extracts, are the most critical parameters of success. The number of spindles that can be recorded is not limited by the pattern itself, but by the acquisition rate of the microscope. Thus using the fastest possible robotic microscope is desirable. The initial printing steps are reproducible, but the method takes ~6 h and surfaces should be used within 7 days and this is time consuming; alternative patterning methods, such as deep UV patterning, would be faster but are less well established (Tarnawska et al., 2014).

Chromatin patterns offer many exciting possibilities; they could, for example, be used to study how chromatin induces the assembly of an actin network in starfish (Lénárt et al., 2005). Another direction is to integrate spindle arrays within microfluidic setups to dynamically change the composition of the cytoplasmic fluid surrounding the immobilized spindles.

ACKNOWLEDGMENTS

We thank Joachim Spatz and members of his group for preparing the resist master, Kresimir Crnokic for maintaining the frog facility. We are thankful to Camilla Godlee for her critical reading of the manuscript. The Nedelec lab acknowledges support from the European Community's Seventh Framework Programme (FP7/2007–2013) under Grant Agreement Nos. 241548 and 258068.

REFERENCES

Athale, C. A., Dinarina, A., Mora-Coral, M., Pugieux, C., Nedelec, F., & Karsenti, E. (2008). Regulation of microtubule dynamics by reaction cascades around chromosomes. *Science (New York, NY)*, *322*(5905), 1243–1247. http://dx.doi.org/10.1126/science.1161820.

Dinarina, A., Pugieux, C., Corral, M. M., Loose, M., Spatz, J., Karsenti, E., et al. (2009). Chromatin shapes the mitotic spindle. *Cell*, *138*(3), 502–513. http://dx.doi.org/10.1016/j.cell.2009.05.027.

Halpin, D., Kalab, P., Wang, J., Weis, K., & Heald, R. (2011). Mitotic spindle assembly around RCC1-coated beads in Xenopus egg extracts. *PLoS Biology*, *9*(12), e1001225. http://dx.doi.org/10.1371/journal.pbio.1001225.

Hannak, E., & Heald, R. (2006). Investigating mitotic spindle assembly and function in vitro using *Xenopus laevis* egg extracts. *Nature Protocols*, *1*(5), 2305–2314. http://dx.doi.org/10.1038/nprot.2006.396.

Heald, R., Tournebize, R., Blank, T., Sandaltzopoulos, R., Becker, P., Hyman, A., et al. (1996). Self-organization of microtubules into bipolar spindles around artificial chromosomes in Xenopus egg extracts. *Nature*, *382*(6590), 420–425. http://dx.doi.org/10.1038/382420a0.

Kumar, A., & Whitesides, G. M. (1994). Patterned condensation figures as optical diffraction gratings. *Science (New York, NY)*, *263*(5143), 60–62. http://dx.doi.org/10.1126/science.263.5143.60.

Lénárt, P., Bacher, C. P., Daigle, N., Hand, A. R., Eils, R., Terasaki, M., et al. (2005). A contractile nuclear actin network drives chromosome congression in oocytes. *Nature*, *436*(7052), 812–818. http://dx.doi.org/10.1038/nature03810.

Murray, A. W. (1991). Cell cycle extracts. *Methods in Cell Biology*, *36*, 581–605.

Noatynska, A., Gotta, M., & Meraldi, P. (2012). Mitotic spindle (DIS)orientation and DISease: Cause or consequence? *The Journal of Cell Biology*, *199*(7), 1025–1035. http://dx.doi.org/10.1083/jcb.201209015.

Petry, S., Pugieux, C., Nedelec, F. J., & Vale, R. D. (2011). Augmin promotes meiotic spindle formation and bipolarity in Xenopus egg extracts. *Proceedings of the National Academy of Sciences of the United States of America*, *108*(35), 14473–14478. http://dx.doi.org/10.1073/pnas.1110412108.

Qin, D., Xia, Y., & Whitesides, G. M. (2010). Soft lithography for micro- and nanoscale patterning. *Nature Protocols*, *5*(3), 491–502. http://dx.doi.org/10.1038/nprot.2009.234.

Tarnawska, K., Pugieux, C., & Nedelec, F. (2014). Mitotic spindle assembly on chromatin patterns made with deep UV photochemistry. *Methods in Cell Biology*, *120*, 3–17. http://dx.doi.org/10.1016/B978-0-12-417136-7.00001-X.

Théry, M., & Piel, M. (2009). Adhesive micropatterns for cells: A microcontact printing protocol. *Cold Spring Harbor Protocols*, *2009*(7), 1–11. http://dx.doi.org/10.1101/pdb.prot5255, pdb.prot5255.

Thiébaud, C., & Fischberg, M. (1977). DNA content in the genus Xenopus. *Chromosoma*, *59*(3), 253–257.

AUTHOR INDEX

Note: Page numbers followed by "*f*" indicate figures and "*np*" indicate footnotes.

D

E

F

G

H

I

J

K

M

N

O

P

S

T

U

V

W

SUBJECT INDEX

Note: Page numbers followed by "*f*" indicate figures.

B

G

M

N

R

S

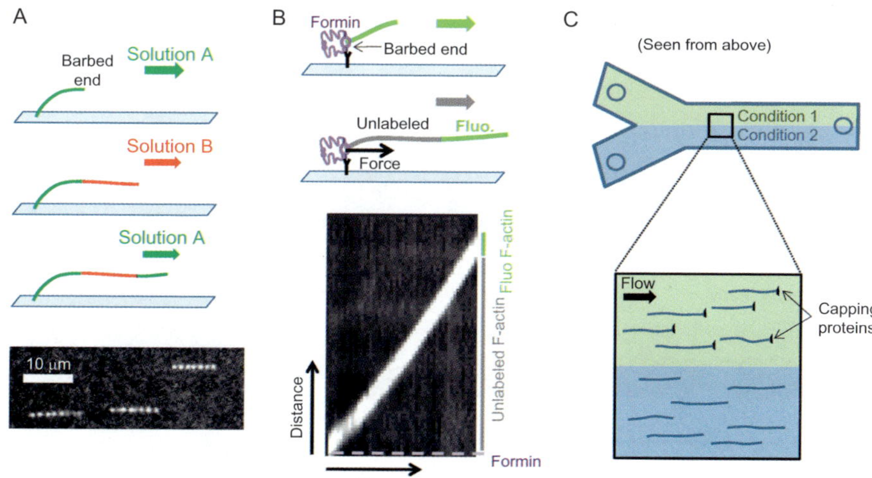

Marie-France Carlier *et al.*, Figure 1.2 Some experimental designs offered by microfluidics. (A) Sequential exposure allows the assembly of hybrid filaments made of actin domains of different nature (label, nucleotide state, or any chemical modification). Bottom: epifluorescence image of "striped filaments" composed of alternating dark and labeled segments. (B) The elongation of filaments from unlabeled actin at surface-bound formins can be monitored tracking a fluorescent segment near the pointed end. Bottom: corresponding kymograph. (C) Two different populations of filaments can coexist in two distinct regions exposed to different incoming solutions.

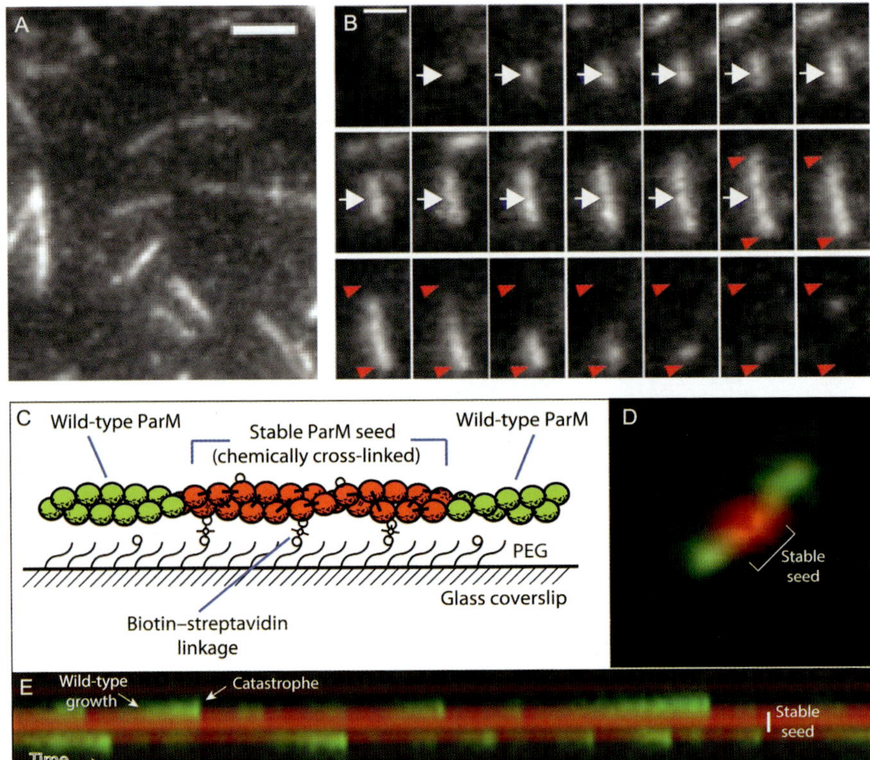

Natalie A. Petek and R. Dyche Mullins, Figure 2.4 Single-filament dynamics by TIRF. (A) A typical TIRF field with GFP-ParM filaments polymerized in AMP-PNP. Scale bar 5 μm. (B) A montage of a single filament over time shows bidirectional growth (white arrow) and catastrophic disassembly (red arrows). Scale bar 2 μm (Garner et al., 2004). (C) Schematic of two-color seeded filament assembly by TIRF microscopy. (D) A representative filament shows bidirectional growth. (E) Kymograph analysis shows bidirectional growth and catastrophic disassembly over many cycles.

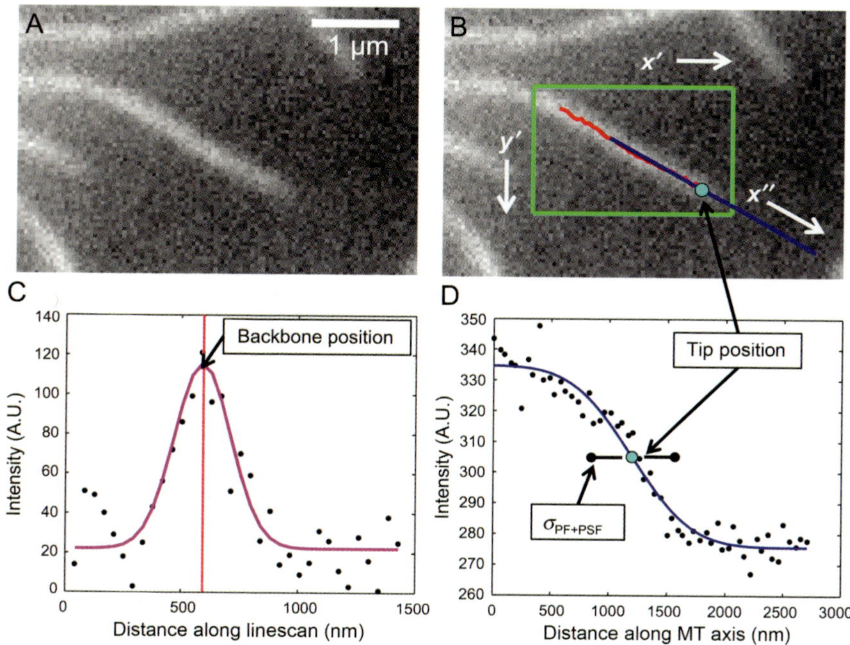

Louis S. Prahl *et al.*, Figure 3.1 Functions used to fit microtubule backbone and tip. (A) An example fluorescence image of an individual microtubule is selected for tip-tracking analysis. (B) Two user-supplied clicks serve as "initial guesses" and define the boundaries of the green box. Within this region, microtubule backbone coordinates (red) are estimated using a Gaussian fit along line scans perpendicular to the major microtubule axis (in this case, the *y*-axis). (C) Plot of the Gaussian function (purple curve) used to fit microtubule backbone coordinates, with black dots representing single pixel intensity values (A.U.) and red vertical line representing the Gaussian mean (backbone coordinate). (D) Plot of the Gaussian survival function (blue curve) used to fit the microtubule tip in (B), with black dots representing intensity values (A.U.) along the x''-axis. Note that TipTracker subtracts the minimum from intensity values for the Gaussian line scan fit (C), but uses the original pixel values to fit the survival function in (D). Black bars mark the standard deviation of the error function (σ_{PF+PSF}), which is used to estimate tip taper.

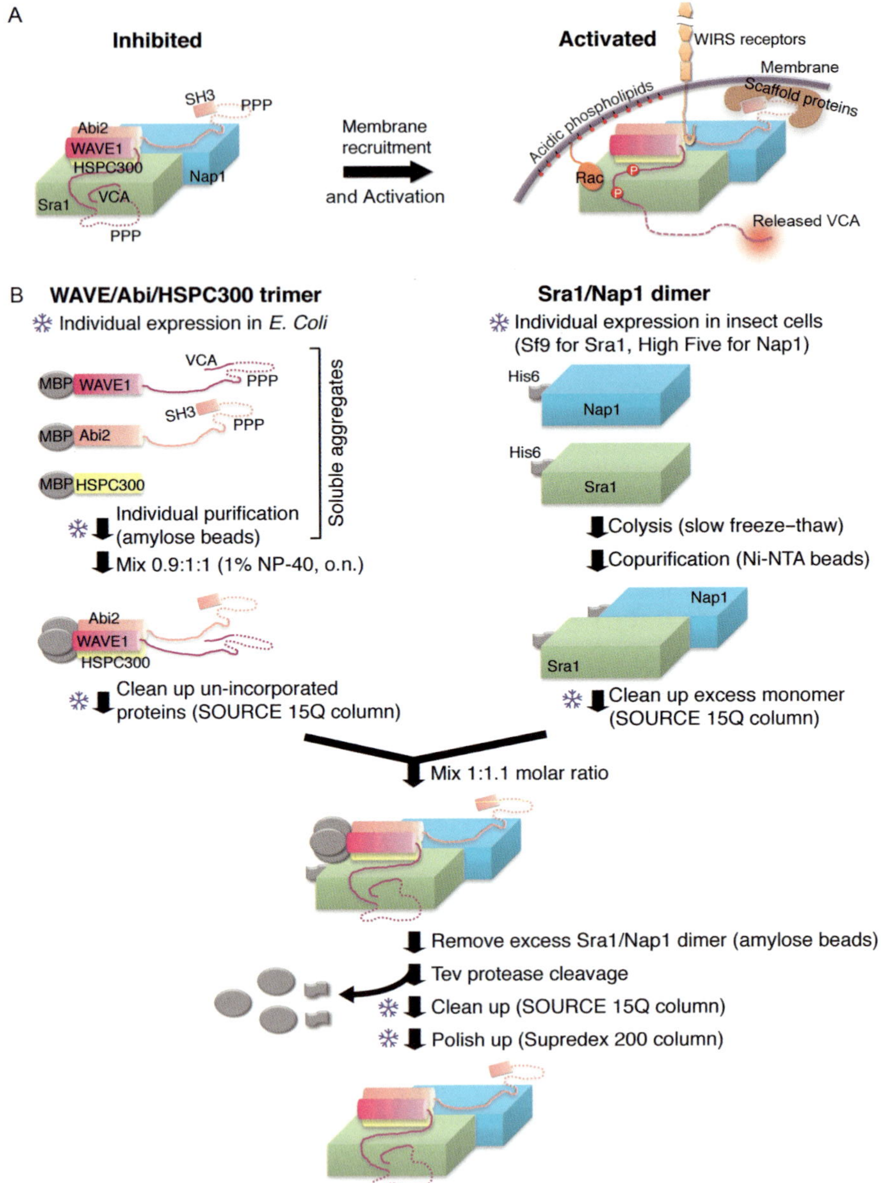

Baoyu Chen *et al.*, Figure 4.1 Activation mechanism and purification strategy of the WRC. (A) Schematic of WRC inhibition, activation, and membrane recruitment. Dotted lines indicate unstructured sequences. (B) Schematic of WRC *in vitro* reconstitution. Snowflake symbols indicate steps after which samples can be frozen and stored for future use.

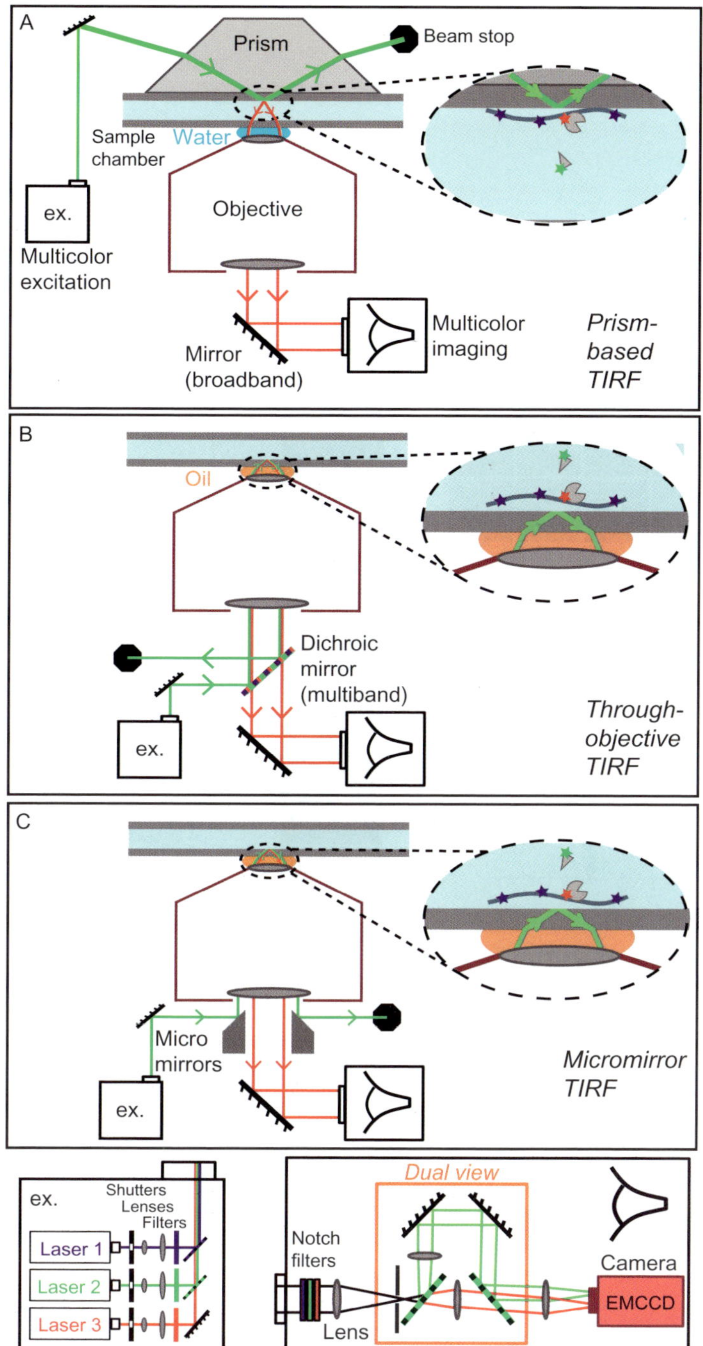

Figure 6.2 Schematics of various multicolor TIRF microscope designs. Prism-based TIRF (A) is in general simpler to set up than through-objective TIRF (B), but is less sensitive since imaging across the full thickness of sample chamber is required. Custom micromirror TIRF (C) further improves the broad spectrum sensitivity by spatially separating excitation (green) and emission (red) optical paths. In each case, excitation (ex.) can be generated by a combination of different color laser beams (bottom left module), and fluorescence emissions can be split into two (or more) color channels, focused on an EMCCD camera, for multicolor image sequence acquisition (bottom right module).

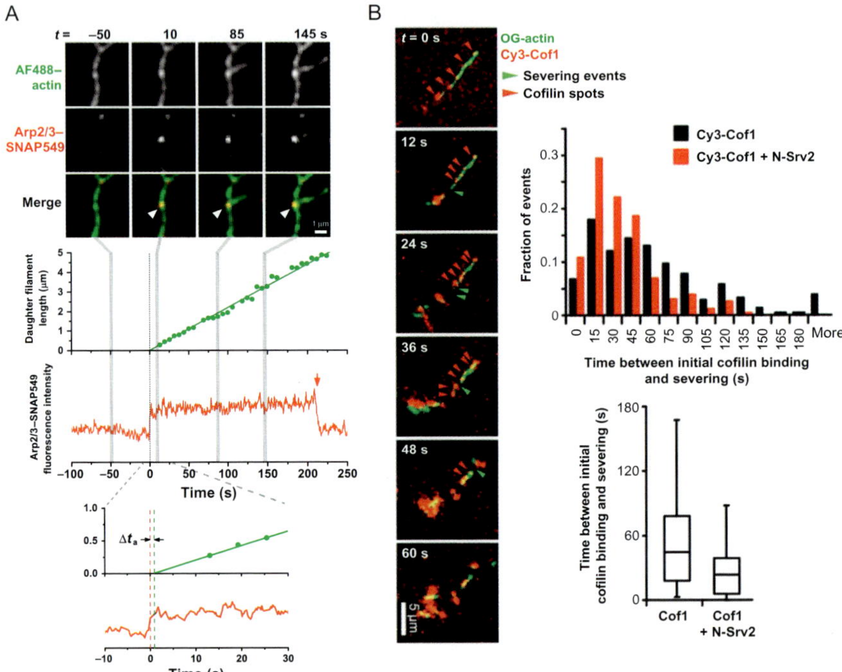

Benjamin A. Smith *et al.*, Figure 6.3 Dual-color TIRF studies of actin filaments and actin-associated proteins. (A) Two-color imaging of actin and individual Arp2/3 complexes showed a short activation time delay ($\Delta t_a \sim 1$ s) between binding of Arp2/3 complex to the side of a filament (at $t=0$) and nucleation of a new filament branch (Smith, Daugherty-Clarke, Goode, & Gelles, 2013). Single-molecule observation was confirmed by single-step photobleaching (red arrow) of the dye label on Arp2/3 complex incorporated into a stable branch junction. (B) Real-time observation of actin and cofilin during cooperative filament severing by Srv2/CAP and cofilin. Srv2 shortened the delay between cofilin binding and severing compared to controls without Srv2 (see histogram and whisker plot). *Reproduced from Chaudhry et al. (2013).*

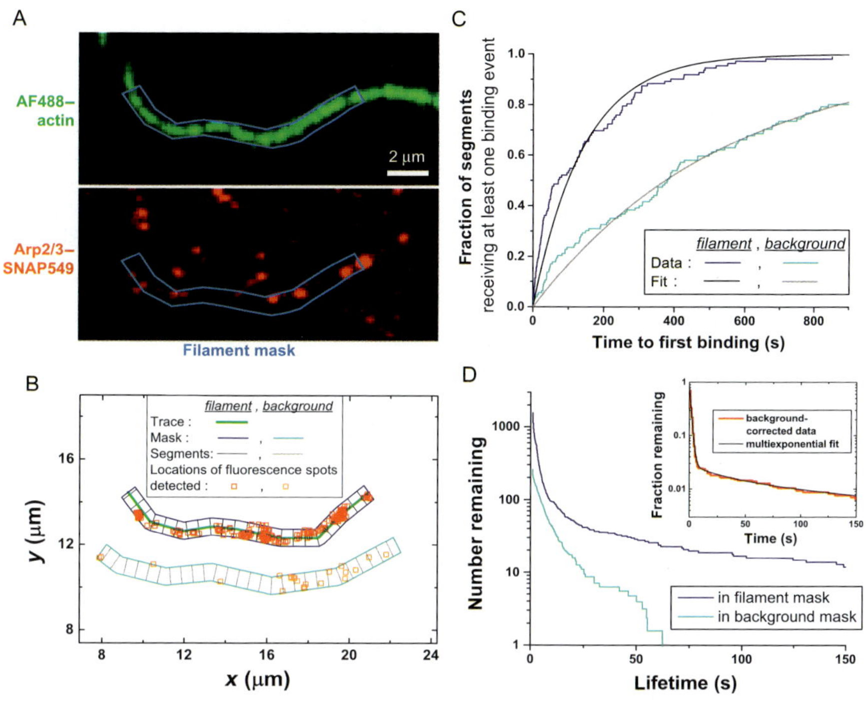

Benjamin A. Smith *et al.*, Figure 6.4 Analysis of the kinetics of a protein binding to and dissociating from filaments. (A) Dual-color observations of surface-tethered actin filaments and Arp2/3 complex were analyzed by first tracing a section of filament and generating a mask of uniform width (blue). (B) Fluorescence spots corresponding to single Arp2/3 complexes were tracked over a ~15-min recording. Spots that appeared in the filament mask (red squares) and in control masks covering regions of the field of view that did not contain filament (orange squares in cyan background mask) were retained for analysis. (C) Cumulative distribution of times of first appearance of Arp2/3 complex in each segment of filament and background masks in B. Single-exponential fits are shown (smooth curves), yielding the binding rate constants. (D) Dissociation rate constants were analyzed by generating survival plots of Arp2/3 complexes in filament and background masks. A specific Arp2/3-filament lifetime distribution was generated (inset: red line) by subtracting the background survival plot from the filament survival plot. Multiexponential fit (black curve) to the background-corrected data reveals that multiple bound states of the Arp2/3-filament complex exist. *Adapted from Smith, Daugherty-Clarke, Goode, and Gelles (2013).*

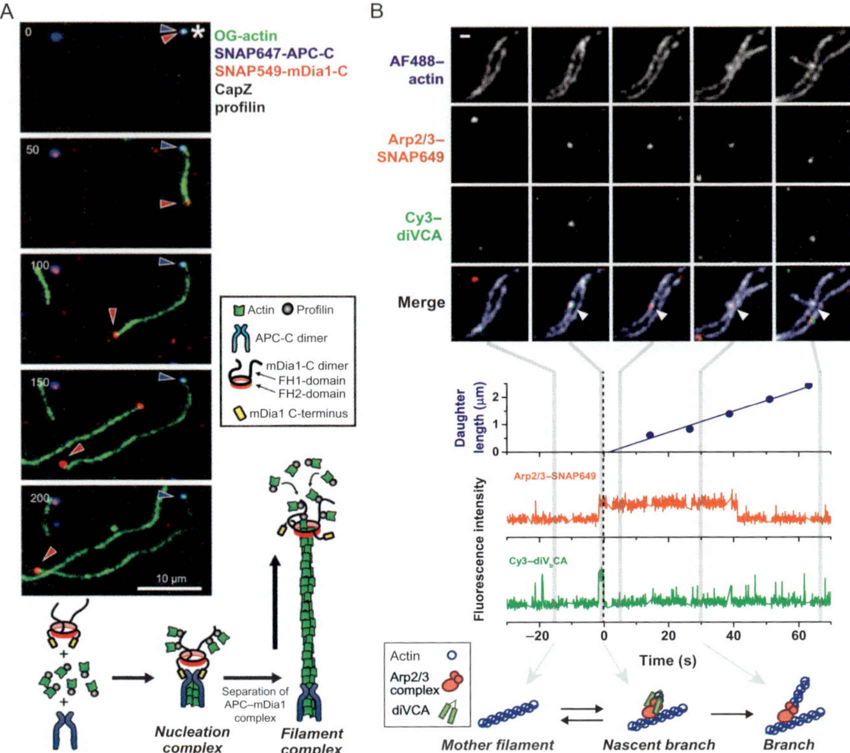

Benjamin A. Smith *et al.*, Figure 6.5 Directly observing the coordination of multiple factors regulating actin filament dynamics. (A) Three-color image sequence (left) showing the coordinated action of APC (blue) and the formin mDia1 (red) in nucleation and elongation of an actin filament (green). Prior to filament elongation, APC and formin colocalized (asterisk), but during elongation APC remained with the filament pointed end whereas formin processively tracked the growing barbed end, defining the "rocket launcher" mechanism shown (cartoon below). (B) Three-color imaging (top) of single molecules of dimeric VCA (green) and Arp2/3 complex (red) binding the side of a preexisting mother filament and generating a new daughter filament (forming an actin branch; blue) (Smith, Padrick, et al., 2013). Fluorescence intensity traces and daughter length records (middle) reveal that Arp2/3 complex and VCA appear on filament sides as a complex, but VCA departs rapidly prior to initiation of daughter filament elongation, indicating that VCA release from the nascent branch may trigger branch formation (simplified kinetic scheme, below). *Panel (A): Reproduced from Breitsprecher et al. (2012). Panel (B): Portions modified from Smith et al. (2013).*

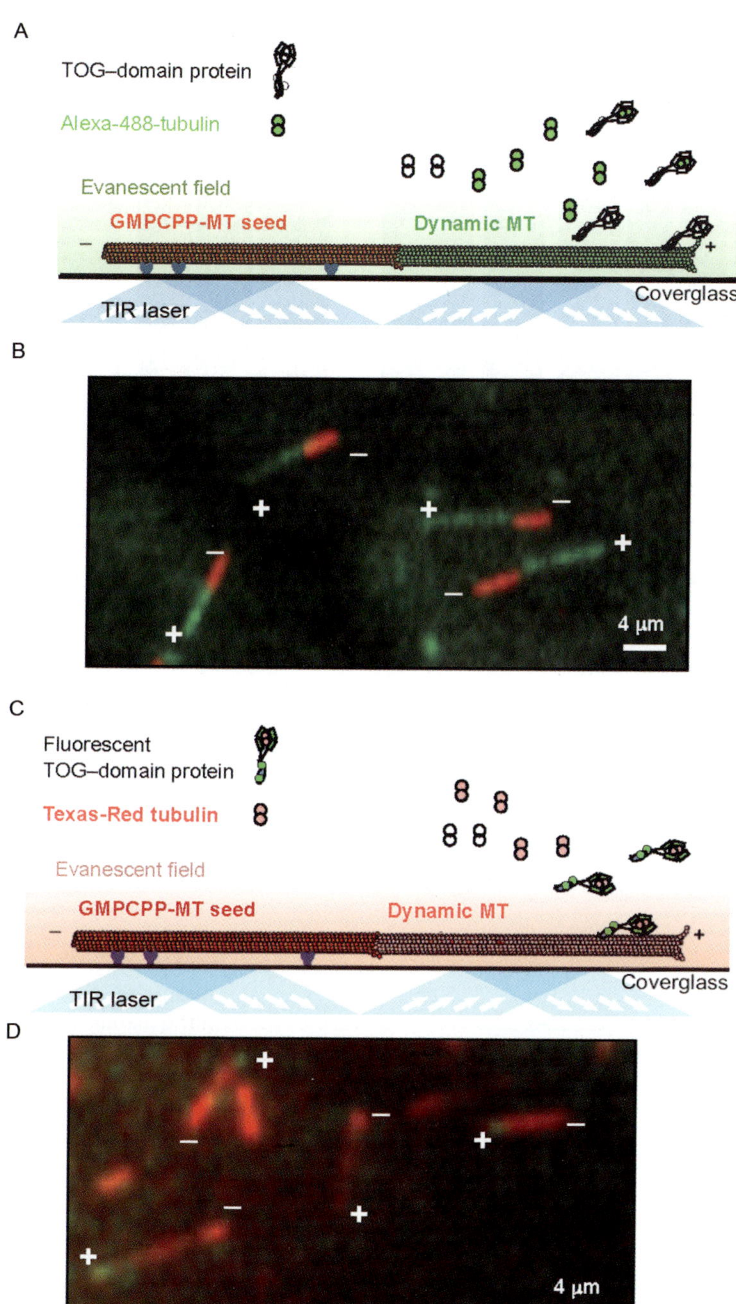

Jawdat Al-Bassam, Figure 8.2 Schematic representation of dynamic MT polymerization reconstitution studies. (A) Scheme described in Section 2.4, for dual fluorescent

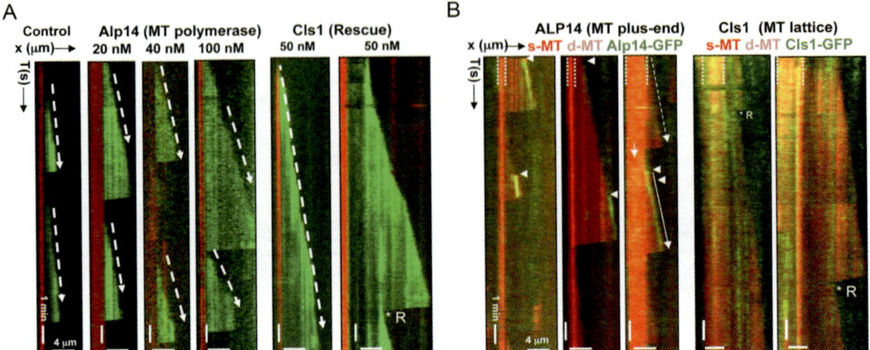

Jawdat Al-Bassam, Figure 8.3 Example MT dynamic polymerization kymographs based on studies described. (A) Kymographs of dynamic MTs produced using ImageJ, using methods described in Section 2.4. Dynamic MTs are shown in green, while stabilized MT seeds are shown in red. Left panel, MTs grow slowly at 6 μM tubulin dimer and depolymerize in frequent catastrophe events. Middle panels, increasing ALp14 concentration incrementally increases MT polymerization, without influencing MT catastrophe frequency (Al-Bassam et al., 2012). Right panels, Cls1 decreases in the frequency of MT catastrophe, and occurrence of MT rescues, labeled R^{*} (Al-Bassam et al., 2010). (B) Kymographs of dynamic MTs produced using ImageJ, using methods described in Section 2.5. Dynamic MTs are shown in light color red, while stabilized MT seeds are shown in intense red. Left panels, Alp14-GFP, shown in green, binds at MT plus ends to increase MT polymerization rate (Al-Bassam et al., 2012). Right panels, Cls1-GFP binds along MT lattices without tracking MT plus ends, where it correlates with absence of MT catastrophes, and its presence coincides with MT rescues, labeled R^{*} (Al-Bassam et al., 2010).All images in this figure are generated with permission from publisher of these references (Al-Bassam et al., 2010, 2012).

MTs to measure MT polymerization dynamics. Dynamic MTs, shown in green, grow from Alexa-Fluor-488-labeled tubulin, while stabilized GMPCPP MT-seeds, shown in red, are polymerized from Texas-Red-labeled tubulin. MT regulators are nonfluorescent. (B) Example data of reconstituted, dynamic MTs, in the scheme described in Section 2.4. (C) Scheme for dynamic MT polymerization reconstitution using fluorescent MT regulator and fluorescent MTs, as described in Section 2.5. Dynamic MTs are shown in faint red grown 10% Texas-Red tubulin, while stabilized MT seeds are shown in red, polymerized from a higher ratio of Texas-Red-labeled tubulin. MT regulators, shown in green, are labeled with GFP-tags. (D) Example data of reconstituted MT dynamics with fluorescent MT regulators at. All images in this figure are generated with permission from publisher of these references (Al-Bassam et al., 2010, 2012). *(B) Reproduced with permission from Al-Bassam et al. (2010). (D) Reproduced with permission from publisher of reference Al-Bassam et al. (2010).*

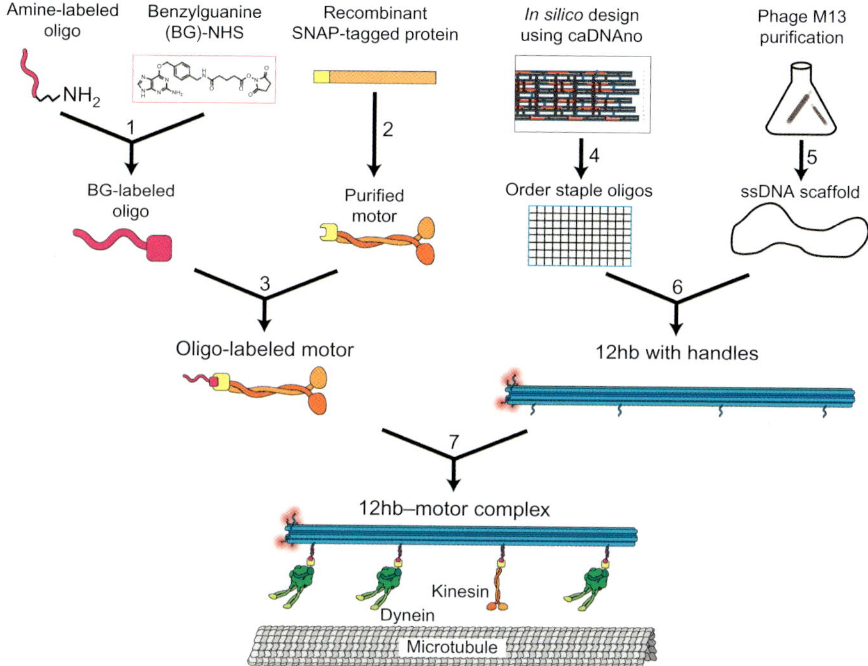

Brian S. Goodman and Samara L. Reck-Peterson, Figure 10.1 Production and assembly of motor–DNA origami structure complexes. (1) Oligo-labeled motors are generated by first coupling amine-labeled oligos to benzylguanine (BG, the SNAP ligand) through NHS ester chemistry. (2) Recombinant SNAP-labeled motor proteins are purified using affinity purification. (3) BG-oligos react with SNAP-tagged motors to generate oligo-labeled motors. (4) The 12hb DNA origami structure is designed using caDNAno software, which generates a list of oligo staples appropriate for a given scaffold sequence. (5) An M13 phage DNA purification generates the ssDNA-scaffolding strand. (6) Staple oligos are mixed with the scaffold strand, and a folding procedure generates the 12hb DNA origami structure. (7) Oligo-labeled motors are mixed with the 12hb DNA origami structure (bearing handle sites complimentary to the motor oligos) to generate a 12hb–motor complex.

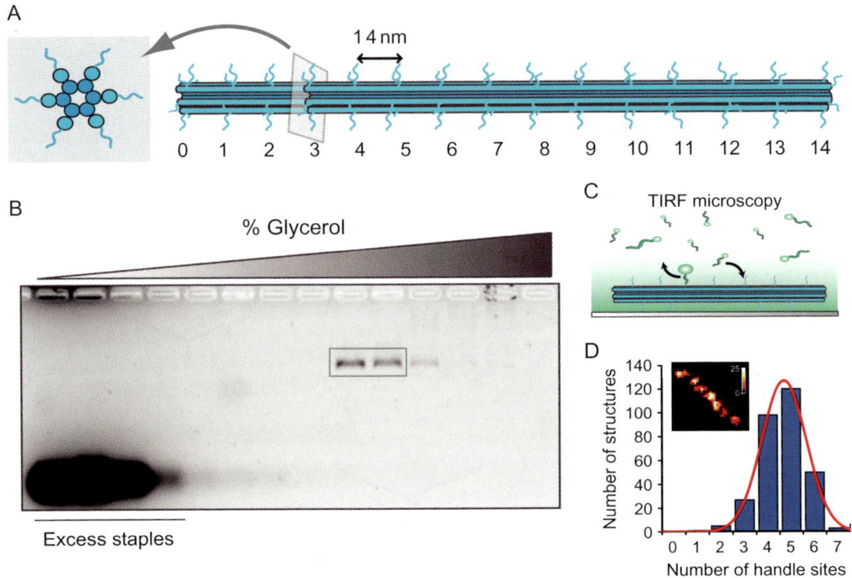

Brian S. Goodman and Samara L. Reck-Peterson, Figure 10.2 Design and analysis of the 12hb DNA origami structure. (A) Design of the 12hb DNA origami structure. The 12hb is composed of six inner helixes and six outer helixes. The structure is 14 nm in diameter and 225 nm in length with optional handle sites every 14 nm. (B) 12hb folding is analyzed by glycerol gradients. After folding, the 12hb is sedimented through a 10–45% glycerol gradient as described (Lin, Perrault, Kwak, Graf, & Shih, 2013) and fractions from the gradient are run on a 2% agarose TBE gel. The excess staples remain in the lower percentage glycerol fractions, while the denser folded 12hb is found in the higher percentage glycerol fractions (box). (C) Handle strand incorporation into the 12hb is assessed by a DNA PAINT experiment (Derr et al., 2012; Jungmann et al., 2010). In this technique, short (10 bp), fluorescent antihandles base pair with the handles of surface-immobilized 12hb structures (Jungmann et al., 2010). Due to the short stretch of homology, stochastic binding of the labeled antihandles generates the ability to perform high-precision localization over many frames of a movie. (D) A DNA PAINT experiment of a 12hb bearing seven handles. A histogram of the number of handle sites observed reveals that the majority of structures have five or more assessable handles. The inset shows a pseudo-image of the handle locations generated from a movie of a single 12hb structure.

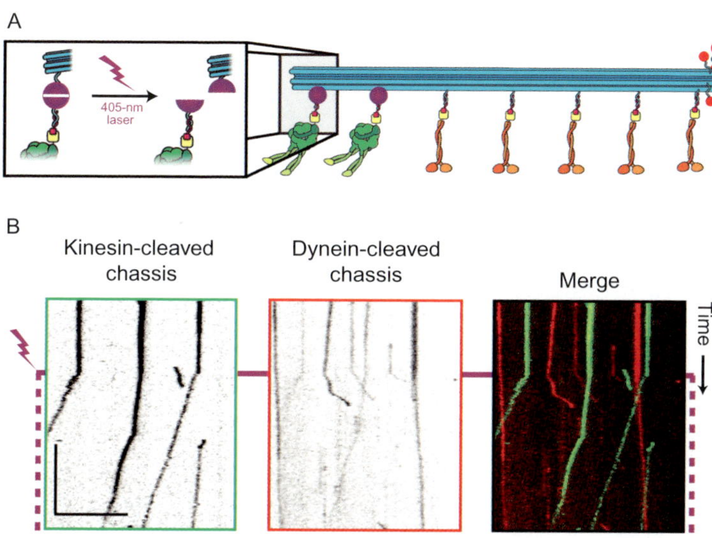

Brian S. Goodman and Samara L. Reck-Peterson, Figure 10.4 Photocleavable oligos allow release of motors from the 12hb. (A) To program the removal of specific motors from the 12hb structure, photocleavable linkers are added to select handle strands (circles). Illumination with 405-nm light induces photocleavage of the motors bearing the photocleavable linker (inset). (B) Two types of 12hb structures with different fluorophores were designed for simultaneous imaging. Kinesin cleavable, mixed-motor 12hb structures (left) were labeled with Cy5 and had photocleavable linkers on the kinesin handle strands. Dynein cleavable, mixed-motor 12hb structures (center) were labeled with TAMRA and contained dynein handle strands with photocleavable linkers. After 1 min of imaging both the TAMRA and Cy5 channels, a 405-nm laser was pulsed in between frames to induce cleavage (dotted line) showing that 12hb structures move in the expected direction when the opposing motor is cleaved. Different colored fluorophores allow the simultaneous observation of the two different motor ensembles during the same experiment (right). Vertical scale bar: 1 min; horizontal scale bar: 10 μm.

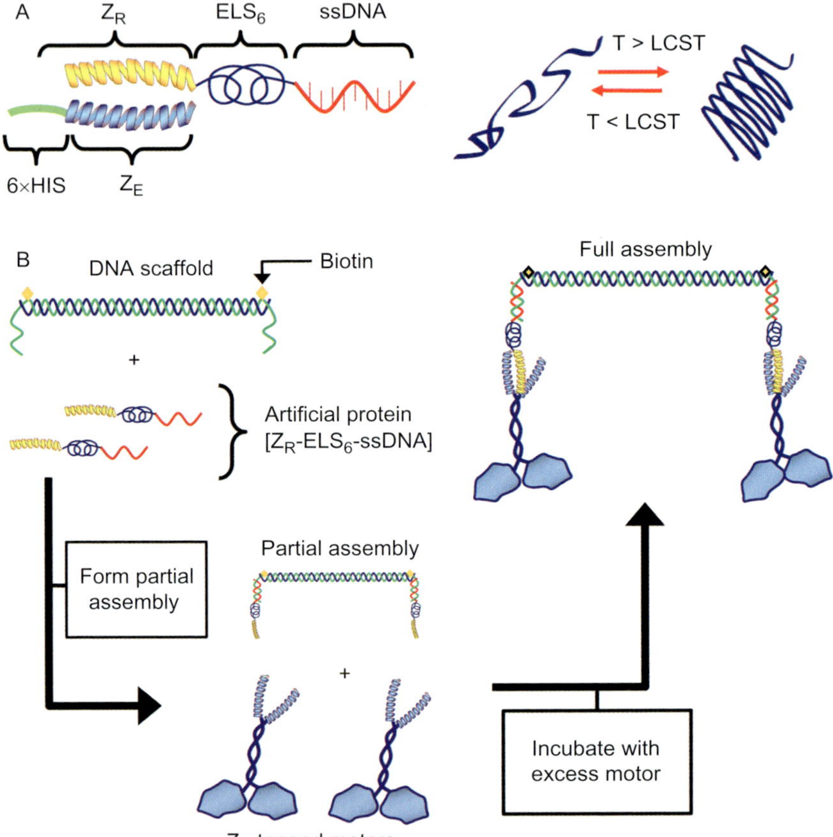

Arthur Rogers *et al.*, Figure 11.1 Self-assembly of DNA-templated multiple-motor complexes. (A) The domain structure of the polymer linkers that connect motors to the DNA scaffold. The thermally driven condensation of the elastin-like ELS_6 polymer domain is depicted at the right. (B) The protocol to assemble two-motor complexes. Once fully formed, these assemblies can be labeled with quantum dots and anchored to bead surfaces.

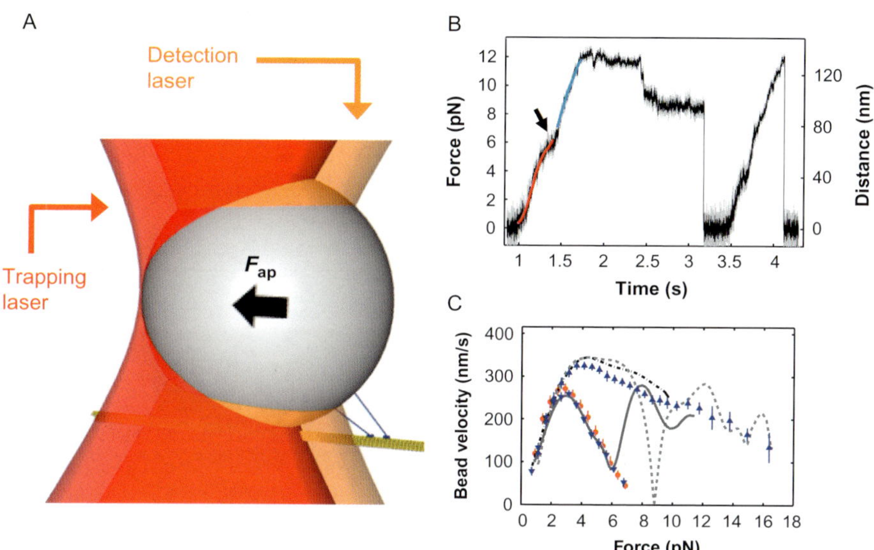

Arthur Rogers *et al.*, Figure 11.3 Analyses of multiple-motor state transitions. (A) Illustration of the optical trapping assay. (B) Bead velocities can change rapidly in an optical trap when a complex transitions between single (red) and two (blue) load-bearing motor states. (C) Configuration-dependent state transitions are also visible in individual force–velocity traces (solid and dashed lines). Thresholds can be defined to identify "slow" and "fast" components of traces to be calculated, allowing the average velocities of each component to be calculated (downward and upward pointing triangles, respectively). See Jamison et al. (2010) for details surrounding the procedures.

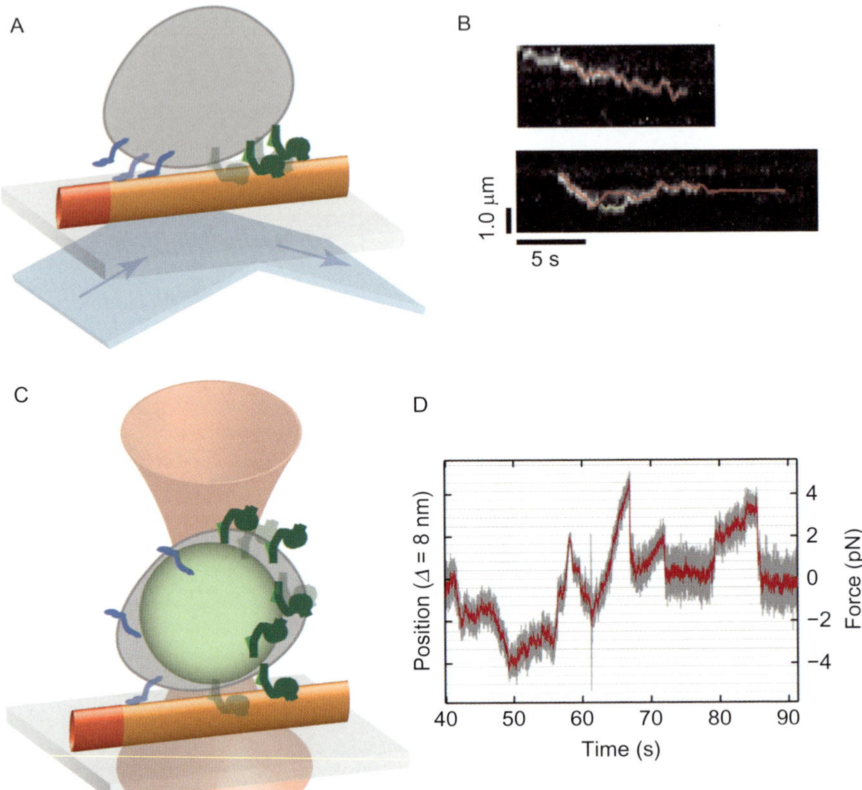

Adam G. Hendricks *et al.*, Figure 14.3 TIRF microscopy and optical trapping show that isolated cargoes move bidirectionally along microtubules *in vitro*, driven by an endogenous complement of stably bound motor proteins. (A) Isolated neuronal transport vesicles are imaged using TIRF microscopy (blue: kinesin-1 and kinesin-2, green: dynein, red/orange: microtubule). Blue arrows indicate incident and totally reflected laser illumination. (B) Vesicles move bidirectionally along microtubules. Red and orange lines indicate subpixel resolution trajectories from automated tracking analysis. (C) The forces exerted by motor proteins on LBCs are measured with an optical trap. (D) Kinesin-1, kinesin-2, and cytoplasmic dynein exert forces on LBCs along microtubules. Plus-end-directed forces are driven by kinesin-1 and kinesin-2 (plotted as positive forces), while dynein drives motility toward the minus-end (negative forces). Thin gray lines indicate 8 nm intervals.

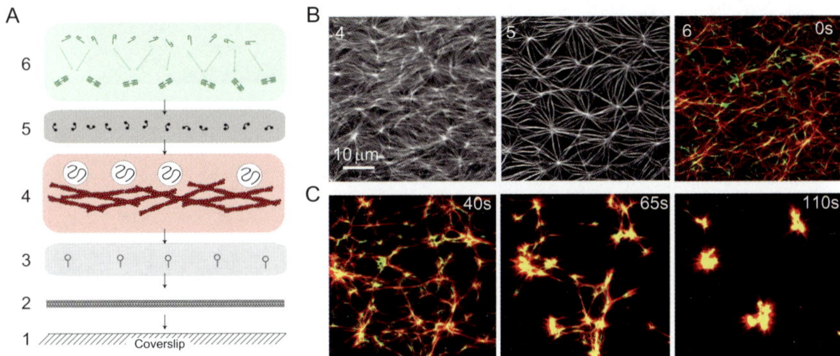

Michael Murrell *et al.*, Figure 15.4 Assembly of a contractile model cortex. (A) Schematic of stepwise cortex assembly: (1) Piranha-treated coverslip, (2) phospholipid bilayer formed on coverslip, (3) incubation with F-actin-membrane attachment factors, (4) crowding of F-actin (F-actin in red, methylcellulose in circles), (5) F-actin cross-linking, and (6) myosin II addition. (B) Fluorescence images of (left) F-actin without cross-linker (Stage 4), (middle) with 30 n*M* α-actinin (Stage 5), and (right) 40 n*M* skeletal muscle myosin II (Stage 6). (C) Time course of contraction, 40, 65, and 110 s after the addition of myosin thick filaments. Red is actin, green is myosin.

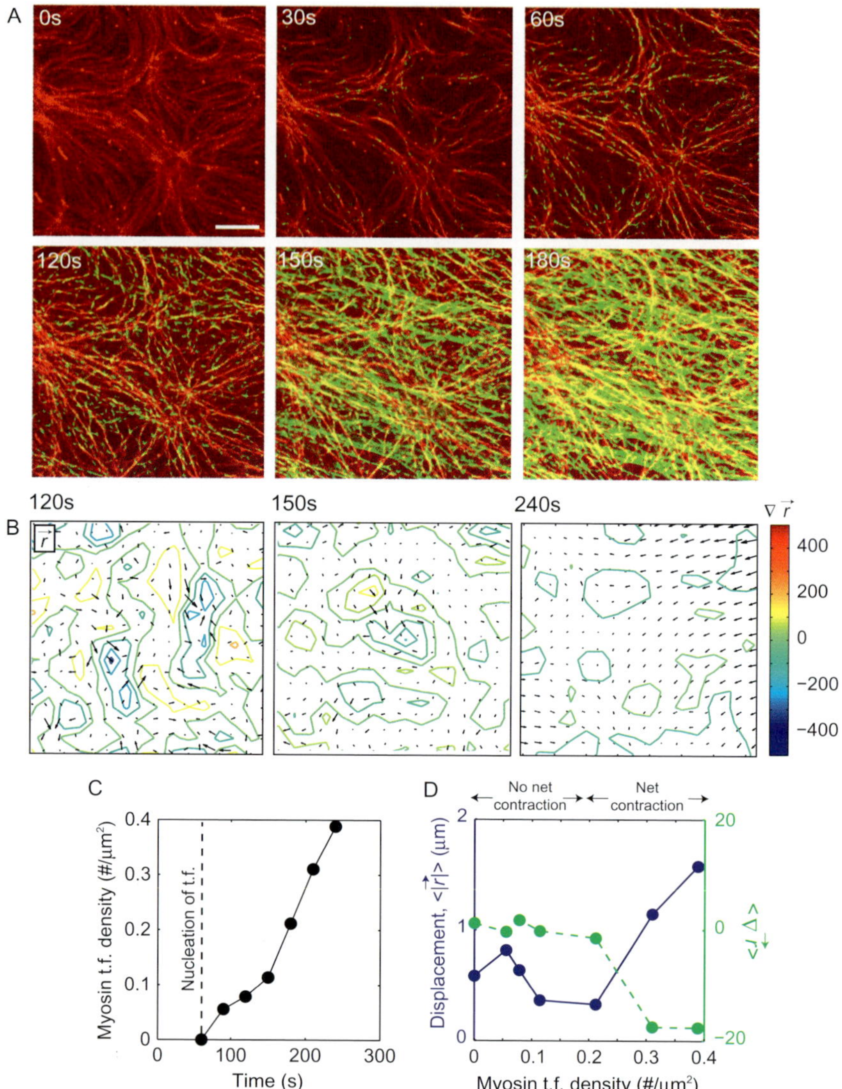

Michael Murrell *et al.*, Figure 15.5 Quantification of F-actin network contractility. (A) F-actin (red) and smooth muscle myosin II (green) within a contracting model cortex. Myosin accumulates over time. Scale bar is 10 μm. (B) Overlay of F-actin displacement (*r*, black arrows) and divergence of F-actin displacement (colored contours) for the contracting network in (A). Hot colors indicate positive divergence (expansion) and cool colors indicate negative divergence (contraction). (C) Myosin thick filament density ρ, over time. (D) Mean divergence (green) and speed (blue) of the F-actin displacement (*r*) as a function of myosin thick filament density, *r*.

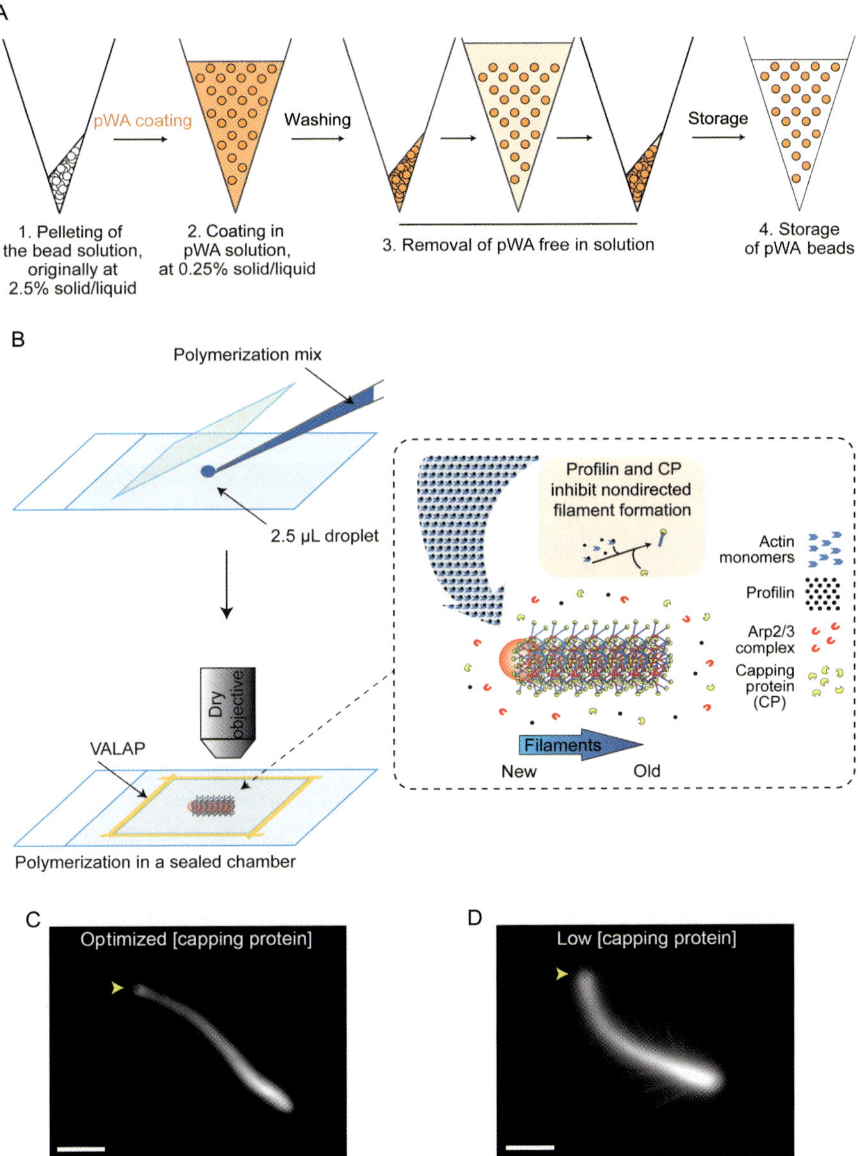

Rajaa Boujemaa-Paterski *et al.*, Figure 16.1 Actin-based motility in a G-actin buffered medium. (A) Schematic representation of the functionalization of polystyrene beads with pWA, an actin nucleation promoting factor. The crucial step of this procedure is the removal of pWA free in solution after the coating in order to prevent nondirected actin assembly in the medium. (B) The polymerization medium is placed between slide and coverslip in a sealed chamber, and bead motility is followed using epifluorescence microscopy. The polymerization medium contains a threefold excess of profilin (black)

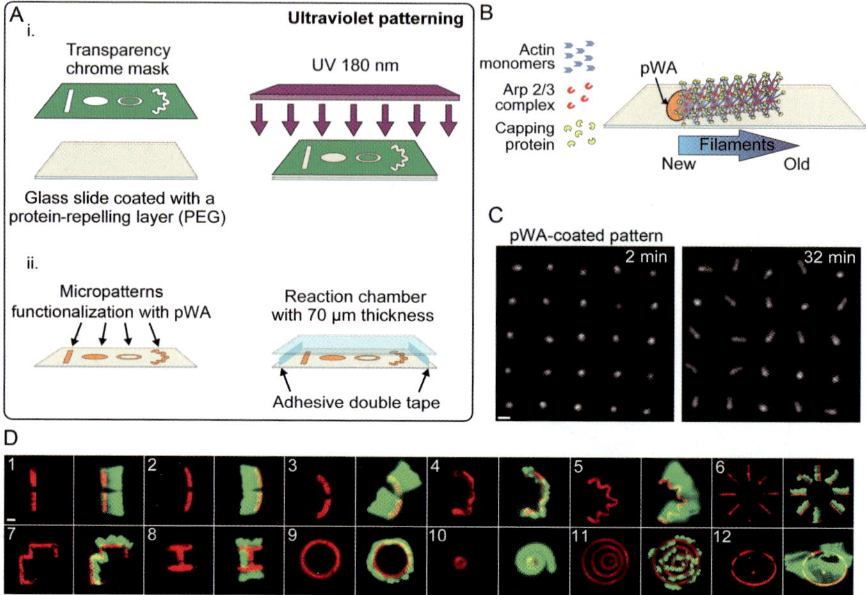

Rajaa Boujemaa-Paterski *et al.*, Figure 16.2 Reconstituted actin-based motility using UV patterning. (A) A chrome photomask containing transparent micropatterns is placed on a PLL-coated glass slide and exposed to ultraviolet light for 5 min (i). The micropatterned glass slide is incubated for 15 min with 1 μ*M* pWA prior to mounting in a reaction chamber (ii). (B) pWA-coated micropatterns are incubated with a motility medium containing actin monomers saturated with profilin (blue), Arp2/3 complex (red), and capping protein (yellow), resulting in the formation of actin filament networks. (C) Dot patterns (3 μm) are imaged by epifluorescence microscopy in a motility medium containing 6 μ*M* actin monomers (Alexa568 labeled), 18 μ*M* profilin, 150 n*M* Arp2/3 complex, and 25 n*M* capping protein. (D) Patterns with different geometries (red patterns, images from 1 to 12) are coated with pWA and fibrinogen 546 and then incubated with a mix containing 4 μ*M* actin monomers (Alexa488 labeled), 12 μ*M* profilin, 150 n*M* Arp2/3 complex, and 25 n*M* capping protein. The second image of the 12 pairs shows the actin network (green) developed after 1 h of polymerization. The width of each pattern line is 3 μm. Scale bars are 5 μm.

---

over actin (blue) in order to efficiently inhibit spontaneous actin nucleation in the medium, and the amount of capping protein (yellow) required for reconstituted motility blocks spontaneous nucleation coming from the small amount of actin monomers not bound to profilin. Actin nucleation and elongation is directed to pWA-coated beads through the recruitment and activation of the Arp2/3 complex (red). (C and D) Reconstituted actin-based motility of 2 μm beads in a medium containing 6 μ*M* actin monomers, 18 μ*M* profilin, 150 n*M* Arp2/3 complex, and 20 n*M* capping protein (C) or 10 n*M* capping protein (D). Scale bars are 10 μm.

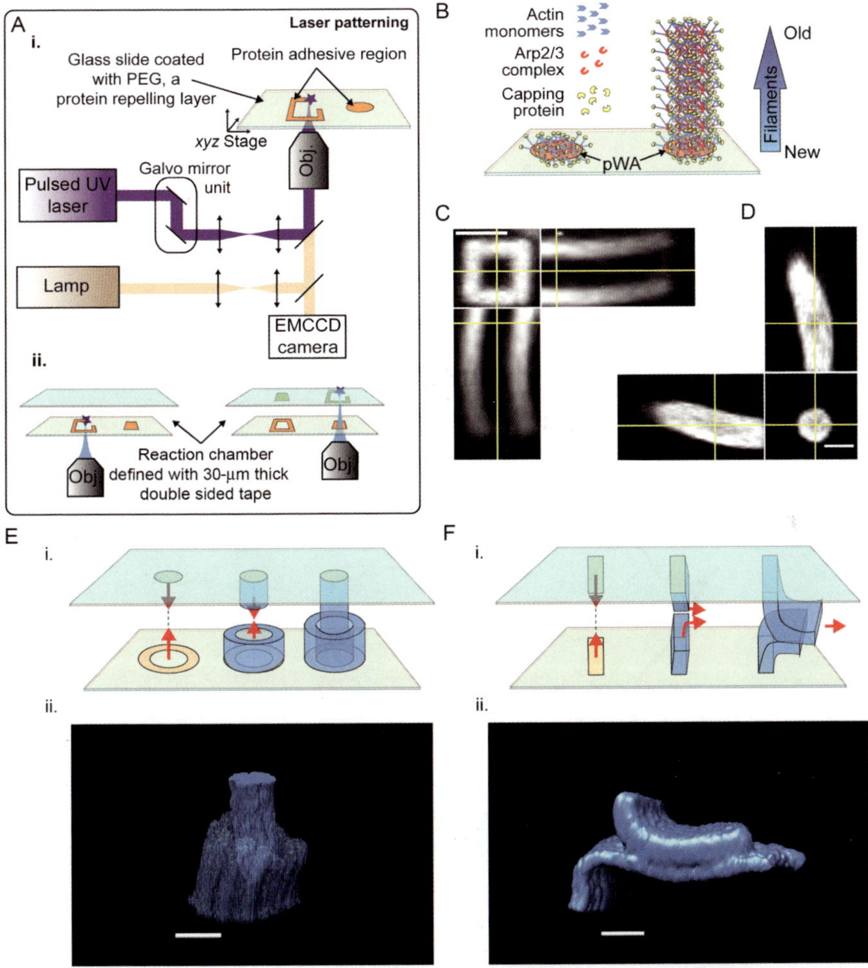

Rajaa Boujemaa-Paterski *et al.*, Figure 16.3 Laser patterning and oriented growth of actin structures in 3D. (A) Schematic representation of the laser-based patterning method. A pulsed UV laser is used to oxidize a protein-repellent layer of polyethylene glycol (PEG) to make it locally adhesive (i). Laser patterning of adhesive regions in a reaction chamber defined by two pegylated coverslips separated by 30–60 μm thick double-sided tape (ii). (B) Schematic representation of the growth of 3D actin structures from a nucleating pattern coated with the pWA in the presence of 3 μ*M* actin monomers (Alexa568 labeled), 9 μ*M* profilin, 50 n*M* Arp2/3 complex, and 60 n*M* capping protein. (C and D) 3D confocal images (planar, horizontal, and vertical cross sections) of actin structures polymerized from an empty square-shaped nucleating pattern (C) and from a disc-shaped nucleating pattern (D). (E) Schematic representation of the interaction of two actin structures growing face-to-face from two complementary-shaped nucleating patterns to form a 3D connection (i) and 3D reconstruction of confocal imaging of such a 3D connection (ii). (F) Schematic representation of the interaction of two actin structures growing face-to-face from two similar rectangular-shaped nucleating patterns that induce the formation of an actin structure in a plane parallel to the coverslip (i) and 3D reconstruction of confocal imaging of such a structure (ii). Scale bars are 10 μm.

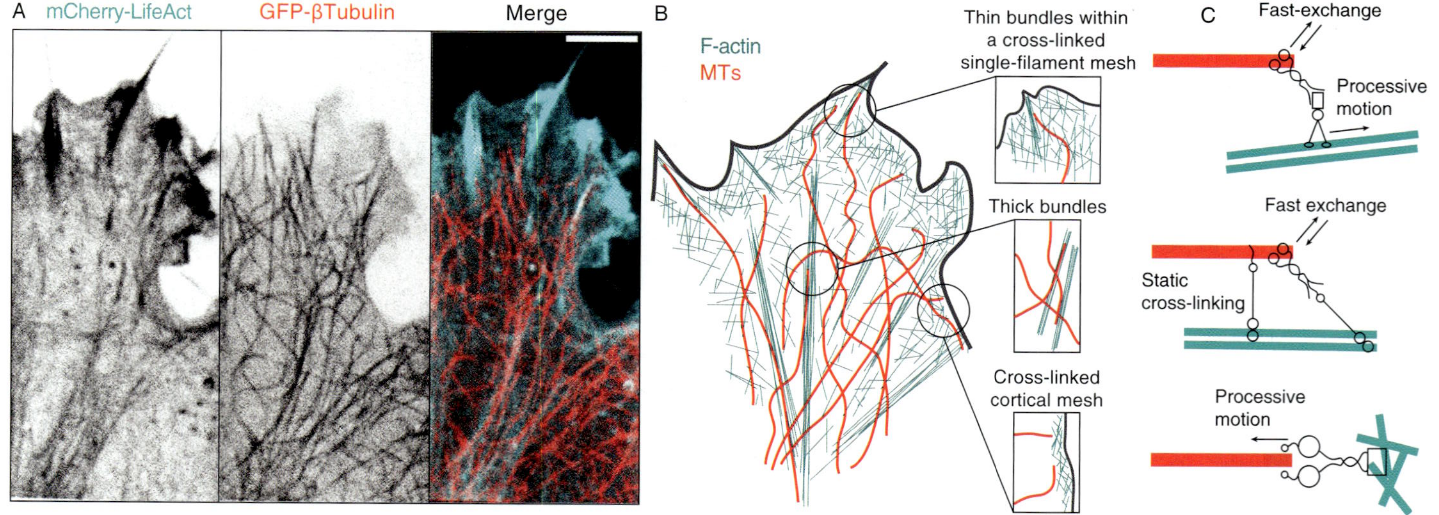

Magdalena Preciado López *et al.*, Figure 17.1 *In vivo* coorganization of F-actin and microtubules. (A) Fluorescence micrograph of part of a 3T3 fibroblast with fluorescent labels showing the distinct organization of F-actin (cyan) and microtubules (red). Actin forms rigid stress fibers and dense networks underneath the plasma membrane (top), whereas microtubules grow toward the cell periphery and form a sparse network. (B) Schematic showing how dynamic microtubules can encounter different F-actin architectures within a cell. (C) Conceptual drawings of the different types of F-actin–microtubule cross-linking proteins and protein complexes, classified based on their activities (e.g., processive motion) and exchange properties (e.g., slow vs. fast). Scale bar: 5 μm.

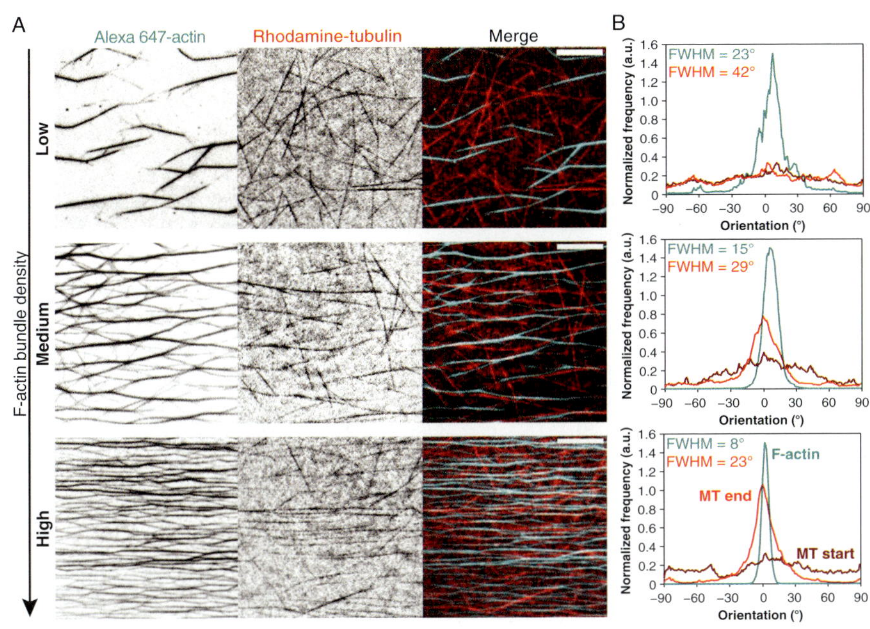

Magdalena Preciado López *et al.*, Figure 17.3 TIRF images of microtubule growth within parallel arrays of F-actin bundles. (A) (Left) Three F-actin bundle arrays of different bundle surface density. (Middle) Resulting microtubule arrays after ~1 of growth within the corresponding F-actin bundle arrays. (Right) Merged panels. (B) Normalized histograms of F-actin bundle and microtubule orientation corresponding to the examples in (A). The cyan curve shows the F-actin histogram, normalized so the maximum frequency is 1.5. The dark-red curve shows the initial distribution of microtubule orientation, which is always random. The light-red curve shows the final microtubule orientation distribution after ~1 of growth, showing that longer microtubules can be better organized by the F-actin bundles. In all graphs, the full width at half maximum (estimated from fitting Gaussian functions to the distributions) is shown for the F-actin bundle array as well as the resulting microtubule (MT) organization (light-red curves). Increasing the F-actin bundle density leads to better actin–microtubule coalignment through steric interactions. Scale bars: 10 μm.

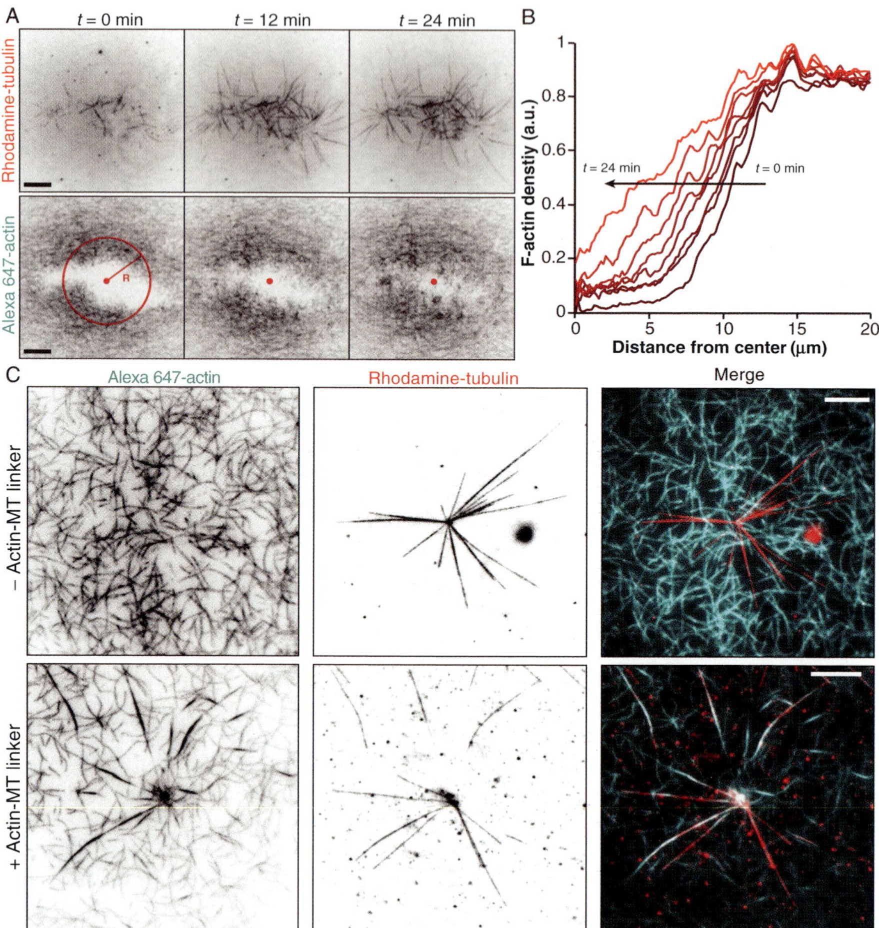

Magdalena Preciado López *et al.*, Figure 17.4 TIRF images of radial microtubule growth from centrosomes within isotropic solutions of actin filaments. (A) Initially, we observe a zone depleted of F-actin around the centrosomes, which typically closes within 15–30 min (in this example, multiple centrosomes are bound to the surface). (B) Radial fluorescence intensity profiles calculated at different time points demonstrate inward diffusion of actin filaments toward the centrosome cluster. (C) Once equilibrated, the actin filaments form a homogeneous entangled network (left) that fully encloses the dynamic centrosome-based microtubule aster (middle). (Right) Merge. (D) In the presence of an engineered F-actin-binding protein that binds to microtubule (MT) plus tips via EB, the actin filaments become radially organized as microtubules grow. Scale bars: 10 μm.

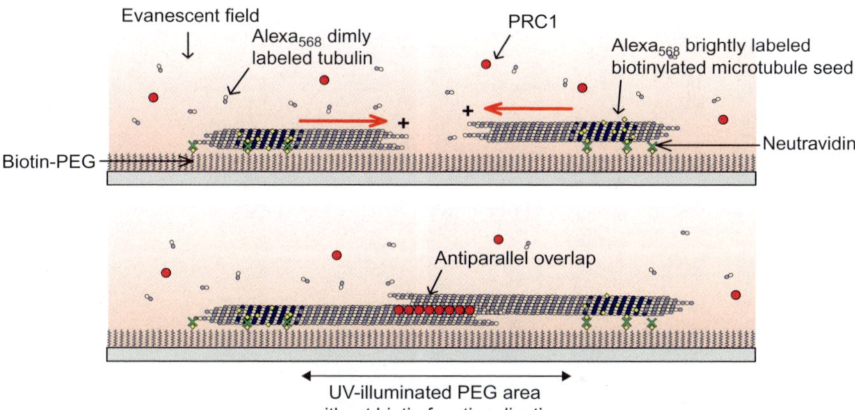

Franck J. Fourniol *et al.*, Figure 19.1 Schematic overview of the dynamic antiparallel microtubule assay on micropatterned coverslips. Brightly labeled microtubule seeds are immobilized on the functionalized areas of the coverslip surface by a biotin–neutravidin sandwich (or alternatively by a streptavidin-Cys directly coupled to the glass—not shown in this scheme) and are visualized using TIRF microscopy. From these seeds, dynamic microtubule extensions are grown with dimly labeled tubulin. When antiparallel microtubules meet, PRC1 accumulates in the overlap region. *(Adapted from Bieling et al. (2010)).*

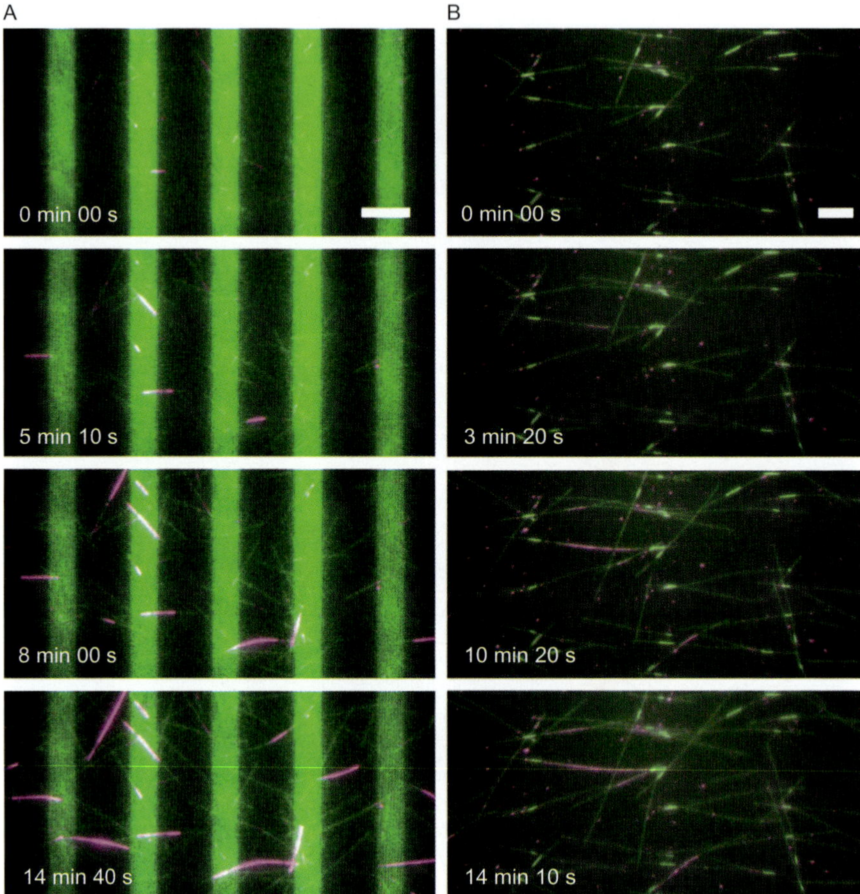

Franck J. Fourniol *et al.*, Figure 19.6 TIRF imaging of antiparallel microtubule nucleated from micropatterned surfaces. (A) Time-course of dynamic microtubules grown from Alexa647-labeled microtubule seeds (dim magenta) aligned on Cys-mCherry (green) and streptavidin-Cys patterns, in the presence of 5 n*M* PRC1-SNAP-Alexa647 (bright magenta) and 22.5 μ*M* Alexa568-tubulin (green). Also see Video 1 (http://dx.doi.org/10.1016/B978-0-12-397924-7.00019-4). (B) Time-course of dynamic microtubules grown from Alexa568 brightly labeled microtubule seeds (bright green) aligned on PEG-biotin patterns, in the presence of 5 n*M* PRC1-SNAP-Alexa647 (magenta) and 17 μ*M* Alexa568-tubulin (green). Scale bars are 10 μm. Also see Video 2 (http://dx.doi.org/10.1016/B978-0-12-397924-7.00019-4).

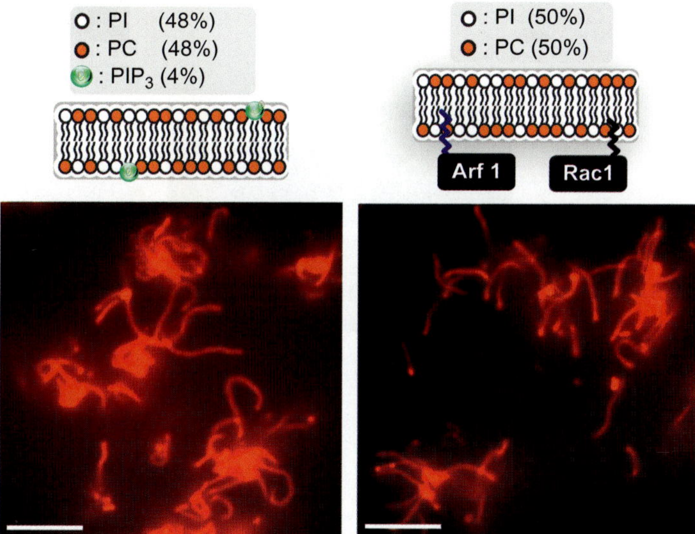

Peter J. Hume *et al.*, Figure 20.3 WAVE-dependent actin-based motility of silica microspheres coated with lipid bilayers composed of PC:PI and either PIP_3 (left) or Arf1 and Rac1 (right). Extract contains rhodamine-labeled G-actin, allowing visualization of comet tails (red). Comet tail formation and consequent microsphere motility triggered by Arf1 and Rac1 (right) are indistinguishable from that induced by PIP_3 (left). Scale bars, 15 μm.

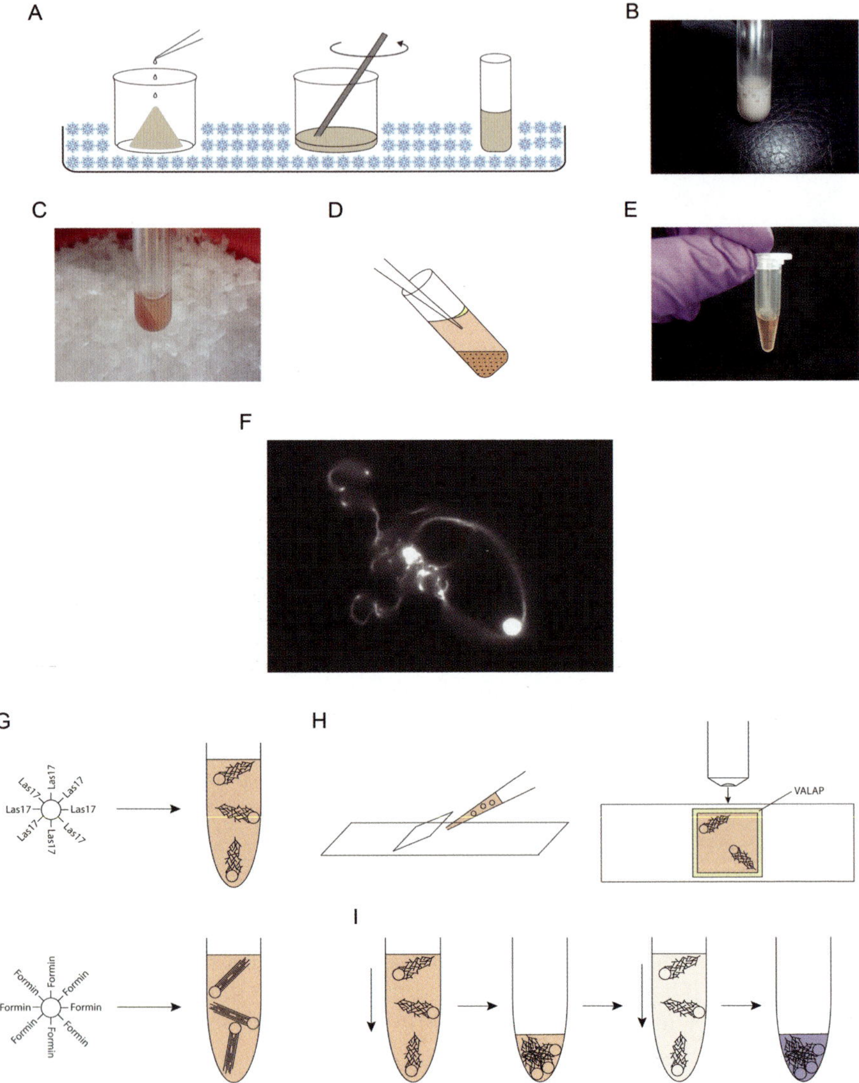

Alphée Michelot and David G. Drubin, Figure 21.2 Preparation and use of yeast extracts for actin network assembly. (A) Illustration of yeast lysate solubilization. Left drawing represents addition of buffers to the yeast powder. Central panel represents the solubilization step. Right panel represents the transfer of the solution to an ultracentrifuge tube. (B) Photograph of solubilized yeast lysates before high-speed centrifugation. (C) Photograph of yeast extract with cell debris pellet after high-speed centrifugation. (D) Illustration of yeast extract collection. Nonsoluble pelleted fraction (dark orange area) and floating lipid layer (yellow area) are to be avoided. (E) Photograph of a yeast extract ready to be used for actin assembly assays. (F) Example of Abp140-3GFP extract observed in fluorescence microscopy 5 h after preparation (right

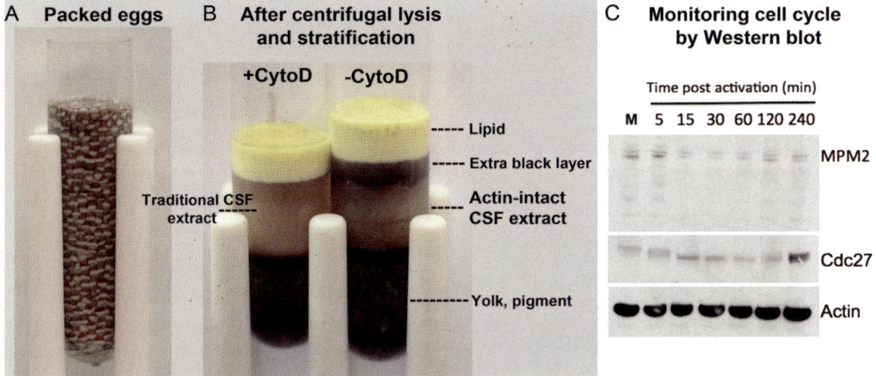

Christine M. Field *et al.*, Figure 22.1 Extract preparation and monitoring cell cycle progression. (A) A centrifuge tube full of unfertilized eggs after the packing spin. Note that the eggs have become slightly deformed as they packed together, forcing out excess buffer. (B) Centrifuge tubes after successful crushing spins. On the left, eggs were pretreated with cytochalasin D to yield traditional M-phase (CSF) extract. The cytoplasmic layer always separates well. On the right, cytochalasin D was omitted to yield actin-intact M-phase extract. Note the presence of an additional band of dark material in the actin-intact system. In this prep, it is well separated from the cytoplasmic layer. This layer is a mixture of mitochondria, smooth ER and lipid droplets as shown by electron microscopy (data not shown). (C) Western blot characterization of cell cycle state of extract. M-phase extract (M) is compared to extract activated by calcium addition. Calcium is added, the extract incubated at 20 °C, and samples taken at times shown. Note the lower intensity of MPM2 starting at 15 min, indicating progression from M-phase arrest and the beginning of I-phase. At 120 min, increased intensity indicates reentry into M-phase. The same samples were also probed with an antibody against Cdc27. Note the shift in electrophoretic mobility following M-phase exit (see 5 min).

---

panel; cable-like actin filament structures of actin filaments may appear). (G) Illustration of the protocol for the assembly of patch-like structures of actin filaments from Las17-coated microbeads and cable-like structures from formin-coated microbeads. (H) Illustration of the protocol for microscopy observation of actin networks. Samples are dropped between a slide and a coverslip, and the sealed with VALAP (in yellow). (I) Illustration of the protocol for the partial purification of actin structures bound to beads for protein composition analysis. Beads are pelleted at low speed, washed carefully multiple times, and quickly pelleted again before resuspension in the buffer of choice.

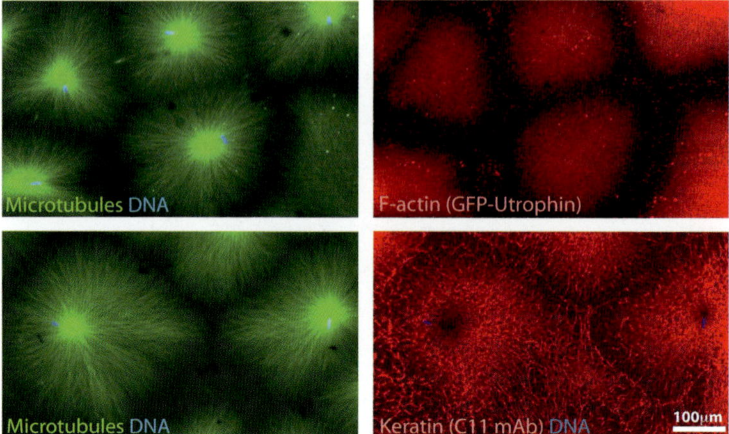

Christine M. Field *et al.*, Figure 22.3 Actin and keratin filaments associate with I-phase microtubule asters. Actin-intact M-phase extract was supplemented with fluorescently labeled tubulin and either GFP–Utrophin to visualize F-actin or directly labeled monoclonal antibody to keratin to visualize keratin filaments. Permeabilized Xenopus sperm (Desai et al., 1999) was added to supply centrosomes, and Ca^{2+} added to initiate I-phase. A volume of 8 μl extract was spread between a whole casein-passivated slide and a 18-mm^2 cover slip. Aster assembly was imaged live with a 10 × objective and widefield fluorescence. These images were taken ~40 min after initiating aster assembly. Note that both F-actin and keratin filaments coorganize into islands associated with each microtubule aster.

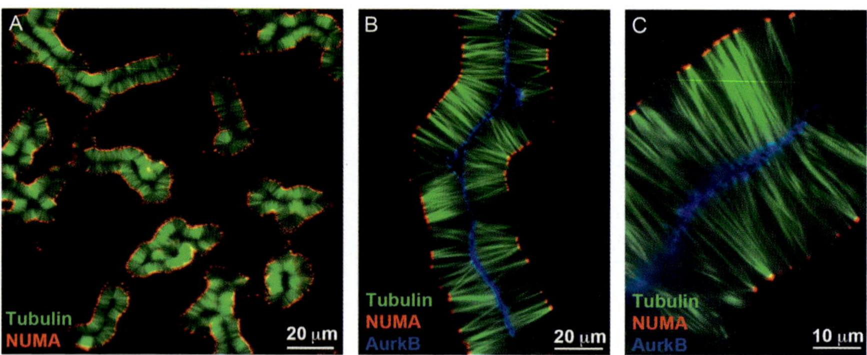

Aaron C. Groen *et al.*, Figure 23.4 Pineapples assembled in HSS + glycogen (fluorescence). These reactions contained tubulin labeled with Alexa 568 (shown green), anti-NUMA IgG labeled with Alexa 488 by conventional in-solution labeling (shown red) and in (B and C) anti-AurkB IgG labeled with Alexa 647 by the on-resin labeling method described above. NUMA accumulates at microtubule minus ends due to dynein-mediated transport. AurkB, as part of the CPC, accumulates at distal plus ends due to transport by one or more midzone kinesins. (A) A field imaged by 20 × widefield after ~30 min of assembly. (B and C) Confocal images after ~120 min of assembly, where the assemblies are wider due to slow plus-end growth.